Ion Exchange Membranes

Ion Exchange Membranes

Design, Preparation, and Applications

Tongwen Xu
Yaoming Wang

WILEY-VCH

Authors

Prof. Tongwen Xu
University of Science & Technology of China
Lab of Functional Membranes
Hefei
China

Prof. Yaoming Wang
University of Science & Technology of China
Lab of Functional Membranes
Hefei
China

Cover Image: © Kateryna Kon/Shutterstock

Printed and bound by CPI Group (UK) Ltd, Croydon, CR0 4YY

C9783527351947_060324

Library of Congress Card No.: applied for

British Library Cataloguing-in-Publication Data
A catalogue record for this book is available from the British Library.

Bibliographic information published by the Deutsche Nationalbibliothek
The Deutsche Nationalbibliothek lists this publication in the Deutsche Nationalbibliografie; detailed bibliographic data are available on the Internet at <http://dnb.d-nb.de>.

Print ISBN: 978-3-527-35194-7
ePDF ISBN: 978-3-527-84142-4
ePub ISBN: 978-3-527-84143-1
oBook ISBN: 978-3-527-84144-8

Typesetting Straive, Chennai, India

Contents

Preface

Ion exchange membranes (IEMs) are membranous ion exchange resins typically made up of a hydrophobic polymer matrix and charged ionic groups. IEMs are one of the earliest types of membranes studied for water treatment and have a long research history. In recent years, IEMs have been widely valued by academia and industry due to the requirements of renewable energy and the upgrading of conventional manufacturing procedures. Notably, progress in advanced IEM has stimulated a multitude of applications of electro-membrane separation processes, such as water desalination, salt concentration, clean separation and production, power generation, and sensitive electrode preparation based on conventional membrane processes, including diffusion dialysis, electrodialysis, bipolar membrane electrodialysis, membrane capacitive deionization, reverse electrodialysis, and fuel cells. In addition, several emerging electro-membrane processes, including alkaline membrane water electrolysis, electrocatalysis, solar-driven water splitting, and acid–base flow batteries, are progressing rapidly and have the potential to reform conventional manufacturing procedures completely. Industrial-scale applications of IEMs will be exemplified from the aspect of process economy and their impact on the environment compared with conventional treatment methods.

Based on the evolution of advanced membrane fabrication technologies and the optimization of rational material design, a desired IEM should be environmentally fabricated with low electrical resistance, high chemical and thermal stability, excellent mechanical properties, and high ion selectivity. Moreover, the progress in advanced IEMs has stimulated a multitude of new applications in electro-membrane separation processes, which have the potential to completely upgrade conventional manufacturing procedures and have been widely employed to address the major challenges related to the environment and energy. Obviously, it is foreseeable that more advanced IEMs will be fabricated toward future demands, especially to fulfill the beautiful vision of "carbon peak and carbon neutral." Of course, it will also continue to transform existing technologies of IEMs, which makes chemical processes energy-efficient and plays an increasingly significant role in sustainable development.

This book attempts to summarize the advanced design and fabrication technologies for IEMs as well as to describe the recent applications of electro-membrane separation processes. The structure of the book consists of an introductory overview followed by eight chapters. The introductory overview is designed to provide a context for relating the history and recent progress on the IEM and the related electro-membrane processes. Next, Chapter 2 provides the fundamental background on the IEM and the basic characterizations of the IEM from the basic physiochemical properties measurement to the electrochemical characterizations. Chapters 3 through 7 feature the latest fabrication technologies for IEM preparation. Specifically, Chapter 3 introduces the manipulation of the architecture topology of side chains for IEM preparation, which are very promising materials used for fuel cells. Chapter 4 describes the polyacylation technique for IEM fabrication. This novel polymerization method provides mild reaction conditions for membrane fabrication with high polymerization activity. Notably, the membrane properties can be precisely tuned by the addition ratio of the ionic monomers. Chapter 5 discusses superacid catalyst polymerization for IEM preparation. Ether-free polymers synthesized by this method were first applied in the field of alkaline membrane fuel cells to solve the alkaline stability problem, which has puzzled this field for many years. Chapter 6 focuses on the modification of intrinsic microporosity polymers for IEM preparation, which is very appropriate for energy storage and conversion processes, such as aqueous organic flow batteries and H_2/O_2 fuel cells. In addition, Chapter 7 describes in situ polymerization for IEM preparation. The in situ polymerization allows the one-step formation of the IEM with little consumption of organic solvent, which provides an effective strategy for the green fabrication of IEMs. Chapter 8 focuses on the fabrication of special IEMs, including metal–organic framework membranes, porous organic cage membranes, covalent organic framework membranes, and electro-nanofiltration membranes. They exhibit excellent properties and broaden the applications of selective ion separations, breaking the trade-off effect of conventional membranes between flux and selectivity. Finally, Chapter 9 discusses the recent applications of IEMs with respect to practical industrial demands, including diffusion dialysis, electrodialysis, bipolar membrane electrodialysis, reverse electrodialysis, metathesis electrodialysis, selective electrodialysis, ion-distillation, and fuel cell, among others.

While previous books have focused on the basics of separation membrane materials and their routine applications, this book focuses on the state-of-the-art development of IEM fabrications and applications. These membrane fabrication methods were recently developed in the past 10 years and are not included in previous related books. The book provides a more comprehensive viewpoint of the recent developments in IEMs that are difficult to obtain solely by reading individual journal papers. It can be seen that this book is suitable as a resource that has a strong guiding role and reference value whether it is for technical research and development, production management practitioners in related fields, or teachers and students in colleges and universities. It can also provide new ideas for further expansion of preparation methods and new applications of IEMs. We hope and expect that this book will sincerely enlighten and benefit a wide range of readers.

In the process of compiling this book, a large number of research results and application examples have emerged in recent years from my laboratory group, which is written with an emphasis on reflecting the most advanced design and preparation methods and application values of IEMs. I strive to combine theory with practice closely, pay attention to connecting academia and industry, and highlight the characteristics of novelty and application. It attempts to show the artistic charm of chemistry in material preparation and the charm of materials in process industry applications, attracting more people to participate in academia or become industrial experts for IEM research. In writing this book, I am deeply grateful to Liang Wu, Yaoming Wang, Zhengjin Yang, Jin Ran, Liang Ge, Xiaolin Ge, Chenxiao Jiang, Xingya Li, Peipei Zuo, and Rongqiang Fu (my group members and former students from the Laboratory of Functional Membranes, University of Science and Technology of China) for their cooperation and contribution. I sincerely thank the National Natural Science Foundation of China (Nos. 22038013, U20A20127, 21720102003, 91534203, and 21490581) and the National Key Research and Development Program of China (Nos. 2020YFB1505601, 2018YFB1502301, and 2012CB932802) for funding support over the past 10 years. Additionally, this book also refers to the literature of domestic and foreign counterparts. Heartfelt thanks to the authors for their original contribution. It must be noted that there are many excellent investigations that are not cited. The references are cited on the author's personal preference. Finally, I am grateful to my wife and our lovely son for their patience and strong support in my research life.

University of Science and Technology of China

Tongwen Xu
Yaoming Wang

1

Overview of Ion Exchange Membranes

1.1 Definition and Classifications

Ion exchange membranes (IEMs) are typically made up of a hydrophobic polymer matrix and ionic groups and can be classified into anion exchange membranes (AEMs) and cation exchange membranes (CEMs) according to the type of ionic groups grafted onto the membrane matrix. CEMs that are fixed with negatively charged groups ($-SO_3^-$, $-COO^-$, etc.) conduct cations but repel anions, while AEMs containing positively charged groups ($-NH_3^+$, $-NRH_2^+$, $-NR_2H^+$, $-NR_3^+$, PR_3^+, $-SR_2^+$, etc.), permit the permeation of anions but retard cations [1, 2]. The typical polymer architectures of IEMs are shown in Figure 1.1a, while the typical groups are shown in Figure 1.1b [3].

According to the connection of ionic groups to the polymer matrix, IEMs can also be categorized as homogeneous and heterogeneous membranes. In homogeneous membranes, the charged groups are chemically bonded to the membrane matrix, and in heterogeneous membranes, they are physically mixed with the membrane matrix [4]. There are many other classification methods, and as a summary, we provide Table 1.1, listing the main categories of IEMs [5].

1.2 Profile of IEMs

Research on IEMs has a long history, dating back to 1890 when Ostwald investigated the properties of semipermeable membranes [6]. It was found that none of the electrolytes could permeate through membranes if these membranes were impermeable to either anions or cations. To explain this phenomenon, the author postulated the "membrane potential" at the boundary between a membrane and its surrounding solution, which was derived from the difference in concentration. The existence of such a boundary was confirmed by Donnan in 1911 [7]. The concentration equilibrium that led to the so-called "Donnan exclusion potential" was described with a mathematical equation. In 1925, Michaelis and Fujita used homogeneous, weak acid collodion membranes to initiate practical basic studies relevant to IEMs [8]. In the 1930s, Sollner demonstrated the idea of a mosaic or amphoteric membrane that had both positively and negatively charged moieties that revealed

Ion Exchange Membranes: Design, Preparation, and Applications,
First Edition. Tongwen Xu and Yaoming Wang.

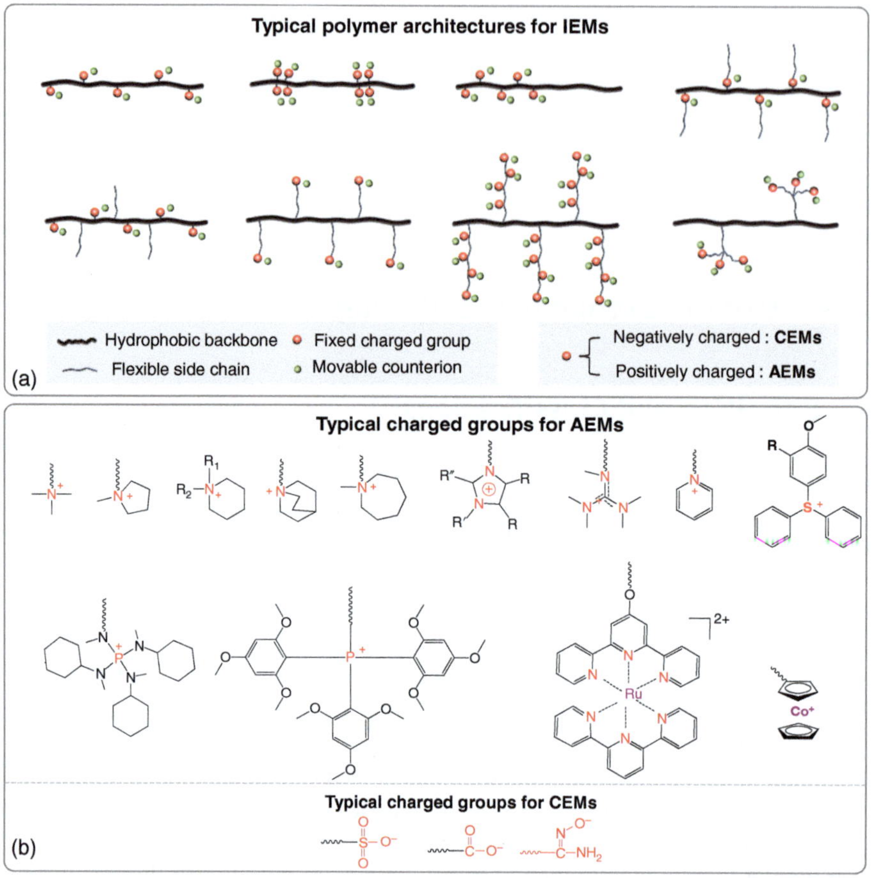

Figure 1.1 Schematic showing the structure of ion exchange membranes (IEMs). (a) Typical polymer architectures, (b) representative positively charged groups of anion exchange membranes (AEMs), while anionic groups, including sulfonate groups, carboxylate groups, and amidoxime groups, are usually introduced for cation exchange membranes (CEMs).

distinctive ion transport phenomena [9]. The increasing interest in IEMs that were used in industrial processes stimulated the development of synthetic IEMs based on phenol–formaldehyde polycondensation around 1940 [10]. Meyer and Strauss proposed an electrodialysis process that was performed in a configured cell [11]. This cell consisted of two compartments formed by an AEM and a CEM placed between two electrodes. In most practical electrodialysis processes, a stack into which multiple electrodialysis cells were arranged was used.

However, at that time, the industrial implementation of electrodialysis was still impeded by the absence of IEMs with good properties, especially low membrane resistance. This situation improved in the 1950s. Juda et al. from Ionics Incorporated [12] and Winger et al. from Rohm and Haas Company [13] promoted the development of IEMs that had improved performance in terms of chemical stability, selectivity, and electric resistance. Electrodialysis on the basis of these membranes

Table 1.1 The main categories of ion exchange membranes.

The basis of classification	Type	Description
According to the type of ionic groups	Cation exchange membrane	Selectively transport cations but repel anions
	Anion exchange membrane	Selectively transport anions but repel cations
According to the arrangement of an ionic group within the membrane	Amphoteric ion exchange membrane	Membranes contain both positively and negatively charged groups and these groups are distributed randomly in membranes
	Zwitterionic ion exchange membrane	Membranes contain an equal number of positively and negatively charged groups.
	Mosaic ion exchange membrane	Membranes are made up of a set of anion and cation exchange elements arranged in parallel
	Bipolar ion exchange membrane	A kind of composition membrane that at least is composed of a cation-selective and an anion-elective layer
	Monovalent anion perm-selective membranes	Membranes can separate monovalent ions from multivalent ions
According to the connection way of ionic groups	Homogeneous ion exchange membrane	Ionic groups are covalently connected to polymer backbone
	Heterogeneous ion exchange membrane	Ionic groups are physically mixed with polymer matrix
	Semi-homogeneous ion exchange membrane	The connections combine above two ways
According to the variety of membrane materials	Inorganic ion exchange membranes	Mainly made up of inorganic materials
	Organic ion exchange membranes	Mainly made up of organic polymer materials
	Inorganic–organic (hybrid) ion exchange membranes	Mixing inorganic materials with organic polymers at the molecular level
According to the cross-sectional structure of membrane	Symmetric ion exchange membranes	Symmetric structure of cross-section
	Asymmetric ion exchange membranes	Asymmetric structure of cross-section
	Composite ion exchange membranes	Covering asymmetric membranes with one or several layers

(Continued)

Table 1.1 (Continued)

The basis of classification	Type	Description
According to pore structure of membranes	Dense ion exchange membranes	Having almost no free volume within membranes
	Microporous ion exchange membranes	Membranes contain pores with size below 2 nm which consist of charged microporous polymers or mixed materials (polymers blended with microporous fillers, such as COFs, MOFs, and graphene)
	Porous ion exchange membranes	Membranes prepared by phase inversion process contain pores with size above 2 nm and are mainly used for diffusion dialysis or electro-nanofiltration

is recommended as an industrial process that demineralizes and concentrates electrolyte solutions. Since then, IEM-based electrodialysis has witnessed wide applications in many fields. Some important examples are listed as follows: (i) In the 1960s, Asahi Co. performed the first salt production from seawater with monovalent ion permselective membranes [14]; (ii) In 1969, electrodialysis reversal (EDR) was invented that could run stably without any precipitation and deposition on both electrodes and membranes [15]; (iii) In the 1970s, DuPont developed a series of CEMs (widely known as Nafion®) made up of sulfonated polytetrafluorethylene [16]. These CEMs are chemically stable and highly conductive, and thus have still been widely used, such as in energy storage (e.g. flow batteries) or conversion systems (e.g. fuel cells); (iv) In 1976, Chlanda et al. presented a new concept of a bipolar membrane that consisted of an AEM layer and a CEM [17], and this concept is still widely studied [18, 19].

Nafion® series combining both excellent electrochemical and physical properties is widely used but hampered by high manufacturing costs. Therefore, alternatives to low-cost hydrocarbon-based polymers have received much attention since around 2000, such as poly(phenylene oxide) (PPO), polyimide (PI), poly(ether sulfone) (PES), poly(ether ketone) (PEK), and polystyrene (PS) [20]. During the membrane-formation process, the charged moieties can self-assemble into a hydrophilic phase within the hydrophobic matrix, and thus, these membranes are called microphase-separated IEMs. This microstructure is crucial to IEM performance in terms of ion conductivity, ion selectivity, mechanical stability, etc. [20–25]. Molecular engineering strategies to enhance polymer self-assembly into highly ordered microstructures are required for high-performance IEMs. Over the past two decades of development, optimal self-assembly has been realized mainly via the following strategies [3]: (i) Densely grafting charged groups to increase the local size of the hydrophilic region (i.e. densely charged or block-type IEMs), (ii) enhancing the mobility of charged segments by introducing flexible spacers between charged groups and the backbone (i.e. side-chain-type IEMs), (iii) introducing secondary interactions such as hydrogen bonding, cation–dipole interactions, and π–π stacking as supplemental driving forces, (iv) introducing the fluorine element with both superhydrophobicity for enhanced contrast of hydrophobic and hydrophilic phases and a strong electron-withdrawing effect for improved dissociation degree of sulfonate moieties, and (v) crystallizing the polymer backbone as robust hydrophobic phase and arranging the charged groups into the continuous ion channels.

Studies on microphase-separated IEMs focus on optimizing the micromorphology from the perspective of enhancing the self-assembly of ionic groups [24]; however, there are very few examples that can precisely form the microstructure at the nanoscale [26]. In addition to the hydrophilic phase acting as ion-conducting channels, micropores are also employed for ion conduction. Constructing micropores within membranes can be achieved via covalent organic frameworks (COFs). For example, in 2018, Kong et al. fabricated positively charged COF membranes through molecular precursor engineering and observed a remarkable hydroxide ion conductivity of over 200 mS/cm [27]. We employed an interfacial growth strategy to successfully construct COF membranes that contained ordered channels with a size of approximately 1.4 nm in size and a large number of hydrogen bonding sites [28]. These membranes enabled very high selectivity of monovalent ions over divalent ions, for example, a selectivity of approximately 217 for Li^+/Mg^{2+}. However, most crystalline COFs suffer from poor processability and low mechanical flexibility [29], which may hinder their widespread use in making large-scale IEMs.

Polymer materials do not have this problem, and the construction of polymers of intrinsic microporosity (PIMs) for microporous membranes has been of great interest. In contrast to polymers for microphase-separated IEMs, PIMs (first developed by Neil B. McKeown et al. in 2004 [30]) usually consist of polymer backbones that are much more rigid and contorted, and the inefficient packing of these backbones can create sufficient intermolecular voids (i.e. micropores) [31–34]. In 2016, our group first observed the fast hydroxide ion transport within membranes from quaternized Tröger's base (TB) polymers. This result stimulated many other studies related to positively charged PIMs, including cross-linked quaternized TB polymers, trimethyl benzyl ammonium-grafted ladder polymers [35], and copolymers of poly(spirobiindane-aryl ether sulfone) with quaternized side chains [36]. Negatively charged PIMs are also employed to construct IEMs. For example, Tan et al. reported a series of amidoxime-based PIMs synthesized from the post-modification of dibenzodioxin-based ladder polymers [37]. Membranes of these polymers are effective in selective ion conduction in alkaline aqueous solutions. However, they may become much less conductive in neutral media, as amidoxime groups are ionizable only in aqueous solutions with pH values exceeding 13. Sulfonated PIMs can avoid this problem, and the current materials developed are synthesized through the post-sulfonation of polyxanthene [38] and ladder polymers based on spirobifluorene [39]. As ion transfer in PIM-based IEMs occurs in micropores, control over pore chemistry and pore architecture is at the center of related research.

Apart from organic polymeric materials, inorganic materials, including zeolites, bentonite, and phosphate salts, can also be exploited to construct IEMs [40–42]. These materials can withstand higher temperatures but are subjected to high costs, relatively poor electrochemical properties, and poor mechanical flexibility [43]. To combine the advantages of both inorganic and organic materials, hybrid IEMs were developed in the late 1990s. They were fabricated by a sol–gel approach that could incorporate inorganic components (e.g. silica) into polymer membranes. These developed membranes were usually applied in severe conditions, such as higher temperatures and strongly oxidizing circumstances [43–45].

Until now, various IEMs, including inorganic–organic (hybrid) IEMs, amphoteric IEMs, mosaic IEMs, and bipolar membranes, have been available, and their position in the IEM development history is schematically shown in Figure 1.2a. With the improved properties of membrane materials in terms of selectivity, membrane resistance, thermal and chemical stability, and mechanical properties, IEM-based processes have also evolved and continuously infiltrated extensive applications (Figure 1.2b). These processes have received increasing interest in the process industries related to food, drug, biotechnology, and wastewater treatment [46–50] and in emerging technologies for coping with environmental and energy issues, such as CO_2 or water electrolyzers [51–53], aqueous organic flow batteries [54, 55], and fuel cells [56].

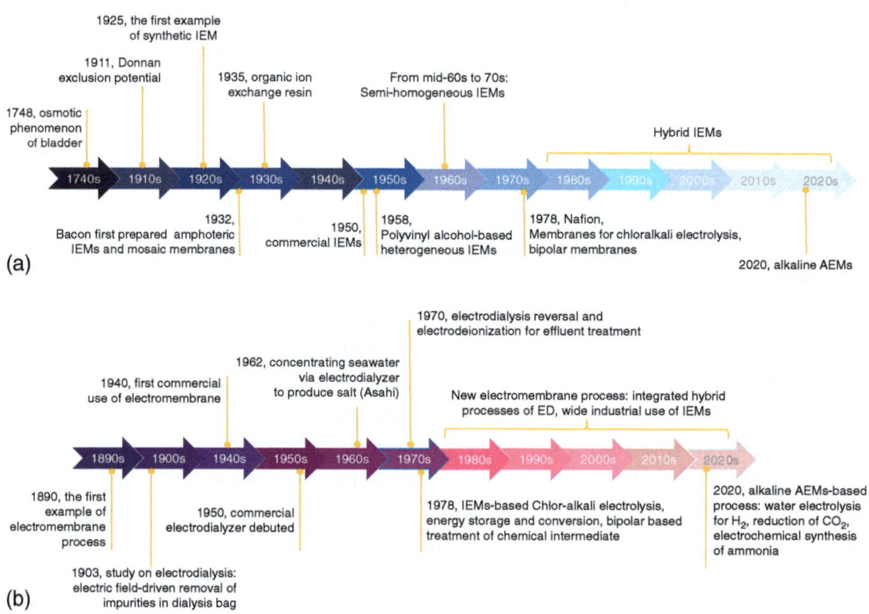

Figure 1.2 (a) The development history of IEMs and (b) IEM-based processes.

1.3 Preparation of IEMs

The synthetic approach of materials and their fabrication process for IEMs are crucial for large-scale preparation of IEMs. The synthetic approaches include the construction of a hydrophobic polymer backbone and introduction of ionic groups and are classified into three categories based on the starting materials [2, 57]: (i) post-modification approach (grafting charged groups onto polymers), (ii) bottom-up approach (direct copolymerization between monomers containing functional groups and nonfunctionalized monomers), and (iii) blend method (blending hydrophobic polymers with hydrophilic polymers with charged groups).

Regarding the post-modification approach, a typical example is the IEM made from polystyrene and its derivatives. These polymers are synthesized starting with styrene and divinylbenzene, combining the advantages of low cost, excellent mechanical properties, and easy modification. They can be made into CEMs from post-sulfonation or AEMs from chloromethylation and subsequent quaternization, as shown in Scheme 1.1 [58–60].

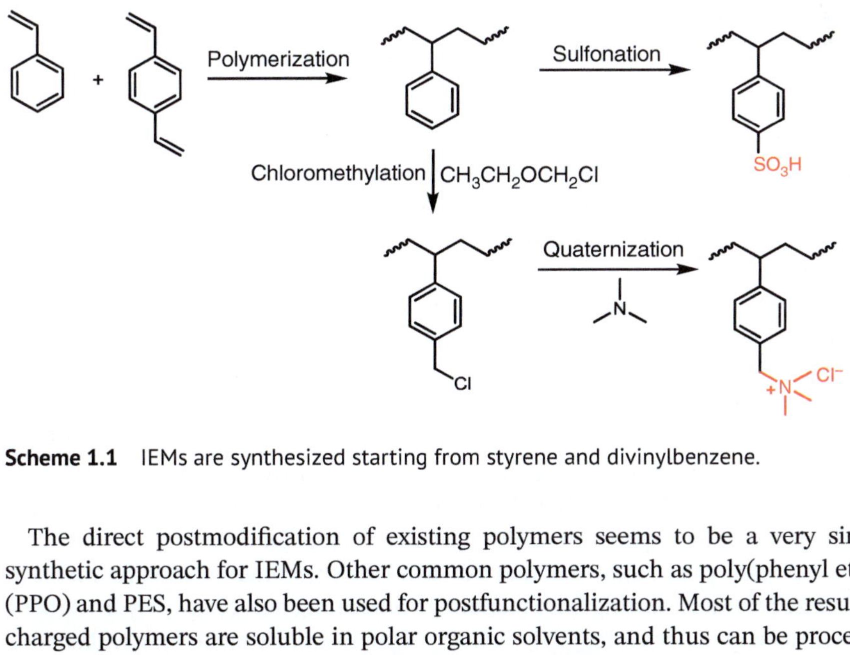

Scheme 1.1 IEMs are synthesized starting from styrene and divinylbenzene.

The direct postmodification of existing polymers seems to be a very simple synthetic approach for IEMs. Other common polymers, such as poly(phenyl ether) (PPO) and PES, have also been used for postfunctionalization. Most of the resulting charged polymers are soluble in polar organic solvents, and thus can be processed into self-supporting films via a solution-casting method [61–69]. To improve the anti-swelling property, crosslinking is sometimes required [70–80]. For example, we triggered the crosslinking of sulfonated PPO membranes through heat treatment and observed that the resulting membranes were insoluble and maintained a high tensile strength of 52.8 MPa [74].

It is noteworthy that, in comparison to the synthesis of CEMs (usually post-sulfonation is used), the preparation of AEMs seems to be more complicated and costly. The process for AEM usually requires chloromethyl methyl ether

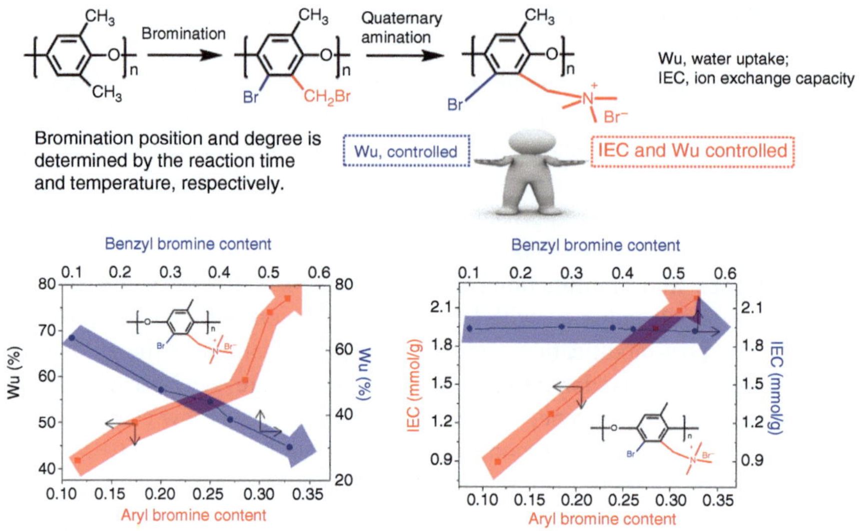

Figure 1.3 A novel synthetic route for anion exchange membrane and water uptake (WU) and evaporated ion exchange capacity (IEC) manipulating strategies based on BPPO and quaternary amination.

during the chloromethylation process, a carcinogen reagent that may cause great harm to human health [81]. Considering the economy and practice, as well as the diversity in membrane formation, direct halogenation of methyl groups that are linked to the benzene ring, instead of direct chloromethylation of the benzene ring, is a better choice due to the cancelation of the use of chloromethyl methyl ether. For example, a simple and practical method for the preparation of AEMs is shown in Figure 1.3. This preparation included the bromination of PPO and subsequent quaternization. Control over the reaction position (benzyl substitution or aryl substitution) and content of bromination, as well as the amination-crosslinking processes, are key to the properties of the resulting membranes [82, 83]. Based on this approach, a series of IEMs have been commercially manufactured and have been widely used in diffusional dialysis to recover inorganic acid under different conditions [84–86], electrodialysis [87], nanofiltration, ultrafiltration [88, 89], fuel cells [90], and in the separation of different anions [91, 92].

Another widely used approach that can avoid the chloromethylation process is to synthesize polymers starting from monomers containing halogenated hydrocarbons (e.g. vinylbenzyl chloride [93–95]). These polymers can be made into AEMs via a simple amination reaction (Scheme 1.2a). Terephthaloyl chloride could be formed into poly(*p*-phenylene dichloroethene) with a π-conjugated system and benzyl chloride groups via the McMurry coupling reaction, and the resulting polymers could be converted into positively charged polymers through quaternization (Scheme 1.2b) [96]. Notably, piperidone and its derivatives as raw materials have recently gained great attention in making poly(arylene piperidinium) AEMs due to their outstanding alkaline stability [97, 98]. These membranes are synthesized from the super acid-catalyzed polycondensation followed by quaternization (Scheme 1.2c),

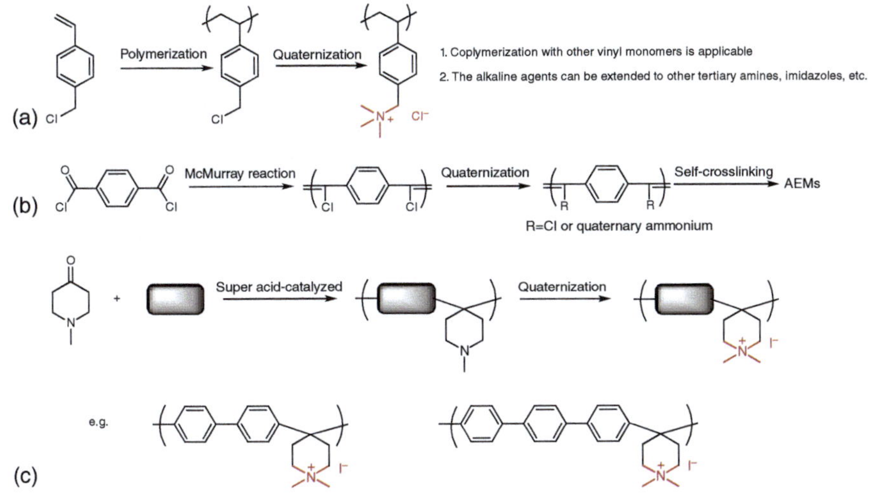

Scheme 1.2 Typical examples of the synthesis of AEMs starting from uncharged functional monomers.

and are now mainly used for AEM fuel cells. Their use in wide applications is highly expected.

For bottom-up synthesis, the copolymerization of commercially available monomers with charged groups and hydrophobic monomers seems to be a simple method. In 2004, Xing et al. reported a series of sulfonated PEK copolymers that were synthesized via aromatic nucleophilic polycondensation of 5,5′-carbonylbis (2-fluorobenzenesulfonate) (a charged monomer) with hexafluoroisopropylidene diphenol and 4,4′-difluorobenzophenone [99]. The major advantage of this approach over post-modification is the precise control over ion exchange capacity by tuning the molar ratio of reactant monomers. The resulting polymers could be processed into robust membranes from a solution-casting approach. To diversify the IEMs from bottom-up synthesis, many other polymerization processes have been employed, such as cyclo-polycondensations of dipiperidines and tetrakis(bromomethyl)benzene for N-spirocyclic quaternary ammonium ionomers [100], and acid-catalyzed polyacylation of pre-quaternized diarene monomers and dibenzoic acid monomers for PEK containing charged side chains [101, 102].

In the above solvent-aided polymerization, solvents were required to obtain the polymer solution, but may bring toxicity risks to the environment. To address this issue, a solvent-free approach via in situ polymerization has been developed [70, 103–105]. The strategy replaced the organic solvents with liquid monomers that were fully incorporated into the resulting membranes. As illustrated in Figure 1.4, brominated PPO was first dissolved in a liquid monomer mixture of vinylbenzyl chloride and styrene, instead of common organic solvents, to perform both crosslinking and quaternization, thus yielding imidazolium-based AEMs [106].

Pore soaking or filling is another approach to introduce charged segments into the pores of polymer films, producing IEMs. Porous films are usually commercially available, cost-effective, and insoluble in common solvents, such as

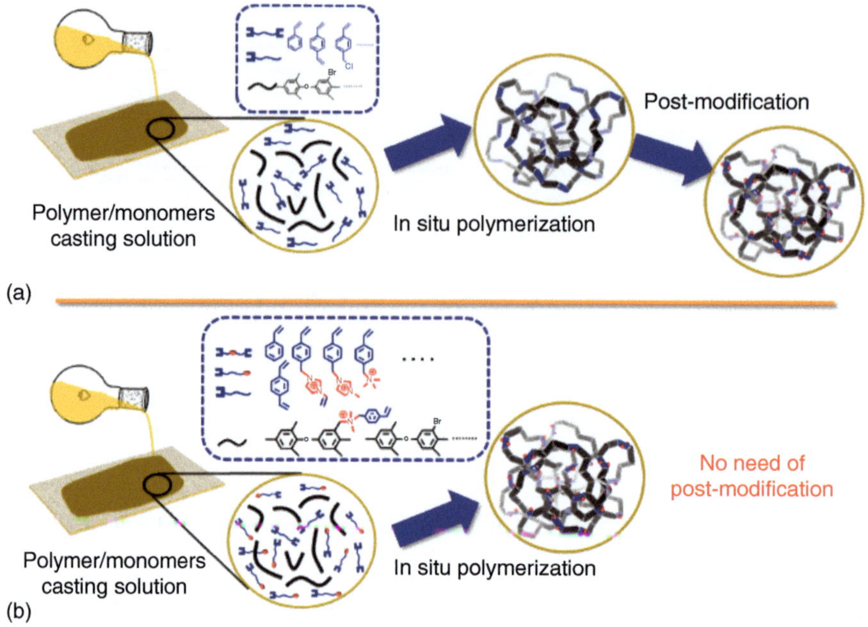

Figure 1.4 Schematic illustrations of (a) conventional, (b) upgraded in situ polymerization strategies for the preparation of IEMs.

films of hydrocarbon or fluorocarbon polymers (polytetrafluoroethylene (PTFE); fluorinated ethylene propylene (FEP), perfluoroalkoxy alkanes (PFA), ethylene tetrafluoroethylene (ETFE), poly (vinylidene fluoride) (PVDF)). Membranes made from this method exhibit excellent mechanical properties and show an advantage in manufacturing cost. For example, Young Moo Lee et al. successfully filled the pores of porous polyethylene (PE) substrates with poly(aryl-*co*-aryl piperidinium), and the reinforced membranes could be applied in alkaline membrane fuel cells [107, 108].

Although various synthetic approaches to IMEs have been developed, not all of the preparations have led to commercialization. There are many obstacles along the way from laboratory synthesis to large-scale production. As a summary of this section, Table 1.2 lists the main commercial IEMs and their main properties.

1.4 Applications

Based on advanced membrane fabrication technologies and rational material design, IEMs with a good combination of low electrical resistance, high chemical and thermal stability, excellent mechanical properties, and high ion selectivity have been developed. The rapid development of materials for IEMs promotes the evolution of traditional IEM-related processes, such as diffusion dialysis, reverser electrodialysis, membrane capacitive deionization, and bipolar membrane electrodialysis. These processes can dispose of wastewater in a wide range of new industrial

Table 1.2 The properties of commercial IEMs from famous membrane companies.

Company	Name	Type	Thickness (mm)	IEC[a] (mmol/g)	R_m [b] (Ω cm^2)	Preparation
FuMA-Tech GmbH, Germany	fumasep® FKB	CEM[c]	0.10–0.13	1.2–1.3	4–6	Polymer dissolution and casting
	fumasep® FAB	AEM[d]	0.10–0.13	1.0–1.1	4–7	
ASTOM Co., Japan	NEOSEPTA® CMX	CEM	0.17	—	3.0	Monomer polymerization
	NEOSEPTA® AMX	AEM	0.14	—	2.4	
Asahi Glass Co., Ltd., Japan	Selemion® CMV	CEM	0.13–0.15	—	2.0–3.5	Desalination
	Selemion® ASV	AEM	0.11–0.15	—	2.3–3.5	Desalination
Chemjoy Co., Ltd., China	Chemjoy® CJMCED	CEM	0.14–0.16	0.9–1.5	<2	In situ polymerization
	Chemjoy® CJMAED	AEM	0.14–0.16	0.9–1.2	<4	Desalination
Tianwei Co., Ltd., China	TWEDC	CEM	0.10–0.13	—	<4	Pore filling
	TWEDA	AEM	0.13–0.16	—	<4	Pore filling
	TWDD	BPM	0.18–0.23	—	-	BPPO matrix
Tingrun Co., Ltd., China	JCM-II	CEM	0.16–0.23	1.8–2.9	1–7	Pore soaking
	JAM-II	AEM	0.16–0.23	1.6–2.2	4–10	Pore soaking

a) IEC: ion exchange capacity.
b) R_m: membrane area resistance (0.5 mol/l NaCl, 25 °C).
c) CEM: cation exchange membrane.
d) AEM: anion exchange membrane.

applications, including salt concentration, cleaner bioproduct production, and salt resource valorization. Notably, many emerging IEM-based technologies have been widely employed to address the issues related to the environment and energy, including fuel cells that convert chemical energy to electric energy, flow batteries that store energy on a large scale, and solar-driven water splitting that produces H_2 and O_2. The detailed applications can be found in the subsequent chapters.

1.5 Potentials

IEMs have broad application prospects in the treatment of high-salinity wastewater to meet the environmental requirement of zero discharge. For example, the transformation processes of coal, such as gasification, liquefaction, and synthesis gas preparation, produce large amounts of high-salinity wastewater that contain toxic and harmful substances, such as phenols, cyanides, oils, and ammonia nitrogen. Wastewater treatment has largely hindered the development of the coal chemical industry. Currently, complex treatment processes, including pretreatment, biochemical treatment, ultrafiltration, reverse osmosis, salt concentration, and evaporation crystallization, still face great challenges of concentrated brine discharge, high investment in evaporation equipment, high operating costs, and low utilization of regenerated salt resources. In contrast, IEM-based electrodialysis technology can concentrate saltwater more economically and effectively, with a high concentration of the resulting NaCl solution that increases to 200,000 mg/l. Therefore, this technology meets the concentration requirements for salt production and offers a new solution for the treatment of coal chemical wastewater, as well as the comprehensive utilization of salt resources. With the increasing improvement of national environmental emission standards and the recognition of national environmental value (i.e. lucid waters and lush mountains are invaluable assets), IEM-based technologies will find their increasingly important role in the reduction and resource recovery of high-salinity wastewater and will transform the industrial production modes of the coal chemical industry, metallurgy, printing and dyeing, and paper-making in terms of reducing both the cost and emissions of carbon dioxide and solid waste.

The future development of IEMs should closely surround the vision of "carbon peaking and carbon neutrality." High-performing IEMs in terms of high ion conductivity, excellent chemical and mechanical stability, low manufacturing cost, and easy production on a large scale are required to promote the evolution of IEM-based technologies that cope with carbon emission issues and steadily improve the conversion efficiency of mass transfer and energy conversion in the chemical industry. These technologies include bipolar membrane electrodialysis for acid/alkali production, CO_2 or water electrolyzers that produce resources, H_2/O_2 or ammonia/O_2 fuel cells that convert chemical energy into electric energy, organic flow batteries that store energy on a large scale, and osmotic energy-conversion devices that are based on the salinity gradient. IEM-based technologies are also receiving attention in ion separation, such as mono-/divalent ion separation, Li^6/Li^7 separation, and F^-/Cl^- separation. Notably, successful membranes may differ greatly in the emphasis placed on certain aspects of their applications. For example, chemical stability is extremely important for AEMs used in alkaline fuel cells, while ion selectivity should be considered first for preparing monovalent anion permselective membranes.

It can be forecasted that an increasing number of advanced membrane materials will be developed and will constantly transform existing membrane-based technologies. The formulation of their industrial standards should be considered urgently. Along with the development of IEM materials, researchers also need to consider IEM application processes and achieve a tailored fit for IEMs. Chemical engineering issues occurring in the related processes should also be given much attention. The following aspects are suggested to be considered for the IEM community:

(1) With the assistance of computer simulation, developing structurally biomimetic membrane stack gaskets helps to optimize the distribution of feed fluids in the

membrane stack and achieve the uniform distribution of fluid streamlines in the chamber, thus effectively reducing fluid flow resistance, pressure drop, and pump loss.

(2) Advanced manufacturing technologies, such as 3D printing, photocuring molding, plasma etching, and micronano processing, are employed to manufacture "framework-style" mesh grids that contribute to regulating fluid disturbances and reducing the thickness of the membrane–solution interface layer. In some specific application scenarios, such mesh grids may also help to reduce mass transfer resistance and improve the limited current density.

(3) Based on micro/nano processing platforms, microfluidic membrane chip technology that combines a chip, membrane materials, and catalytic materials, is developed to realize synergistic coupling between mass conversion and energy conversion.

(4) On the basis of quantum chemical calculations and machine learning techniques, develop advanced intelligent membrane devices that integrate intelligent synthesis, in situ characterization, intelligent acquisition, and intelligent response. Building an artificial intelligence management and control system and interconnecting an intelligent synthetic cloud database will help membrane devices to match well with both the front-end feeding units and the back-end processing units, and thus realize chemical processes that are intelligent, intensive, and of low carbonization.

In short, it is important to understand and solve scientific problems, technical issues, and engineering matters occurring during the development, production, and application of IEMs. The future development of IEMs should closely surround both the industrial demand and the vision of "carbon peaking and carbon neutrality" and promote them into ones having better performance in terms of low resistance, high chemical and mechanical stability, and high ion selectivity. These membranes will transform IEM-based technologies, which can make many chemical processes more energy-efficient and environmentally friendly.

References

1 Hideo, K., Tsuzura, K., and Shimizu, H. (1991). Ion exchangers. In: *Ion Exchange Membranes* (ed. K. Dorfner). Berlin: Walter de Gruyter.

2 Strathmann, H. (1995). Electrodialysis and related processes. In: *Membrane Separation Technology—Principles and Applications* (ed. R.D. Stern and R.D. Noble), 214–278. Elsevier Science BV.

3 Zuo, P., Xu, Z., Zhu, Q. et al. (2022). Ion exchange membranes: constructing and tuning ion transport channels. *Advanced Functional Materials* 32: 2207366.

4 Risen, J.W. (1996). Applications of ionomers. In: *Ionomers—Characterization, Theory and Applications* (ed. S. Schuck). New Jersey: CRC Press.

5 Xu, T.W. and Huang, C.H. (2008). *Preparation and Application Technology of Ion Exchange Membrane*. Beijing: Chemical Industry Press.

6 Ostwald, W. (1890). Elektrische Eigenschaften halbdurchlässiger Scheidewände. *Zeitschrift für Physikalische Chemie* 6: 71.

7 Donnan, F.G. (1911). The theory of membrane equilibrium in presence of a non-dialyzable electrolyte. *Zeitscrift für Elektrochemie* 17: 572–581.

8 Michaelis, L. and Fujita, A. (1911). The electric phenomen and ion permeability of membranes. II. Permeability of apple peel. *Biochemische Zeitschrift* 158: 28–37.

9 Söllner, K. (1932). Uber mosaikmembranen. *Biochemische Zeitschrift* 244: 390.

10 Wassenegger H. and Jaeger K. (1940). Effecting cation-exchange in removing calcium from hard waters.

11 Meyer, K.H. and Straus, W. (1940). La perméabilité des membranes VI. Sur le passage du courant électrique à travers des membranes sélectives. *Helvetica Chimica Acta* 23: 795–800.

12 Juda, W. and McRae, W.A. (1950). Coherent ion-exchange gels and membranes. *Journal of the American Chemical Society* 72: 1044.

13 Winger, A.G., Bodamer, G.W., and Kunin, R. (1953). Some electrochemical properties of new synthetic ion exchange membranes. *Journal of the Electrochemical Society* 100: 178–184.

14 Nishiwaki, T. (1972). Concentration of electrolytes prior to evaporation with an electromembrane process. In: *Industrial Process with Membranes* (ed. R. Lacey and S. Loeb), 83–106. New York: Wiley-Interscience.

15 Mihara, K. and Kato, M. (1969) Polarity reversing electrode units and electrical switching means therefore.

16 Grot, W.G. (1973). Laminates of support material and fluorinated polymer containing pendant side chains containing sulfonyl groups. U.S.

17 Chlanda, F.P., Lee, L.T.C., and Liu, K.J. (1976). Bipolar membranes and method of making same. U.S.

18 Sebastian, Z., Oener, M.J.F., and Boettcher, S.W. (2020). Accelerating water dissociation in bipolar membranes and for electrocatalysis. *Science* 369: 1099–1103.

19 Shehzad, M.A., Yasmin, A., Ge, X. et al. (2021). Shielded goethite catalyst that enables fast water dissociation in bipolar membranes. *Nature Communications* 12: 9.

20 Hickner, M.A., Ghassemi, H., Kim, Y.S. et al. (2004). Alternative polymer systems for proton exchange membranes (PEMs). *Chemical Reviews* 104: 4587–4612.

21 Li, N. and Guiver, M.D. (2014). Ion transport by nanochannels in ion-containing aromatic copolymers. *Macromolecules* 47: 2175–2198.

22 Ge, X., Zhang, F., Wu, L. et al. (2022). Current challenges and perspectives of polymer electrolyte membranes. *Macromolecules* 55: 3773–3787.

23 Peckham, T.J. and Holdcroft, S. (2010). Structure-morphology-property relationships of non-perfluorinated proton-conducting membranes. *Advanced Materials* 22: 4667–4690.

24 Shin, D.W., Guiver, M.D., and Lee, Y.M. (2017). Hydrocarbon-based polymer electrolyte membranes: importance of morphology on ion transport and membrane stability. *Chemical Reviews* 117: 4759–4805.

25 Luo, X., Kushner, D.I., Li, J. et al. (2021). Anion exchange ionomers: impact of chemistry on thin-film properties. *Advanced Functional Materials* 31: 2008778.

26 Trigg, E.B., Gaines, T.W., Marechal, M. et al. (2018). Self-assembled highly ordered acid layers in precisely sulfonated polyethylene produce efficient proton transport. *Nature Materials* 17: 725–731.

27 Kong, Y., He, X., Wu, H. et al. (2021). Tight covalent organic framework membranes for efficient anion transport via molecular precursor engineering. *Angewandte Chemie, International Edition* 60: 17638–17646.

28 Sheng, F., Wu, B., Li, X. et al. (2021). Efficient ion sieving in covalent organic framework membranes with sub-2-nanometer channels. *Advanced Materials* 33: e2104404.

29 Yuan, S., Li, X., Zhu, J. et al. (2019). Covalent organic frameworks for membrane separation. *Chemical Society Reviews* 48: 2665–2681.

30 Budd, P.M., Ghanem, B.S., Makhseed, S. et al. (2004). Polymers of intrinsic microporosity (PIMs): robust, solution-processable, organic nanoporous materials. *Chemical Communications* (2): 230–231.

31 Low, Z.X., Budd, P.M., McKeown, N.B., and Patterson, D.A. (2018). Gas permeation properties, physical aging, and its mitigation in high free volume glassy polymers. *Chemical Reviews* 118: 5871–5911.

32 McKeown, N.B. (2017). The synthesis of polymers of intrinsic microporosity (PIMs). *Science China Chemistry* 60: 1023–1032.

33 McKeown, N.B. and Budd, P.M. (2006). Polymers of intrinsic microporosity (PIMs): organic materials for membrane separations, heterogeneous catalysis and hydrogen storage. *Chemical Society Reviews* 35: 675–683.

34 Wang, Y., Ghanem, B.S., Ali, Z. et al. (2021). Recent progress on polymers of intrinsic microporosity and thermally modified analogue materials for membrane-based fluid separations. *Small Structures* 2: 2100049.

35 Tong Huang, J.Z., Pei, Y., Liu, X. et al. (2021). Mechanically robust microporous anion exchange membranes with efficient anion conduction for fuel cells. *Chemical Engineering Journal* 418: 129311.

36 Yang, Q., Cai, Y.Y., Zhu, Z.Y. et al. (2020). Multiple enhancement effects of crown ether in Troger's base polymers on the performance of anion exchange membranes. *ACS Applied Materials and Interfaces* 12: 24806–24816.

37 Tan, R., Wang, A., Malpass-Evans, R. et al. (2020). Hydrophilic microporous membranes for selective ion separation and flow-battery energy storage. *Nature Materials* 19: 195–202.

38 Zuo, P., Li, Y., Wang, A. et al. (2020). Sulfonated microporous polymer membranes with fast and selective ion transport for electrochemical energy conversion and storage. *Angewandte Chemie, International Edition* 59: 9564–9573.

39 Ye, C., Wang, A., Breakwell, C. et al. (2022). Development of efficient aqueous organic redox flow batteries using ion-sieving sulfonated polymer membranes. *Nature Communications* 13: 3184.

40 Bishop, H.K., Bittels, J.A., and Gutter, G.A. (1969). Investigation of inorganic ion-exchange membranes for electrodialysis. *Desalination* 6: 369–380.

41 Bregman, J.I. and Braman, R.S. (1965). Inorganic ion exchange membranes. *Journal of Colloid Science* 20: 913–922.

42 Srivastava, S.K., Jain, A.K., Agrawal, S., and Singh, R.P. (1978). Studies with inorganic ion-exchange membranes. *Talanta* 25: 157–159.

43 Ohya, H., Paterson, R., Nomura, T. et al. (1995). Properties of new inorganic membranes prepared by metal alkoxide methods. Part I: a new permselective cation exchange membrane based on Si/Ta oxides. *Journal of Membrane Science* 105: 103–112.

44 Kogure, M., Ohya, H., Paterson, R. et al. (1997). Properties of new inorganic membranes prepared by metal alkoxide methods. Part II: new inorganic-organic anion-exchange membranes prepared by the modified metal alkoxide methods with silane coupling agents. *Journal of Membrane Science* 126: 161–169.

45 Ohya, H., Masaoka, K., Aihara, M., and Negishi, Y. (1998). Properties of new inorganic membranes prepared by metal alkoxide methods. Part III: new inorganic lithium permselective ion exchange membrane. *Journal of Membrane Science* 146: 9–13.

46 Dammak, L., Fouilloux, J., Bdiri, M. et al. (2021). A review on ion-exchange membrane fouling during the electrodialysis process in the food industry, part 1: types, effects, characterization methods, Fouling mechanisms and interactions. *Membranes* 11: 789.

47 Yang, X., Merenda, A., Al-Attabi, R. et al. (2022). Towards next generation high throughput ion exchange membranes for downstream bioprocessing: a review. *Journal of Membrane Science* 647: 120325.

48 Åkerman, S., Viinikka, P., Svarfvar, B. et al. (1998). Transport of drugs across porous ion exchange membranes. *Journal of Controlled Release* 50: 153–166.

49 Swanckaert, B., Geltmeyer, J., Rabaey, K. et al. (2022). A review on ion-exchange nanofiber membranes: properties, structure and application in electrochemical (waste)water treatment. *Separation and Purification Technology* 287: 120529.

50 Barros, K.S., Marti-Calatayud, M.C., Scarazzato, T. et al. (2021). Investigation of ion-exchange membranes by means of chronopotentiometry: a comprehensive review on this highly informative and multipurpose technique. *Advances in Colloid and Interface Science* 293: 102439.

51 Salvatore, D.A., Gabardo, C.M., Reyes, A. et al. (2021). Designing anion exchange membranes for CO_2 electrolysers. *Nature Energy* 6: 339–348.

52 Lees, E.W., Mowbray, B.A.W., Parlane, F.G.L., and Berlinguette, C.P. (2021). Gas diffusion electrodes and membranes for CO_2 reduction electrolysers. *Nature Reviews Materials* 7: 55–64.

53 Lagadec, M.F. and Grimaud, A. (2020). Water electrolysers with closed and open electrochemical systems. *Nature Materials* 19: 1140–1150.

54 Zhang, L., Feng, R., Wang, W., and Yu, G. (2022). Emerging chemistries and molecular designs for flow batteries. *Nature Reviews Chemistry* 6: 524–543.

55 Xiong, P., Zhang, L., Chen, Y. et al. (2021). A chemistry and microstructure perspective on ion-conducting membranes for redox flow batteries. *Angewandte Chemie, International Edition* 60: 24770–24798.

56 Jiao, K., Xuan, J., Du, Q. et al. (2021). Designing the next generation of proton-exchange membrane fuel cells. *Nature* 595: 361–369.

57 Nasefa, M.M. and Hegazy, E.-S.A. (2004). Preparation and applications of ion exchange membranes by radiation-induced graft copolymerization of polar monomers onto non-polar films. *Progress in Polymer Science* 29: 499–561.

58 Kusumoto, K., Sata, T., and Mizutani, Y. (1976). Modification of anion exchange membranes with polystyrenesulfonic acid. *Polymer Journal* 8: 225–226.

59 Meares, P. (1983). Trends in ion-exchange membrane science and technology. In: *Ion-exchange Membranes* (ed. D.S. Flett). Chichester: Horwood Ltd.

60 McRae, W.A. and Alexander, S.S. (1960) Sulfonating reagent and its use in preparing cation exchange membranes. U.S.

61 Kreuer, K.D. (2001). On the development of proton conducting polymer membranes for hydrogen and methanol fuel cells. *Journal of Membrane Science* 185: 29–39.

62 Wang, F., Hickner, M., Kim, Y.S. et al. (2002). Direct polymerization of sulfonated poly(arylene ether sulfone) random (statistical) copolymers: candidates for new proton exchange membranes. *Journal of Membrane Science* 197: 231–242.

63 Kang, M.-S., Choi, Y.-J., Choi, I.-J. et al. (2003). Electrochemical characterization of sulfonated poly(arylene ether sulfone) (S-PES) cation-exchange membranes. *Journal of Membrane Science* 216: 39–53.

64 Kerres, J., Cui, W., and Reichle, S. (1996). New sulfonated engineering polymers via the metalation route. I. Sulfonated poly(ethersulfone) PSU Udel® via metalation-sulfination-oxidation. *Journal of Polymer Science Part A: Polymer Chemistry* 34: 2421–2438.

65 Hwang, G.-J. and Ohya, H. (1998). Preparation of anion exchange membrane based on block copolymers. Part II: the effect of the formation of macroreticular structure on the membrane properties. *Journal of Membrane Science* 149: 163–169.

66 Hwang, G.-J., Ohya, H., and Nagai, T. (1999). Ion exchange membrane based on block copolymers. Part III: preparation of cation exchange membrane. *Journal of Membrane Science* 156: 61–65.

67 Zschocke, P. and Quellmalz, D. (1985). Novel ion exchange membranes based on an aromatic polyethersulfone. *Journal of Membrane Science* 22: 325–332.

68 Nagarale, R.K., Gohil, G.S., Shahi, V.K. et al. (2004). Preparation and electrochemical characterization of cation- and anion-exchange/polyaniline composite membranes. *Journal of Colloid and Interface Science* 277: 162–171.

69 Sarrazin, J., Persin, M., and Cretin, M. (2002). Conductive polymer membranes. *Macromolecular Symposia* 188: 1–12.

70 Lin, X., Liu, Y., Poynton, S.D. et al. (2013). Cross-linked anion exchange membranes for alkaline fuel cells synthesized using a solvent free strategy. *Journal of Power Sources* 233: 259–268.

71 Liu, Y., Pan, Q., Wang, Y. et al. (2015). In-situ crosslinking of anion exchange membrane bearing unsaturated moieties for electrodialysis. *Separation and Purification Technology* 156: 226–233.

72 Wu, L., Pan, Q., Varcoe, J.R. et al. (2015). Thermal crosslinking of an alkaline anion exchange membrane bearing unsaturated side chains. *Journal of Membrane Science* 490: 1–8.

73 Ertem, S.P., Tsai, T.-H., Donahue, M.M. et al. (2015). Photo-cross-linked anion exchange membranes with improved water management and conductivity. *Macromolecules* 49: 153–161.

74 Wu, D., Wu, L., Woo, J.-J. et al. (2010). A simple heat treatment to prepare covalently crosslinked membranes from sulfonated poly(2,6-dimethyl-1,4-phenylene oxide) for application in fuel cells. *Journal of Membrane Science* 348: 167–173.

75 Liu, M., Wang, Z., Mei, J. et al. (2016). A facile functionalized routine for the synthesis of imidazolium-based anion-exchange membrane with excellent alkaline stability. *Journal of Membrane Science* 505: 138–147.

76 Zhu, L., Zimudzi, T.J., Li, N. et al. (2016). Crosslinking of comb-shaped polymer anion exchange membranes via thiol–ene click chemistry. *Polymer Chemistry* 7: 2464–2475.

77 Si, K., Wycisk, R., Dong, D. et al. (2012). Rigid-rod poly(phenylenesulfonic acid) proton exchange membranes with cross-linkable biphenyl groups for fuel cell applications. *Macromolecules* 46: 422–433.

78 Ko, T., Kim, K., Jung, B.-K. et al. (2015). Cross-linked sulfonated poly(arylene ether sulfone) membranes formed by in Situ casting and click reaction for applications in fuel cells. *Macromolecules* 48: 1104–1114.

79 Xu, W., Zhao, Y., Yuan, Z. et al. (2015). Highly stable anion exchange membranes with internal cross-linking networks. *Advanced Functional Materials* 25: 2583–2589.

80 Sata, T. (1996). Anion exchange membrane with viologen moiety as anion exchange groups and generation of photo-induced electrical potential from the membrane. *Journal of Membrane Science* 118: 121–126.

81 Wright, M.E., Toplikar, E.G., and Svejda, S.A. (1991). Details concerning the chloromethylation of soluble high-molecular-weight polystyrene using dimethoxymethane, thionyl chloride, and a Lewis acid: a full analysis. *Macromolecules* 24: 5879–5880.

82 Tongwen, X. and Weihua, Y. (2001). Fundamental study of a new series of anion exchange membranes membrane preparation and characterization.pdf. *Journal of Membrane Science* 190: 159–166.

83 Tongwena, X. and Zha, F.F. (2002). Fundamental studies on a new series of anion exchange membranes: effect of simultaneous amination-crosslinking processes on membranes ion-exchange capacity and dimensional stability. *Journal of Membrane Science* 199: 203–210.

84 Tongwen, X. and Weihua, Y. (2001). Sulfuric acid recovery from titanium white (pigment) waste liquor using diffusion dialysis with a new series of anion exchange membranes—static runs. *Journal of Membrane Science* 183: 193–200.

85 Tongwen, X. and Weihua, Y. (2003). Industrial recovery of mixed acid ($HF + HNO_3$) from the titanium spent leaching solutions by diffusion dialysis

with a new series of anion exchange membranes. *Journal of Membrane Science* 220: 89–95.

86 Tongwen, X. and Weihua, Y. (2004). Tuning the diffusion dialysis performance by surface cross-linking of PPO anion exchange membranes–simultaneous recovery of sulfuric acid and nickel from electrolysis spent liquor of relatively low acid concentration. *Journal of Hazardous Materials* 109: 157–164.

87 Tongwen, X. and Weihua, Y. (2002). Effect of cell configurations on the performance of citric acid production by a bipolar membrane electrodialysis. *Journal of Membrane Science* 203: 145–153.

88 Tongwen, X. and Weihua, Y. (2003). A novel positively charged composite membranes for nanofiltration prepared from poly(2,6-dimethyl-1,4-phenylene oxide) by in situ amines crosslinking. *Journal of Membrane Science* 215: 25–32.

89 Tang, B., Xu, T., Gong, M., and Yang, W. (2005). A novel positively charged asymmetry membranes from poly(2,6-dimethyl-1,4-phenylene oxide) by benzyl bromination and in situ amination: membrane preparation and characterization. *Journal of Membrane Science* 248: 119–125.

90 Ran, J., Wu, L., Varcoe, J.R. et al. (2012). Development of imidazolium-type alkaline anion exchange membranes for fuel cell application. *Journal of Membrane Science* 415–416: 242–249.

91 Xu, T. and Hu, K. (2004). A simple determination of counter-ionic permselectivity in an ion exchange membrane from bi-ionic membrane potential measurements: permselectivity of anionic species in a novel anion exchange membrane. *Separation and Purification Technology* 40: 231–236.

92 Xu, T., Liu, Z., and Yang, W. (2005). Fundamental studies of a new series of anion exchange membranes: membrane prepared from poly(2,6-dimethyl-1,4-phenylene oxide) (PPO) and triethylamine. *Journal of Membrane Science* 249: 183–191.

93 Herman, H., Slade, R.C.T., and Varcoe, J.R. (2003). The radiation-grafting of vinylbenzyl chloride onto poly(hexafluoropropylene-co-tetrafluoroethylene) films with subsequent conversion to alkaline anion-exchange membranes: optimisation of the experimental conditions and characterisation. *Journal of Membrane Science* 218: 147–163.

94 Sata, T., Yamane, Y., and Matsusaki, K. (1998). Preparation and properties of anion exchange membranes having pyridinium or pyridinium derivatives as anion exchange groups. *Journal of Polymer Science Part A: Polymer Chemistry* 36: 49–58.

95 Varcoe, J.R., Slade, R.C., and Lam How Yee, E. (2006). An alkaline polymer electrochemical interface: a breakthrough in application of alkaline anion-exchange membranes in fuel cells. *Chemical Communications* 1428–1429.

96 Zhang, F., Zhang, Y., Sun, L. et al. (2023). A π-conjugated anion-exchange membrane with an ordered ion-conducting channel via the McMurray coupling reaction. *Angewandte Chemie International Edition* 62: e202215017.

97 Olsson, J.S., Pham, T.H., and Jannasch, P. (2018). Poly(arylene piperidinium) hydroxide ion exchange membranes: synthesis, alkaline stability, and conductivity. *Advanced Functional Materials* 28: 1702758.

98 Wang, J., Zhao, Y., Setzler, B.P. et al. (2019). Poly(aryl piperidinium) membranes and ionomers for hydroxide exchange membrane fuel cells. *Nature Energy* 4: 392–398.
99 Xing, P., Robertson, G.P., Guiver, M.D. et al. (2004). Sulfonated poly(aryl ether ketone)s containing the hexafluoroisopropylidene diphenyl moiety prepared by direct copolymerization, as proton exchange membranes for fuel cell application. *Macromolecules* 37: 7960–7967.
100 Pham, T.H., Olsson, J.S., and Jannasch, P. (2017). N-Spirocyclic quaternary ammonium ionenes for anion-exchange membranes. *Journal of the American Chemical Society* 139: 2888–2891.
101 Zhang, Z., Wu, L., Varcoe, J. et al. (2013). Aromatic polyelectrolytes via polyacylation of pre-quaternized monomers for alkaline fuel cells. *Journal of Materials Chemistry A* 1: 2595.
102 Zhang, Z. and Xu, T. (2014). Proton-conductive polyimides consisting of naphthalenediimide and sulfonated units alternately segmented by long aliphatic spacers. *Journal of Materials Chemistry A* 2: 11583.
103 Xu, T., Woo, J.-J., Seo, S.-J., and Moon, S.-H. (2008). In situ polymerization: a novel route for thermally stable proton-conductive membranes. *Journal of Membrane Science* 325: 209–216.
104 Wu, L., Zhou, G., Liu, X. et al. (2011). Environmentally friendly synthesis of alkaline anion exchange membrane for fuel cells via a solvent-free strategy. *Journal of Membrane Science* 371: 155–162.
105 Li, Y., Lin, X., Wu, L. et al. (2015). Quaternized membranes bearing zwitterionic groups for vanadium redox flow battery through a green route. *Journal of Membrane Science* 483: 60–69.
106 Pan, Q., Hossain, M.M., Yang, Z. et al. (2016). One-pot solvent-free synthesis of cross-linked anion exchange membranes for electrodialysis. *Journal of Membrane Science* 515: 115–124.
107 Hu, C., Park, J.H., Kim, H.M. et al. (2022). Robust and durable poly(aryl-co-aryl piperidinium) reinforced membranes for alkaline membrane fuel cells. *Journal of Materials Chemistry A* 10: 6587–6595.
108 Wang, H.H., Hu, C., Park, J.H. et al. (2022). Reinforced poly(fluorenyl-co-terphenyl piperidinium) anion exchange membranes for fuel cells. *Journal of Membrane Science* 644.

2

Fundamentals and Characterizations

2.1 Donnan Equilibrium

The thermodynamic equilibrium between charged colloids and electrolytes was named by Frederic Donnan, who first published it in 1911 [1]. Donnan pointed out that the charged colloids located on one side of the semipermeable membrane restrict ions with the same charge on the other side from passing through the semipermeable membrane. Based on the theory of Donnan's equilibrium, we can understand the principle of selectivity in ion exchange membranes [2].

The two-phase ion exchange membrane and salt solution system can be abstracted as a simplified model in Figure 2.1. The interface between the ion exchange membrane and the salt solution is equivalent to a semipermeable membrane, allowing salt ions to pass through. On the right, the NaCl solution dissociates into Na^+ and Cl^-. The phase of the ion exchange membrane is on the left. Suppose the membrane is a cation exchange membrane, and its ion pair can be noted as NaR. In an aqueous environment, the Na^+ ions can move freely on both sides, but functional groups (R^-) are limited on the left by the semipermeable membrane (Figure 2.1a). Driven by the concentration difference on both sides, Cl^- and Na^+ ions migrate across the semipermeable membrane, causing a concentration change by ion migration denoted as x (Figure 2.1b).

When the migration of ions on both sides reaches a dynamic equilibrium, the chemical potential of NaCl on both sides of the membrane becomes equal, that is:

$$\mu_{\text{NaCl},1} = \mu_{\text{NaCl},2} \tag{2.1}$$

Then:

$$RT \ln a_{\text{NaCl},1} = RT \ln a_{\text{NaCl},2} \tag{2.2}$$

Because the activity coefficient of ions in a dilute solution is 1, the formula above can be simplified as:

$$[\text{Na}^+]_1 \cdot [\text{Cl}^-]_1 = [\text{Na}^+]_2 \cdot [\text{Cl}^-]_2 \tag{2.3}$$

Substituting the value in Figure 2.1b into the equation above, we obtain:

$$(c_1 + x)x = (c_2 - x)^2 \tag{2.4}$$

Ion Exchange Membranes: Design, Preparation, and Applications,
First Edition. Tongwen Xu and Yaoming Wang.

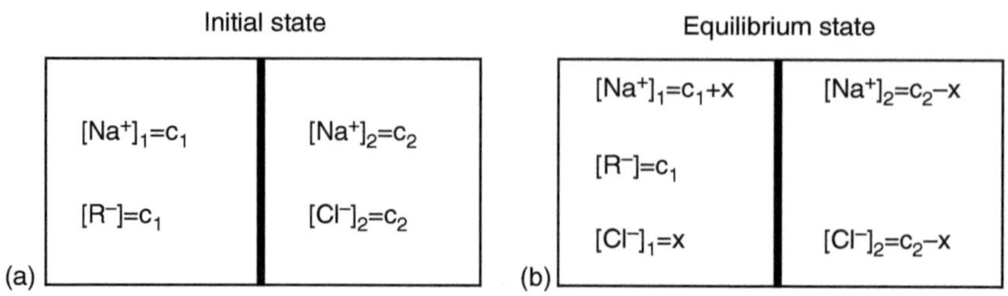

Figure 2.1 Scheme of Donnan equilibrium in the (a) initial state and (b) equilibrium state.

Solve as:

$$x = \frac{c_2^2}{c_1 + 2c_2} \text{ and } \frac{x}{c_2} = \frac{c_2}{c_1 + 2c_2} \tag{2.5}$$

x is the concentration change of Cl^- in the equilibrium state compared to the initial state, so $\frac{x}{c_2}$ is the proportion of Cl^- migrating to the left. When the initial concentration of ions on the left is much lower than the solution on the right, for example, we let $c_1 = 0.01c_2$. Then, $\frac{x}{c_2} = \frac{c_2}{c_1+2c_2} = \frac{1}{2.01} \approx \frac{1}{2}$. This result indicates that approximately half of the NaCl passes through the semipermeable membrane to the left. There is little inhibition of the transport of Cl^- and Na^+ ions, which are evenly distributed on both sides of the semipermeable membrane.

When the initial concentration of ions on the left is much higher than that on the right, the distribution of ions will be different. We let $c_2 = 0.01c_1$. Then, $\frac{x}{c_2} = \frac{c_2}{c_1+2c_2} = \frac{1}{102} \approx 0.01$, which indicates that less than 1% of Cl^- ions penetrate the semipermeable membrane. The high concentration of R^- on the left significantly hinders the migration of Cl^- ions to the left chamber. Thus, only the Na^+ ions can move freely between the two sides of the semipermeable membrane.

Based on this principle, charged groups are grafted onto the polymer backbones of membranes, endowing the membranes with ion-selective permeability. These membranes are called ion exchange membranes. Membranes that are negatively charged and can selectively penetrate cations are called cation exchange membranes. Similarly, the positively charged ion exchange membrane, called the anion exchange membrane, allows anions to penetrate. Applying an electric field to the alternately arranged anion and cation exchange membranes allows the salt solution between the membranes to be desalinated or concentrated, which is the principle of electrodialysis.

2.2 Membrane Potential

When the electrolyte concentration on both sides of the ion exchange membrane differs, the concentration difference drives the counterions to flow across the membrane, generating a potential difference on both sides. The potential difference is called the membrane potential.

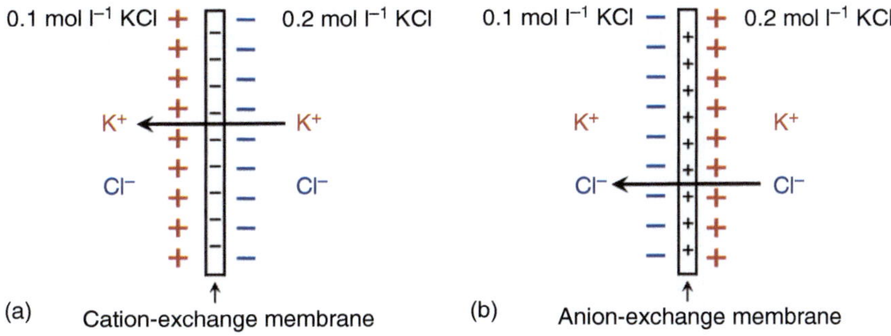

Figure 2.2 Membrane potential of (a) cation exchange membrane and (b) anion exchange membrane.

In Figure 2.2a, the concentration of the right chamber is higher than that of the left chamber. K^+ ions pass across the cation exchange membrane driven by the concentration gradient. The side where the K^+ ions migrate in becomes positive, and the side where the K^+ ions migrate out becomes negative. The situation of the anion exchange membrane is the opposite. The value of the membrane potential can be determined by the following equation:

$$E_m = (\bar{t}_+ - \bar{t}_-)\frac{RT}{zF}\ln\frac{a_1}{a_2} \tag{2.6}$$

E_m is the membrane potential. $\bar{t}_+$ and $\bar{t}_-$ are the transference numbers of cations and anions in the membrane, respectively. a_1 and a_2 are the activities of the electrolytic solutions ($a_1 > a_2$). z is the valence of the electrolyte, and F is the Faraday constant. Because $\bar{t}_+ - \bar{t}_- = 1$, combined with equation (2 – 6), we can obtain the value of $\bar{t}_+$ by the following equation:

$$E_m = (2\bar{t}_+ - 1)\frac{RT}{zF}\ln\frac{a_1}{a_2} \tag{2.7}$$

If the cation exchange membrane is the ideal membrane that only allows cations to permeate, that is, $\bar{t}_+ = 1$ and $\bar{t}_- = 0$. Then, the membrane potential reaches its maximum value, called the Nernst potential:

$$E_0 = \frac{RT}{zF}\ln\frac{a_1}{a_2} = (E_m)_{max} \tag{2.8}$$

In the experiment, the membrane potential can be measured by a lab-made two-compartment cell in Figure 2.3. These two chambers were filled with KCl solutions at different concentrations. The solutions are maintained using peristaltic pumps to stabilize their concentration and avoid concentration polarization. The potentiometer is connected to two chambers by two AgCl/Ag electrodes. It is worth noting that the E_m value obtained by the potentiometer is more accurate than that obtained by the voltmeter. When we measure E_m by a voltmeter, the voltmeter and the membrane form a circuit, which causes a decrease in the membrane potential. In addition, the voltmeter can only measure part of the membrane potential due to its resistance. To improve the accuracy, some researchers use a source meter to record the potential value at zero current on the I–V curve as the membrane potential [3].

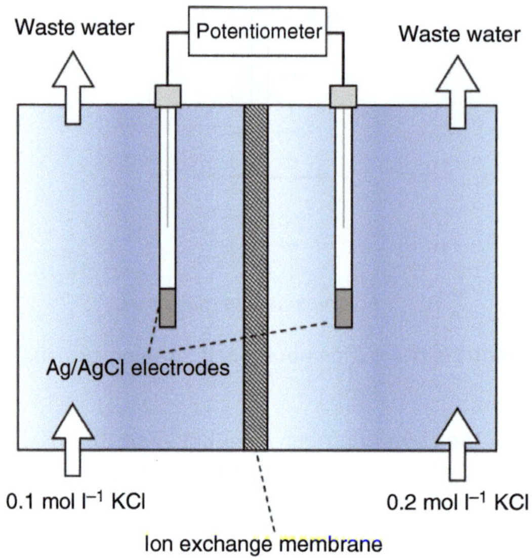

Figure 2.3 The equipment used to measure the membrane potential.

Some factors affect the value of membrane potential. According to the theory of Donnan equilibrium, the selectivity of the ion exchange membrane is based on the concentration difference between the inside and outside of the membrane. When the concentration of solutions in the chambers increases, the membrane's selectivity decreases. More coions pass through the membrane and offset the membrane potential. In addition, a higher concentration ratio of the two chambers may lead to excessive equilibrium time for measurement. Therefore, the ratio should be within the range of 2–10. Higher temperatures accelerate the transport of ions and increase the membrane potential; however, the pores of some membranes may change under elevated temperatures, leading to reduced selectivity.

2.3 Transference Number

In ion exchange membranes, ions are the primary carriers. The transference number of ions is often used to estimate the selectivity of a membrane to the ions. If I_i is the current carried by ion i, then the ratio of I_i to the total current (I) is the transference number of ion i ($\bar{t}_i$):

$$\bar{t}_i = \frac{I_i}{I} \tag{2.9}$$

The selectivity of ion exchange membranes is determined by the relative concentration of ion exchange groups within the membrane and ions in the solution. In general, the concentration of the ion exchange group in the membrane is higher than that in the external solution, and the counterion is the primary carrier in the membrane.

The expression for the transport numbers of anions and cations passing through the membrane can be derived as follows:

$$\bar{t}_+ = \frac{I_+}{I} = \frac{z_+J_+}{z_+J_+ + z_-J_-} \quad (2.10)$$

$$\bar{t}_- = \frac{I_-}{I} = \frac{z_-J_-}{z_+J_+ + z_-J_-} \quad (2.11)$$

$$\bar{t}_+ + t_- = 1 \quad (2.12)$$

where z_+ and z_- are the charges of cations and anions, respectively, and J_+ and J_- are the fluxes of cations and anions, respectively. The flux cannot be obtained directly in the experiment, so we always calculate the transference number by the membrane potential as discussed above:

$$E_m = (2\bar{t}_+ - 1)\frac{RT}{zF}\ln\frac{a_1}{a_2} \quad (2.13)$$

2.4 Diffusion Coefficient and Ion Permeability

The diffusion coefficient is an indicator of the mobility of species in the system, which includes cations, anions, and solvent molecules. The value of the diffusion coefficient depends on the movement frequency and step length of the species. In the current study, the diffusion behavior of ions within membranes varies. The structure and properties of the membrane materials have a significant impact on the forces that restrict ion transport. Diffusion coefficient measurement is crucial for mechanistic research and electrodialysis engineering.

The diffusion coefficient of the ion exchange membrane can be measured by the apparatus shown in Figure 2.4. The donating chamber on the left is filled with 0.1 M NaCl solution. The donating chamber can also be filled with other salt solutions to evaluate the diffusion coefficient of other ions. The receiving chamber on

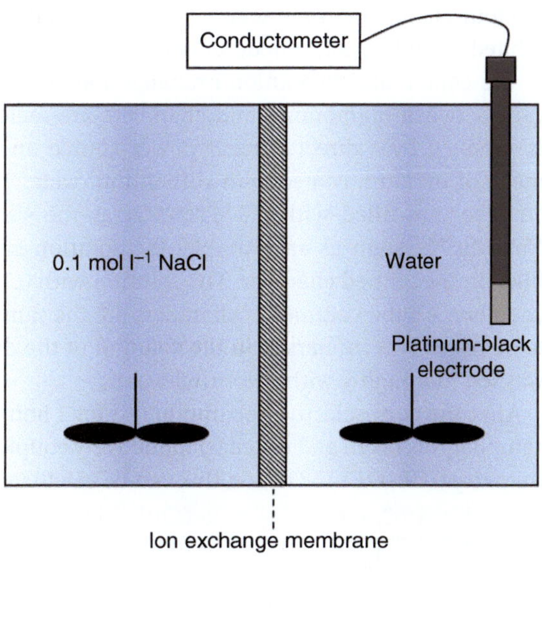

Figure 2.4 Schematic diagram of the diffusion coefficient measuring equipment. Source: Adapted from Zhou et al. [4].

the right is filled with deionized water. The two chambers are under vigorous stirring to reduce the concentration polarization and maintain uniform concentration. The conductivity of the receiving chamber can be measured by a conductometer (DDS-307A) connected to a platinum-black electrode immersed in the receiving solution. The conductivity can be converted to concentration (c) using a standard curve.

We can calculate the diffusion coefficient (D) by the following formula [4]:

$$D = \frac{\mathrm{d}c_t}{\mathrm{d}t}\frac{VL}{A(c_0 - c_t)} \approx \frac{\mathrm{d}c_t}{\mathrm{d}t}\frac{VL}{Ac_0} \tag{2.14}$$

where c_t and c_0 are the molar concentrations of NaCl measured at the end and the beginning of the measurement. A is the effective area of the membrane, and L is the thickness of the membrane. The volumes (V) of both chambers are the same. The ion permeability (P) is calculated by the following equation:

$$P = \frac{D}{L} \tag{2.15}$$

2.5 Ion Flux and Permselectivity

In the previous section, we described the ion transport in ion exchange membranes driven by the concentration gradient. However, in electrodialysis, the cations and anions in the solvent are driven by electrostatic forces, causing them to move spontaneously toward the cathode and anode. When an electric field is applied, ion transport becomes different. Therefore, the measurement methods for ion transport in ion exchange membranes have also changed markedly.

Ion flux and permselectivity are crucial indicators that demonstrate the performance of ion exchange membranes in electrodialysis. The equipment shown in Figure 2.5 is an electrodriven cation permselective single cell. From left to right, the four chambers in the scheme are the cathode chamber, concentrated chamber, diluted chamber, and anode chamber. The membrane to be measured is placed in the center, and two anion exchange membranes are placed on both sides. All gaps between membranes and chambers are sealed by silicone pads. The current is supplied by a direct current power source and applied on the two electrodes made of titanium coated with ruthenium oxide. The anode chamber and cathode chamber are filled with 0.3 M Na_2SO_4 as the supporting electrolyte. Taking the Na^+/Mg^{2+} system as an example, the solution with 0.1 M NaCl and 0.1 $MgCl_2$ is filled in the diluted chamber. The solution with 0.1 KCl is filled in the concentrated chamber. All the volumes of chambers are the same. Before the test, the membrane to be tested was immersed in the solution of the diluted chamber for 24 hours and washed thoroughly with deionized water.

After the permselective separation test for 1 hour, the solvent in the concentrated chamber was taken and tested by inductively coupled plasma–optical emission spectrometry (ICP–OES) . The ion flux can be calculated by the following equation:

$$J = \frac{V(C_t - C_0)}{A_{\mathrm{m}}t} \tag{2.16}$$

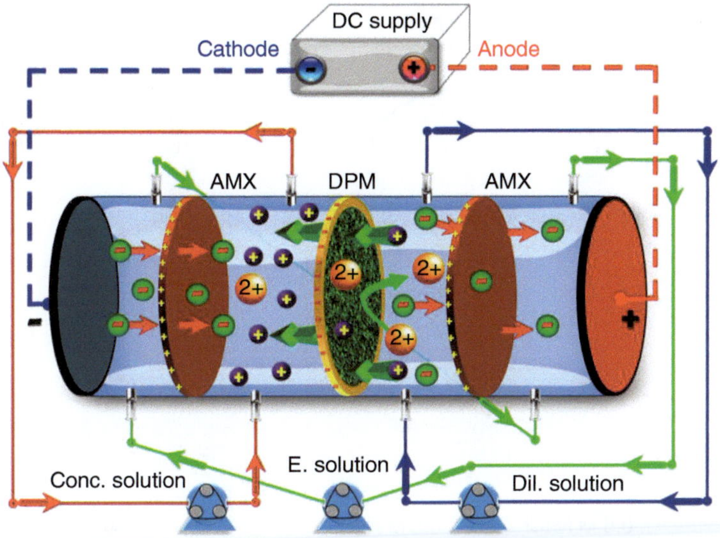

Figure 2.5 Schematic diagram of an electrodriven cation permselectivity single-cell configuration. Source: Shehzad and coworkers [5]/John wiley & Sons.

where J (mol h^{-1} m^{-2}) is the ion flux and C_0 and C_t (mol l^{-1}) are the concentrations of this ion in the concentrated chamber at the beginning and at the end, respectively. A_m is the effective area of the membrane, V is the volume of the concentrated chamber, and t is the time of the measurement. With the known flux and concentration of ion A and ion B, the permselective nature of this membrane for the two kinds of ions ($P_{A/B}$) can be calculated by the following equation:

$$P_{A/B} = \frac{J_A/C_A}{J_B/C_B} \tag{2.17}$$

2.6 Area Resistance

The ion exchange membrane is a polyelectrolyte, which can be regarded as a conductor conducting ions in the solvent. For common conductors, the resistance R can be expressed by the following formula:

$$R = \rho \frac{L}{A} \tag{2.18}$$

where ρ is the resistivity, and L and A are the length and cross-sectional area of the conductor, respectively. Generally, we can evaluate the conductive performance of the conductor by the value of ρ. Nevertheless, the situation of the ion exchange membrane is quite different. For ion exchange membranes of the same model, their thickness (L) is the same. Therefore, the constant obtained by the product of ρ and L represents the conductive performance of this model of the membrane, and this constant is the so-called area resistance. Ion exchange membranes with the same

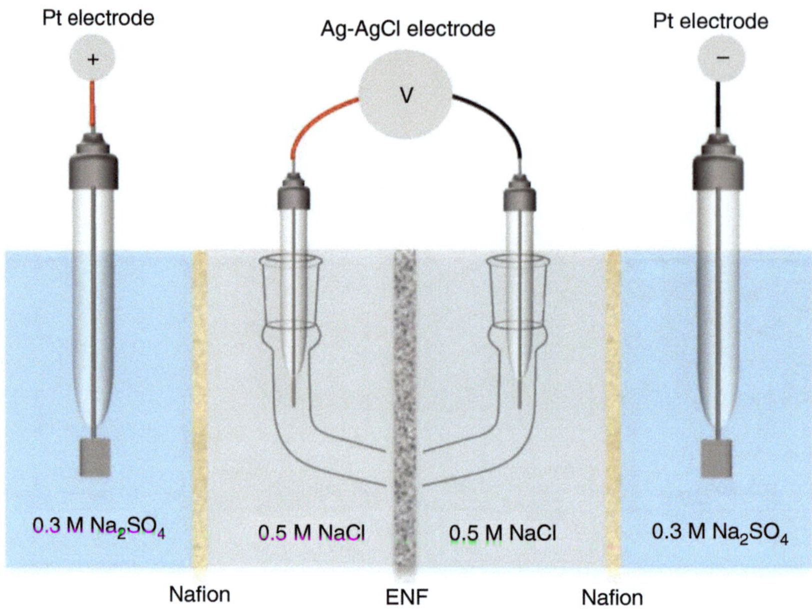

Figure 2.6 Schematic illustration of the equipment used to measure the area resistance of membranes. Source: Zhao et al. [6]/American chemical Society.

area resistance (r_m) can be considered to have the same conductive performance, regardless of how different their ρ and L are.

$$r_m = RA = \rho L \tag{2.19}$$

A four-chamber apparatus is applied to measure the area resistance of the ion exchange membrane (Figure 2.6). The two chambers in the middle are filled with 0.5 M NaCl, and the other two outside chambers are filled with 0.3 M Na_2SO_4. Two platinum electrodes in the outer chambers supply stable direct current for the system. The multimeter connected to two Ag/AgCl electrodes in the middle chambers records the transmembrane voltage drop. The Ag/AgCl electrodes are located in the Haber–Luggin capillaries almost one millimeter away from the membrane surface to reduce the interference of the solution resistance. To achieve equilibrium, the ion exchange membrane to be tested was immersed in 0.5 M NaCl overnight. In addition, two Nafion cation ion exchange membranes are placed between the middle chamber and the electrode chamber to reduce the effect of gas bubbles from the electrode reaction on the data measured [7]. During the measurement, peristaltic pumps circulate the solution in the middle chambers to avoid concentration polarization. The area resistance (r_m) can be calculated by the following equation:

$$r_m = R_m A_m = \frac{U_m}{i} \tag{2.20}$$

where r_m ($\Omega\,cm^2$) equals the resistance of the membrane (R_m) multiplied by the area of the membrane (A_m). U_m is the potential across the membrane, and i is the current density. To minimize equipment error, we calculate U_m by taking the difference between the potential with the membrane (U_{m+c}) and that without the membrane (U_c). Therefore, the equation of area resistance can be rewritten as follows:

$$r_m = r_{m+c} - r_c = \frac{U_{m+c} - U_c}{I/A_m} \tag{2.21}$$

r_{m+c} is the area resistance with the membrane, and r_c is the area resistance without the membrane. I is the current of the system.

2.7 Concentration Polarization

Concentration polarization is a special phenomenon at the interface between the membrane and the solution, and its essence is the imbalance of ion input and output at this interface. If the input number of ions exceeds the output, the ions will be exhausted at the interface. The depletion and concentration of ions generally occur in pairs, which are located on both sides of the membrane. Concentration polarization significantly impacts the electroosmosis process because it requires electrodialysis to operate at a low current density, thus limiting the processing capacity of the unit equipment.

The key factor in concentration polarization is the current density. Figure 2.7 shows the changes in the concentration of ions in the boundary layer under different current densities. Taking an anion exchange membrane as an example, all the membranes are immersed in NaCl aqueous solution in these systems, and the cathode is placed on the right. The dotted lines represent the edges of the boundary layers, and the full lines represent the concentration of ions in the solution. In Figure 2.7a, without the external electric field, the concentration of Cl^- on both sides of the membrane remains constant. When current is applied, chloride ions move to the anode on the right (Figure 2.7b). In the bulk solution, both Na^+ and Cl^- are driven by the electric field and become carriers. In the anion exchange membrane, only Cl^- ions pass through and serve as the only carrier. Therefore, the transference number of Cl^- in the membrane is twice that of the bulk solution. Excessive consumption results in a decreased Cl^- concentration on the side near the anode and an increased Cl^- concentration on the side near the cathode. When the current density continues to increase, the Cl^- concentration on the cathode side drops close to 0, and the OH^- ions become another carrier and pass through the membrane, which leads to the hydrolyzation of more water molecules to maintain the ionization balance of water (Figure 2.7c).

2.8 Limiting Current Density and Current–Voltage Curves

In electrodialysis, the salt concentration of the boundary layer of the desalination chamber decreases due to concentration polarization and reduces to 0 at the limiting current density. We can analyze the ion flux on the membrane surface on the desalination side to obtain an expression for the ultimate current density.

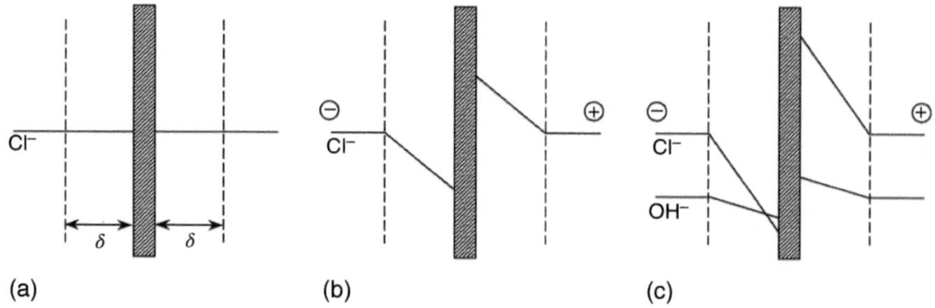

Figure 2.7 Scheme of the concentration polarization of the anion exchange membrane. (a) Membrane without current, (b) with moderate current, and (c) with excessive current.

In the boundary layer on the desalinated side, ions move from the boundary layer to the membrane driven by the electric field, forming the electrical flux of ion A in the membrane ($\bar{J}_{A(e)}$, mol cm^{-2} s^{-1}):

$$\bar{J}_{A(e)} = \frac{\bar{t}_A i}{z_A F} \tag{2.22}$$

where F is the Faraday constant, i is the current density (A cm^{-2}), $\bar{t}_A$ is the transference number of ion A in the membrane, and z_A is the valence of ion A.

The electrical flux of ion A in the boundary layer ($J_{A(e)}$) is similar to $\bar{J}_{A(e)}$:

$$J_{A(e)} = \frac{t_A i}{z_A F} \tag{2.23}$$

where t_A is the transference number of ion A in the solution.

In the solution of the boundary layer, the concentration of ion A is significantly lower than that of the bulk solution. Thus, some ions A in bulk solution diffuse to the boundary layer driven by the concentration gradient. The diffusion flux in the solution of ion A ($J_{A(d)}$) follows Fick's law:

$$J_{A(d)} = -D\frac{dc}{dx} \tag{2.24}$$

where D is the diffusion coefficient and $\frac{dc}{dx}$ is the concentration gradient.

When the system reaches a steady state, the three fluxes reach equilibrium at the interface of the boundary layer and the membrane. The fluxes of ions entering and exiting the interface are equal. Then:

$$\bar{J}_{A(e)} = J_{A(d)} + J_{A(d)} \tag{2.25}$$

$$\frac{\bar{t}_A i}{z_A F} = \frac{t_A i}{z_A F} - D\frac{dc}{dx} \tag{2.26}$$

When i reaches the limiting current density, we can integrate the formula above. In the integral, the range of x is $0 \sim \delta$, and the range of c is $0 \sim c_A$, where δ is the thickness of the boundary layer and c_A is the concentration of ion A in the solution. Then, we obtain the expression of limiting current density [8]:

$$i_{lim} = \frac{z_A F D c_A}{\delta(\bar{t}_A - t_A)} \tag{2.27}$$

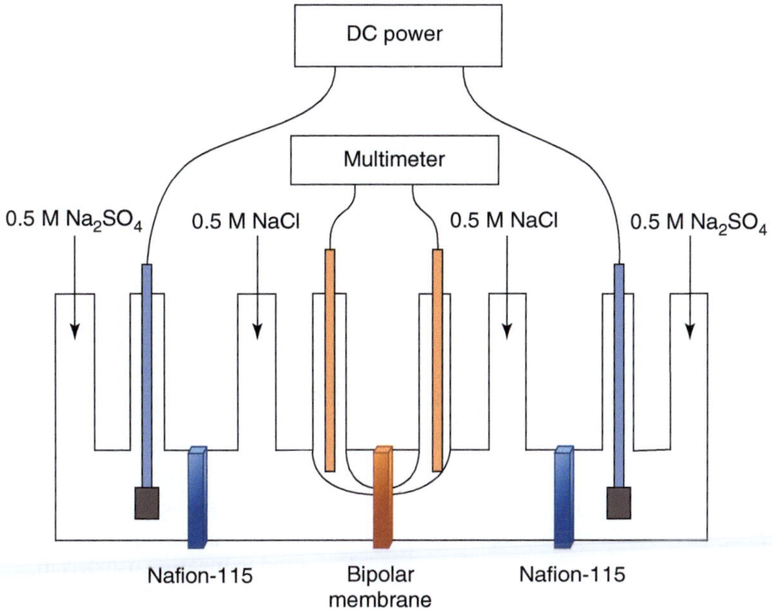

Figure 2.8 The equipment for measuring I-V curves in the bipolar membrane. Source: Adapted from Shehzad and coworkers [9].

In most cases, $\bar{t}_i$ is close to 1, and t_i is close to 0.5. In the absence of accurate transference numbers, $\bar{t}_A - t_A$ can be roughly estimated as 0.5. However, accurately measuring the thickness of the boundary layer (δ) is challenging.

In the experiment, there is a simple method to estimate the value of limiting current density with a four-compartment cell, as shown in Figure 2.8. During the measurement process, the direct current power source supplies stable current to the cells through the two platinum electrodes. The transmembrane voltage drop is measured by a multimeter connected to a pair of Ag/AgCl electrodes placed on both sides of the membrane. Then, we record the value of the transmembrane voltage drop under the current with stepwise increases and obtain the current–voltage curve, as shown in Figure 2.9. The current at the first inflection point of the curve corresponds to the limiting current density. Generally, researchers divide the I–V curve into three parts according to its slope. The first part is called the ohmic region because the current of this curve increases linearly with increasing voltage, which is in accordance with Ohm's law. In the second part, the current exceeds the limiting current, and the curve becomes flat. At this stage, chloride ions in the boundary layer have been completely exhausted, and hydroxides become the only carrier to pass across the membrane. As a result, it is difficult for the current to increase with increasing voltage, forming a plateau region in the curve. When the voltage is further increased, the current finally breaks through the value of the limiting current and continues to increase with the increasing voltage. This part of the curve is called the overlimiting region. In this region, electricity is mainly used to dissociate the water molecules to generate more hydroxides as carriers.

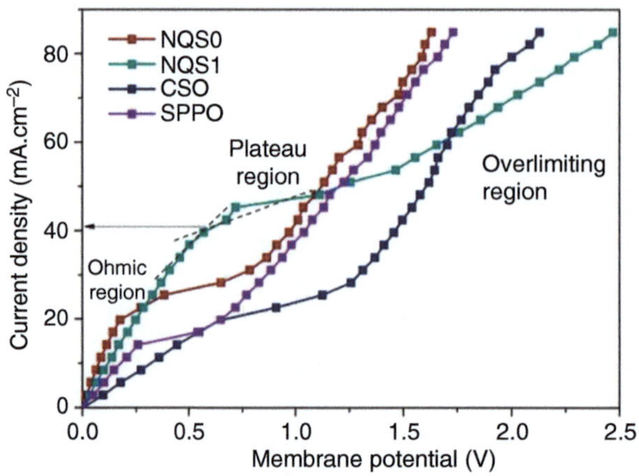

Figure 2.9 A typical current–voltage curve of an ion exchange membrane and the position of the limiting current density (the current density at the intersection of the ohmic region extension line and plateau region extension line). Source: Hou et al. [10]/with permission of Elsevier.

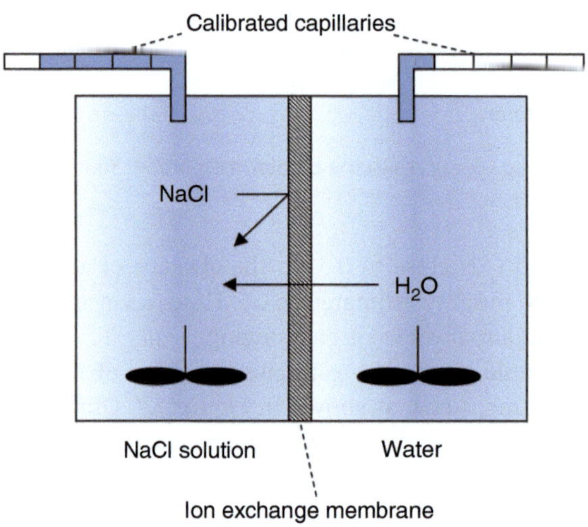

Figure 2.10 Schematic diagram of the equipment used to measure osmosis in the ion exchange membrane. Source: Adapted from Giorno et al. [11].

2.9 Water Transport

Water molecules pass through the ion exchange membrane by osmosis and electroosmosis. Without an external electric field, water molecules are transported through ion exchange membranes mainly by osmosis. In Figure 2.10, the left chamber is filled with a NaCl solution, and the right chamber is filled with water. Therefore, the concentration difference between the two chambers results in osmotic pressure, forcing water molecules to move toward the left chamber. Calibrated capillaries can measure the change in solvent volume. Osmotic pressure is directly proportional to salt concentration, and higher osmotic pressure leads to higher osmotic water volume.

The equipment used to measure electroosmosis is shown in Figure 2.11. Under the external electric field of direct current, anions and cations migrate toward the

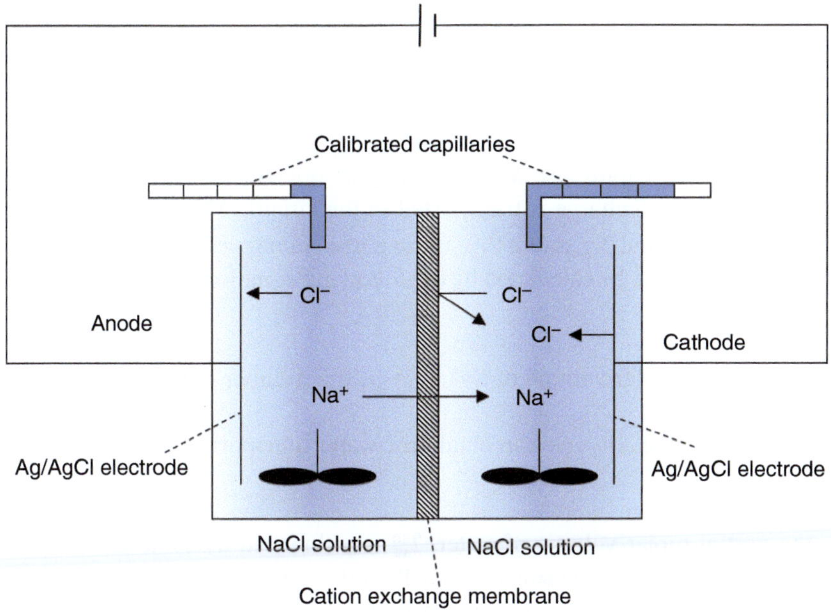

Figure 2.11 Schematic diagram of the equipment used to measure electroosmosis in the cation exchange membrane Source: Adapted from Giorno et al. [11].

anode and cathode, respectively. If a cation exchange membrane separates the two-compartment cell, Na^+ ions are transported through the membrane, while Cl^- ions are blocked. In an aqueous solution, water molecules form a solvation shell around the ions. When Na^+ migrates through the ion exchange membrane, the water molecules in the solvation shell also cross the membrane along with Na^+ ions. The transport of electroosmotic water can be calculated by the following equation:

$$T_{\mathrm{w}}^{\mathrm{m}} = \frac{FJ_{\mathrm{w}}}{i} \tag{2.28}$$

$T_{\mathrm{w}}^{\mathrm{m}}$ is the transport number of water through the membrane, and J_{w} is the water flux in moles per unit area and time. F is the Faraday constant, and i is the current density.

The change in water volume in electroosmosis can also be measured by the calibrated capillaries, but there is a difference as compared to osmosis. We assume that the membrane in Figure 2.11 is an ideal cation exchange membrane, and only Na^+ ions pass through it. Despite this, the measurement is also influenced by the changes in the volume of the Ag/AgCl electrodes. The reactions on the electrodes are shown below:

$$\text{Anode: Ag} + \text{Cl}^- \rightarrow \text{AgCl} + \text{e}$$

$$\text{Cathode: AgCl} + \text{e} \rightarrow \text{Ag} + \text{Cl}^-$$

With the current applied, the two reactions occur, and Na^+ ions migrate to the cathode with water in the solvation shell.

$$\Delta V_F = \Delta V_{tot} + \Delta \overline{V}_{AgCl} - \Delta \overline{V}_{Cl^-} \tag{2.29}$$

ΔV_F is the electroosmotic water flux, and ΔV_{tot} is the total volume change measured by calibrated capillaries. ΔV_F and ΔV_{tot} are not equal for the reactions on the electrodes. When 1 F charge is transported in the circuit, $\Delta \overline{V}_{AgCl}$ is the partial molar volume of AgCl and Ag, and $\Delta \overline{V}_{Cl^-}$ is the partial molar volume of Cl^-.

Similarly, ΔV_F can also be calculated by the following equation:

$$\Delta V_F = T^{m}_{Na^+} n^{Na^+}_{w} + \overline{V}_{Na^+} \tag{2.30}$$

$\overline{V}_{Na^+}$ is the partial molar volume of Na^+. $n^{Na^+}_{w}$ is the number of water molecules carried by Na^+ ions.

Obtaining the value of ΔV_F, we can obtain the water transport number, T^{m}_{w}:

$$\Delta V_F = T^{m}_{w} \overline{V}_{w} \tag{2.31}$$

$\overline{V}_{w}$ is the partial molar volume of water. T^{m}_{w} is influenced by some factors; for example, Na^+ ions have a large solvation shell, and their water transport number (T^{m}_{w}) is 6–8. The solvation shell of a proton is smaller, and its T^{m}_{w} is less than 3. In solutions with high salt concentrations, the solvation shells of ions are limited, and the T^{m}_{w} of ions is lower than the T^{m}_{w} of ions at low salt concentrations.

2.10 Membrane Scaling and Fouling

Ion exchange membranes have been increasingly applied in modern industry, including energy conversion, seawater desalination, and chemical production. The complex application environment leads to various scaling and fouling on the membrane surface. The cost of the membrane constitutes approximately 50% of the total cost of the apparatus [12]. Thus, prevention and control of these problems are crucial for the efficiency and cost of the membrane process.

The composition of pollutants on the membrane surface is complex and varies due to the application environment; however, the mechanisms reducing the membrane performance are similar. In most natural water sources, there are many negatively charged colloids or particles, including iron oxide, aluminum oxide, manganese oxide, carbohydrates, proteins, and aromatic substances [13–15]. These substances remain suspended in solution by electrostatic repulsion. During the electrodialysis process, they are adsorbed on the membrane surface to form a fouling layer. In addition, excess hydroxides generated by concentration polarization precipitate calcium and magnesium ions in the solution to form the scale adhering to the membrane. Generally, pH, temperature, membrane properties, and salt concentration also affect the formation of the scaling layer [16].

The ions driven by the electric field have to pass through the fouling layer and the scaling layer before entering the membrane during their migration. Figure 2.12 shows the current–voltage curves of the ion exchange membrane before and after

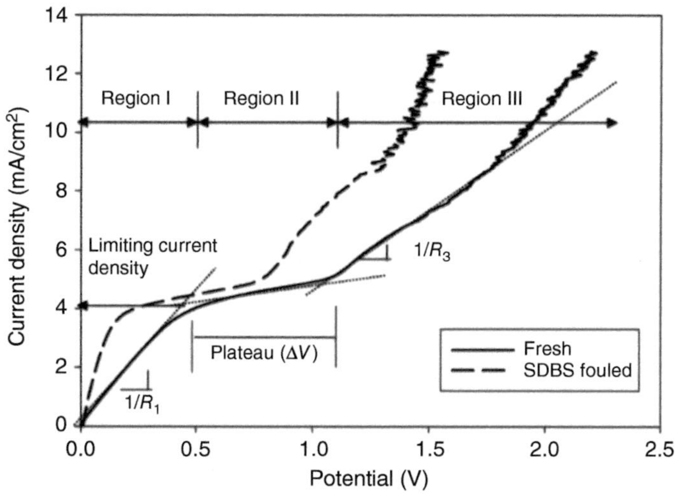

Figure 2.12 Current–voltage curves of fresh and fouled ion exchange membranes. Source: Lee et al. [17]/Elsevier.

fouling. In Region I and Region III, the slope of the fouled membrane curve is much higher than that of the fresh membrane. Thus, the fouling layer increases the resistance of the membrane and reduces the ionic conductivity. Moreover, the limiting current density drops from 4.20 to 3.90 mA cm^{-2} [17]. The fouling layer inhibits the supplement of salt ions from the bulk solution, lowering the limiting current density.

In addition to fouling and scaling, metal ions can poison ion exchange membranes and cause the degradation of membrane performance [18]. Some metal ions, especially high-valence ions, have stronger binding forces with the functional groups in ion exchange membranes. Therefore, these ions have lower transference numbers and are difficult to elute from membranes. Schalenbach et al. found that ion exchange membranes can accumulate cations from salt impurities in the feed liquid, which causes severe voltage increases in the electrolysis cells, as shown in Figure 2.13. Under a high current density, the voltage of the proton exchange membrane increases rapidly with only a tiny amount of sodium ions [19].

2.11 Zeta Potential

Zeta potential can be used to characterize the stability of catalyst colloids [20], the surface charge of the ion exchange membranes [6], and reactions occurring on the membrane surface [21]. The measurement of zeta potential is based on the Gouy [22]–Chapman [23]–Stern [24] electrical double layer theory. For negatively charged particles, the negative charges on their surface attract positive ions in the solution. The positive ions are tightly adsorbed on the particle surface, forming a positive layer called the Stern layer. Outside the Stern layer, anions and cations are not tightly bound to the particles. This layer is called the diffuse layer. There is a slipping plane

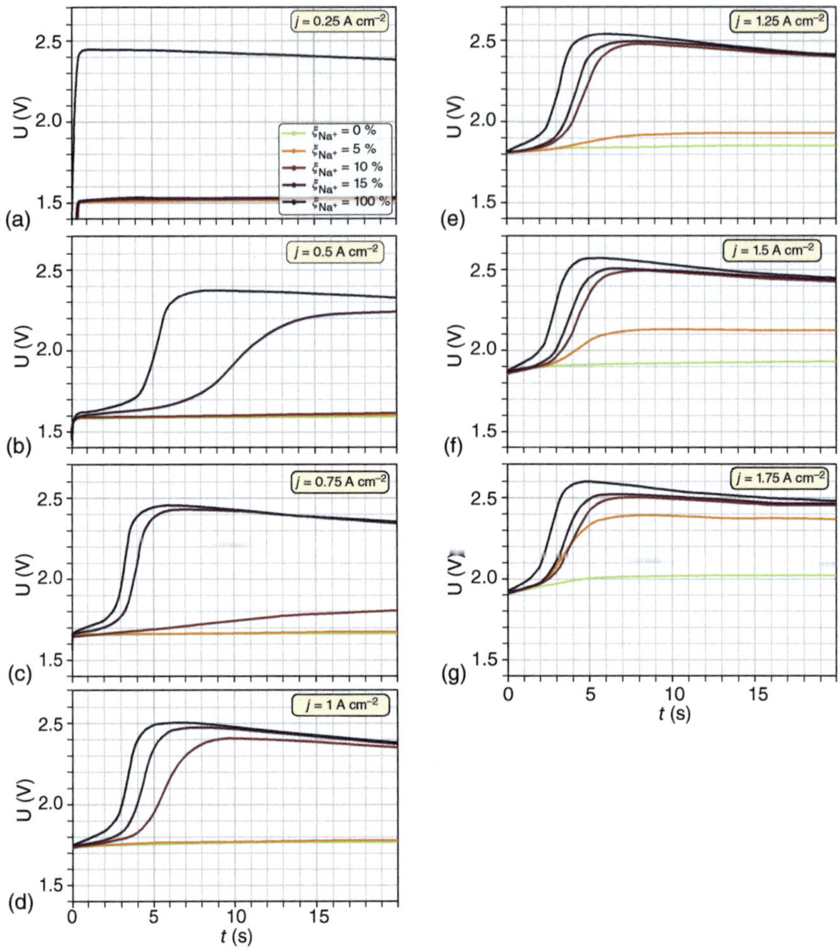

Figure 2.13 Time progression of the voltage of the electrolysis cell for various sodium ion contents of 0%, 5%, 10%, 15%, and 100%. Source: Schalenbach et al. [19]/Institute of Physics/Public domain CC-BY-4.0.

between the Stern layer and the diffuse layer. The potential difference between the slipping plane and the bulk solution is the zeta potential (Figure 2.14).

Zeta potential is crucial for the characterization of ion exchange membranes, especially during the modification process. In the study of Ge et al. $Fe(OH)_3$ catalyst colloid particles with a high zeta potential have more potent Coulomb repulsive forces and maintain good dispersion. After immersion in the catalyst colloid, the colloid attaches to the surface of the cation exchange membrane, and its zeta potential increases significantly [20].

The value of the zeta potential is affected by many factors, and pH is the most prominent factor. Let us consider negative particles as an example. When we add alkali to the solution, the particles become more negative in the presence of increased hydroxides attached to them. In contrast, if acid is added to the system, the negative charge of particles will be neutralized, which results in a higher zeta potential.

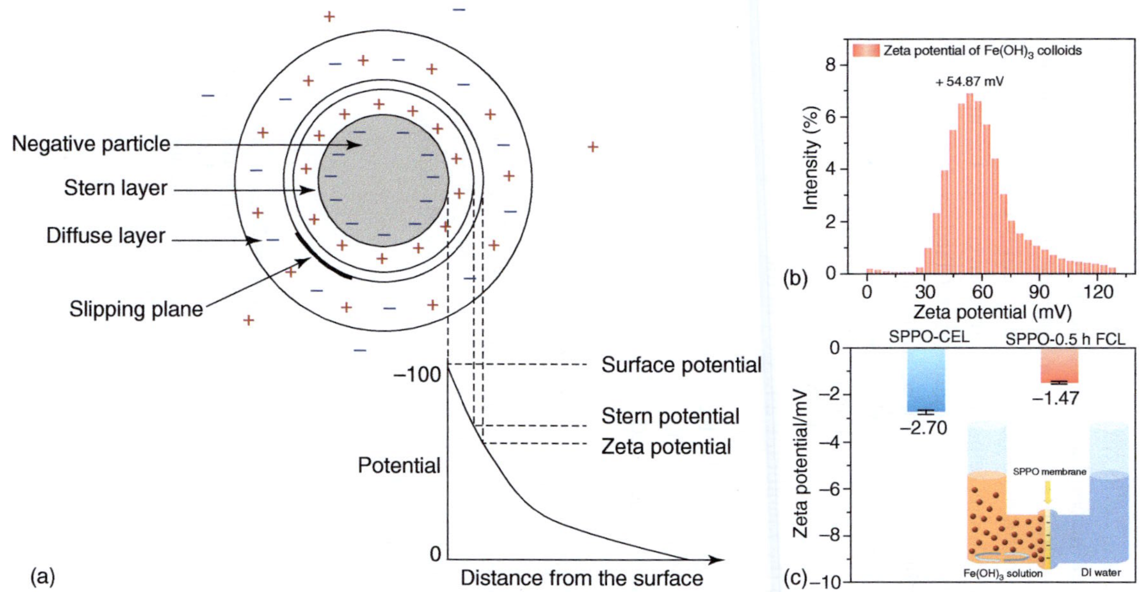

Figure 2.14 (a) Scheme of the electrical double layer (EDL) of a charged particle and its zeta potential. Zeta potential of (b) $Fe(OH)_3$ colloid and (c) cation exchange membrane before and after modification (the inset is the apparatus scheme of the membrane modification). Source: Ge et al. [20]/Elsevier.

2.12 Other Conventional Characterization

2.12.1 Conductivity

In a dry state, ion exchange membranes are almost insulated. When the membranes are immersed in aqueous solutions, functional groups in membranes dissociate into charged fixed groups and act as carriers. The faster the migration of ions within the membrane, the higher the conductivity of the membrane.

To measure the conductivity of the membrane, the membrane should be pretreated first. The membrane sample of 4 cm × 1 cm was immersed in 1 M aqueous NaOH for 12 hours at room temperature to convert the membrane to the OH^- form. Then, the extra water on the membrane surface are removed by washing and immersing the membrane in deionized water sevaral times. After this, the membrane is placed in a Teflon fixture, which has two current-collecting electrodes and two potential sensing electrodes (Figure 2.15). The fixture and the membrane are immersed in a deionized water bath. During these processes, it is crucial to reduce the contact between the membrane and CO_2 in the air to prevent the conversion of OH^- within the membrane to CO_3^{2-}[25]. The conductivity measurement is based on the standard four-point probe technique under the galvanostatic mode with a frequency range of 1 MHz to 300 Hz.

The conductivity (σ) is calculated by the following formula:

$$\sigma = \frac{1}{R} \times \frac{L}{S} = \frac{L}{R \times T \times W}\ (\mathrm{S \cdot cm^{-1}})$$

where L is the distance between potential sensing electrodes (1 cm) and R is the ohmic resistance measured from the Nyquist plot. L/S is the conductivity cell constant. S, T, and W are the cross-sectional area, thickness, and width of the membrane sample, respectively.

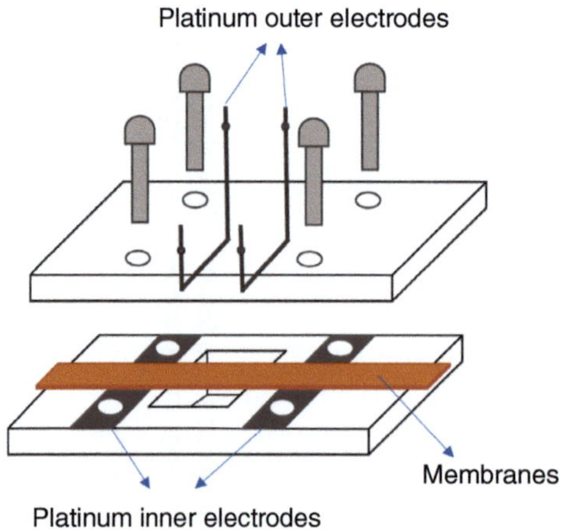

Figure 2.15 Schematic illustration of the experimental setup to measure the conductivity of ion exchange membranes. Source: Adapted from Zuo et al. [3].

2.12.2 Ion Exchange Capacity

When the ion exchange membranes are immersed in a salt solution, the ions loosely bound within the membrane are gradually displaced by counterions in the solution. To evaluate the ion exchange property of these membranes, an indicator, ion exchange capacity (IEC), is defined based on the concentration of active ion exchange groups within the membrane. IEC is represented by milliequivalents of ions exchanged per gram of dry ion change membrane (meq g^{-1}), milliequivalents of ions exchanged per cubic centimeter of the membrane (meq cm^{-3}), or millimoles of ions exchanged per gram of dry ion change membrane (mmol g^{-1}).

IEC significantly impacts the conductivity, stability, and selectivity of ion exchange membranes. The values of IEC are mainly related to the chemical structure of ion exchange membranes. When the structural units of the backbone have a high molecular weight or there are fewer functional groups grafted on the backbone, the IEC of the membrane is lower. In general, membranes with higher IEC have higher water uptake and conductivity; however, their mechanical strength will decrease accordingly. Based on the Donnan equilibrium, a higher IEC enhances the concentration of functional groups within the membrane, which is beneficial for improving selectivity. However, excessive IEC also leads to swelling of the membrane. A more porous structure intensifies the coion transport and reduces the selectivity of the membrane. Therefore, keeping the value of IEC within a reasonable range is crucial for the performance of ion exchange membranes.

Generally, the IEC is achieved by titration of ion exchange solutions of membranes. Take an example of a cation exchange membrane with quaternary ammonium functional groups in the Cl^- form. When the membrane is immersed in a 1 M NaOH aqueous solution for 24 hours, the Cl^- ions in the membrane can be exchanged for OH^- and are retained in the solution by the following chemical equation:

$$\mathrm{R-N(CH_3)_4Cl + OH^- \rightarrow R-N(CH_3)_4OH + Cl^-}$$

Then, we titrated the solution obtained with 0.01 M $AgNO_3$ and judged the titration endpoint by a K_2CrO_4 colorimetric indicator. The formulas to calculate IEC are as follows:

$$\mathrm{IEC(mmol/g)} = \frac{V_{\mathrm{AgNO_3}} \cdot C_{\mathrm{AgNO_3}}}{W_{\mathrm{dry}}} \tag{2.32}$$

where $V_{\mathrm{AgNO_3}}$ is the volume of $AgNO_3$ solution consumed, $C_{\mathrm{AgNO_3}}$ is the concentration of $AgNO_3$, and W_{dry} is the weight of the dry anion exchange membrane.

The principle of IEC measurement for cation exchange membranes is similar. To measure it, we converted the cation exchange membrane to the H^+ form and exchanged the H^+ ions within the membrane with a 1 M NaCl solution. Then, the obtained solution was titrated by NaOH solution with a phenolphthalein colorimetric indicator. It is worth noting that the IEC obtained by the same membrane under different ion forms is different, and we also cannot compare the IEC values of membranes in various ion forms.

2.12.3 Water Uptake and Swelling Ratio

The dry ion exchange membranes have to absorb water to achieve complete swelling. Water uptake describes the water required for swelling, which is defined as the ratio of water mass to dry membrane mass in percentage. The formula is as follows:

$$\mathrm{WU}(\%) = \frac{W_{\mathrm{wet}} - W_{\mathrm{dry}}}{W_{\mathrm{dry}}} \times 100\% \tag{2.33}$$

The process of measuring water uptake involves the following steps: the ion exchange membrane should be immersed in 1 M selected salt solution for six hours (repeat four times) to exchange the ion form completely. Residual salt was removed by washing and immersing the membrane in deionized water several times. After wiping off the water on the membrane surface, the wet mass of the membranes (W_{wet}) was obtained. To obtain the dry mass of the membrane (W_{dry}), we placed the membrane in a vacuum oven at 60 °C for 24 hours. Many factors affect the water uptake value, including the concentration and ion form of the solution, the concentration and dissociation ability of the functional group in the ion exchange membrane, and the crosslinking degree of the membrane.

Before and after being immersed in the solution, the ion exchange membranes absorb water and undergo dimensional changes in length, area, or volume, which is called swelling. The swelling ratio is usually calculated by measuring the size change of the membrane before and after drying. The formula is as follows:

$$\mathrm{SR}(\%) = \frac{X_{\mathrm{wet}} - X_{\mathrm{dry}}}{X_{\mathrm{dry}}} \times 100\% \tag{2.34}$$

where X_{wet} and X_{dry} are the lengths of the membranes in the wet and dry states, respectively. The calculations of the swelling ratio of area or volume are performed in similar ways.

2.12.4 Mechanical Strength

During practical applications, ion exchange membranes may face the impact of water flow and mechanical compression from the devices. The common indicators used to evaluate the mechanical strength of ion exchange membranes are burst strength and tensile strength.

Burst strength refers to the maximum vertical pressure that the membrane can withstand and is measured in MPa. The Mullen test can measure the burst strength of membranes, as shown in Figure 2.16a. During the test, the membrane is placed between two ring-like clamps. Then, compressed air flows into the air chamber with the control of the mass flow controller. The flow rate and the pressure are continuously recorded during this process. The pressure of the air chamber at the membrane rupture is the bursting pressure. Tensile strength refers to the maximum

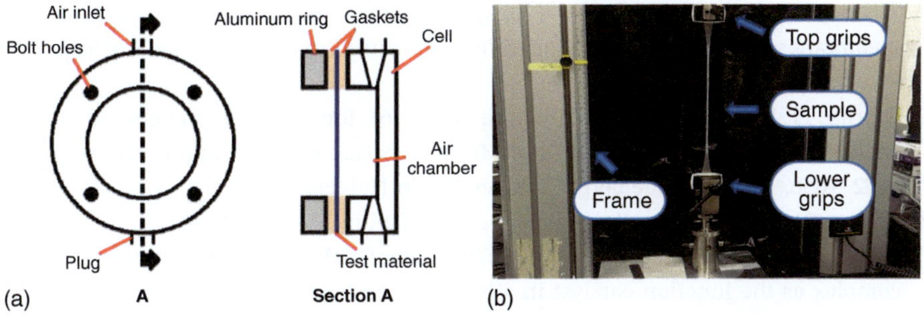

Figure 2.16 The equipment for measuring the burst strength and tensile strength of membranes. Source: Herman et al. [26]/Springer Nature/CCBY 4.0/Public domain.

tensile force that the membrane can withstand per unit cross-sectional area in MPa or N mm^{-2}. The material of the membrane should be shaped into a standard shape and placed in the materials testing machine, as shown in Figure 2.16b. During the test, the stress–strain curves are recorded. The tensile strength (σ) can be calculated by the following formula:

$$\sigma = \frac{F_b}{S_0} \tag{2.35}$$

where F_b is the maximum force that the specimen bears when it breaks and S_0 is the original cross-sectional area of the specimen.

Mechanical strength relates to the internal structures and chemical composition of membranes. Higher IEC or water uptake may lead to a loose membrane structure, reducing mechanical strength. Increasing the crosslinking degree will increase the strength but reduce the processability of membranes. Using nonwoven fabric as a substrate can significantly improve the mechanical strength, but it sacrifices part of the IEC of the membrane.

References

1 Donnan, F.G. (1911). The theory of membrane equilibrium in oresence of a non-dialyzable electrolyte. *Zeitschrift fuer Elektrochemie* 17: 572–581.

2 Donnan, F.G. (1995). Theory of membrane equilibria and membrane potentials in the presence of nondialyzing electrolytes. A contribution to physical-chemical physiology. *Journal of Membrane Science* 100: 45–55.

3 Zuo, P., Ye, C., Jiao, Z. et al. (2023). Near-frictionless ion transport within triazine framework membranes. *Nature* 617: 299–305.

4 Zhou, J., Liu, Y., Zuo, P. et al. (2021). Highly conductive and vanadium sieving Microporous Tröger's Base Membranes for vanadium redox flow battery. *Journal of Membrane Science* 620: 118832.

5 Xu, T., Shehzad, M.A., Yu, D. et al. (2019). Highly cation permselective metal–organic framework membranes with leaf-like morphology. *ChemSusChem* 12: 2593–2597.

6 Zhao, Z., Li, X., Zhang, H. et al. (2022). Polyamide-based electronanofiltration membranes for efficient anion separation. *Industrial and Engineering Chemistry Research* 61: 9869–9878.

7 Giorno, L., Drioli, E., and Strathmann, H. (2016). Ion-exchange membrane characterization. In: *Encyclopedia of Membranes* (ed. E. Drioli and L. Giorno), 1052–1056. Berlin, Heidelberg: Springer Berlin Heidelberg.

8 Zhang, W. (1995). *Elecrodialysis Engineering*. Beijing: Science Press.

9 Ge, Z., Shehzad, M.A., Ge, L. et al. (2020). Beneficial use of a coordination complex as the junction catalyst in a bipolar membrane. *ACS Applied Energy Materials* 3: 5765–5773.

10 Hou, L., Pan, J., Yu, D. et al. (2017). Nanofibrous composite membranes (NFCMs) for mono/divalent cations separation. *Journal of Membrane Science* 528: 243–250.

11 Giorno, L., Drioli, E., and Strathmann, H. (2015). Water transport in ion-exchange membranes. In: *Encyclopedia of Membranes* (ed. E. Drioli and L. Giorno), 1–4. Berlin, Heidelberg: Springer Berlin Heidelberg.

12 Grebenyuk, V.D., Chebotareva, R.D., Peters, S., and Linkov, V. (1998). Surface modification of anion-exchange electrodialysis membranes to enhance anti-fouling characteristics. *Desalination* 115: 313–329.

13 Lee, H., Park, J., Kang, M., and Moon, S. (2003). Effects of silica sol on ion exchange membranes: electrochemical characterization of anion exchange membranes in electrodialysis of silica sol containing-solutions. *Korean Journal of Chemical Engineering* 20: 889–895.

14 Banasiak, L.J. and Schäfer, A.I. (2009). Removal of boron, fluoride and nitrate by electrodialysis in the presence of organic matter. *Journal of Membrane Science* 334: 101–109.

15 Guo, H., Xiao, L., Yu, S. et al. (2014). Analysis of anion exchange membrane fouling mechanism caused by anion polyacrylamide in electrodialysis. *Desalination* 346: 46–53.

16 Mikhaylin, S. and Bazinet, L. (2016). Fouling on ion-exchange membranes: Classification, characterization and strategies of prevention and control. *Advances in Colloid and Interface Science* 229: 34–56.

17 Lee, H., Hong, M., Han, S. et al. (2008). Analysis of fouling potential in the electrodialysis process in the presence of an anionic surfactant foulant. *Journal of Membrane Science* 325: 719–726.

18 Momose, T., Higuchi, N., Arimoto, O. et al. (1991). Effects of low concentration levels of calcium and magnesium in the feed brine on the performance of a membrane chlor-alkali cell. *Journal of the Electrochemical Society* 138: 735–741.

19 Schalenbach, M., Keller, L., Janotta, B. et al. (2022). The effect of ion exchange poisoning on the ion transport and conduction in polymer electrolyte membranes (PEMs) for water electrolysis. *Journal of the Electrochemical Society* 169: 94510.

20 Ge, Z., Shehzad, M.A., Yang, X. et al. (2022). High-performance bipolar membrane for electrochemical water electrolysis. *Journal of Membrane Science* 656: 120660.

21 Shehzad, M.A., Yasmin, A., Ge, X. et al. (2021). Shielded goethite catalyst that enables fast water dissociation in bipolar membranes. *Nature Communications* 12.

22 Gouy, M. (1910). Sur la constitution de la charge électrique à la surface d'un électrolyte. *Journal de Physique Théorique et Appliquée* 9: 457–468.

23 Chapman, D.L. (1913). LI. *A* contribution to the theory of electrocapillarity. *The London, Edinburgh, and Dublin Philosophical Magazine and Journal of Science* 25: 475–481.

24 Stern, O. (1924). Zur theorie der elektrolytischen doppelschicht. *Zeitschrift für Elektrochemie und Angewandte Physikalische Chemie* 30: 508–516.

25 Zhu, Y., Ding, L., Liang, X. et al. (2018). Beneficial use of rotatable-spacer side-chains in alkaline anion exchange membranes for fuel cells. *Energy & Environmental Science* 11: 3472–3479.

26 Herman, A., Duraiswamy, N., Nandy, P. et al. (2023). Mechanical and leakage integrity testing considerations for evaluating the performance of tissue containment systems. *Scientific Reports* 13.

3

Side-chain Engineering for Ion Exchange Membrane Preparation

3.1 Principles of Side-chain Engineering

With the increasing requirements for the performance of homogeneous ion-exchange membrane (IEMs), designing IEMs at the molecular level has become an important means for obtaining access to desirable properties. The molecular design of IEMs involves the regulation of compositions and the design of structures. The composition regulation of IEMs needs to take into account several factors, such as processability and chemical stability, so the materials that can be used to prepare homogeneous IEMs are still limited to polyether sulfone, polyether ketone, polyimide, polyolefin, and other polymers. In contrast, the structural tailoring of polymers can further enrich the types of IEMs and improve their properties. In terms of molecular structure, polymeric IEMs can be divided into main-chain and side-chain architectures. The main-chain IEM refers to the ion exchange group that is directly connected with the backbone unit, while the side-chain IEM refers to the ion exchange group that is connected with the backbone through the spacer unit or that is distributed on the graft chain. In the early development of homogeneous IEMs, membrane materials mainly relied on the synthesis of main-chain polymers, but the performance of main-chain IEMs needs to be further improved, especially in terms of ion conductivity and chemical stability. In recent years, scientists in the field of IEMs have gradually focused on the exploration of side-chain structures. Specifically, side-chain IEMs are mainly developed based on Nafion-like and graft structures. Furthermore, through side-chain engineering, the length, distribution, and connection modes of side chains are regulated to further optimize the performance of IEMs and help to deeply understand the relationship between molecular structure and macroscopic performance in the field of IEMs. The theoretical basis of side-chain engineering comes from the study of Nafion and the theory of polymer microphase separation.

3.1.1 Inspiration of Nafion

The early development of side-chain structure IEMs was inspired by the perfluorosulfonic acid polyolefin IEMs. Although the perfluorinated sulfonate cation

Ion Exchange Membranes: Design, Preparation, and Applications,
First Edition. Tongwen Xu and Yaoming Wang.

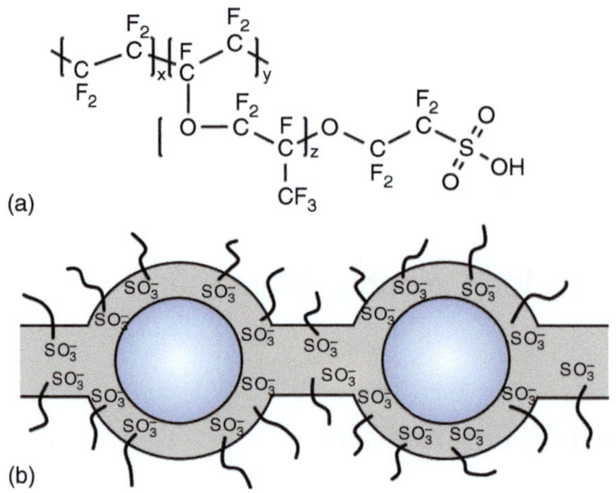

Figure 3.1 Chemical structure of Nafion and its ionic cluster morphology.

exchange membrane (CEM) developed by the DuPont company has been used for more than 50 years, it is still the benchmark of IEMs. Perfluorinated sulfonate CEMs are composed of perfluorinated olefins containing sulfonic acid groups, which are usually obtained by free radical copolymerization and the hydrolysis of tetrafluoroethylene and perfluorinated olefins containing "presulfonic acid groups" (Figure 3.1a, sulfonyl fluoride or sulfonate ester form) [1]. Perfluorinated polyolefin CEMs have excellent chemical and thermal stability and high proton conductivity. At the beginning of the study, it was believed that the excellent performance of the membrane was due solely to its perfluorinated chemical composition. On the one hand, the high bond energy of the C—F bond (497 kJ/mol) gives the membrane excellent chemical and thermal stability. On the other hand, the strong electron absorption of the CF_2 group gives its adjacent sulfonic acid group super acidity, which in turn endows the membrane with good proton conductivity. With the deepening of research, it has been found that the performance of perfluorosulfonate polyolefin cation exchange membranes is derived not only from their composition but also from their microscopic morphology [2, 3].

Gierker and his collaborators used small-angle X-ray diffraction (SAXS) and wide-angle X-ray diffraction (WAXD) to observe the membrane microstructure of Nafion (a kind of perfluorinated sulfonate CEM) manufactured by DuPont. A "cluster-network" model has been proposed to describe the microstructure of Nafion membranes (Figure 3.1b) [4, 5]. In this model, sulfonated perfluoroalkyl ether groups are aggregated to form ion clusters with a size of approximately 4 nm, and these ion clusters are connected by channels with a size of approximately 1 nm to form connective ion transport channels. Due to the existence of vehicular and Grotthuss mechanisms, protons can be transported at high speed. In addition, Roche et al. found that Teflon chain segments can form crystalline regions to confer excellent mechanical and chemical stability to Nafion membranes [6, 7]. In short, the ideal hydrophilic and hydrophobic phase separation structure of Nafion gives it excellent performance. Although the Nafion membrane has excellent comprehensive properties, its complex synthesis process and the usage of highly toxic fluorinated monomers make it too expensive. Based on the in-depth study of Nafion membranes, developers of homogeneous IEMs began to imitate the structure of Nafion and prepare short side-chain IEMs with similar structures to Nafion using side-chain engineering to find alternatives to Nafion.

3.1.2 Microphase Separation of Grafted Polymers

Through the in-depth investigation of Nafion, it can be confirmed that the beneficial microphase separation morphology is the key factor in gaining excellent performance for homogeneous IEMs. In addition to the side-chain structure of Nafion, grafted side-chain-based structures are also expected to self-assemble to form an ideal microphase separation morphology. The graft structure refers to one or more polymer chains connected to repetitive units of the linear polymer skeleton, forming a brush-like structure [8]. Due to the different compositions of the main chain and side chain, they are incompatible and form a microphase separation structure with aggregations of ionic moieties. The resulting microphase separation morphology can be tailored by changing the composition of main and side chains, the length of chain segments, the volume of each chain segment, and the crystallization states of each chain segment [9]. By adjusting these parameters, diverse self-assembly morphologies can be constructed, such as spherical, cylindrical, biconcontinuous, and layered structures. The order of the constructed morphology is determined by the regularity of the polymer structure, which requires that the grafted chains have the same degree of polymerization as possible. Therefore, "grafting-from," "grafting-through," and "grafting-onto" strategies based on active polymerization technology are often used to synthesize grafted IEMs [10].

3.2 Construction of Side-chain Architecture

At present, the design of side-chain IEMs is mainly realized based on aromatic polymers. Aromatic polymers mainly consist of aromatic rings or aromatic heterocyclic units. The aromatic backbone endows these polymers with excellent chemical, thermal, and mechanical properties. Commercial aromatic polymers include polyphenylene ether (PPO), polyphenylene sulfide (PPS), polyaryl ether ketone (PEK), polyether ether ketone (PEEK), polyaromatic ether sulfone (PES), polyimide (PI), polybenzimidazole (PBI), and polybenzoxazole (PBO). Through chemical modification of these commercial aromatic polymers, the preparation of side-chain IEMs can be completed. Alternatively, starting from the design of sulfonated monomers, the synthesis of aromatic side-chain IEMs can also be achieved (Figure 3.2).

3.2.1 Design of Side-chain CEMs

3.2.1.1 Design of Side-chain CEMs with Similar Nafion Structures

Synthesis of Nafion-like Structures via Post-functionalization. When the aromatic monomer with aromatic rings in its side chain is selected for polymerization, the

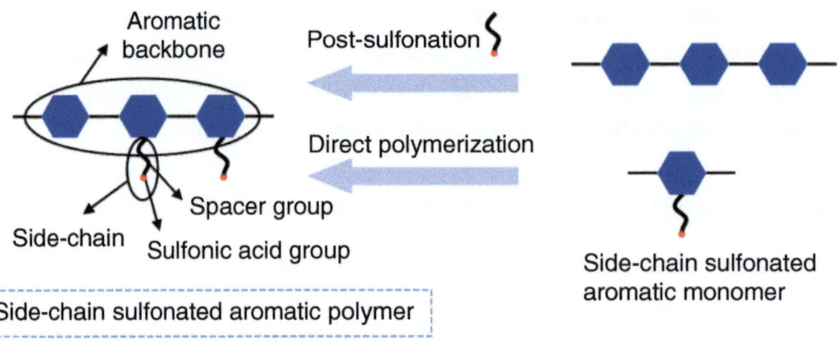

Figure 3.2 The available strategies for preparing side-chain CEMs with similar Nafion structures.

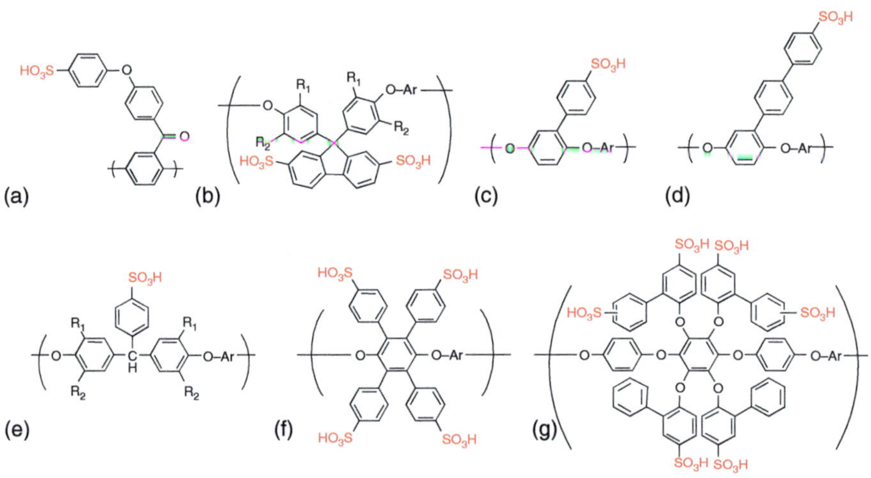

Figure 3.3 Side-chain sulfonated aromatic polymers synthesized by the electrophilic sulfonation reaction.

corresponding side-chain aromatic polymers can be obtained. Then, the appropriate sulfonation reagent can be selected through the electrophilic sulfonation reaction to obtain sulfonated side chains. This method was first applied to commercial polymers. The poly(4-phenoxy-benzoyl-1,4-phenylene) was dissolved in concentrated sulfuric acid under a nitrogen atmosphere and sulfonated at room temperature. The benzene ring on the side chain was sulfonated, but the benzene ring on the main chain was not (Figure 3.3a). The final degree of sulfonation can reach more than 80% by regulating the sulfonation time [11].

Commercialization of polymers with aromatic ring side chains is limited, so most side-chain CEM synthesis starts with monomer design. The monomer containing an aromatic ring on the side chain was first synthesized and then reacted with the proper monomer via polycondensation to obtain targeted polymers with aromatic side chains. Finally, the side-chain benzene rings were sulfonated to fabricate side-chain CEMs. The key to this method is the design and synthesis of monomers containing aromatic side chains.

Diphenol monomers containing fluorene can polymerize with dihalide monomers under alkaline conditions to obtain high molecular weight polymers, and then sulfonation is completed by dropping low-concentration chlorosulfonic acid (0.1 mol/l) into the polymer solution (Figure 3.3b). The sulfonation process can only sulfonate the 2 and 7 positions of fluorene without affecting the main chain. For example, side-chain PES CEMs containing fluorene with molecular weights over 40,000 can be obtained by the nucleophilic reaction of 9,9-bis(4-hydroxyphenyl) fluorene and bis(tetrafluorophenyl) sulfone and sulfonated by chlorosulfonic acid [12]. Similarly, the polycondensation reaction of diphenol monomers containing fluorene with 4,4′-difluorophenone and subsequent sulfonation with chlorosulfonic acid can prepare PEEK side-chain CEMs containing fluorene, and its sulfonation degree can reach up to 150% [13].

Biphenyl-2,5-diol is another monomer that can be used to prepare side-chain CEMs by post-sulfonation (Figure 3.3c) [14]. Such a monomer reacted with 4,4′-difluorophenone via the polycondensation reaction and was further sulfonated by chlorosulfonic acid. It was found that only the benzene ring on the side chain was sulfonated, while the benzene ring on the main chain was not affected. Only 1 hour of sulfonation reaction can make the final product sulfonation degree nearly 100%. The phenyl group on the side chain of the polymer can also be sulfonated by concentrated sulfuric acid, but after 20 hours of reaction, no sulfonation products were detected. After 15 days of reaction, only 40% of the side-chain benzene rings were successfully sulfonated. In contrast, the synthesis of diphenyl-structured PEEK can significantly improve the sulfonation activity (Figure 3.3d) [15]. Three days of sulfonation can increase the degree of sulfonation to 100%. Bisphenol monomers containing triphenyl or tetraphenyl groups can also be used to synthesize side-chain CEMs (Figure 3.3e and f) [16]. However, both bisphenol monomers were easily oxidized in the process of polymerization with perfluorinated biphenyls, even in an atmosphere of nitrogen. The degradation of bisphenol monomers can be avoided by reacting bisphenol monomers with isonitrile esters to obtain carbamate, which can be redecomposed to react with perfluorinated biphenyl at 150 °C to obtain polyaryl ether. The obtained polyaryl ether was sulfonated by chlorosulfonic acid, and sulfonation only occurred at the para position on the benzene ring of the side chain [17].

To further improve the sulfonation degree of side-chain sulfonated CEMs and the aggregation degree of sulfonic acid groups, researchers designed a diphenol monomer containing eight sulfonation sites (Figure 3.3g) [18]. This monomer can be directly polycondensed with dichlorodiphenylsulfoxide to form PES. Similarly, the eight benzene ring units on the designed monomer can be completely sulfonated by chlorosulfonic acid, while the main chain will not be sulfonated due to the presence of the electron-withdrawing sulfoxide group. In the polymerization process, 2,2-bis(4-p-hydroxybenzene) hexafluoropropane containing fluorine can also be added to participate in the polymerization [19]. The introduction of fluorinated monomers can improve the hydrophobicity, which is beneficial to the aggregation of ion regions and the formation of proton transport channels.

The above work is based on the design of diphenol monomers to prepare side-chain CEMs. In addition, the design of dihalide monomers can also realize the development of side-chain CEMs. The bis(4-fluorophenyl) sulfone monomer can be polycondensed with sodium sulfide to give a high molecular weight polymer. However, the reaction of bis(4-fluorophenyl) sulfone containing a phenyl moiety on the side group with sodium sulfide yields only low molecular weight polymers. This is due to the steric effect of the ortho-substituted phenyl group, which prevents thiols from attacking fluorine atoms [20]. Bis(4-fluorophenyl) sulfone with benzene rings on its side group can only be polymerized with bisphenol monomers such as 2,2-bis(4-p-hydroxybenzene) hexafluoropropane by the nucleophilic substitution condensation reaction to obtain high molecular weight polymers. The side-chain sulfonated CEM was prepared by sulfonating the side-phenyl ring with chlorosulfonic acid. Only two benzene rings on the side group can be sulfonated. To further increase the number of side group sulfonation sites, researchers designed a monomer with four side group sulfonation sites [21]. The monomer can be polymerized with bis(tetrahydroxy benzene) sulfone to form a high molecular weight polymer, which is then sulfonated by chlorosulfonic acid to prepare a side-chain CEM with four sulfonation sites in the structural unit.

In addition to electrophilic sulfonation reactions, nucleophilic reactions can also attach sulfonic acid groups onto the side chains of aromatic polymers. The key to the success of this method still lies in the design and synthesis of polymer precursors. Commercial bisphenol A-type PES was reacted first with butyl lithium to metallize it and then with 4-fluorobenzyl chloride to introduce fluorophenyl side chains into the polymer. Fluorophenyl in the sulfoxide ortho position can covalently bond with various phenolic monomers containing sulfonic acid groups through the etherification reaction, such as 2-naphthol-6,8-disulfonate and 8-hydroxy-1,3,6-pyrene trisulfonate, thus completing the preparation of side-chain CEMs (Figure 3.4) [22]. Since no electron-donating ether is located in the para or ortho position of the sulfonic acid group, the sulfonated products obtained are stable. Brominated poly (2,6-diphenyl-1,4-phenylene oxide) (BPPO) produced by Shandong Tianwei Membrane Co., Ltd., can also be used to prepare side-chain CEMs by etherification. Since BPPOs possess active halogenated hydrocarbons, they do not need pretreatment for metallization, such as bisphenol A-type PES [23]. Bromomethyl can react with sodium 2-naphthol 6-sulfonate under a mild potassium carbonate alkaline environment, and the degree of sulfonation can be regulated by changing the amount of sodium 2-naphthol 6-sulfonate added. In addition, metallized bisphenol A-type PES can react with p-o-sulfobenzoic anhydride via a one-step ring-opening reaction to introduce side-chain aryl sulfonic acid groups [24].

Commercial polybenzimidazole can also be used as polymer precursor for the synthesis of side-chain CEMs. The active hydrogen on the imidazole unit nitrogen of the polybenzimidazole backbone can be metallized first, and then the side-chain CEM was prepared by nucleophilic reactions with sodium 4-bromomethylbenzenesulfonic acid or 1,3-propanesulfonic lactone (Figure 3.5) [25, 26]. Due to the high activity of polybenzimidazole, 100% metallization can be achieved by reacting with lithium hydride. Therefore, the sulfonation degree of the side-chain CEM based on polybenzimidazole can be easily regulated by changing the added mount of sulfonated monomers.

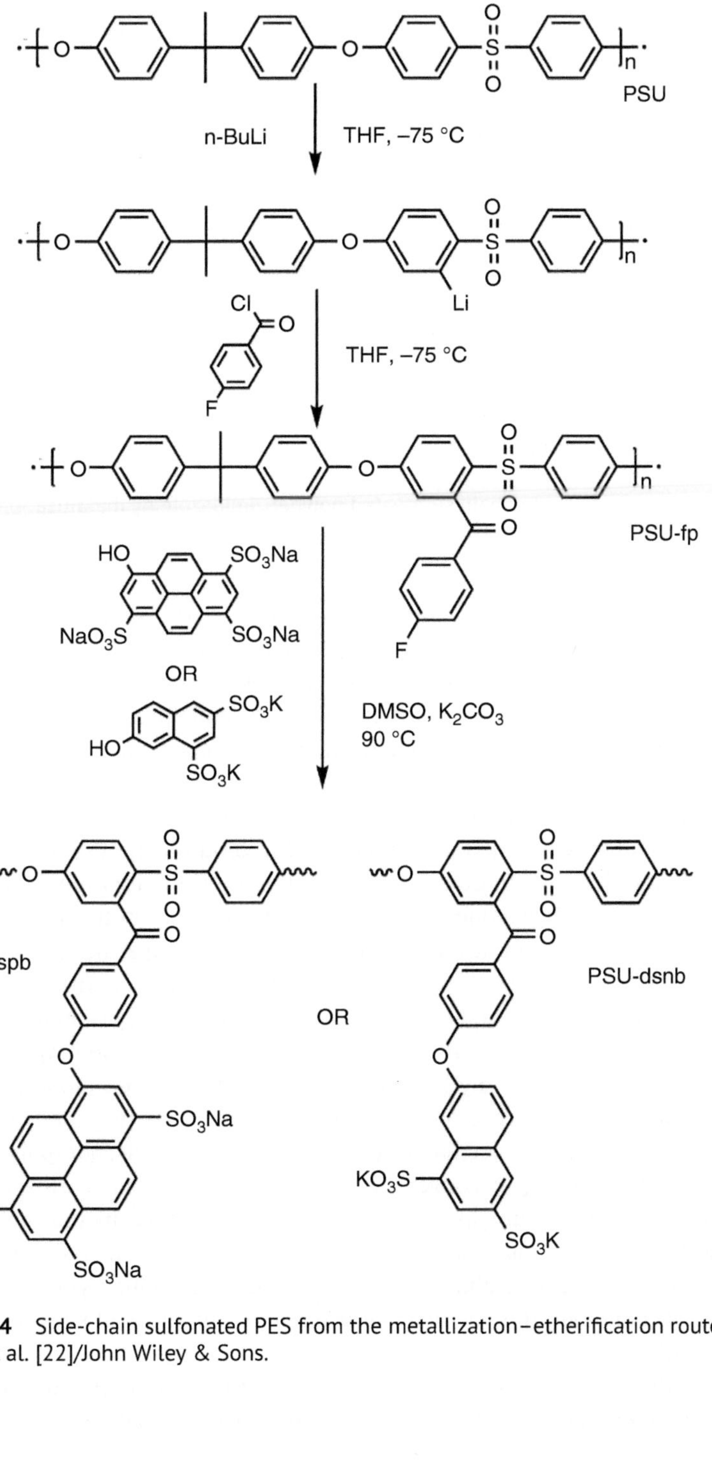

Figure 3.4 Side-chain sulfonated PES from the metallization–etherification route. Source: Lafitte et al. [22]/John Wiley & Sons.

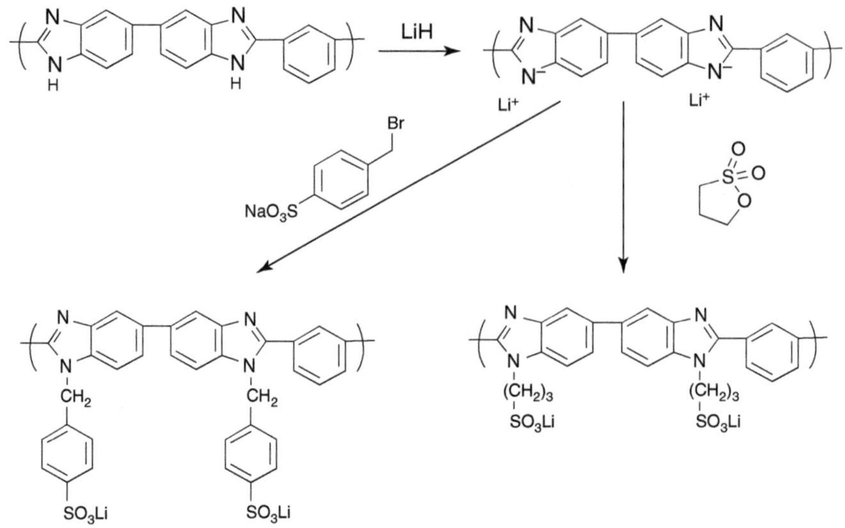

Figure 3.5 Synthesis of side-chain sulfonated polybenzimidazole by metallization. Source: Glipa et al. [25], Kawahara et al. [26].

Polymers containing hydroxyl functional groups in the aromatic ring of the backbone can also be employed to introduce side-chain sulfonic acid groups by nucleophilic substitution reactions. When diphenol monomers with methoxy in their side group are polymerized with dihalide monomers, methoxy moieties can be attached to the backbone. Then, methoxy groups can be converted into hydroxyl groups through the demethoxy reactions, and the side-chain CEM can be prepared by further etherization reactions with various sulfonated monomers [27]. For example, methoxy hydroquinone and 4,4′-difluorodiphenyl sulfone undergo a typical nucleophilic substitution polycondensation reaction to synthesize PES with methoxy groups in its backbone. Under the influence of boron tribromide, all the methoxy groups on PES can be converted into hydroxyl groups, and the sulfonated monomers can be etherized to synthesize side-chain CEMs. To control the degree of sulfonation and the distribution of sulfonated chain segments, bisphenol A can be added to participate in the polymerization process. Moreover, the usage of 9,9-bis (3,5-dimethoxy-4-hydroxyphenyl) fluorene further improves the sulfonation degree [28]. Such monomers can react with dihalide monomers via polycondensation to obtain polymers with a molecular weight of 50,000. Similarly, the polymer was demethylated to completely convert the methoxy group into hydroxyl groups. Under the action of sodium hydroxide, the hydroxyl groups in the main chain can cause the ring-opening reaction of 1,4-butane sulfonate lactone, resulting in the preparation of the side-chain CEM . Since there are four hydroxyl sites on the fluorenyl monomer, this method can bring four sulfonic acid sites on a single structural unit.

Although the above post-sulfonation methods can be used to prepare side-chain CEMs, these methods are only able to introduce aromatic sulfonic acid or alkyl sulfonic acid groups. However, aromatic or alkyl sulfonic acids are relatively weak, with a pK_a value of approximately −1, while the perfluoroalkyl sulfonic acid on Nafion is extremely acidic, with a pK_a value of approximately −6. Therefore, it is necessary to develop perfluorinated sulfonic acid side-chain cation exchange membranes with similar cationic group compositions to Nafion. The introduction of perfluorinated sulfonic acid side chains by the post-sulfonation method mainly relies on Ullmann coupling reactions. Perfluorinated sulfonic acid side-chain CEMs can be prepared by halogenating the benzene ring of the aromatic polymer or by synthesizing aromatic monomers with halogens, which are then coupled to 1,1,2,2-tetrafluoro-2-(1,1,2,2-tetrafluoro-2-iodoethoxy) ethane sulfonate (Figure 3.6) [29].

Bisphenol A type-PES can be treated with bromine to anchor bromine elements in the ortho-position of its ether bond, followed by the Ullmann coupling reaction with 1,1,2,2-tetrafluoro-2-(1,1,2,2-tetrafluoro-2-iodoethoxy) ethane sulfonate potassium catalyzed by copper powders [30]. Since the reaction is extremely sensitive to humidity, all raw materials and solvents need to be thoroughly dried. After the Ullmann coupling, all brominated sites can be converted to fluorinated sulfonic acid side chains. Perfluorinated sulfonic acid (PEES) containing naphthalene ring structures can also be used in a similar way to prepare CEMs with perfluorinated sulfonic acid side chains [31]. The bromination site was introduced into the naphthalene ring through the abovementioned bromination process, and the side-chain CEM was prepared by a similar coupling reaction with fluorinated sulfonic acid monomers. The strong hydrophobicity of naphthalene rings facilitates the formation of microphase separation structures.

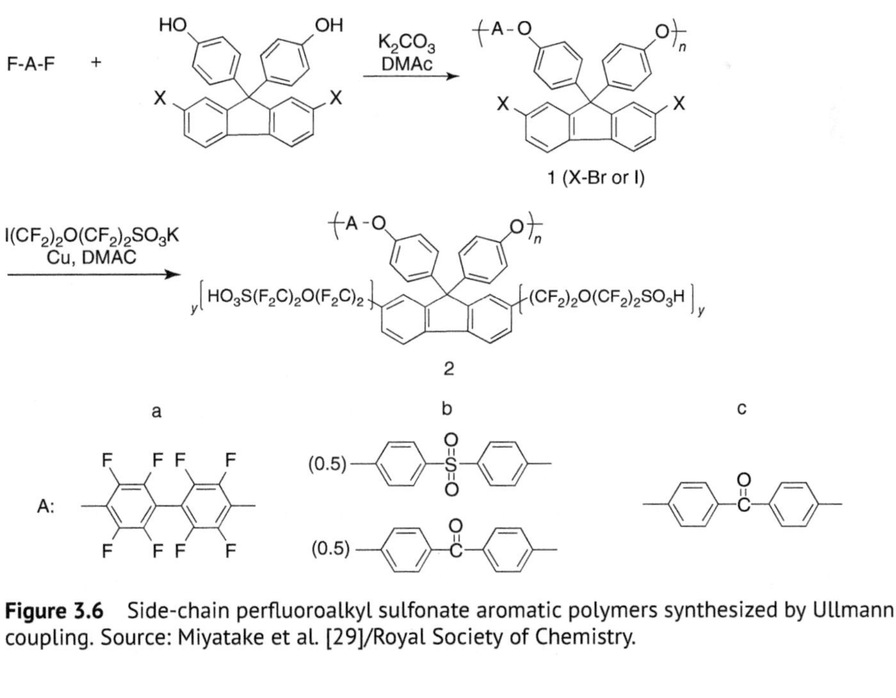

Figure 3.6 Side-chain perfluoroalkyl sulfonate aromatic polymers synthesized by Ullmann coupling. Source: Miyatake et al. [29]/Royal Society of Chemistry.

Fluorene can also be used to synthesize CEMs with perfluorinated sulfonic acid side chains. The brominated or iodinated diphenol monomer of fluorene is reacted with various dihalide monomers to generate a high molecular weight polymer, and then the target polymer is synthesized by the same Ulmann coupling reaction. By comparing the reactivity of iodized and brominated fluorene groups, it can be found that the iodine functional group has higher activity in the Ullmann reaction. Through the introduction of fluorene, a structural unit can be made with two perfluorinated sulfonic acid groups. To further enhance the aggregation degree of sulfonic acid groups on the polymer chain, oligomers containing fluorene iodide and oligomers without fluorene iodide can be synthesized [32]. The two oligomers terminate with dihalide monomer and diphenol monomer, respectively, so that they can be further polypolymerized into high molecular weight multiblock polymers. Similarly, the perfluorinated sulfonic acid side chain is introduced by the Ulmann reaction. 1,4-Diphenol-2,5-diphenylbenzene is another structural unit that can be used to synthesize fluorinated sulfonic acid side chains containing CEMs [33]. The diphenol monomer was polymerized with perfluorobiphenyl or perfluorobenzene and subjected to an Ulmann coupling reaction with potassium 1,1,2,2-tetrafluoro-2-(1,1,2,2-tetrafluoro-2-iodoethoxy) ethane sulfonate. Due to the high activity of the reaction, the degree of sulfonation can be facilely regulated by changing the amount of added fluorinated sulfonic acid side chains.

Synthesis of Nafion-like Structures Starting from Side-chain Sulfonated Monomers There is a great choice of reaction and side-chain categories in the post-sulfonation synthesis of side-chain sulfonated aromatic polymers, but it is still difficult to finely control the amount of sulfonated groups. When using the presulfonated aromatic monomers directly participating in polymerization, according to the principle of the polycondensation reaction "functional equivalence," the final IEC can be precisely monitored by controlling the addition of the sulfonated monomer proportion. In addition, the location of the sulfonic acid groups in the resulting polymer is directly determined by the structure of the sulfonated monomer. Therefore, the design of side-chain sulfonated aromatic monomers is the key to the synthesis of side-chain aromatic polymers. The following will be divided into four categories: side-chain sulfonated aromatic dihalide monomers, side-chain sulfonated aromatic diphenol monomers, side-chain sulfonated aromatic diamine monomers, and side-chain sulfonated aromatic dicarboxylic monomers.

The aromatic dihalide monomer and diphenol monomer can undergo polyetherized polycondensation. It is very simple to adjust the content of the sulfonic acid group in the resulting CEMs only by controlling the ratio of sulfonated aromatic dihalide monomers and nonsulfonated dihalide monomers. The polymerization is carried out under alkaline conditions (usually potassium carbonate as the base), and the reaction temperature needs to be above 160 °C to obtain a high molecular weight polymer. Aromatic dihalide monomers also undergo polycondensation reactions with "quasidiphenol" monomers [34]. Thiophenol is more nucleophilic than phenol and more easily participates in aromatic halogen reactions. In addition, diazanaphone can also react with aromatic halides to generate aromatic polymers [35].

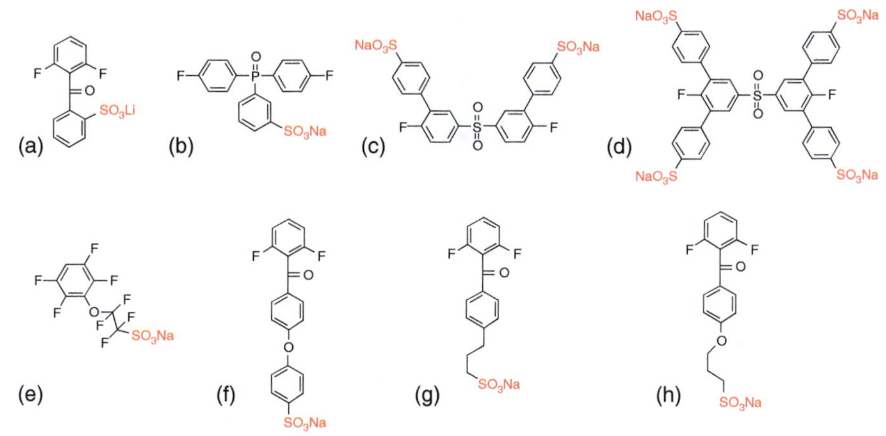

Figure 3.7 Side-chain sulfonated aromatic dihalide monomers.

Side-chain sulfonated aromatic dihalide monomers include sulfonic acid groups connected to the main chain through rigid spacer groups, flexible spacer groups, rigid/flexible spacer groups, and other structures. Starting from 1,3-difluorobenzene, it was metallized by butyl lithium and reacted with o-sulfobenzoic anhydride to synthesize side-chain sulfonated difluoride monomers connected to the main chain by a rigid benzene ring (Figure 3.7a) [36]. The monomer can be polymerized with diphenol or dithiol monomers to fabricate side-chain CEMs. Another synthesis route for the side-chain sulfonated difluoride monomer connected to the main chain through the benzene ring is based on the Grignard reaction between fluorophenyl magnesium bromide and phenylphosphonyl dichloride and then treated by fuming sulfuric acid (Figure 3.7b) [37]. The sulfonic acid group is introduced in the ortho position of the phosphonic oxygen group. Such monomers can polymerize with 4-(4-hydroxyphenyl) diazanaphthalene phenolone to obtain high molecular weight ionic polymers. The degree of sulfonation of the final CEM can be controlled by varying the ratio of sulfonated difluoride monomers and nonsulfonated difluoride monomers. As more sulfonated difluoride monomers participate in the polymerization, the required polymerization time will be prolonged accordingly.

The above-mentioned sulfonated monomer has only one sulfonated group per structural unit. To enhance the sulfonation density, a special monomer design is needed. The double (4-fluorophenyl) sulfone with benzene rings on its side group, as described above, can be sulfonated with fuming sulfuric acid to incorporate two or four sulfonic acid groups into a difluoride monomer, and the sulfonic acid group is separated from the main chain by a benzene ring (Figure 3.7c and d) [20]. Due to steric hindrance and electronic effects, the paracentric activity of the side-group benzene ring is relatively higher, so the sulfonation substitution reaction is mainly performed at this position. The monomer can then be reacted with diphenol to produce high molecular weight PEES to gain side-chain CEMs.

It is relatively difficult to design such structures linked to the main chain by flexible spacer groups, and only dihalide monomers with perfluorinated sulfonic acid side chains have been reported with such a structure. The nucleophilic substitution reaction of 1,2-dibromotetrafluoroethane reacts with phenols containing halogen functional groups in the presence of potassium hydroxide, and further oxidation using sodium disulfite to generate perfluorinated sulfonic acid side chains (Figure 3.7e) [38]. Aiming to make the nucleophilic substitution reaction proceed smoothly in the first step, the selected phenol-containing halogen should be either a four-substituted structure with fluorine atoms in both ortho and interposition of hydroxyl groups or a disubstituted structure with chlorine (or bromine) in both ortho and interposition of hydroxyl groups. However, only the four-substituted structure with fluorine atoms can react with various diphenol monomers to form PES, PEEK, and PEES. Due to the strong electron absorption effect of fluorine, the polymerization activity of this monomer is lower than that of common dihalide monomers, so the reaction still needs a high-temperature environment above 160 °C to obtain high-molecular-weight polymers.

It is relatively difficult to synthesize sulfonated aromatic dihalides with side chains connected to the main chain by completely flexible spacer groups; however, completely rigid spacer groups will limit the movement ability of side-chain sulfonic acid groups, which is not favorable for self-assembly into penetrating ion channels. Therefore, another large class of side-chain sulfonated aromatic dihalide monomers have flexible/rigid spacer groups connected to the backbone. Under the catalysis of dried aluminum chloride, 2,6-difluorobenzyl chloride and diphenyl ether were subjected to the Friedel–Crafts acylation reaction. It was further sulfonated by concentrated sulfuric acid to synthesize the target monomer (Figure 3.7f) [39]. Due to steric and electronic effects, sulfonation occurs only at the para-position site where the phenyl ring at the end of the side chain is connected to the ether bond. To improve the flexibility of the side chain, diphenyl ether can also be replaced with 1-bromo-3-phenylpropane and then sulfonated by sodium sulfite to introduce a sulfonic acid group at the end (Figure 3.7g) [40]. In addition, diphenyl ether can also be substituted with 1-bromo-3-phenoxy propane, and an ether bond is introduced in the terminal alkyl chain to further improve the flexibility of the side chain (Figure 3.7h) [41]. However, when these monomers were reacted with hydroquinone or biphenyl, the resulting polymers had low molecular weights and could not form complete membranes. Only with the use of more active diphenol monomers, such as 4,4′-dihydroxy-diphenyl ether, high molecular weight polymers can be obtained. However, due to the relatively low activity of the designed dihalide monomer, the polymerization temperature still needs to reach more than 180 °C.

Side-chain sulfonated aromatic diphenol monomers can undergo polyetherized polycondensation with dihalide monomers. However, the side-chain sulfonated aromatic diphenol monomers reported thus far are limited. One example is the commercial 2,8-m-dinaphthol 6-sulfonate, which contains both sulfonic acid groups and polymerizable diphenol functional groups. The monomer can be reacted with 4,4′-difluorophenophenone, 1,3-bis(4-fluorophenoyl) benzene, difluorophenonitrile, and heteronaphenyl phenolone to synthesize various compositions of high molecular weight polymers [42–46]. Different from linear diphenol monomers, 2,8-m-dinaphthol-6-sulfonate easily forms low molecular weight circular polymers in the polymerization process. Therefore, it is necessary to add other linear diphenol monomers for copolymerization to avoid the appearance of circular oligomers. Another side-chain sulfonated diphenol monomer is 2,2-bis-(*p*-hydroxyphenyl)-pentafluoropropane sulfonate with side-chain sulfonic acid synthesized through a multistep reaction [47]. The synthesis of this monomer starts from hexafluoroacetone and then reacts with triethyl phosphite and phenol to synthesize the target monomer. The sulfonic acid group in the monomer is connected to the main chain by a fluorinated side chain, so it is more acidic than benzene sulfonic acid. In addition, the monomer can be polymerized with dihalide monomers to form high molecular weight polymers for the preparation of side-chain CEMs.

By the polycondensation reaction of aromatic diamine and dianhydride monomers, the content of sulfonic acid groups can be precisely controlled by adjusting the ratio of sulfonated aromatic diamine and non-sulfonated diamine. This polymerization occurs at high temperatures, usually above 200 °C, to obtain fully polyimylated polymers, and the solvent used is cresol, which is irritating and difficult to remove [48, 49]. However, because sulfonated PI membranes have good properties, especially mechanical properties and swelling resistance, side-chain sulfonated aromatic diamine monomers are the most studied. Similar to sulfonated aromatic dihalide monomers, the structural design of sulfonated aromatic diamine monomers is mainly based on sulfonic acid groups connected to the backbone through rigid spacer groups, flexible spacer groups, or a combination of both groups.

Diamine monomers with sulfonated groups on fluorene are typical side-chain sulfonated monomers connected to the main chain by rigid spacer groups (Figure 3.8a) [50]. The synthesis of the monomer first requires chlorination of fluorene and further coupling with aniline compounds. Further sulfonation of fluorene can synthesize diamine monomers with side-chain sulfonation acid. The monomer can be polycondensed with 1,4,5,8-naphthalene tetramethyl anhydride. To regulate the content of sulfonic acid groups in the polymer as well as the composition of the polymer, other nonsulfonated diamine monomers can be added for copolymerization, such as 1,10-decatenyl diamine, bis(4-(3-aminophenoxy) phenyl) sulfoxide, and 4,4′-bis(4-aminophenoxy) biphenyl [51]. Sulfonation of 2,2′-diphenyl benzidine by fuming sulfuric acid can introduce sulfonic acid groups into the side-phenyl ring, that is, synthesis of a side-chain sulfonated diamine monomer attached to the main chain through a rigid spacer group (Figure 3.8b) [52]. Due to the strong electron absorption effect of protonated amine groups, the sulfonation site mainly appears in the para site of side-chain phenyl groups. The biphenyl unit in this monomer has strong rigidity. To improve its flexibility, ether bonds can be added to the monomer design to synthesize 4,4′-bis(4-aminophenoxy)-3,3′-bis(4-sulfophenyl) biphenyl. In this monomer, the biphenyl group and the benzene ring are separated by ether bonds, which is beneficial for improving the flexibility of the synthesized polymer (Figure 3.8c) [53]. Similarly, the sulfonation site of the polymer is located on the phenyl group of the side chain. Both monomers can be polymerized with 1,4,5,8-naphthalene tetra-methylene anhydride to synthesize side-chain CEMs.

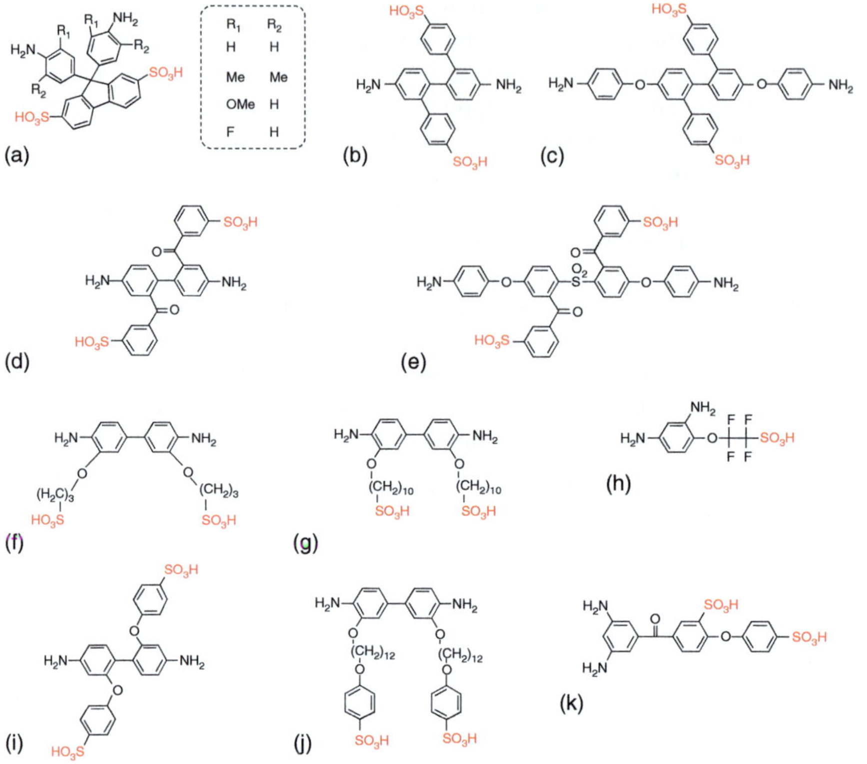

Figure 3.8 Side-chain sulfonated aromatic diamine monomers.

The spacer in a side-chain sulfonated diamine monomer connected to the main chain by a rigid spacer group can also be a benzoyl functional group, such as 2,2′-bis(benzoyl 3-sulfonate) benzidine and bis[4-(4-aminophenoxy)-2-(benzoyl-3-sulfonate)] phenyl sulfone (Figure 3.8d and e) [54, 55]. The synthesis of the former begins with bibenzoic acid, which is treated with fuming nitric acid to introduce nitro groups at the intersites of the carboxyl groups. Furthermore, benzoyl was introduced into the side chain through the action of Friedel–Crafts benzoyl chloride catalyzed by aluminum trichloride. Under the action of palladium and hydrazine hydrate, the nitro moiety is reduced to an amine group. Sulfonic acid groups can then be introduced on the benzoyl of the side chain by fuming sulfuric acid sulfonation. Different from 2,2′-bis(benzoyl 3-sulfonate) benzidine, the synthesis of bis[4-(4-aminophenoxy)-2-(3-sulfonate benzoyl)] phenyl sulfone with a sulfone functional group starts with 4,4′-dichlorodiphenyl sulfone, which in turn interacts with butyllithium and carbon dioxide to introduce carboxyl groups. Under the action of dichlorosulfoxide, the carboxyl group can be further converted to acyl chloride. Similarly, the introduction of the side-chain benzoyl is accomplished by the Friedel–Crafts alkylation reaction. The side chains are sulfonated with fuming sulfuric acid and then further reacted with p-nitrophenol to convert the halogens at both ends to amino groups. Similar to the diamine monomers described previously, these two monomers can also be polycondensed with 1,4,5,8-naphthalene tetramethylene anhydride for the synthesis of side-chain CEMs. In addition, the rigid spacer can also be a benzimidazole group, such as 2-(3,5-diaminophenyl) benzimidazole-5-sulfonic acid [56]. The synthesis of the monomer begins with o-phenylenediamine and 3,5-dinitrobenzoyl chloride, and then catalyzed by palladium and hydrazine, the nitro group can be reduced to the amino group. The benzene ring of benzimidazole can be sulfonated by concentrated sulfuric acid. The monomer can be polycondensed with 1,4,5,8-naphthalene tetra methylene anhydride to synthesize side-chain CEMs.

Sulfonated side-chain aromatic diamine monomers, connected to the main chain by flexible spacer groups are mainly designed based on the biphenediamine structure. The phenolic hydroxyl group can be converted into sulfonic acid ether functional groups by the interaction of nitrophenol with sodium bromoalkyl sulfonate. The compound can be converted into azo catalyzed by zinc powders under alkaline conditions. Under the action of weak acid, azo compounds can be further transformed into hydrazide propyl compounds. In the presence of hydrochloric acid, hydrazo compounds can be converted into biphenediamines with alkyl sulfonic acid side chains (Figures 3.8f and 8 g) [57]. The length of the side chain of the compound can be adjusted by the number of carbons in the alkyl group used in the first-step reaction [58]. With the increasing number of alkyl carbons in the side chain, the flexibility of the side chain will be further enhanced. The compound can also be polycondensed with 1,4,5,8-naphthalene tetramethylene anhydride to form a high molecular weight polymer. Another diamine monomer with a flexible spacer is designed based on the perfluorinated sulfonic acid side chain. The amino group of 2,4-diaminophenol hydrochloride is protected by acetyl groups. The protected monomer can undergo the Williamson reaction with 1,2-dibromo-tetrafluoroethane, and further sulfonic acid groups were introduced by sodium dithionite. After deprotection, the synthesis of the target monomer is completed (Figure 3.8 h) [59]. CEMs with flexible side chains were synthesized by reacting with 1,4,5,8-naphthalene tetanethoxylanhydride.

Another class of side-chain sulfonated diamine monomers has a structure connected to the backbone through rigid/flexible spacer groups. On the basis of the 2,2′-diphenylbenzidine monomer mentioned above, an ether bond can be introduced between the benzene rings of the main chain and the side chain to improve the flexibility of the side chain. The synthesis of this monomer starts from the reaction of fluorobenzene with fuming sulfuric acid to synthesize sodium 4-fluorobenzene sulfonate, which reacts with 3,3′-dihydroxybenzidine by the etherification reaction under alkaline conditions to obtain targeted monomers (Figure 3.8i) [60]. To enhance the flexibility of the side chain, alkyl ether bonds can be further introduced. The functionalization is achieved by etherification of dihydroxybenzidine with halogenated hydrocarbon side chains with sodium 4-hydroxybenzenesulfonic acid (Figure 3.8j) [61]. The synthesis of dihydroxybenzidine with halogenated hydrocarbon side chains requires the protection of the amine group of the dihydroxybenzidine precursor by acetyl, and then the phenol hydroxyl group interacts with the dibromoalkane to acquire the alkyl ether side chain.

By adjusting the carbon number of the dibromoalkane, the length and flexibility of the side chain can be adjusted. Both monomers can be synthesized by the polycondensation reaction with 1,4,5,8-naphthalene tetramethylene anhydride to complete high molecular weight polymers, and the composition of IEMs can be regulated by introducing special elements containing fluorene, sulfone, biphenyl, fluorine, and so on. Another monomer, 3,5-diamino-3′-sulfonic acid-4′-(4-sulfonic acid phenoxy) benzophenone, is also a side-chain sulfonated diamine monomer (Figure 3.8k) [62]. The synthesis of the monomer is initiated from 3,5-dinitrobenzoyl chloride by the introduction of diphenyl ether on the side chain through a Friedel–Crafts alkylation reaction. Through the sulfonation of fuming sulfuric acid, sulfonic acid groups can be introduced into the two benzene rings connected with the ether bond, and then nitro can be reduced into amine to synthesize the target monomer. The side-chain sulfonated monomers connected to the main chain by the rigid/flexible spacer can be polycondensed with the dianhydride monomer to synthesize the side-chain CEM.

Aromatic carboxylic monomers can polymerize with di(o-phenylendiamine), di(o-hydroxyanaline), or di(2-aminothiophenol) via polycondensation to generate PBI, PBO, or PBT, respectively. When the polymer reaction is carried out in concentrated phosphoric acid, the temperature is very high, usually above 200 °C, to obtain fully cyclized polymers, whereas when the polymerization reaction is carried out in the Eaton reagent, polymerization can occur below 150 °C [63, 64]. These heterocyclic sulfonated aromatic polymers have good properties, but di(o-phenylendiamine), di(o-hydroxyanaline), or di(2-aminothiophenol) monomers are difficult to purify and store, so they are expensive compared to other aromatic polymers. At present, there are few reports and studies on side-chain sulfonated dicarboxylic groups. 4′-Sulfonic acid-2,5-dicarboxyphenyl sulfone is a side-chain sulfonated dicarboxylic monomer whose sulfonic acid group is bonded to the main chain through the structure of the sulfoxide [65]. The monomer was synthesized by the nucleophilic substitution reaction of 2,5-dimethyl phenobenzene with sodium 4-fluorobenzene sulfonate, followed by oxidation of methyl and thioether bonds to carboxyl and sulfone groups, respectively, under the strong oxidation of potassium permanganate. The monomer can be reacted with 3,3′-diaminobenzidine at more than 200 °C to produce PBI with sulfonic acid side chains.

The side-chain sulfonated monomers designed above usually need to undergo polycondensation with other monomers at high temperatures to obtain polymers with high molecular weights, and most of the reaction mechanisms involved are based on the nucleophilic substitution mechanism. However, it can also be polycondensation of dicarboxylic derivatives with electron-rich diaryl monomers under the catalysis of Lewis acid or protic acid. The reaction mechanism lies in the electrophilic substitution process, and the reaction temperature is low, even close to room temperature. To achieve low-temperature synthesis of side-chain sulfonated IEMs, it is necessary to design dicarboxylic or diaryl monomers containing side-chain sulfonic acid groups.

Morphologies of Nafion-similar Side-chain CEMs. The microphase separation in side-chain CEMs is usually related to the movement, composition, distribution, and acidity of side chains. Studies on the morphology of a series of side-chain polyaryl ether CEMs show that longer side chains are more conducive to the formation of ion clusters, so it is more beneficial to enable CEMs with high proton conductivity [66]. In addition, by designing special groups on the side chains, the aggregation of ionic groups will also be facilitated. For example, CEM containing naphthalene sulfonic acid in the side chain can be orderly packed between adjacent naphthalene rings through π–π interactions, which will also lead to orderly packing of sulfonic acid groups to form ion channels [23]. The distribution of side chains is also a key factor affecting the morphology of microphase separation. Compared with the membrane containing only one sulfonic acid group side chain, increasing the density of the sulfonic acid group on the structural unit will be beneficial to enhance the degree of microphase separation. For example, for a polyaryl ether containing four sulfonic acid groups on its structural unit, transmission electron microscope (TEM) characterization confirmed that the structure could contribute to connective ionic channels with widths of 2–5 nm [20]. In contrast, PEK containing eight sulfonic acid groups on a structural unit has a better microphase separation structure [19]. Strengthening the acidity of the side chain and thus increasing the polarity gap between the side chain and the main chain will also be conducive to enhancing phase separation [38]. Similar to the commercial Nafion membrane, the side-chain CEM containing the fluorosulfonic acid chain has a high microphase separating structure.

3.2.1.2 Design of Side-chain CEMs with Grafted Structures

"Grafting Through." The synthesis of side-chain CEMs by this strategy requires the presynthesis of polymerizable macromolecular monomers with sulfonated side chains. However, there are few types of such monomers, and only successful examples have been completed by anionic active polymerization technology reported thus far [67]. Butyl lithium was used to initiate the polymerization of methyl styrene, and diphenyl ethylene with a tert-butyl dimethyl silane (TBDMS) functional group was added to make the end of the polymer with a special functional group. Further treatment with methanol and potassium iodide converted the ends of the polymer to methyl groups. In the presence of weak acidity, the TBDMS functional group can be converted to a hydroxyl group, thus completing the preparation of diphenol monomers with methyl styrene long side chains. The amount of methyl styrene can be changed by regulating the ratio of initiator to monomer. The synthesized diphenol monomer can be polymerized with perfluorinated biphenyls. To control the proportion of the monomer in the polymer, bisphenol A monomers can be added for copolymerization. To ensure that sulfonation only occurs on the benzene ring of the side chain, the obtained polymer needs to be sulfonated with acetyl sulfate at a low temperature (50 °C) to complete the preparation of the grafted side-chain CEM.

"Grafting Onto." The key to synthesizing grafted CEMs using this strategy is to rationally design the polymeric backbone and side chains separately so that the high molecular weight backbone and long side chains can complete the covalent bond. Copper-catalyzed azide–acetylene ring addition and click reactions are often used to conveniently create grafted polymers by "grafting onto." The click reaction is extremely active and can be carried out under mild conditions. Such a reaction has no byproducts and is suitable for compounds with various functional groups. Commercialized bisphenol A-type PES was metallized and then reacted with 3-(chloromethyl)benzoyl chloride and sodium azide. The azide functional group was successfully introduced into the main chain. Atom transfer radical polymerization (ATRP) technology was used in the design of side chains to ensure the preparation of well-defined side chains. To guarantee the presence of an alkyne group on the side chain, the ATRP initiator used is 3-bromo-1-trimethylsilyl-1-propionate to initiate the polymerization of 2,3,4,5,6-pentafluorostyrene. During the ATRP reaction course, the length of the side chain can be controlled by adjusting the ratio of monomers relative to initiators. The grafting polymer can be synthesized by the copper-catalyzed click reaction between alkynyl-containing side chains and the azide-functionalized main chain. Furthermore, trimethylsilyl phosphate was used to phosphorylate fluorine on the paracentric site of pentafluorostyrene to complete the preparation of grafted side-chain CEMs [68].

The side chains prepared by ATRP polymerization are usually aliphatic. To obtain aromatic side chains, aromatic oligomers need to be synthesized in advance. 2,6-Diphenyl phenol can be polymerized into polyphenyl ether oligomers by the copper-catalyzed oxidation coupling reactions, and its number average molecular weight can be controlled in the range of 2000–4000. The end of the polyphenyl ether oligomer is a hydroxyl group, which can be converted into a fluorine group by reacting with perfluorobenzene. For polyphenyl ether oligomers, the nucleophilic substitution reactions can be carried out with various hydroxy-containing main chains to form grafted polymers. For the synthesis of hydroxy-containing backbones, the relevant route has been provided previously. The resulting graft-type polymer was further functionalized using chlorosulfonic acid, which can sulfonate the side phenyl on the side chain. Via this route, side-chain CEMs with both Nafion-like and graft-like structures were synthesized (Figure 3.9) [69].

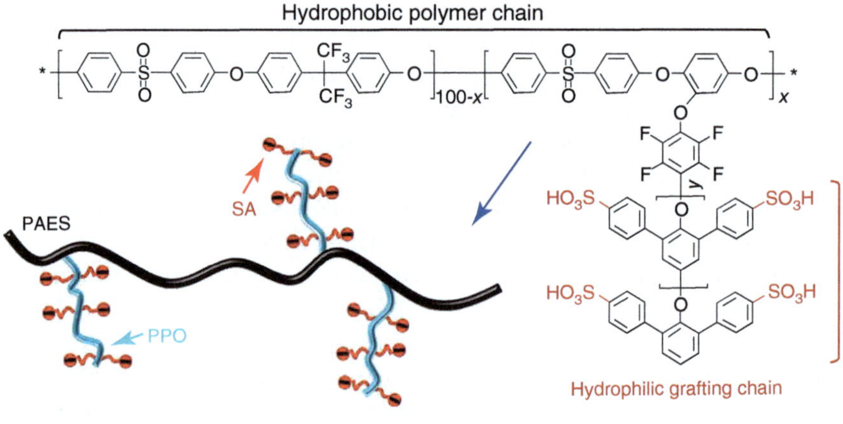

Figure 3.9 Synthesis of side-chain PES CEMs through "grafting onto". Source: Li et al. [69]/John Wiley & Sons.

"Grafting From." The synthesis of side-chain CEMs using this strategy usually relies on the use of anionic or ATRP, and other active polymerization techniques to induce suitable monomer polymerization to complete the preparation of cation-exchange membranes with regular side chains. Therefore, this strategy requires that the polymeric backbone be functionalized in advance so that it has initiating sites. To prepare side-chain CEMs, the metallization of bisphenol A-type PES can initiate monomer polymerization via anionic polymerization, but this initiation point is not effective for polymerizing vinyl diethyl phosphate monomers. To give the host chain the ability to initiate anionic polymerization, the metallized PES reacted with 1, 1-stilbene to form an initiation point of the 1,1-diphenyl alkyl anion, which then induced vinyl diethyl phosphate polymerization. The length of the side chain can be regulated by regulating the amount of monomer. Under acidic conditions, side-chain polymers bearing phosphate cation exchange groups can be obtained by hydrolyzing diethyl esters on the side chains to hydroxyl groups [70]. In contrast to PES, PPO does not contain a sulfone group, which is a strong electron-absorbing site. Therefore, it is convenient to introduce anion polymerization initiation sites on PPO by one-step metallization. This initiation site can also initiate the polymerization of vinyl diethyl phosphate and further prepare grafted side-chain CEMs through hydrolysis [71].

Anionic active polymerization technology can be achieved through "grafting from" strategies for the preparation of grafting-type CEMs. Nevertheless, the applicability of anionic polymerization with monomers has certain limitations, so the preparation of IEMs is solely based on phosphate groups, and common sulfonic acid anchoring IEMs through synthetic technology remains difficult. In contrast, ATRP technology can be used to synthesize sulfonic acid-type side-chain CEMs [72]. Similar to anionic active polymerization, the successful use of ATRP requires the introduction of an ATRP initiator on the backbone polymer. Through the copolymerization of vinylidene fluoride and chlorotrifluoride, an ATRP initiation site with a C–Cl functional group can be synthesized, which can subsequently initiate the polymerization of styrene. The grafted CEM can be prepared by sulfonating the styrene on the side chain with sulfuric acid acetyl [73, 74].

Morphologies of Side-chain CEMs with Grafted Structures. The disparity in the composition of the main chain and side chain is conducive to microphase separation to build up well-connected channels. For example, the fluorine-containing backbone and the sulfonated styrene side chain have a large polarity difference, so the membrane material of the polymer has a hydrophilic/hydrophobic microphase separation morphology, and the hydrophilic region is composed of ion clusters with a size of 2–4 nm. For long side-chain CEMs, the length of the side chains has a significant effect on the morphology. A longer side chain is much more beneficial to the aggregation of ionic groups for the formation of ion clusters of larger size (Figure 3.10) [73]. The side-chain CEM with aromatic main chains and sulfonated styrene side chains can also realize the construction of a high-degree microphase separation morphology. TEM and small-angle neutron scattering (SANS) characterization confirmed that highly continuous, worm-like ion channels were uniformly distributed in the

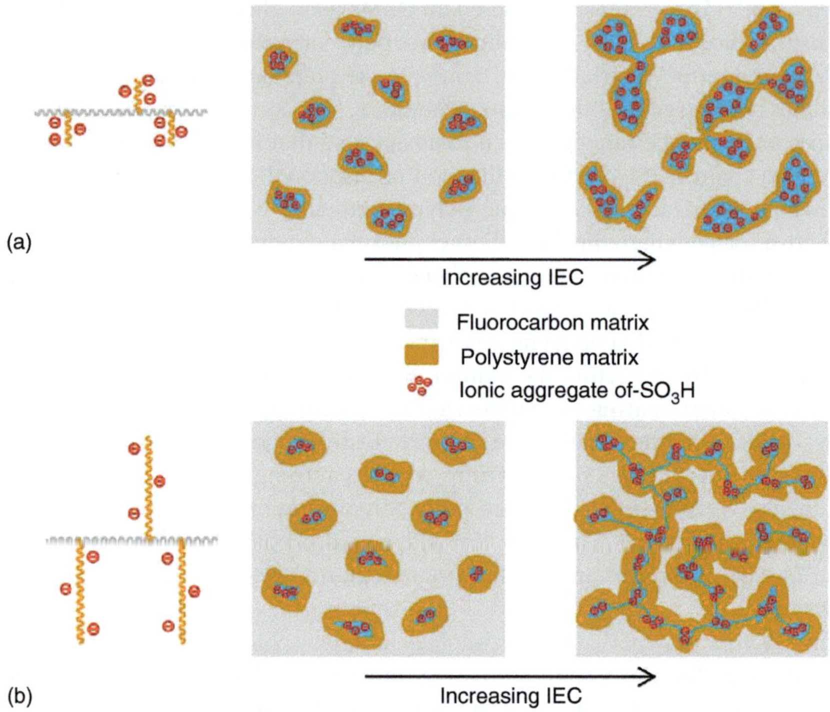

Figure 3.10 Schematic representation of the postulated ionic aggregation in grafted membranes with (a) short and (b) long graft lengths at different ion contents. Source: Tsang et al. [73]/American Chemical Society.

hydrophobic region. Microphase separation can be further enhanced by increasing the density of side-chain sulfonic acid groups. In addition, the main chain is a sulfonated aromatic polymer, and the side chain is a fluoropolymer, which also shows a desired phase-separated structure. It shows a highly ordered microphase separation morphology and a layered or cylindrical microstructure depending on the length of the side chains [75].

3.2.2 Design of Side-chain AEMs

3.2.2.1 Design of Side-chain AEMs with Similar Nafion Structures

Synthesis of Nafion-like Structures via Post-functionalization. When the side chain of the polymer has special groups such as chloromethyl, bromomethyl, and chloroacetyl, it can react with various basic functionalized monomers to prepare the side-chain anion exchange membrane (AEM). Similarly, when the side chains of polymers have primary, secondary, or tertiary amines, they can be further reacted with alkylating reagents to introduce anion exchange groups to prepare side-chain AEMs. Another way to synthesize side-chain AEMs through post-functionalization is to synthesize side chains with anion exchange groups and other active groups in advance and to prepare side-chain AEMs by bonding onto the main chain. In the following, according to the synthesis method adopted, the route of the side-chain AEM prepared by the post-functionalization method is introduced.

The Friedel–Crafts reaction is one of the most common methods to prepare side-chain AEMs. Under the catalysis of aluminum trichloride, bromo-chlorinated alkyl side chains can be introduced by the one-step reaction between the main polymer chain containing the electron-rich benzene ring and the halogenated alkyl group (or halogenated alkyl chloride). Through the Friedel–Crafts reaction, the side chain connected to the main chain by the flexible or rigid spacer group can both be designed. The alkyl side chain containing benzyl bromide can be successfully introduced into the benzene ring of the PPO main chain by the Friedel–Crafts acylation reaction with brominated alkyl chloride. The AEM with flexible side chains can be prepared by interacting benzyl bromide groups with common alkaline functionalized reagents (Figure 3.11) [76]. Polyaryl ether with a phenyl side group can be synthesized by polycondensation of phenyl hydroquinone with perfluorinated biphenyl. Under the catalysis of tin tetrachloride, the polyaryl ether can undergo the Friedel–Crafts alkylation reaction with chloromethyl ether, which then can introduce the chloromethyl group onto the side phenyl group. Due to the high activity of the reaction, the reaction temperature and the amount of catalyst should be strictly controlled to avoid crosslinking. The resulting side-chain polymers can interact with basic-functionalized reagents to prepare AEMs with rigid side chains [77].

Nucleophilic substitution reactions can also be used to prepare side-chain AEMs. Similar to the preparation of side-chain CEMs, polymers with hydroxyl functional groups can be further subjected to nucleophilic substitution reactions to complete the preparation of side-chain AEMs. PES (PEK) polymers with methoxy groups can be prepared by nucleophilic polycondensation of bisphenol monomers with methoxy groups and difluoride monomers. Using boron tribromide and other Lewis acids, methoxyl can be reduced to phenolic hydroxyl and then react with bromoalkyl quaternary ammonium (QA) salts in an alkaline environment to prepare side-chain AEMs [78]. The bromoalkyl QA salt is mainly prepared by dibromoalkane and trimethylamine (TMA). Imidazole-type side-chain AEMs can also be prepared by replacing TMA with imidazole [79].

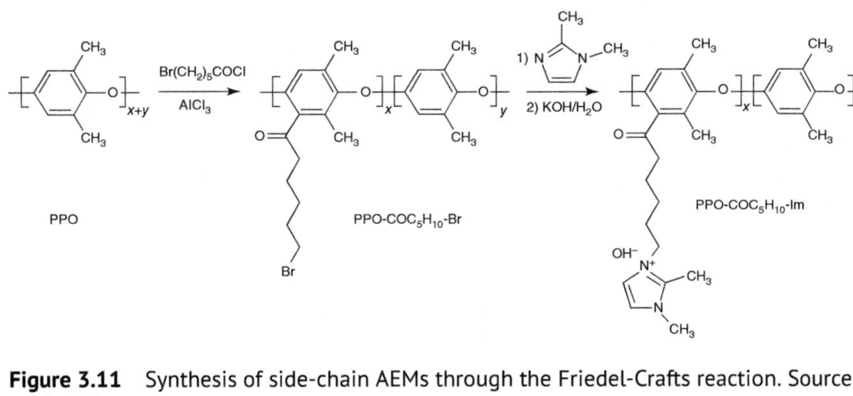

Figure 3.11 Synthesis of side-chain AEMs through the Friedel-Crafts reaction. Source: Gao et al. [76].

By using dibromoalkanes with different carbon chain lengths, the length of the side chains can be adjusted. For the preparation of side-chain AEMs, the density of side chains can be further regulated by using different bisphenol monomers with methoxy groups. Only one side chain can be introduced into one structural unit for the polymer based on methoxy-hydroquinone. However, for polymers based on 9,9-bis(3,5-dimethoxy-4-hydroxyphenyl) fluorene, four side chains can be introduced into a structural unit, which is favorable for further aggregation of ionic functional groups [80, 81]. In addition, side-chain AEMs can be prepared by a one-step ring-opening reaction with epoxide QA salts. However, this reaction needs to be completed under the catalysis of palladium chloride [82].

In addition to the phenolic hydroxyl group, polymers with nitrogen heterocyclic rings can also undergo the nucleophilic substitution reaction with electrophiles, and then the side-chain AEM can be prepared. Carbazolyl PES can be prepared by polycondensation of bisphenol monomers with benzocarazole and difluorodiphenyl sulfone monomers. PES containing carbazole groups can be successively reacted with bromochlorane and 1,2-dimethylimidazole to obtain imidazole-type side-chain AEMs [83]. However, because of the weak nucleophilicity of the nitrogen atom on carbazole, the reaction with bromoalkane can be carried out only after sodium hydroxide activation. In contrast, the nitrogen atoms in benzimidazole have high nucleophilicity, and polybenzimidazole can directly react with bromoalkyl QA salts to prepare side-chain AEMs [84]. In addition, side-chain AEMs can be synthesized from polymers containing amino functional groups by a similar reaction route. A commercial polymer containing polyphenylphosphonine oxygen can react with bromoalkyl QA salts in the presence of an acidic agent. By regulating the proportion of monomers added, one or two side chains can be attached to the amino moieties to regulate the density of side chains [85].

Nucleophilic substitution reactions are also available between fluorinated polyaryl ethers and phenols containing tertiary amine groups to prepare side-chain AEMs. First, perfluorinated biphenyls were reacted with bisphenol A for the synthesis of aryl ether-containing fluorine groups. Then, the polymer was functionalized by 2,4,6-tri(dimethylamine)-phenol in the presence of cesium carbonate via the nucleophilic reaction process to introduce tertiary amine groups on the main chain. Subsequently, the side-chain AEM was prepared with bromoalkane or bromoalkyl QA salts. In particular, the reaction treated by bromoalkane can introduce an anion exchange group at each tertiary amine site of the side-group benzene ring, and this ion exchange group is connected to the main chain by the rigid benzene ring spacer group. If it reacts with bromoalkyl QA salts, two anion exchange groups can be introduced into each tertiary amine site, and they are separated by flexible side chains (Figure 3.12) [86]. This route can also be realized on modified polyphenyl ethers containing fluorine. Similarly, tertiary amines can be introduced by the reaction with 2,4,6-tri(dimethylamine)-phenol, which further interacts with basic functionalized reagents to complete side-chain AEM preparation [87].

The Suzuki reaction can also be used to prepare side-chain AEMs. BPPO can be prepared by the reaction of liquid bromine with PPO at low temperatures. The Suzuki reaction of BPPO with bromoalkylbenzene esterified with boric acid gave the polymer precursor with bromoalkylbenzene side chains. Side-chain AEMs can

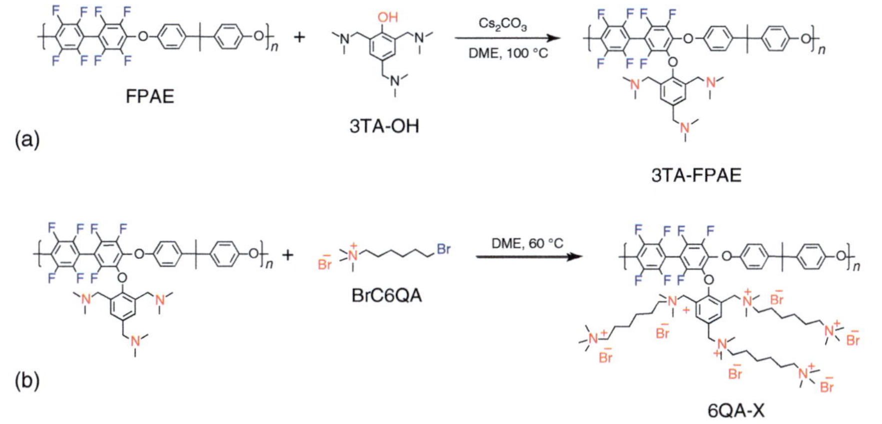

Figure 3.12 Synthesis of side-chain AEMs through the nucleophilic substitution reaction. Source: He et al. [86].

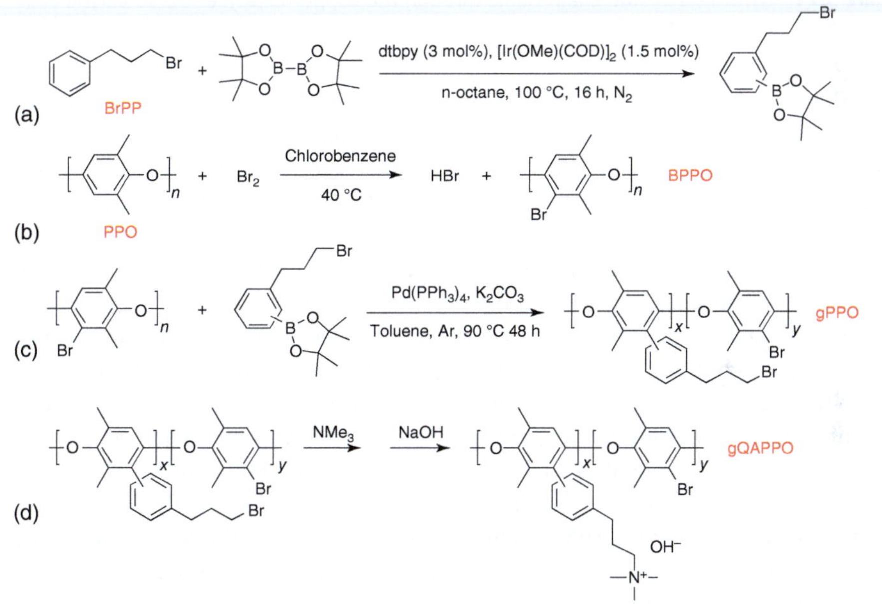

Figure 3.13 Synthesis of side-chain AEMs through the Suzuki reaction. Source: Yang et al. [88].

be prepared by further treatment with basic functionalized reagents (Figure 3.13) [88]. If BPPO reacts with 9,9-bis(6′-bromohexyl) fluorene esterified with boric acid and further reacts with TMA or N,N-dimethylhexylamine, a double side-chain AEM can be prepared [89]. In the future, the Suzuki reaction can be used in the design of AEMs to introduce richer functional groups due to its carbon–carbon coupling process, strong substrate applicability, and group tolerance. However, the main drawback is that the preparation of borate ester reagent is complicated, and purity is strictly needed.

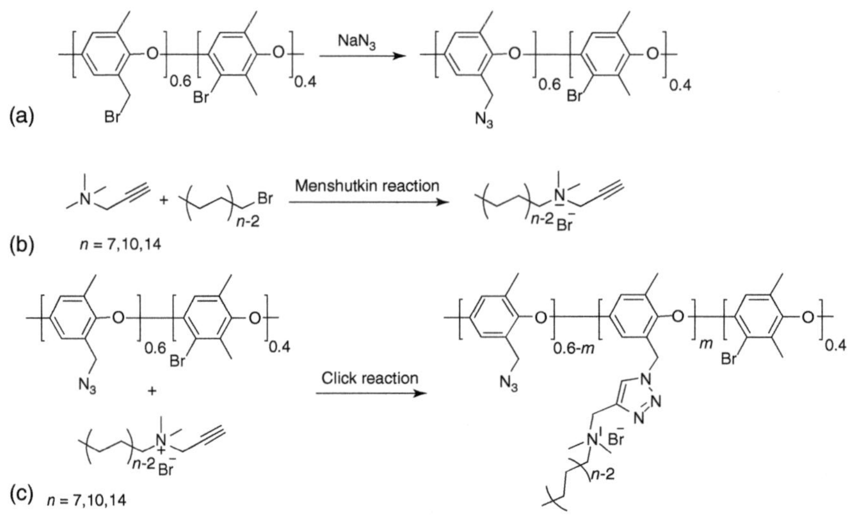

Figure 3.14 Synthesis of side-chain AEMs through the copper-catalyzed azide-alkyne click reaction . Source: Ge et al. [91].

The copper-catalyzed azide-alkyne click reaction can also be used to prepare side-chain AEMs. Similar to the preparation of side-chain CEM, the polymer backbone with azide groups was obtained by the reaction of BPPO with sodium azide. The ionized side chains were prepared by the reaction of alkyl bromide with N,N-dimethyl-2-propynylamine. Finally, the side-chain AEM was obtained through the high-efficiency click reaction [90]. The conditions of this reaction route are mild; however, the click reaction requires stricter anaerobic conditions. The triazole group generated by the reaction can form a hydrogen bond network with water molecules, which is beneficial to the improvement of ionic conductivity. However, due to the electron absorption effect of the triazole group, the alkaline resistance of the anion exchange group will be reduced. To improve alkaline resistance, ionized side chains can be reacted with bromoalkanes via 1-dimethylamino-2-propyne. Subsequently, it was subjected to the click reaction with azide polyphenyl ether to prepare the side-chain AEM (Figure 3.14) [91]. Due to the steric hindrance and hydrophobic effect of the alkyl chain at the end of the QA group, the alkaline resistance can be effectively improved.

In addition, based on the click reaction, side-chain AEMs containing imidazole, guanidine, and other anion exchange groups can also be prepared. The ionized side chain with alkyne can be synthesized by reacting clopentylene with 1,2-dimethylimidazole, and the imidazole-type AEM can be prepared by the click reaction of polyphenyl ether with azide [92]. The imidazole functional group in the AEM is separated from the main chain by a flexible alkyl chain, which is beneficial to avoid the adverse influence of the strong electron absorption effect from the triazole group. In addition, to improve the basic stability of the AEM, adamantane functional groups can be introduced in the N3 position of imidazole cations [93]. Because of the large volume and hydrophobic effect of adamantane, it can hinder the attack of alkali and improve the alkaline resistance of AEMs. The specific reaction route involves the reaction of 1-bromoadamantane with imidazole or 1-methylimidazole, which then acts with chloroprene to synthesize ionized side chains with alkyne and adamantane. The imidazole-type side-chain AEM was synthesized by the click reaction with polyphenyl ethers containing azide groups. Similarly, to prepare guanidine-based side-chain AEMs based on the click reaction, guanidine-based ionic side chains with alkynes should be prepared first, and then the guanidine-based side-chain AEMs can be synthesized by click course with polymers containing azide groups [94]. Notably, because guanidine is unstable in the environment of click reactions, it needs to be protected. After the click reaction, guanidine-type anion exchange groups can be obtained by removing the protection.

The Wittig reaction is another route used to synthesize side-chain AEMs. BPPO was reacted with triphenyl phosphine and treated with sodium hydroxide or sodium ethanol to produce phosphorus ylides. Then, the bromoalkyl side chain was successfully introduced by the Wittig reaction with 4-(3-bromoalkyl oxygen) benzaldehyde. After reacting with nucleophiles such as trimethylamine and imidazole, side-chain AEMs can be finally obtained (Figure 3.15) [95]. The main advantage of this method is that AEMs with different side chain lengths can be prepared by adjusting the molecular structure of bromoalkoxy benzaldehyde. The main disadvantage of this route is that the preparation of bromoalkoxy benzaldehyde is complicated, and it needs to be strictly purified.

After the main polymer chain is activated with the butyl lithium reagent, it can be further reacted with electrophiles such as carbonyl and bromoalkane to prepare side-chain AEMs. Bisphenol A-type PES was activated with butyl lithium, which

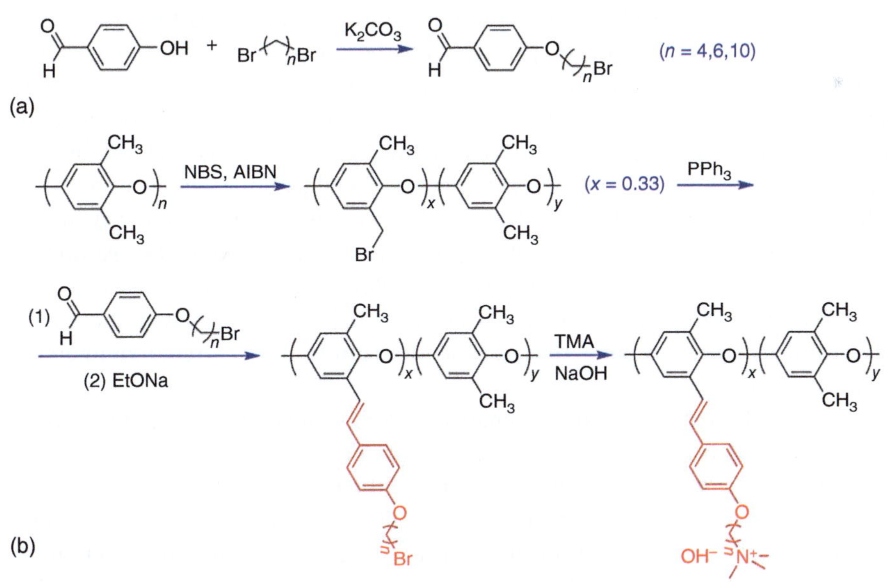

Figure 3.15 Synthesis of side-chain AEMs through the witting reaction. Source: Hou et al. [95].

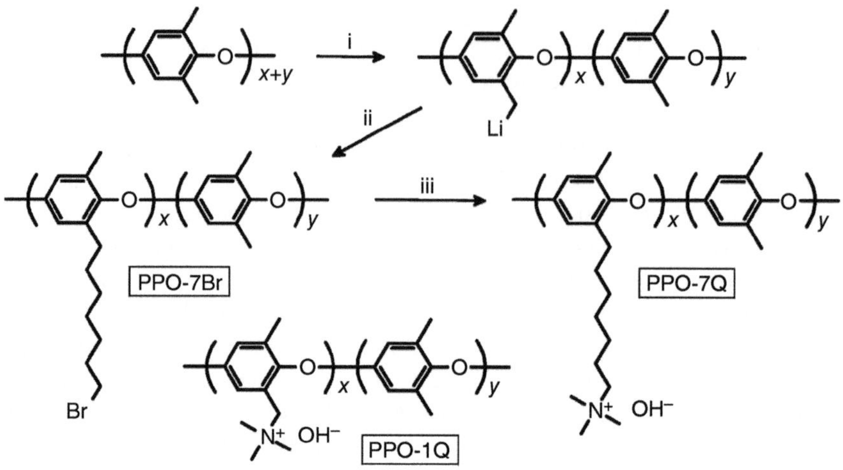

Figure 3.16 Introduction of side chains for AEMs through butyl lithium activation. Source: Dang et al. [97].

was then reacted with di-dimethylaminophenyl ketone to successfully introduce a functionalized side chain with two tertiary amine groups. After quaternizing with iodide, side-chain AEMs can be obtained [96]. However, because the QA group is directly connected with the benzene ring, the electron absorption effect greatly reduces the alkali stability of the membrane material. Therefore, the side-chain composition needs to be further regulated. Commercial polyphenyl ethers were treated with butyllithium to introduce a carboanion in the position of its bimethyl group and then reacted with dibromohexane and TMA to successfully prepare a long side-chain AEM (Figure 3.16) [97]. Because the QA group is bonded to the main chain through the aliphatic chain, the alkaline stability is improved. To develop highly basic-resistant AEMs, TMA can also be substituted with adipose circular basic functionalized reagents, such as 1-methylpiperidine. If 4,4′-tris-methylene bis(1-methylpiperidine) is used as a basic functionalization reagent, two anion exchange groups separated by aliphatic carbon chains can be introduced into a side chain, which is conducive to the aggregation of ionic groups [98]. Although the preparation of side-chain AEMs with butyllithium reagents is feasible and allows the composition and structure of the side chains to be regulated, the use of butyllithium is extremely dangerous and therefore limits the large-scale commercial application of this route.

Reversible addition-fragmentation chain transfer polymerization (RAFT) is a kind of active controllable polymerization that can also be used to synthesize side-chain AEMs. The use of RAFT initiators thus initiates the polymerization of styrene and 4-(3-butenyl) styrene. 4-(3-Butenyl) styrene is obtained by the coupling reaction of chloromethyl styrene and allyl magnesium bromide. By the addition reaction of the obtained polymer with hydrogen bromide, the double bond at the end of the side chain can be converted into bromoalkane, and further reacting with TMA can obtain the side-chain AEMs [99]. Since the polymer backbone prepared by RAFT technology is a block polymer, the AEM has both block and graft structures, which is conducive to the formation of a microphase separation morphology. However, the strong rigidity of the polystyrene backbone leads to a lack of sufficient entanglement of the polymer chain, so unlike aromatic polymers, the membrane materials made from this aliphatic polymer have poor mechanical properties. In a later study, isolated double bonds can be grafted onto a polymer matrix with good mechanical properties to obtain an AEM with excellent comprehensive properties. In addition, RAFT polymerization, despite its mild regulation, is sensitive to oxygen and requires strict deoxygenation during the process.

The Menshutkin reaction occurs between tertiary amine and halogenated hydrocarbon to form QA salts, which are often used in the preparation of side-chain AEMs. The synthesis of ionized monomers mentioned above involves the Menshutkin reaction, but it still depends on other reactions to bond the ionized monomers to the backbone. In this paragraph, the preparation of side-chain AEMs using only the Menshutkin reaction is described. With BPPO as the main chain, the ionized side chain with a tertiary amine group is connected to the main chain through the Menshutkin reaction, which can complete the introduction of the side chain and increase the density of ionic groups on the side chain in one step. The ionic side chain containing tertiary amine groups usually starts from N,N,N′,N′-methyl-1,6-hexyl amine as the functional monomer, which then reacts with bromoethane through the Menshutkin reaction. One end of the tertiary amine can be transformed into QA groups. Such molecules can continue to be attached to BPPO via the Menshutkin reaction to gain AEMs with two QA moieties on a single side chain. If bromoethane is replaced by 4-bromo-N,N,N-trimethylbutane-1-ammonium bromide, a side-chain AEM with three QA groups on a single side chain can be obtained (Figure 3.17) [100]. The synthesis of 4-bromo-N,N,N-trimethylbutane-1-ammonium bromide was obtained by the reaction of dibromobutane with TMA. Since the synthesis of the above-functionalized side chains is based on the bifunctional monomer, the ratio of the two monomers should be carefully controlled to ensure that only one end of the functional group is functionalized, while the other end of the functional group is reserved for further functionalization. The above reaction route can also be adjusted to preferentially react BPPO with N,N,N′,N′-tetramethyl-1,6-hexanamine via the Menshutkin process, and the terminal tertiary amine was converted to QA with halogenated hydrocarbon to obtain side-chain AEMs [101]. Side-chain AEMs can also be prepared by using BPPO to react with N,N-dimethylalkyl amines via the Menshutkin reaction [102]. On this basis, the membrane can be prepared by further reaction with alkyl amine. The side-chain membrane is not strictly anion exchange groups connected to the main chain by spacer groups. However, studies have shown that the hydrophobic alkyl side chain is also conducive to the aggregation of ionic groups on the main chain and suppressing water swelling.

Synthesis of Nafion-like Structures Starting from Functionalized Monomers. While the preparation route mentioned previously enables most AEMs with excellent properties, the synthesis of Nafion-like structures starting from functionalized monomers has the advantage of precise control of ionic group content and positions. However,

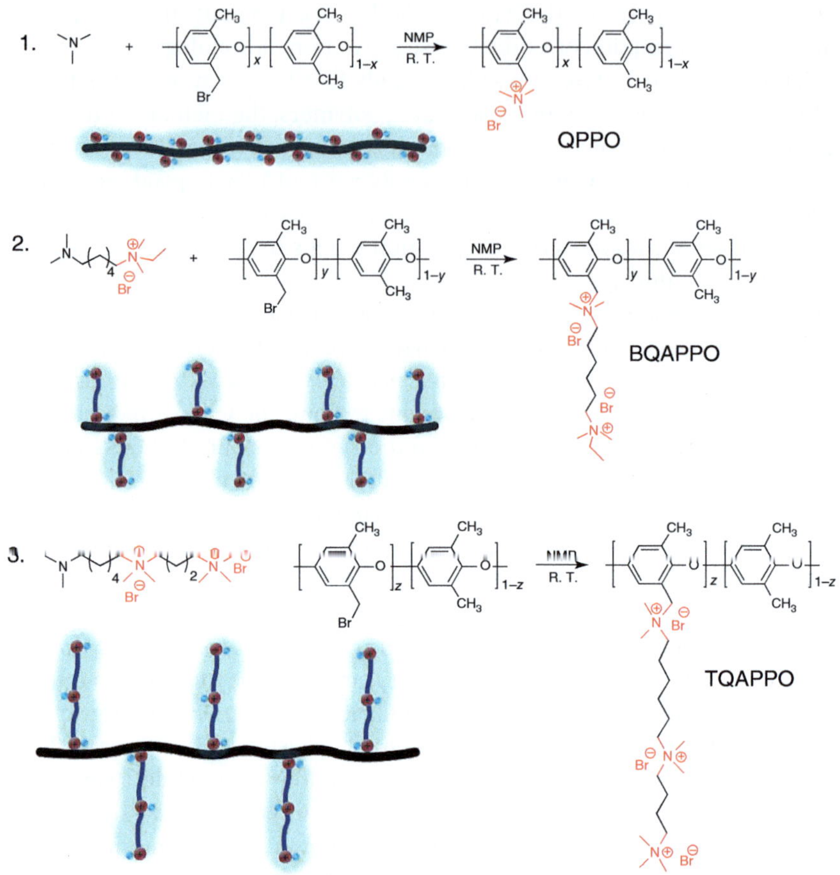

Figure 3.17 Synthesis of side-chain AEMs through the Menshutkin reaction. Source: He et al. [100].

because anion exchange groups such as QA are usually not very stable under high temperatures or basic polymerization conditions, there are relatively few reports on the direct polymerization of ionized side-chain monomers to synthesize side-chain AEMs.

Dicarboxylic monomers and diaryl monomers can be polyacylated using Eaton reagents to form high molecular weight polymers. The reaction can be completed under mild conditions at 60 °C, so it can be used to prepare side-chain AEMs from functionalized monomers. With 2,2′-dihydroxybiphenyl and 4-bromobutyltrimethyl ammonium bromide, the side-chain QA diaromatic monomer can be obtained under the action of potassium carbonate. Polyacylation of this monomer with the non-QA diamine monomers 2,2′-dimethoxy-biphenyl and 4,4′-diphenyl ether dicarboxylic acid in trifluoromethanesulfonic acid can afford AEMs with flexible side chains. In addition, the acid-catalyzed Friedel–Crafts polycondensation reaction can also be utilized to synthesize side-chain AEMs, which can be completed at room temperature under nitrogen. Under the catalysis of trifluoromethanesulfonic acid, biphenyl,

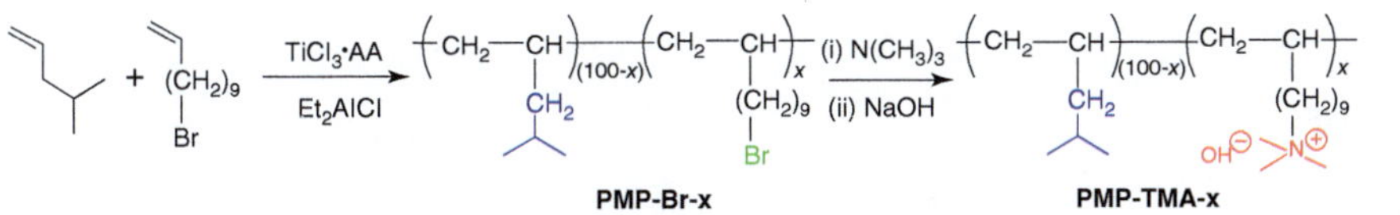

Figure 3.18 Synthesis of side-chain AEMs from functionalized monomers. Source: Zhang et al. [104].

and trifluoromethyl bromoalkyl ketone can undergo a Friedel–Crafts polycondensation reaction, which is followed by interaction with TMA to generate side-chain AEMs [103]. Although supermolecular weight bromoalkyl side-chain polymers can be prepared under very mild conditions in this route, the preparation of the precursor trifluoromethyl bromoalkyl ketone monomer is quite complicated.

Under the catalysis of the Ziegler–Natta catalyst, polymerization of olefin and bromoalkane olefin monomers can also be used to prepare side-chain AEMs. For example, the polymerization of 4-methyl-1-pentene with 11-bromo-1-carbene can yield ultrahigh molecular weight polymers with molecular weights exceeding 110,000 (Figure 3.18) [104]. As the polymers obtained by olefin polymerization are entirely aliphatic chains, the resulting membranes are usually not strong enough. Therefore, divinylbenzene can also be added to the polymerization process to participate in copolymerization and cross-linking to improve the mechanical properties of the obtained membrane material [105]. Because of the ultrahigh molecular weight, the polymer obtained by this method has poor solubility, so it needs to be made into membranes by melting hot pressing. Then, the membrane was immersed in TMA solution to complete the preparation of side-chain AEMs. Because effective microphase separation is difficult to achieve during hot pressing, the conductivity of the obtained membrane is usually not high. However, this route is very suitable for industrial large-scale production and has good application prospects.

Morphologies of Side-chain AEMs with Similar Nafion Structures. Similar to CEMs, flexible side chains containing ionic groups can self-assemble into well-defined channels for efficient hydroxide transport. The motility of the side chains greatly affects the phase separation. For example, in the side-chain AEM prepared by the Suzuki reaction, the QA group is bonded to the main chain through the flexible alkyl chain and the benzene ring, which can achieve a highly hydrophilic/hydrophobic microphase separation morphology [88]. If the ether bond is introduced into the side chain, the flexibility of the side chain can be enhanced to improve the degree of microphase separation [95]. In addition, the aggregation degree of ionic groups can be further improved by increasing the density of ionic groups on a single structural unit of the main chain. For example, an AEM containing six QA groups on a backbone benzene ring structure unit was prepared by a nucleophilic reaction [86]. The AFM observation shows that the morphology of the ion channels is perforative and distributed in a continuous hydrophobic region, which can complete efficient hydroxide conduction. Similarly, increasing the density of side-chain ion exchange groups is beneficial for constructing continuous channels. For example, two, three, or more ion exchange groups can be introduced into the side chain through the

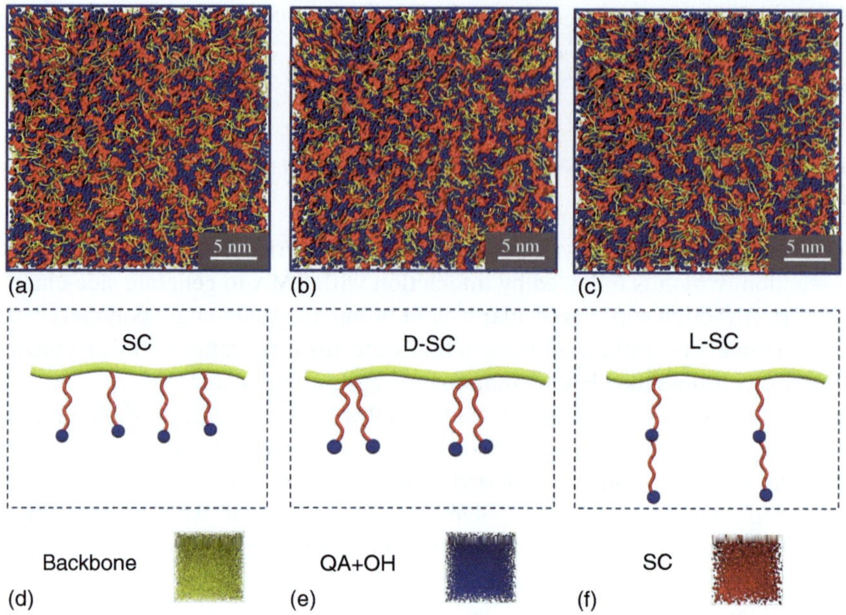

Figure 3.19 Microphase separation comparisons for different side-chain architectures (coarse-grained MD simulation results). Source: Ran et al. [85]/With permission of Royal Society Chemistry.

Menshutkin reaction using polyphenyl ether as the skeleton [100]. AFM images confirmed the successful construction of the highly microphase separation morphology. For the two strategies to improve the degree of phase separation, AFM, TEM, and coarse-grained molecular dynamics (MD) simulation were used to study them. It can be proven that increasing the density of ionic groups on a single structural unit of the backbone is more conducive to the construction of long-range ordered and wide ion channels. In contrast, increasing the density of ionic groups on the side chain is conducive to the construction of smaller ion clusters, which are less connective than the former (Figure 3.19) [85]. Therefore, the strategy of increasing the density of ion exchange groups on the structural unit is more conducive to speeding up the conduction of hydroxide, while increasing the density of ion exchange groups on the side chain is more conducive to improving the stability of membrane materials.

Hydrophobic side chains can also induce microphase separation, whether they are located at the end of anion exchange groups or directly connected to the main chain [91, 102]. Usually, the longer the side chain length is, the higher the degree of microphase separation, but a side chain that is too long can lead to macroscopic inhomogeneity of the membrane. Compared with the hydrophobic structure of the main chain and the hydrophilic side chain, the microstructure completed by the hydrophilic structure of the main chain and the hydrophobic side chain not only has well-established ionic groups but also has excellent continuity in the hydrophobic area. Therefore, the membrane can effectively balance the hydroxyl conductivity and stability.

3.2.2.2 Design of Side-chain AEMs with Grafted Structures

"Grafting from" Synthesis of Side-chain AEMs. The fabrication of grafted side-chain AEMs is mainly due to the "grafting from," "grafting onto" and "grafting through" methods, which are similar to the act of grafting side-chain CEMs. The "grafting from" method is most commonly reported for the fabrication of grafted side-chain AEMs. Grafted AEMs can be obtained by irradiation grafting of p-chloromethylstyrene monomers onto membranes based on perfluorinated or partially fluorinated aliphatic polymers, followed by immersion in aqueous TMA [106–109]. This method is suitable for large-scale production, but because the grafted chain is obtained by a heterogeneous reaction, the grafted chain has no ability to aggregate into ion channels. In addition, the high-energy rays used in irradiation grafting may damage the backbone of the polymer.

Therefore, it is necessary to find grafting strategies that can be carried out in homogeneous systems. The ATRP polymerization technique described above can be used to synthesize grafted side-chain AEMs. The C—F bond on commercial PVDF or polyperfluoroethylene propylene can be used as an initiator of ATRP to initiate the polymerization of ionized alkene monomers, such as QA chloromethylstyrene [110]. It should be noted that the nonionized chloromethylstyrene monomer is not suitable for the preparation of grafted AEMs by ATRP. This is because chloromethyl itself is also an ATRP initiator, and its presence will lead to uncontrollable structure. The grafted AEM can be prepared by one-step ATRP grafting. In addition, due to the high applicability of the ATRP reaction to monomers, other monomers can be added for graft polymerization to give the membrane material richer properties. However, due to the high C—F bond energy, weak ATRP initiation ability, and weak alkaline resistance of aliphatic fluorinated alkenes, this route is only suitable for research and difficult to popularize in practice.

BPPO is also the macromolecular initiator of the ATRP reaction, and the ATRP initiation ability of benzyl bromides is much better than that of the C—F bond, which can successfully initiate the polymerization of alkaline functionalized monomers (Figure 3.20) [111–113]. By regulating the amount of monomer added, the length of the graft chain can be changed, thus inducing different microphase aggregation patterns. By changing the amount of CuBr used in the catalyst, the grafting density can be changed, and the morphology of microphase aggregation can also be affected. By comparing the comb structure of high graft density and short graft chain with the rod-linear structure of low graft density and long graft chain, it can be found that the latter has lower water content and higher hydroxide conductivity, although the IEC is similar. In addition, unlike fluorinated aliphatic polymers, the BPPO backbone has excellent alkaline resistance and can be used in real-world hydroxide conduction environments. In future research, BPPO can be used as a development platform to complete the development of grafted AEMs suitable for different fields by regulating the composition of grafted chains.

Morphologies of Side-chain AEMs With Grafted Structures. The side-long AEM not only has a high degree of microphase separation morphology but can also be further constructed accurately by adjusting the graft density and graft length. Long side

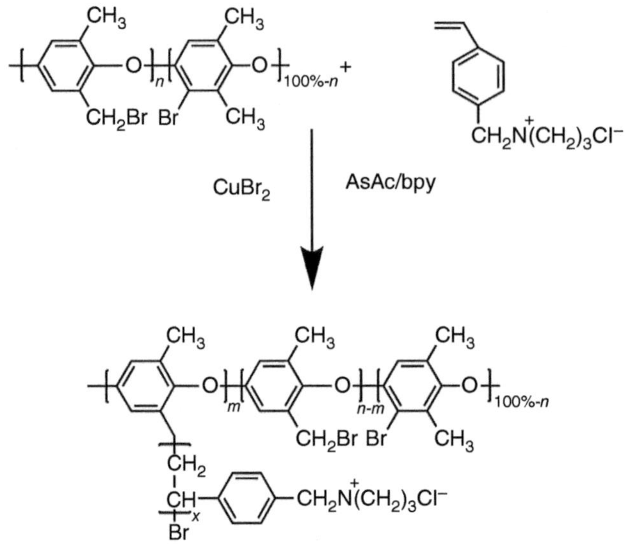

Figure 3.20 Synthesis of side-chain AEMs with graft structures through "grafting from". Source: Ran et al. [111].

chain AEMs prepared by ATRP with BPPO as the main chain. When the grafting density is high and the grafting chain is short, ion channels with larger sizes and better penetration can be constructed. When the grafting density is low and the grafting chain is long, ion channels with smaller sizes and more uniform distributions in the hydrophobic region can be constructed. In addition, the reduction in graft density was beneficial to improve the basic resistance of side-chain AEMs (Figure 3.21) [112].

3.2.2.3 Design of Functional Groups for Side-chain AEMs

Compared with cationic exchange groups, the types of side-chain anion exchange groups are much richer. The hydroxyl dissociation constants of "quaternary salt" with different structures are different, and the larger the hydroxyl dissociation constants are, the more basic they are, and the higher the anion conduction ability is in the case of "quaternary salt" with the same concentration. However, the more important aspect is that the "quaternary salt" structure is different, leading to different alkali resistance stabilities. The performance of different quaternary salt groups will be introduced below.

Quaternary Ammonium (QA) Anion Exchange Groups. The QA group is widely used in the preparation of side-chain AEMs due to its advantages, such as easy preparation and inexpensive raw materials (Figure 3.22a) [114, 115]. However, in general, the QA-type IEM is still limited by its own existence of two fatal shortcomings: poor ion conduction ability and poor alkaline resistance. The poor ion conduction ability is due to the weak dissociation ability between the QA group and its accompanying hydroxide ion. Another reason is the poor solubility of QA salts in organic

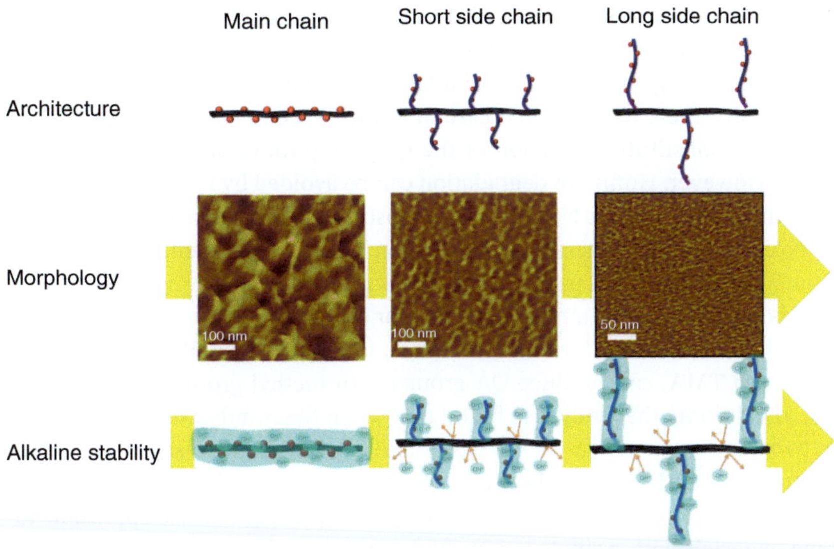

Figure 3.21 The structure–property relationship for side-chain AEMs with grafted structures.

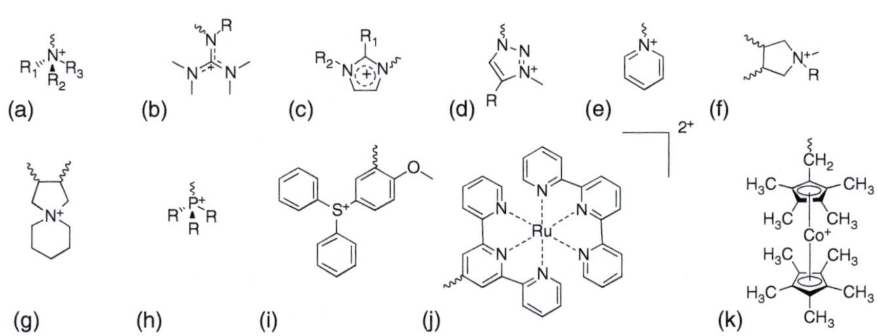

Figure 3.22 Common anion exchange groups for AEMs.

solvents. Therefore, the common QA functionalization route is generally based on first solidifying into membranes and then soaking in an alkaline solution. Due to the introduction of ionic groups through functionalization, it is not beneficial to microphase separation and the formation of ion clusters in the process of membrane formation. Although many side-chain AEMs have been reported previously, they are limited by the QA functionalization route, so it is not beneficial to obtain high-ion conduction AEMs to a certain extent. However, the poor alkaline stability is due to degradation reactions such as nucleophilic substitution, Hofmann elimination, and ylide formation in high-temperature and strongly alkaline environments. Specifically, the carbon in the α carbon position around the N^+ center of the QA group will be attacked by hydroxide, in which case the QA group will degrade to form tertiary amines (which have no ion-exchange capacity) and alcohols (which detangle from the membrane). However, if there is hydrogen in the β carbon position of the QA group, it will also be attacked by hydroxide and cause Hofmann degradation of the quaternary ammonium group. In this case, the QA group will undergo elimination reactions and generate olefins and tertiary amines (without ion exchange ability). Generally, there must be an α carbon in the N^+ center of the QA group, so the nucleophilic substitution reaction of the QA group under alkaline conditions is inevitable. However, Hofmann degradation can be avoided by removing the presence of β-hydrogen around the N^+ center. The most common method is to occupy the β carbon position with a benzene ring, which can avert the occurrence of Huffman degradation. The degradation path can also be avoided by substituting all hydrogen in the β carbon position with a methyl group. For example, the use of the acylation reagent 3-chloro-2,2-dimethylpropionyl chloride to functionalize PES, followed by interaction with TMA, can produce QA groups with methyl groups in the β position [116]. In addition, the study also found that when the number of carbon atoms between two QA groups increased, the base stability of QA could also be improved. When there are six carbon atoms between the two QA groups, the Hofmann elimination reaction can be effectively inhibited [117]. This principle is also applicable to membrane materials with a single ion-exchange group in the side chain; that is, in the design of the polymer structure, the interval between the QA group and another electron-absorbing group (carbonyl group on the polymer main chain or side chain, sulfoxide group, etc.) should be greater than six carbon atoms.

Guanidinium Anion Exchange Groups. Guanidine is a common organic base for organic synthesis (Figure 3.22b). After the guanidinium cation is formed with nucleophilic groups such as bromine, the central carbon atom, and its three connected nitrogen atoms are planar conjugated structures. The positive charge on the carbon atom can be efficiently dispersed to the three nitrogen atoms by charge delocalization, which results in a strong dissociation between the guanidine cation and its accompanying anion. For example, the pK_a value of pentamethyl-substituted guanidine is 13.8, which can be used to synthesize quaternary guanidine anion exchange groups [118, 119]. Among them, pentamethylguanidinium was obtained by reaction of tetramethylurea with oxaloyl chloride to obtain Vilsmeyer salt, and then the product was obtained by the further reaction of methylamine and sodium hydride. Although the guanidine salt has excellent ionic conduction ability, its alkali resistance is not excellent. Under alkaline resistance test conditions of 60 °C and 1 mol/l KOH, its hydroxide conductivity decays rapidly within 48 hours [120]. At present, the degradation process of guanidine in alkaline environments is not sufficiently clear.

Heterocyclic Anion Exchange Groups. AEMs based on azacyclic compounds have also been widely studied. The imidazole-based AEM is the most promising one. Compared with QA groups, imidazolium anion exchange groups have similar ionic conduction ability, and their preparation method is simple using inexpensive raw materials (Figure 3.22c) [121]. In addition, due to the good solubility of the imidazole molecule and imidazolium in organic solvents, imidazolium-functionalized AEMs can be prepared by the prefunctionalization course. In the process of membrane formation, ionic groups with imidazole side chains can aggregate and form well-established ion channels. Therefore, the ionic conductivity of imidazolium-type AEMs reported in most literature is maintained at a high level [122]. The alkaline resistance of imidazole AEMs is closely related to the composition and position of substituents on the imidazole five-membered ring.

For the AEM functionalized with 1-methylimidazole, the imidazolium-type anion exchange group can remain stable, usually in alkaline solutions with concentrations less than 1 mol/l at room temperature. However, when the alkaline concentration is higher than 1 mol/l or the temperature is increased, the imidazolium-type AEM is unstable. This is because the two N atoms around the C2 site on the imidazole ring absorb electrons, which makes the C2 proton more acidic and vulnerable to attack by hydroxide under alkaline conditions, resulting in ring opening of the imidazole cation at the C2 site and loss of ion exchange ability [123]. Even using polybenzimidazole as the alkaline functional reagent preparation of imidazolium, there is also an existing C2 position open-ring degradation phenomenon [124].

Therefore, the introduction of substituents at the imidazole C2 position is one of the main ways to improve the alkali resistance, and imidazolium groups with different C2 substituents also have great differences in alkali resistance [125]. For example, when a tert-butyl group is introduced at the C2 position, the stability of the imidazole group decreases rather than increases. This is mainly because the tert-butyl cation generated by hydroxide attacking the 2-position carbon atom is relatively stable, so it effectively promotes the occurrence of ring-opening degradation. Therefore, the steric hindrance effect cannot completely determine the imidazolium stability, and the electronic effect must also be considered to reduce the C2 substituent on the proton acid. For example, the use of trimethoxy phenyl as the C2 substituent of the imidazole group effectively improves the basic stability of imidazole cation due to the steric hindrance effect of trimethoxy phenyl and the strong electron donor effect of methoxy [126]. In addition, the introduction of substituents at C4 and C5 can further improve the basic stability. Compared with methyl, isopropyl and 3,5-dimethylphenyl have stronger steric effects and can more effectively shield the nucleophilic attack of hydroxide ions [127].

In addition to the C2, C4, and C5 substitutions, the N3 substituents of the imidazole ring also affect the stability of the imidazole cation. The effects of methyl, ethyl, isopropyl, heptyl, octyl, phenyl, and diphenyl methyl groups on the stability of imidazole cations were studied by experiments and density functional theory (DFT) calculations. Isopropyl-substituted imidazolium has the best alkali resistance, which is better than that of QA groups. AEMs developed based on imidazole substituted with isopropyl at the N3 position also showed excellent alkaline stability [128].

The triazole group is similar to the imidazole group structure. As mentioned above, triazole groups can be obtained by the click reaction of azide and alkyne groups (Figure 3.22d). Triazole-type AEMs can be obtained by using electrophilic reagents such as iodide methane to react with triazole. Its ionic conduction ability is similar to that of the QA group, and its alkaline stability is lower than that of the QA-type AEM but higher than that of the C2 nonsubstituted imidazole AEM [129].

Currently, there are limited studies on triazoles as anion exchange groups and suitable substituents can be further introduced at the C4 and C5 positions to improve the alkaline stability.

In addition to imidazole and triazole groups, pyridine can also be used as ion conduction groups for AEMs (Figure 3.22e). Due to the strong basicity and the positive charge delocalization effect of the conjugated pyridine ring, the alkali resistance of pyridine-type AEMs is comparable to that of QA-type AEMs, and further optimization can be expected [130, 131]. Pyrrole can also be used as a basic functionalization reagent to prepare AEMs. Similar to imidazole-type AEMs, the basic resistance of the pyrrole-based AEM is closely related to the type of N substituent. By studying the effect of a series of substituents, such as methyl, ethyl, butyl, octyl, isopropyl, and phenyl groups, on the basic stability of pyrrolidine salt, the N,N-ethylmethyl substituted pyrrolidine cations were found to be the most stable (Figure 3.22f) [132]. In addition, pyrrolidinium can also act as the anion exchange group (Figure 3.22g) [133]. Almost no degradation of polydiallyl piperidine was observed in 1 mol/l KOH solution (80 °C), indicating its excellent stability.

Quaternary Phosphine Anion Exchange Groups. By reacting trialkyl phosphorus with haloalkanes, a quaternary phosphorus AEM can be obtained (Figure 3.22h). However, due to the severe nucleophilic reaction, the alkaline stability of the obtained membrane material is poor [134]. By using tri(2,4,6-trimethoxyphenyl) phosphine as a basic functionalized reagent, the alkaline stability can be greatly improved. This is because the electron donor and steric hindrance effects of methoxy groups can effectively reduce the nucleophilic attack of hydroxide [135]. However, due to the high molecular weight of trimethoxylphosphine, the resulting AEM IEC is usually low, which limits its ionic conduction ability. In another type of quaternary phosphorus AEM, phosphorus atoms are linked to four dialkyl-substituted nitrogen atoms. Similar to the guanidine group, the positive charge on the phosphorus atom can be dispersed on the nitrogen atom by charge delocalization, resulting in strong dissociation between the quaternary phosphorus and its accompanying anion. On the other hand, the methyl and cyclohexyl substituents on nitrogen atoms have strong steric hindrance, which can effectively shield the nucleophilic attack of hydroxide ions. Although the IEC of the membrane is still very low, its ion conduction ability has been strikingly improved, and it shows excellent alkaline stability. At present, there are few reports on quaternary phosphorus groups, but their excellent ionic conductivity and alkaline stability have been proven. In future work, further simplifying the synthesis steps, and reducing the molecular weight of functional groups to improve the IEC, is expected to develop a quaternary phosphorus AEM with practical application potential.

Tertiary Sulfur Anion Exchange Groups. Tertiary sulfur cations can also be used as functional groups in AEMs, but triphenyl or trialkyl tert-sulfur cations are not stable under alkaline conditions and are easily attacked by hydroxide to undergo nucleophilic substitution reactions. By using tertiary sulfur cations with trimethoxylbenzene as anion exchange groups, the electron-donating and steric effects of trimethoxylbenzene can effectively improve the alkaline stability of membranes (Figure 3.22i) [136].

Metal Cationic Anion Exchange Groups. By coordinating rhodium chloride with tripyridine, anion exchange groups based on metal rhodium can be synthesized (Figure 3.22j) [137]. Since Ru is positively divalent in this compound, each Ru center can be accompanied by two anions. In addition, the metal rhodium-based AEM exhibited excellent alkaline stability and did not show any change with respect to its chemical structure after immersion in 1 mol/l NaOH solution (at room temperature) for 6 months. Cobalt diamene can also act as an anion exchange group (Figure 3.22k) [138]. Five methyl substituents were introduced into each benzene ring of cobalt diamene to shield the nucleophilic attack of hydroxide ions through steric effects, and the resulting AEM showed excellent alkaline stability.

At present, the QA group is still the most commonly used ion exchange group in the preparation of AEMs. In the reported research work, many ion exchange groups have excellent performances in ionic conductivity or chemical stability. However, considering factors such as ionic conductivity, alkali stability, and synthesis complexity, these ionic groups are still limited for practical use. Most likely to replace the QA group is the imidazolium group. It has similar ionic conduction ability to QA groups, the synthesis process is simple, and the raw material is inexpensive and easy to obtain. The introduction of alkyl substituents in the C2, N3, C4, and C5 positions of imidazole rings can achieve much higher alkaline stability than the QA group.

3.3 Construction of the Cross-linking Side Chain

High-performance side-chain IEMs can be constructed by adjusting the composition, structure, and graft density (or length). However, for some side-chain IEMs, the performance of the membrane may have a remarkable "trade-off" effect during the process of regulation. For example, if the ionic conductivity increases, the swelling of the separation membrane will become severe. Therefore, it is necessary to find ways to improve the comprehensive performance of side-chain IEMs. Among them, cross-linking is an effective means. Cross-linking the side-chain IEM with excellent conductivity and high ion transport efficiency can effectively avoid the problem of excessive swelling to obtain an IEM with excellent comprehensive performance.

Cross-linking by Unsaturated End Groups. By introducing unsaturated and highly reactive end groups in the design process of IEMs, a cross-linking reaction can occur accompanying membrane formation. For example, during the course of membrane solidification, vinyl double bonds can exhibit free radical polymerization to complete the cross-linking of IEMs [139, 140]. The constructed cross-linking network can greatly improve the membranes' mechanical properties, such as tensile strength, and effectively inhibit methanol permeability. In addition to radical polymerization, highly reactive Friedel–Crafts acylation also enables cross-linking of IEMs. Chloracetylation of PPO was carried out to introduce chloracetyl groups. PPO containing chloroacetyl groups was blended with BPPO, and the chloroacetyl group can be cross-linked by Friedel–Crafts acylation with hydrogen on the phenyl ring of PPO during membrane formation [141, 142]. Due to the high reactivity of Friedel–Crafts acylation, the cross-linking process can be completed at room temperature without any extra catalysts. Similar to the double-bond cross-linking effect, Friedel–Crafts acylation cross-linking can significantly improve the thermal stability and mechanical strength of the membrane material and reduce the methanol permeability. In addition, due to the occurrence of Friedel–Crafts acylation cross-linking on the benzene ring structure of adjacent polymers, it is beneficial to induce the orderly arrangement of polymer chains, thus promoting the regular arrangement of ionic groups and realizing efficient ion conduction. Similarly, bromomethyl on BPPO can react with hydrogen on the benzene ring to prepare cross-linked IEMs. As mentioned above, the side-chain IEM can be prepared by using BPPO as the macromolecular initiator of ATRP. On this basis, BPPO was added for cross-linking, which greatly improved the thermal stability of the membrane. At the same time, the cross-linked structure also improves the alkaline stability of the membrane [143]. After long alkali immersion, the conductivity of the crosslinked membrane was 1.5 times higher than that of the non-crosslinked membrane.

Cross-linking by Hydrogen Bonds. As mentioned above, the ATRP reaction provides a platform for the preparation of side-chain IEMs. Through the ATRP reaction, not only ionized side chains but also other functionalized polymer chains can be introduced. With PVDF as the polymerization backbone, through the ATRP reaction, N-isopropyl acrylamide and quaternary ammonium p-chloromethyl styrene polymerization are successively initiated to prepare side chains containing temperature-sensitive groups and ion exchange groups [144]. When the temperature-sensitive group is above the phase transition temperature, hydrogen bonds will be formed between molecular chains, which is conducive to improving the swelling resistance of IEMs under high temperatures. Above 40 °C, the water content and swelling property of side-chain AEMs containing temperature-sensitive groups are significantly smaller than those without temperature-sensitive groups (Figure 3.23).

Cross-linking by Unsaturated Groups of Main Chains. The bromomethyl in BPPO can be converted into a double bond by the Wittig reaction, which can be further polymerized with p-chloromethyl styrene by radical polymerization and quaternary ammonium functionalization, thus completing the preparation of cross-linked IEMs with interpenetrating networks [145]. The swelling of the IEM is greatly inhibited by the presence of an interpenetrating network. The water content of the quaternized poly(phenylene) oxide (QPPO) membrane is as high as 335%, and that of the PPO-based IEM with cross-linking networks is only 78.6% at the same IEC level. In addition, the introduction of an interpenetrating network effectively improves the thermal stability of the membrane.

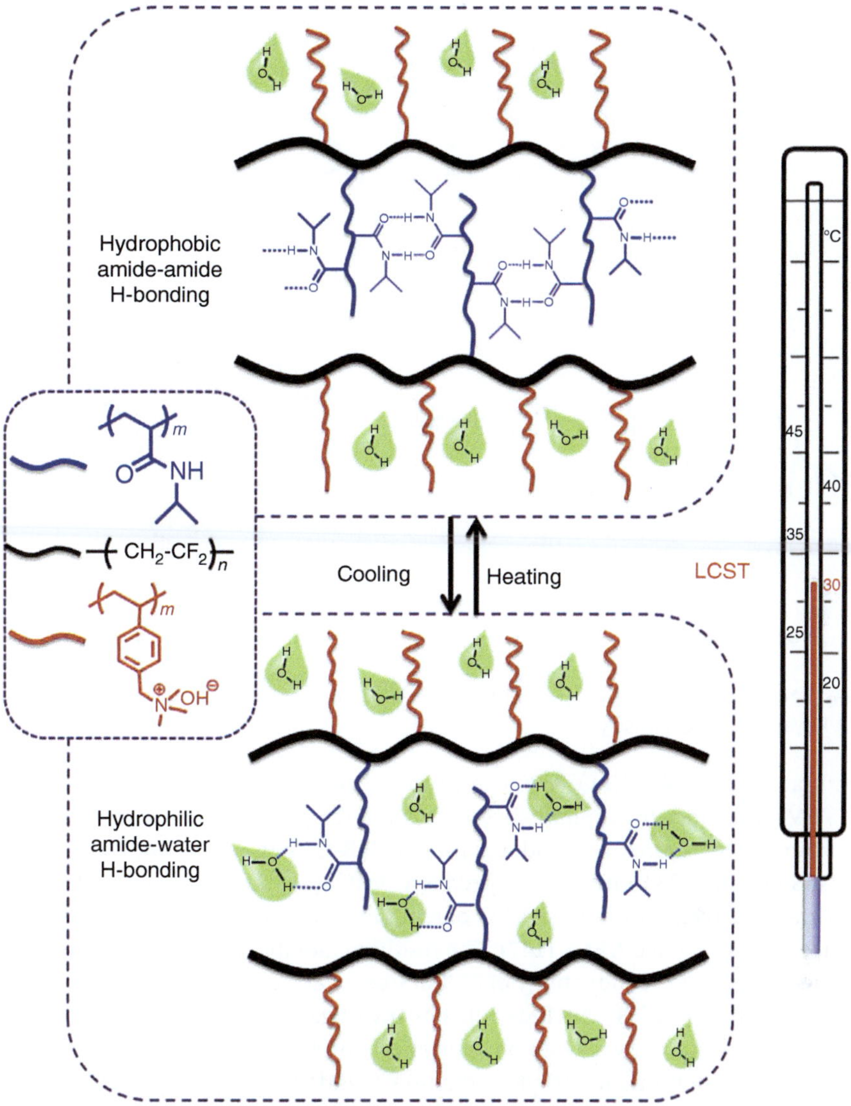

Figure 3.23 The temperature-responsive working principle of side-chain AEMs cross-linked by hydrogen bonds. Source: Wu et al. [144].

3.4 Construction of Hyperbranched Networks

Compared with dense SC AEMs, hyperbranched structures have higher cationic group aggregation degrees, which may favor the collection of ionic groups. In our group, we developed two routes to prepare hyperbranched AEMs. One uses ATRP, and the other employs RAFT polymerization. 4-(Chloromethyl) styrene (VBC) was a good candidate to trigger ATRP reactions and residue chloromethyl could be further functionalized to be cationic groups (Figure 3.24) [146]. The IECs of this

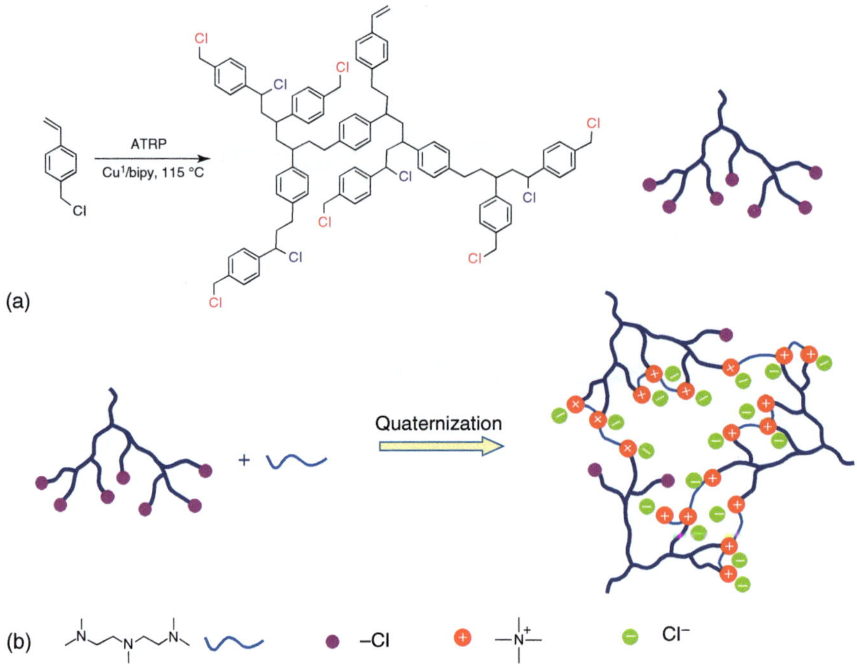

Figure 3.24 Synthesis of side-chain AEMs hyperbranched networks. Source: Ge et al. [146]/Royal Society chemistry.

hyperbranched AEM (HB-QPVBC) can reach 2.36 mmol/g. As observed in AFM and TEM, a distinct hydrophilic–hydrophobic microphase-separated morphology was visible. The hydroxide conductivities of HB-QPVBC at 20 and 80 °C were 39.57 mS/cm and 123.01 mS/cm, respectively. VBC was functionalized with RAFT chain transfer agents to initiate VBC polymerization, which was another way to produce hyperbranched AEMs [147]. The resulting membrane (IEC = 1.26 mmol/g) exhibited a hydroxide conductivity of 85.1 mS/cm at 80 °C.

Regarding hyperbranched AEMs, further investigations found that the compositions of ammonium agents had obvious effects on the microphase morphologies [148]. N,N,N′,N′-tetramethyl ethylene diamine (TEMEDA), N,N,N′,N′-tetramethyl-1,6-hexanediamine (TEMHDA), and N,N,N′,N′-tetramethyl-1,10-decyldiamine were chosen as functionalized agents to prepare three kinds of hyperbranched AEMs, termed HBM-2C, HBM-6C, and HBM-10C, respectively. With TEM observations, HBM-6C displayed distinct microphase separation with interconnected hydrophilic domains with an average width of 3 nm. In comparison, HBM-2C and HBM-10C showed fewer interconnected ionic channels.

3.5 Construction of Dynamic Transfer Regions

A molecular machine is an assembly of different numbers of molecules that can perform machine-like motions with appropriate external stimuli. Molecular machines have attracted great interest and attention in the supramolecular field, but there is no report regarding their use in IEMs. This situation has encouraged researchers to introduce the idea of molecular machines into the design of IEM structures and explore how to use molecular machines to further improve membrane performance. In recent years, our research group introduced polyrothane into the preparation of IEMs for the first time, and the results showed that polyrothane-based IEMs possess dynamic transfer regions, thus greatly improving the ion conductivity of the membrane without changing or even reducing IECs.

Dynamic Transfer Region Construction in CEMs. As shown in Figure 3.25, the polymer skeleton was prepared by copolymerizing dibenzo 24-crown ether-8, diphenyl ether, and sebacic acid to crown ether containing main chains. Then, the guest side chain with secondary ammonium salt was prepared, and the Schiff base was obtained by the reaction of methyl p-formylbenzoate and methyl 4-aminomethylbenzoate. The tertiary ammonium in the Schiff base was reduced to form a secondary amine.

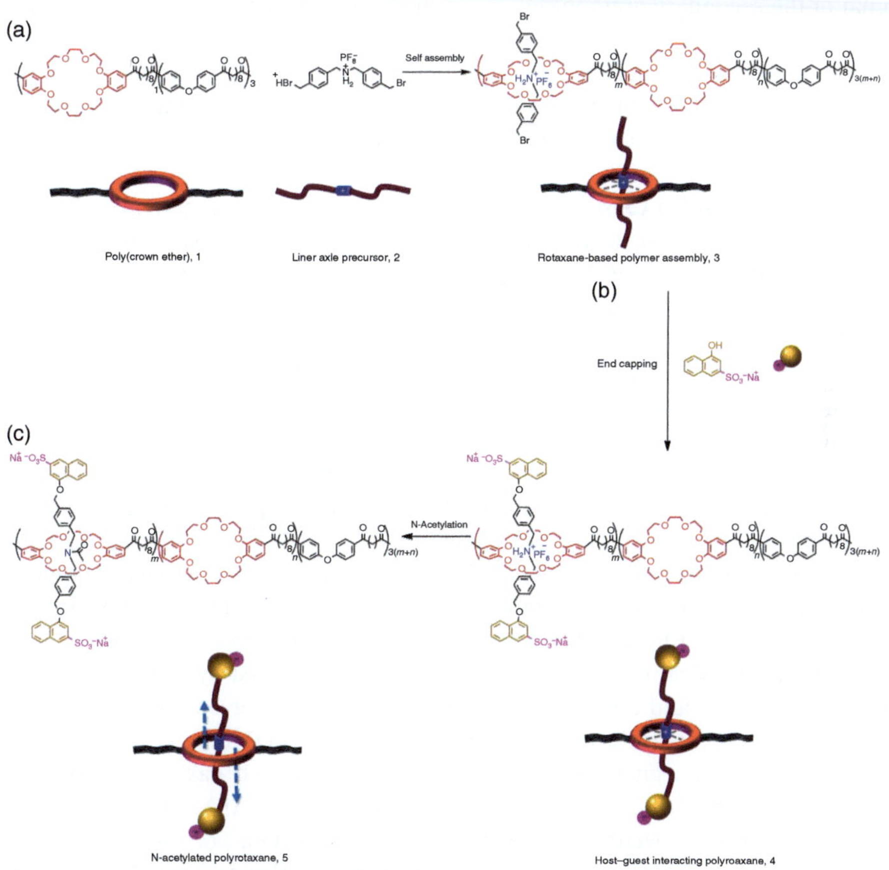

Figure 3.25 Synthesis of side-chain CEMs with dynamic transfer regions. Source: Ge et al. [149]/Springer Nature/CCBY 4.0/Public domain.

Finally, the secondary ammonium salt was protonated by hydrobromic acid and exchanged with ammonium hexafluorophosphates to obtain secondary ammonium salt associated with hexafluorophosphates. Through the host–guest interaction between the electron-rich crown ether donor and the electron-deficient secondary ammonium salt guest axis, the linear axis was inserted into the poly(crown ether) skeleton to assemble the cyclorotaxane structure polymer. Then, the two ends of the polymer were sealed with the ionic group 1-naphthol-3-sulfonate, and the polyrotaxane PEM was obtained [149].

To verify the successful construction of polyrotaxane in the IEM, the structure of the products obtained at each step was first determined by 1H-NMR spectroscopy, which confirmed the formation of the polyrotaxane structure. Due to the strong host–guest interaction, when the linear axis is surrounded by macrocyclic crown ether, the typical proton signal peak of the secondary ammonium group in the axis moves to the lower field. The complexation peak between the amino hydrogen and the crown ether ring was verified by two-dimensional NMR, and the proton signal in the macrocyclic crown ether showed a strong spatial correlation with the proton signal of the secondary ammonium group in the axis, which further verified that the crown ether polymer and the secondary ammonium salt axis had a host–guest interaction. Due to the successful construction of mobile transfer regions, the membranes exhibit unprecedented ion conduction. When the IEC is 0.73 mmol/g and the temperature is 85 °C, the conductivity is as high as 280 mS/cm.

Dynamic Transfer Region Construction in AEMs. Shared with the experience of PEM preparation, the polymer skeleton containing crown ether prepared in the previous section is continued for AEM development. The guest axis is changed into a small molecular chain with two secondary ammonium salt sites. The reaction steps and conditions in the previous section are adopted. Then, based on the specific identification characteristics between the polymer framework containing crown ether and the guest double-site secondary ammonium salt, tri(2-ethylhexyl) amine was introduced into the quritaxane as the end group (Figure 3.26) [150].

The hydroxide conductivity increases sharply when the temperature exceeds 60 °C, indicating that the polyrotaxanes has obvious temperature-corresponding characteristics. The membrane conductivity (IEC = 0.68 mmol/g) is as high as 180 mS/cm when the temperature is 90 °C. Comparing the conductivity of polyrotaxanes with the results reported in the literature, it can be found that this kind of membrane has better hydroxide transfer abilities, especially under low IECs.

The AEM with floating OH^- counterions is easily attacked by CO_2 in ambient air for applications and then converted into carbonate. However, due to the large ionic radius of carbonate, its movement ability is weak, and the degree of hydration is low, resulting in a great reduction in conductivity, which has become one of the main problems for AEM applications. In addition, the AEM in some energy storage devices itself needs to conduct bulk ion conduction. For instance, in the flow battery, the AEM needs to transport sulfate or chloride ions. These large anions cannot form hydrogen bond networks with water and therefore lack the means of transport through the Grotthuss mechanism. To address this challenge,

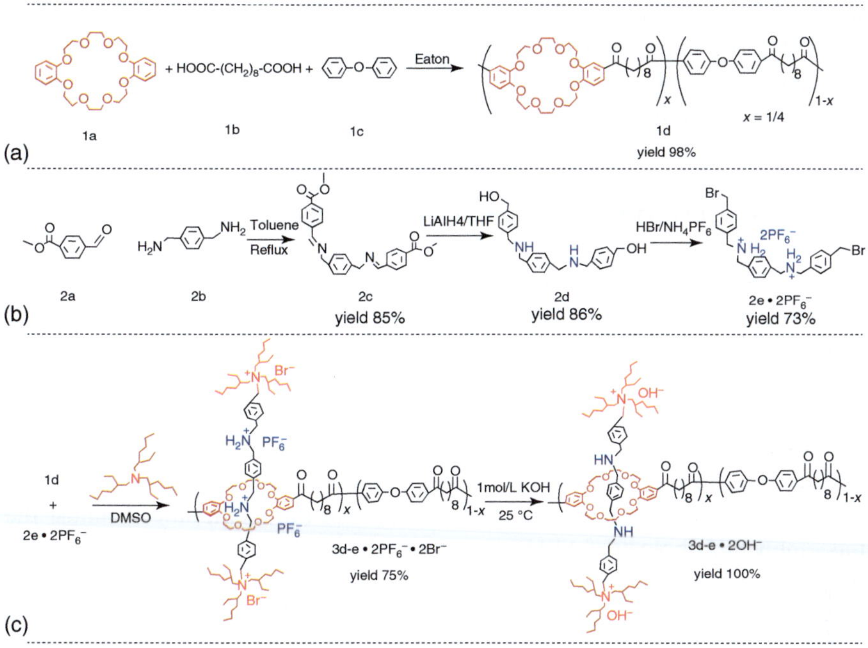

Figure 3.26 Synthesis of side-chain CEMs with dynamic transfer regions. Source: Ge et al. [150].

our laboratory proposed the diffusion conduction mechanism of triggering motion and rotation in the molecular machine to transfer large anions on the basis of the work of the polyrotaxanes.

The carbonate-type AEM preparation method is similar to that in previous work. The guest small molecule was first penetrated the crown ether ring, and the end was sealed by bulk quaternary phosphorus groups. The prepared IEMs were soaked in sodium carbonate solution to obtain mobile carbonate AEMs. Two-dimensional NMR was used to prove that the side chain successfully penetrated the crown ether ring, and the spatially correlated coupling peak between the secondary amine group in the side chain and the crown ether ring could be observed [151].

MD simulations of the polyrotaxanes system visually elucidate the mobility and range of motion of free-shuttling cationic groups. With increasing temperature, the motion characteristics of traditional AEMs do not change significantly, and the range of relative motion positions is small. These results show that the ionic conducting groups are mainly fixed in a specific position due to the limitation of covalent bonds. In contrast, due to its unique shuttle characteristics, the relative position range of the polyrotaxanes is larger, and the distribution of possibility density is more uniform. This free shuttle behavior with a wide range of motion is thought to provide an additional solvated shell fluctuation and reduce the conduction energy barrier during the transfer between two cationic groups. Similarly, the conductivity test results showed that the side chain began to slide in the crown ether ring of the main chain, driving the movement of water molecules and ions, resulting in a significant increase in conductivity. At 80 °C, the conductivity of the carbonate type can reach 105 mS/cm and that of the bromine type can reach 112.1 mS/cm. Compared with the carbonate- or bromine-type AEMs reported thus far, polyrothane AEMs have obvious advantages in terms of conductivity.

3.6 Construction of Cation–Dipole Interactions

Ion–dipole interactions are electrostatic interactions between ions and neutral dipole molecules that contribute significantly to the self-assembly of biological macromolecules such as DNA and proteins. By mimicking the ion–dipole interaction functions in biological systems to promote the directed aggregation and self-assembly of ion-exchange groups in AEMs, highly ordered ion conduction channels can be expected to be built. The introduction of neutral macromolecular fragments into the polymer framework will help promote the directional aggregation of cationic groups and dipole molecules within the membrane to adjust the self-assembly morphology of AEMs and construct regular ion transport channels.

Ion Conduction Channel Establishment by Ion–Dipole Interactions. Poly(ethylene glycol) (PEG) is an archetypical polar homopolymer that contains abundant uniformly distributed dipoles owing to the strongly electron-withdrawing oxygen heteroatoms in the long polymer main chain. Thus, the slightly electronegative oxygen atoms can provide a strong driving force for cation movement (e.g. lithium and QA) via cation–dipole interactions. Cation–dipole interactions in Li-ion battery polyelectrolytes have been demonstrated to provide attractive ion conduction. The incorporation of PEG into AEM manufacturing will direct the self-assembly of covalently attached quaternary ammonium cations, forming enhanced highways for ion conduction. Furthermore, such cation–dipole pairs behave as noncovalent cross-linkers, preventing dimensional deformation of the ion channels. Finally, the ethylene glycol groups would form an efficient, continuous hydrogen-bonded network to enhance the OH^- conduction (Figure 3.27) [152].

The self-assembled geometry and formation of ion-conducting channels have been revealed by SAXS and nanoscale microscopy analysis (TEM and AFM). As shown in Figure 3.28, the SAXS profile of QA38PPO-P shows a distinct scattering peak ($q = 1.35$ n/m, in contrast to the SAXS profile of QA38PPO), consistent with periodically distributed clusters composed of ionic groups. TEM and AFM phase images also provide supporting evidence for the presence of ionic clusters with a nanoscale phase-segregated geometry. MD simulation of the polymers (QAxPPO-P versus QAxPPO) was carried out to determine the benefits of cation-dipole interactions for QA aggregation. Quantitative information on the spatial location of the QA groups relative to the polymer main chain is gained from the N O* distance (d) and the N O O* angle (θ). Figure 3.29 shows this information in joint probability density distribution contour maps. Compared to QAxPPO, QAxPPO-P shows a deeper red zone (a strong probability of spatial distribution) at a larger θ and smaller d. This demonstrates a mutual affinity between the ether oxygen and the QAs, which

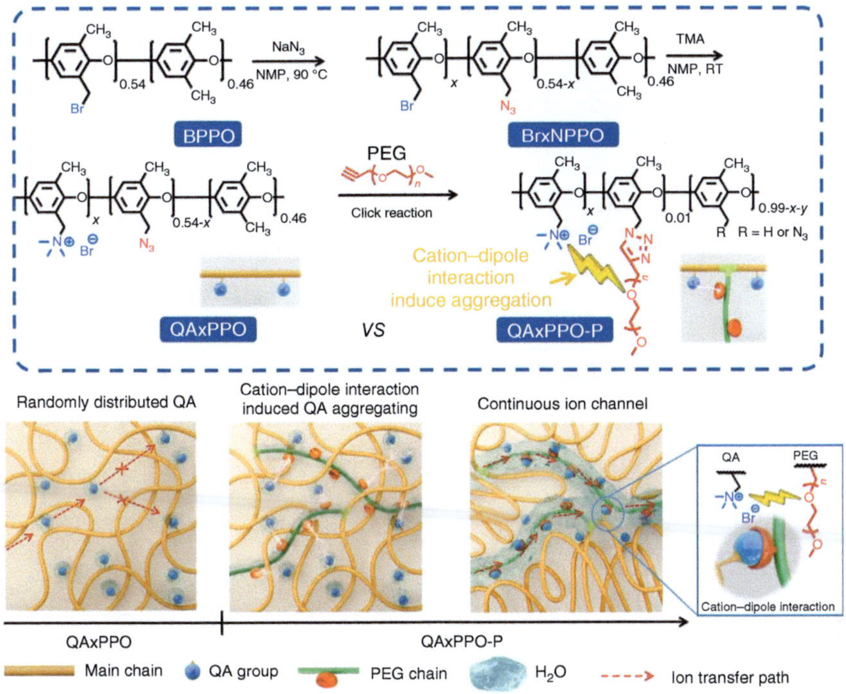

Figure 3.27 Synthesis of side-chain membranes with ion–dipole interactions for ion conduction channel establishment. Source: Zhang et al. [152]/John Wiley & Sons.

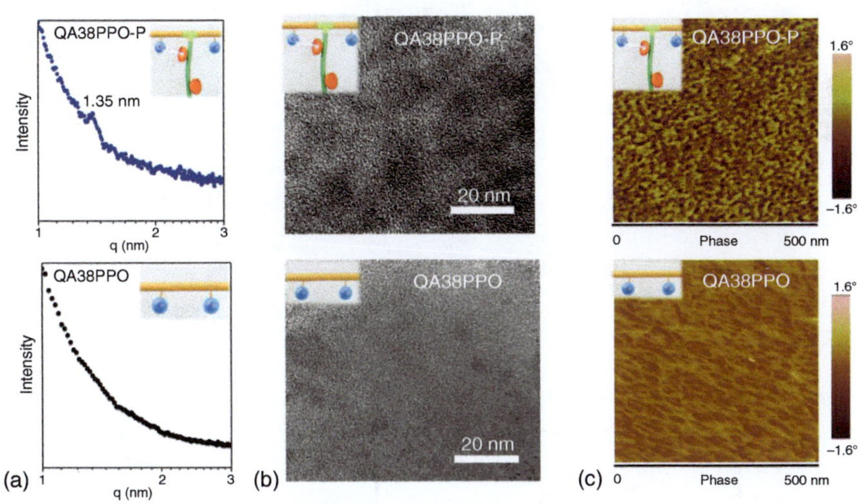

Figure 3.28 Micromorphologies of the QA38PPO-P and QA38PPO membranes [152]. Jianjun Zhang et al. 2020 / With permission of American Institute of Chemical Engineers.

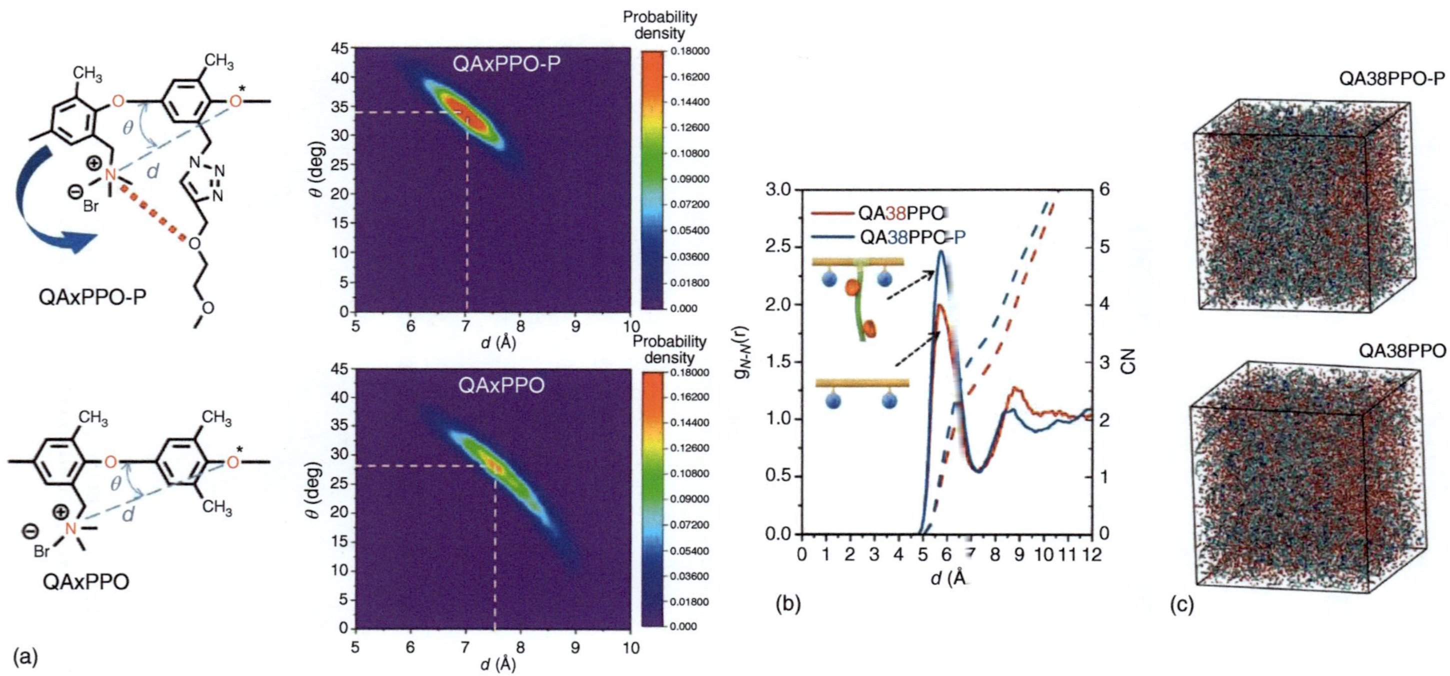

Figure 3.29 MD simulations for observing the microphase separation of QAxPPO-P and QAxPPO. Source: Zhang et al. [152]/John Wiley & Sons.

endows a significant motion of the QA cations toward the PEG side chain, in support of the expected aggregation behavior. The radial distribution function (RDF) and its corresponding CN allow the quantification of such aggregation behavior. The first peak represents the contribution of the N–N correlation between two polymer chains. QA38PPO-P has a total density probability g(N–N)(r) of 2.49 at $r = \sim 5.75$ Å, while PEG-free QA38PPO exhibits a smaller g(N–N)(r) of 2.00 at the same given distance (r). The second peak corresponds to the adjacent N–N correlation in the same chain. The correlation peak shifts inwards from 8.8 (QA38PPO) to 8.4 Å (QA38PPO-P), suggesting a decreased distance between QA groups in the presence of PEG grafts. Snapshots of equilibrated configurations provide an intuitive verification. QA38PPO-P presents a more concentrated configuration, while QA38PPO yields a more random distribution. The more self-assembled cationic aggregates will percolate through the hydrophobic polymer matrix to form continuous nanoscale ion channels for enhanced H_2O and OH^- transport.

Alkaline Stable Ion Conduction Channel Establishment by Ion–Dipole Interactions. The above-mentioned works preliminarily verified that ion–dipole interactions can provide an additional driving force for the aggregation of ion-exchange groups, enhancing the degree of molecular chain segment self-assembly and thus constructing efficient ion conduction channels. However, in previous work, there are still shortcomings in the alkaline stability and ion conductivity of the membrane. To further optimize this design idea, our research group prepared an AEM with both high conductivity and high alkali resistance by using the ether-free main chain with better alkaline resistance and the cyclic piperidine group [153]. The representative synthesis route is shown in Figure 3.30 [154]. The energy barrier of alkoxy molecule rotation is much smaller than that of ordinary alkyl spacers, so the side chain containing alkoxy has better flexibility. The flexibility can be used to greatly increase the freedom of movement of the ionized side chain to promote the mutual aggregation of groups. In addition, the strong electron-donating effect of alkoxy groups can weaken the electrophilicity of cationic groups, thus promoting hydroxide ion dissociation and accelerating ion conduction in the membrane.

Two AEMs, abbreviated as O-PDQA and PDQA, containing different cationic side chains were synthesized. O-PDQA has ethylene oxide spacers between the backbone and cationic groups, while PDQA has an aliphatic chain as the spacer. MD simulations were conducted to investigate the flexibility of O-DQA and DQA. The spatial distributions of O-DQA and DQA were transferred to joint probability density distribution contour maps. As shown in Figure 3.31, vectors d and q describe the extension and bending of the cationic side chain. DQA shows a deeper red zone than O-DQA, implying a more concentrated spatial probability density distribution. This demonstrates that the more rigid alkyl side chains in DQA lead to undesired constraints and inhibit the self-assembly process. Conversely, the EO spacer in O-DQA provides a less concentrated probability of chain distribution, which can trap the adjacent cationic side chains more efficiently, thus promoting the self-assembly process of the O-DQA-containing membrane.

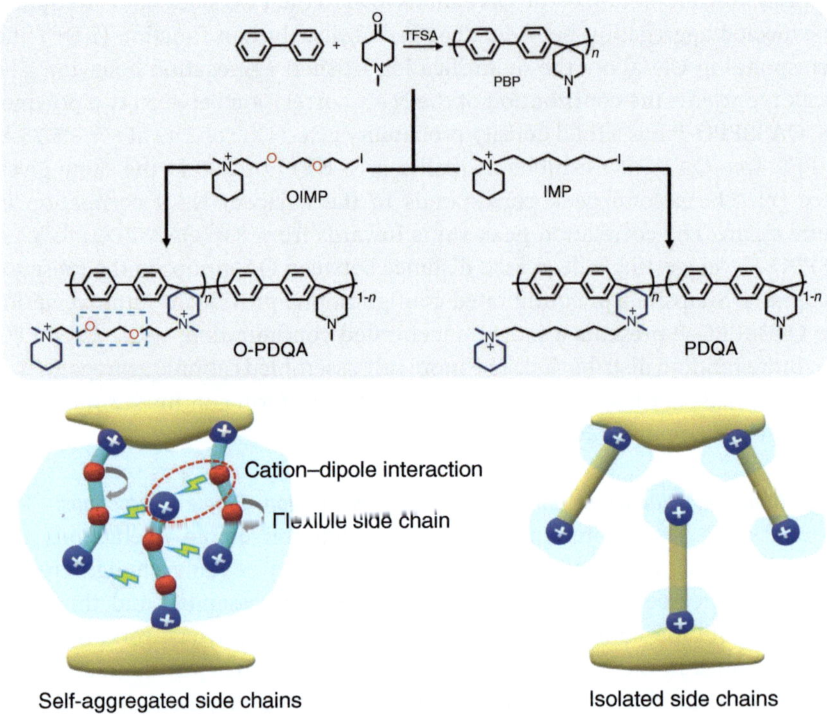

Figure 3.30 Synthesis of side-chain membranes with ion–dipole interactions for alkaline stability enhancement. Source: Zhang et al. [154]/Royal Society chemistry.

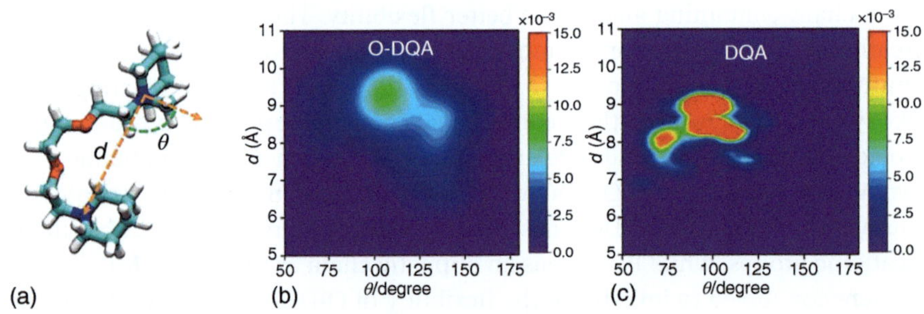

Figure 3.31 MD simulations for observing the microphase separation of O-DQA and DQA Source: Zhang et al. [154]/Royal Society chemistry.

References

1 Souzy, R. and Ameduri, B. (2005). Functional fluoropolymers for fuel cell membranes. *Progress in Polymer Science* 30: 644–687.

2 Mauritz, K.A. and Moore, R.B. (2004). State of understanding of Nafion. *Chemical Reviews* 104: 4535–4586.

3 Elliott, J.A. and Paddison, S.J. (2007). Modelling of morphology and proton transport in PFSA membranes. *Physical Chemistry Chemical Physics* 9: 2602–2618.

4 Hsu, W.Y. and Gierke, T.D. (1983). Ion transport and clustering in nafion perfluorinated membranes. *Journal of Membrane Science* 13: 307–326.

5 Gierke, T.D., Munn, G.E., and Wilson, F.C. (1981). The morphology in nafion perfluorinated membrane products, as determined by wide- and small-angle X-ray studies. *Journal of Polymer Science Polymer Physics Edition* 19: 1687–1704.

6 Roche, E.J., Pineri, M., Duplessix, R., and Levelut, A.M. (1981). Small-angle scattering studies of nafion membranes. *Journal of Polymer Science Polymer Physics Edition* 19: 1–11.

7 Roche, E.J., Pineri, M., and Duplessix, R. (1982). Phase separation in perfluorosulfonate ionomer membranes. *Journal of Polymer Science Polymer Physics Edition* 20: 107–116.

8 Li, Z., Tang, M., Liang, S. et al. (2021). Bottlebrush polymers: from controlled synthesis, self-assembly, properties to applications. *Progress in Polymer Science* 116: 101387.

9 Perin, F., Motta, A., and Maniglio, D. (2021). Amphiphilic copolymers in biomedical applications: synthesis routes and property control. *Materials Science and Engineering: C* 123: 111952.

10 Peckham, T.J. and Holdcroft, S. (2010). Structure-morphology-property relationships of non-perfluorinated proton-conducting membranes. *Advanced Materials* 22: 4667–4690.

11 Kobayashi, T., Rikukawa, M., Sanui, K., and Ogata, N. (1998). Proton-conducting polymers derived from poly(ether-etherketone) and poly(4-phenoxybenzoyl-1,4-phenylene). *Solid State Ionics* 106: 219–225.

12 Miyatake, K., Chikashige, Y., and Watanabe, M. (2003). Novel sulfonated poly(arylene ether): a proton conductive polymer electrolyte designed for fuel cells. *Macromolecules* 36: 9691–9693.

13 Shang, X., Fang, S., and Meng, Y. (2007). Synthesis and characterization of poly(arylene ether ketone) with sulfonated fluorene pendants for proton exchange membrane. *Journal of Membrane Science* 297: 90–97.

14 Jeong, M.-H., Lee, K.-S., Hong, Y.-T., and Lee, J.-S. (2008). Selective and quantitative sulfonation of poly(arylene ether ketone)s containing pendant phenyl rings by chlorosulfonic acid. *Journal of Membrane Science* 314: 212–220.

15 Liu, B., Kim, Y.S., Hu, W. et al. (2008). Homopolymer-like sulfonated phenyl- and diphenyl-poly(arylene ether ketone)s for fuel cell applications. *Journal of Power Sources* 185: 899–903.

16 Wang, L., Meng, Y.Z., Wang, S.J. et al. (2004). Synthesis and sulfonation of poly(aryl ethers) containing triphenyl methane and tetraphenyl methane moieties from isocynate-masked bisphenols. *Macromolecules* 37: 3151–3158.

17 Miyatake, K. and Hay, A.S. (2001). Synthesis and properties of poly(arylene ether)s bearing sulfonic acid groups on pendant phenyl rings. *Journal of Polymer Science Part A: Polymer Chemistry* 39: 3211–3217.

18 Matsumoto, K., Higashihara, T., and Ueda, M. (2009). Locally sulfonated poly(ether sulfone)s with highly sulfonated units as proton exchange membrane. *Journal of Polymer Science Part A: Polymer Chemistry* 47: 3444–3453.

19 Matsumoto, K., Higashihara, T., and Ueda, M. (2009). Locally and densely sulfonated poly(ether sulfone)s as proton exchange membrane. *Macromolecules* 42: 1161–1166.

20 Li, N., Shin, D.W., Hwang, D.S. et al. (2010). Polymer electrolyte membranes derived from new sulfone monomers with pendent sulfonic acid groups. *Macromolecules* 43: 9810–9820.

21 Seo, D.W., Lim, Y.D., Lee, S.H. et al. (2012). Preparation and characterization of sulfonated poly(tetra phenyl ether ketone sulfone)s for proton exchange membrane fuel cell. *International Journal of Hydrogen Energy* 37: 6140–6147.

22 Lafitte, B. and Jannasch, P. (2007). Proton-conducting aromatic polymers carrying hypersulfonated side chains for fuel cell applications. *Advanced Functional Materials* 17: 2823–2834.

23 Zhang, Z., Wu, L., and Xu, T. (2011). Synthesis and properties of side-chain-type sulfonated poly(phenylene oxide) for proton exchange membranes. *Journal of Membrane Science* 373: 160–166.

24 Lafitte, B., Puchner, M., and Jannasch, P. (2005). Proton conducting polysulfone ionomers carrying sulfoaryloxybenzoyl side chains. *Macromolecular Rapid Communications* 26: 1464–1468.

25 Glipa, X., El Haddad, M., Jones, D.J., and Rozière, J. (1997). Synthesis and characterisation of sulfonated polybenzimidazole: a highly conducting proton exchange polymer. *Solid State Ionics* 97: 323–331.

26 Kawahara, M., Rikukawa, M., Sanui, K., and Ogata, N. (2000). Synthesis and proton conductivity of sulfopropylated poly(benzimidazole) films. *Solid State Ionics* 136–137: 1193–1196.

27 Kim, D.S., Robertson, G.P., and Guiver, M.D. (2008). Comb-shaped poly(arylene ether sulfone)s as proton exchange membranes. *Macromolecules* 41: 2126–2134.

28 Wang, C., Li, N., Shin, D.W. et al. (2011). Fluorene-based poly(arylene ether sulfone)s containing clustered flexible pendant sulfonic acids as proton exchange membranes. *Macromolecules* 44: 7296–7306.

29 Miyatake, K., Shimura, T., Mikami, T., and Watanabe, M. (2009). Aromatic ionomers with superacid groups. *Chemical Communications* 6403–6405.

30 Yoshimura, K. and Iwasaki, K. (2009). Aromatic polymer with pendant perfluoroalkyl sulfonic acid for fuel cell applications. *Macromolecules* 42: 9302–9306.

31 Nakagawa, T., Nakabayashi, K., Higashihara, T., and Ueda, M. (2011). Polymer electrolyte membrane based on poly(ether sulfone) containing binaphthyl units with pendant perfluoroalkyl sulfonic acids. *Journal of Polymer Science Part A: Polymer Chemistry* 49: 2997–3003.

32 Mikami, T., Miyatake, K., and Watanabe, M. (2011). Synthesis and properties of multiblock copoly(arylene ether)s containing superacid groups for fuel cell membranes. *Journal of Polymer Science Part A: Polymer Chemistry* 49: 452–464.

33 Nakabayashi, K., Higashihara, T., and Ueda, M. (2011). Polymer electrolyte membranes based on poly(phenylene ether)s with pendant perfluoroalkyl sulfonic acids. *Macromolecules* 44: 1603–1609.

34 Jutemar, E.P. and Jannasch, P. (2010). Influence of the polymer backbone structure on the properties of aromatic ionomers with pendant sulfobenzoyl side chains for use as proton-exchange membranes. *ACS Applied Materials and Interfaces* 2: 3718–3725.

35 Chen, H.G., Wang, S.J., Xiao, M., and Meng, Y.Z. (2007). Novel sulfonated poly(phthalazinone ether ketone) ionomers containing benzonitrile moiety for PEM fuel cell applications. *Journal of Power Sources* 165: 16–23.

36 Jutemar, E.P., Takamuku, S., and Jannasch, P. (2010). Facile synthesis and polymerization of 2,6-difluoro-2′-sulfobenzophenone for aromatic proton conducting ionomers with pendant sulfobenzoyl groups. *Macromolecular Rapid Communications* 31: 1348–1353.

37 Ma, X., Zhang, C., Xiao, G. et al. (2008). Synthesis and characterization of sulfonated poly(phthalazinone ether phosphine oxide)s by direct polycondensation for proton exchange membranes. *Journal of Polymer Science Part A: Polymer Chemistry* 46: 1758–1769.

38 Xu, K., Oh, H., Hickner, M.A., and Wang, Q. (2011). Highly conductive aromatic ionomers with perfluorosulfonic acid side chains for elevated temperature fuel cells. *Macromolecules* 44: 4605–4609.

39 Pang, J., Zhang, H., Li, X., and Jiang, Z. (2007). Novel wholly aromatic sulfonated poly(arylene ether) copolymers containing sulfonic acid groups on the pendants for proton exchange membrane materials. *Macromolecules* 40: 9435–9442.

40 Pang, J., Zhang, H., Li, X. et al. (2007). Low water swelling and high proton conducting sulfonated poly(arylene ether) with pendant sulfoalkyl groups for proton exchange membranes. *Macromolecular Rapid Communications* 28: 2332–2338.

41 Pang, J., Zhang, H., Li, X. et al. (2008). Poly(arylene ether)s with pendant sulfoalkoxy groups prepared by direct copolymerization method for proton exchange membranes. *Journal of Power Sources* 184: 1–8.

42 Kim, D.S., Kim, Y.S., Guiver, M.D., and Pivovar, B.S. (2008). High performance nitrile copolymers for polymer electrolyte membrane fuel cells. *Journal of Membrane Science* 321: 199–208.

43 Gao, Y., Robertson, G.P., Guiver, M.D. et al. (2004). Synthesis of poly(arylene ether ether ketone ketone) copolymers containing pendant sulfonic acid groups bonded to naphthalene as proton exchange membrane materials. *Macromolecules* 37: 6748–6754.

44 Jeong, M.-H., Lee, K.-S., and Lee, J.-S. (2009). Synthesis and characterization of sulfonated poly(arylene ether ketone) copolymers containing crosslinking moiety. *Journal of Membrane Science* 337: 145–152.

45 Kim, D.S. and Guiver, M.D. (2008). Comparative effect of phthalazinone units in sulfonated poly(arylene ether ether ketone ketone) copolymers as proton exchange membrane materials. *Journal of Polymer Science Part A: Polymer Chemistry* 46: 989–1002.

46 Lee, J.K., Li, W., and Manthiram, A. (2009). Poly(arylene ether sulfone)s containing pendant sulfonic acid groups as membrane materials for direct methanol fuel cells. *Journal of Membrane Science* 330: 73–79.

47 Yip, N.Y., Tiraferri, A., Phillip, W.A. et al. (2011). Thin-film composite pressure retarded osmosis membranes for sustainable power generation from salinity gradients. *Environmental Science & Technology* 45: 4360–4369.

48 Deligöz, H., Vantansever, S., Koç, S.N. et al. (2008). Preparation of sulfonated copolyimides containing aliphatic linkages as proton-exchange membranes for fuel cell applications. *Journal of Applied Polymer Science* 110: 1216–1224.

49 Li, Y., Jin, R., Cui, Z. et al. (2007). Synthesis and characterization of novel sulfonated polyimides from 1,4-bis(4-aminophenoxy)-naphthyl-2,7-disulfonic acid. *Polymer* 48: 2280–2287.

50 Miyatake, K., Yasuda, T., and Watanabe, M. (2008). Substituents effect on the properties of sulfonated polyimide copolymers. *Journal of Polymer Science Part A: Polymer Chemistry* 46: 4469–4478.

51 Guo, X., Fang, J., and Watari, T. (2002). Novel sulfonated polyimides as polyelectrolytes for fuel cell application. 2. Synthesis and proton conductivity of polyimides from 9,9-Bis(4-aminophenyl)fluorene-2,7-disulfonic acid. *Macromolecules* 35: 6707–6713.

52 Hu, Z., Yin, Y., Kita, H. et al. (2007). Synthesis and properties of novel sulfonated polyimides bearing sulfophenyl pendant groups for polymer electrolyte fuel cell application. *Polymer* 48: 1962–1971.

53 Chen, K., Chen, X., Yaguchi, K. et al. (2009). Synthesis and properties of novel sulfonated polyimides bearing sulfophenyl pendant groups for fuel cell application. *Polymer* 50: 510–518.

54 Li, N., Cui, Z., Zhang, S. et al. (2007). Preparation and evaluation of a proton exchange membrane based on oxidation and water stable sulfonated polyimides. *Journal of Power Sources* 172: 511–519.

55 Chen, S., Yin, Y., Tanaka, K. et al. (2006). Synthesis and properties of novel side-chain-sulfonated polyimides from bis[4-(4-aminophenoxy)-2-(3-sulfobenzoyl)]phenyl sulfone. *Polymer* 47: 2660–2669.

56 Álvarez-Gallego, Y., Nunes, S.P., Lozano, A.E. et al. (2007). Synthesis and properties of novel polyimides bearing sulfonated benzimidazole pendant groups. *Macromolecular Rapid Communications* 28: 616–622.

57 Yin, Y., Fang, J., Watari, T. et al. (2004). Synthesis and properties of highly sulfonated proton conducting polyimides from bis(3-sulfopropoxy)benzidine diamines. *Journal of Materials Chemistry* 14: 1062–1070.

58 Asano, N., Aoki, M., Suzuki, S. et al. (2006). Aliphatic/aromatic polyimide ionomers as a proton conductive membrane for fuel cell applications. *Journal of the American Chemical Society* 128: 1762–1769.

59 Saito, J., Tanaka, M., Hirai, M. et al. (2011). Polyimide ionomer containing superacid groups. *Polymers for Advanced Technologies* 22: 1305–1310.

60 Savard, O., Peckham, T.J., Yang, Y., and Holdcroft, S. (2008). Structure–property relationships for a series of polyimide copolymers with sulfonated pendant groups. *Polymer* 49: 4949–4959.

61 Yasuda, T., Li, Y., Miyatake, K. et al. (2006). Synthesis and properties of polyimides bearing acid groups on long pendant aliphatic chains. *Journal of Polymer Science Part A: Polymer Chemistry* 44: 3995–4005.

62 Hu, Z., Yin, Y., Chen, S. et al. (2006). Synthesis and properties of novel sulfonated (co)polyimides bearing sulfonated aromatic pendant groups for PEFC applications. *Journal of Polymer Science Part A: Polymer Chemistry* 44: 2862–2872.

63 Kang, S., Zhang, C., Xiao, G. et al. (2009). Synthesis and properties of soluble sulfonated polybenzimidazoles from 3,3′-disulfonate-4,4′-dicarboxylbiphenyl as proton exchange membranes. *Journal of Membrane Science* 334: 91–100.

64 Jouanneau, J., Gonon, L., Gebel, G. et al. (2010). Synthesis and characterization of ionic conducting sulfonated polybenzimidazoles. *Journal of Polymer Science Part A: Polymer Chemistry* 48: 1732–1742.

65 Tan, N., Xiao, G., Yan, D., and Sun, G. (2010). Preparation and properties of polybenzimidazoles with sulfophenylsulfonyl pendant groups for proton exchange membranes. *Journal of Membrane Science* 353: 51–59.

66 Li, N. and Guiver, M.D. (2014). Ion transport by nanochannels in ion-containing aromatic copolymers. *Macromolecules* 47: 2175–2198.

67 Norsten, T.B., Guiver, M.D., Murphy, J. et al. (2006). Highly fluorinated comb-shaped copolymers as proton exchange membranes (PEMs): improving PEM properties through rational design. *Advanced Functional Materials* 16: 1814–1822.

68 Dimitrov, I., Takamuku, S., Jankova, K. et al. (2012). Polysulfone functionalized with phosphonated poly(pentafluorostyrene) grafts for potential fuel cell applications. *Macromolecular Rapid Communications* 33: 1368–1374.

69 Li, N., Wang, C., Lee, S.Y. et al. (2011). Enhancement of proton transport by nanochannels in comb-shaped copoly(arylene ether sulfone)s. *Angewandte Chemie International Edition* 50: 9158–9161.

70 Parvole, J. and Jannasch, P. (2008). Polysulfones grafted with poly(vinylphosphonic acid) for highly proton conducting fuel cell membranes in the hydrated and nominally dry state. *Macromolecules* 41: 3893–3903.

71 Ingratta, M., Elomaa, M., and Jannasch, P. (2010). Grafting poly(phenylene oxide) with poly(vinylphosphonic acid) for fuel cell membranes. *Polymer Chemistry* 1: 739–746.

72 Ran, J., Wu, L., Zhang, Z., and Xu, T. (2014). Atom transfer radical polymerization (ATRP): a versatile and forceful tool for functional membranes. *Progress in Polymer Science* 39: 124–144.

73 Tsang, E.M.W., Zhang, Z., Yang, A.C.C. et al. (2009). Nanostructure, morphology, and properties of fluorous copolymers bearing ionic grafts. *Macromolecules* 42: 9467–9480.

74 Tsang, E.M.W., Zhang, Z., Shi, Z. et al. (2007). Considerations of macromolecular structure in the design of proton conducting polymer membranes: graft

versus diblock polyelectrolytes. *Journal of the American Chemical Society* 129: 15106–15107.

75 Ingratta, M., Jutemar, E.P., and Jannasch, P. (2011). Synthesis, nanostructures and properties of sulfonated poly(phenylene oxide) bearing polyfluorostyrene side chains as proton conducting membranes. *Macromolecules* 44: 2074–2083.

76 Gao, L., He, G., Pan, Y. et al. (2016). Poly(2,6-dimethyl-1,4-phenylene oxide) containing imidazolium-terminated long side chains as hydroxide exchange membranes with improved conductivity. *Journal of Membrane Science* 518: 159–167.

77 Zhou, J., Ünlü, M., Anestis-Richard, I. et al. (2011). Solvent processible, high-performance partially fluorinated copoly(arylene ether) alkaline ionomers for alkaline electrodes. *Journal of Power Sources* 196: 7924–7930.

78 Lin, C.X., Huang, X.L., Guo, D. et al. (2016). Side-chain-type anion exchange membranes bearing pendant quaternary ammonium groups via flexible spacers for fuel cells. *Journal of Materials Chemistry A* 4: 13938–13948.

79 Zhuo, Y.Z., Lai, A.L., Zhang, Q.G. et al. (2015). Enhancement of hydroxide conductivity by grafting flexible pendant imidazolium groups into poly(arylene ether sulfone) as anion exchange membranes. *Journal of Materials Chemistry A* 3: 18105–18114.

80 Guo, D., Lai, A.N., Lin, C.X. et al. (2016). Imidazolium-functionalized poly(arylene ether sulfone) anion-exchange membranes densely grafted with flexible side chains for fuel cells. *ACS Applied Materials and Interfaces* 8: 25279–25288.

81 Hu, E.N., Lin, C.X., Liu, F.H. et al. (2018). Poly(arylene ether nitrile) anion exchange membranes with dense flexible ionic side chain for fuel cells. *Journal of Membrane Science* 550: 254–265.

82 Wang, X.Q., Lin, C.X., Zhang, Q.G. et al. (2017). Anion exchange membranes from hydroxyl-bearing poly(ether sulfone)s with flexible spacers via ring-opening grafting for fuel cells. *International Journal of Hydrogen Energy* 42: 19044–19055.

83 Zuo, P., Su, Y., and Li, W. (2016). Comb-like poly(ether-sulfone) membranes derived from planar 6,12-diaryl-5,11-dihydroindolo[3,2-b]carbazole monomer for alkaline fuel cells. *Macromolecular Rapid Communications* 37: 1748–1753.

84 Li, S., Zhu, X., Liu, D., and Sun, F. (2018). A highly durable long side-chain polybenzimidazole anion exchange membrane for AEMFC. *Journal of Membrane Science* 546: 15–21.

85 Ran, J., Ding, L., Chu, C. et al. (2018). Highly conductive and stabilized side-chain-type anion exchange membranes: ideal alternatives for alkaline fuel cell applications. *Journal of Materials Chemistry A* 6: 17101–17110.

86 He, Y., Ge, X., Liang, X. et al. (2018). Anion exchange membranes with branched ionic clusters for fuel cells. *Journal of Materials Chemistry A* 6: 5993–5998.

87 Li, Q., Liu, L., Miao, Q. et al. (2014). A novel poly(2,6-dimethyl-1,4-phenylene oxide) with trifunctional ammonium moieties for alkaline anion exchange membranes. *Chemical Communications* 50: 2791–2793.

88 Yang, Z., Zhou, J., Wang, S. et al. (2015). A strategy to construct alkali-stable anion exchange membranes bearing ammonium groups via flexible spacers. *Journal of Materials Chemistry A* 3: 15015–15019.

89 Zhu, L., Pan, J., Christensen, C.M. et al. (2016). Functionalization of poly(2,6-dimethyl-1,4-phenylene oxide)s with hindered fluorene side chains for anion exchange membranes. *Macromolecules* 49: 3300–3309.

90 Li, N., Guiver, M.D., and Binder, W.H. (2013). Towards high conductivity in anion-exchange membranes for alkaline fuel cells. *ChemSusChem* 6: 1376–1383.

91 Ge, Q., Ran, J., Miao, J. et al. (2015). Click chemistry finds its way in constructing an ionic highway in anion-exchange membrane. *ACS Applied Materials and Interfaces* 7: 28545–28553.

92 Wang, J., Wei, H., Yang, S. et al. (2015). Constructing pendent imidazolium-based poly(phenylene oxide)s for anion exchange membranes using a click reaction. *RSC Advances* 5: 93415–93422.

93 Wang, J., Wang, X., Zu, D. et al. (2018). N3-adamantyl imidazolium cations: alkaline stability assessment and the corresponding comb-shaped anion exchange membranes. *Journal of Membrane Science* 545: 116–125.

94 Chen, Y., Tao, Y., Wang, J. et al. (2017). Comb-shaped guanidinium functionalized poly(ether sulfone)s for anion exchange membranes: effects of the spacer types and lengths. *Journal of Polymer Science Part A: Polymer Chemistry* 55: 1313–1321.

95 Hou, J., Wang, X., Liu, Y. et al. (2016). Wittig reaction constructed an alkaline stable anion exchange membrane. *Journal of Membrane Science* 518: 282–288.

96 Li, N., Zhang, Q., Wang, C. et al. (2012). Phenyltrimethylammonium functionalized polysulfone anion exchange membranes. *Macromolecules* 45: 2411–2419.

97 Dang, H.-S., Weiber, E.A., and Jannasch, P. (2015). Poly(phenylene oxide) functionalized with quaternary ammonium groups via flexible alkyl spacers for high-performance anion exchange membranes. *Journal of Materials Chemistry A* 3: 5280–5284.

98 Dang, H.-S. and Jannasch, P. (2016). Alkali-stable and highly anion conducting poly(phenylene oxide)s carrying quaternary piperidinium cations. *Journal of Materials Chemistry A* 4: 11924–11938.

99 Wang, L. and Hickner, M.A. (2016). Highly conductive side chain block copolymer anion exchange membranes. *Soft Matter* 12: 5359–5371.

100 He, Y., Pan, J., Wu, L. et al. (2015). A novel methodology to synthesize highly conductive anion exchange membranes. *Scientific Reports* 5: 13417.

101 He, Y., Wu, L., Pan, J. et al. (2016). A mechanically robust anion exchange membrane with high hydroxide conductivity. *Journal of Membrane Science* 504: 47–54.

102 Ran, J., Fu, C., Ding, L. et al. (2018). Dual hydrophobic grafted chains facilitating quaternary ammonium aggregations of hydroxide conducting polymers: a theoretical and experimental investigation. *Journal of Materials Chemistry A* 6: 5714–5723.

103 Lee, W.-H., Kim, Y.S., and Bae, C. (2015). Robust hydroxide ion conducting poly(biphenyl alkylene)s for alkaline fuel cell membranes. *ACS Macro Letters* 4: 814–818.

104 Zhang, M., Shan, C., Liu, L. et al. (2016). Facilitating anion transport in polyolefin-based anion exchange membranes via bulky side chains. *ACS Applied Materials and Interfaces* 8: 23321–23330.

105 Zhang, M., Liu, J., Wang, Y. et al. (2015). Highly stable anion exchange membranes based on quaternized polypropylene. *Journal of Materials Chemistry A* 3: 12284–12296.

106 Danks, T.N., Slade, R.C.T., and Varcoe, J.R. (2003). Alkaline anion-exchange radiation-grafted membranes for possible electrochemical application in fuel cells. *Journal of Materials Chemistry* 13: 712–721.

107 Varcoe, J.R. and Slade, R.C.T. (2006). An electron-beam-grafted ETFE alkaline anion-exchange membrane in metal-cation-free solid-state alkaline fuel cells. *Electrochemistry Communications* 8: 839–843.

108 Danks, T.N., Slade, R.C.T., and Varcoe, J.R. (2002). Comparison of PVDF- and FEP-based radiation-grafted alkaline anion-exchange membranes for use in low temperature portable DMFCs. *Journal of Materials Chemistry* 12: 3371–3373.

109 Slade, R.C.T. and Varcoe, J.R. (2005). Investigations of conductivity in FEP-based radiation-grafted alkaline anion-exchange membranes. *Solid State Ionics* 176: 585–597.

110 Ran, J., Wu, L., Lin, X. et al. (2012). Synthesis of soluble copolymers bearing ionic graft for alkaline anion exchange membrane. *RSC Advances* 2: 4250–4257.

111 Ran, J., Wu, L., and Xu, T. (2013). Enhancement of hydroxide conduction by self-assembly in anion conductive comb-shaped copolymers. *Polymer Chemistry* 4: 4612–4620.

112 Hu, M., Ding, L., Shehzad, M.A. et al. (2019). Comb-shaped anion exchange membrane with densely grafted short chains or loosely grafted long chains? *Journal of Membrane Science* 585: 150–156.

113 Ran, J., Wu, L., Wei, B. et al. (2014). Simultaneous enhancements of conductivity and stability for anion exchange membranes (AEMs) through precise structure design. *Scientific Reports* 4: 6486.

114 Jasti, A. and Shahi, V.K. (2013). Multi-block poly(arylene ether)s containing pre-choloromethylated bisphenol: anion conductive ionomers. *Journal of Materials Chemistry A* 1: 6134–6137.

115 Nie, G., Li, X., Tao, J. et al. (2015). Alkali resistant cross-linked poly(arylene ether sulfone)s membranes containing aromatic side-chain quaternary ammonium groups. *Journal of Membrane Science* 474: 187–195.

116 Chen, J., Li, C., Wang, J. et al. (2017). A general strategy to enhance the alkaline stability of anion exchange membranes. *Journal of Materials Chemistry A* 5: 6318–6327.

117 Komkova, E.N., Stamatialis, D.F., Strathmann, H., and Wessling, M. (2004). Anion-exchange membranes containing diamines: preparation and stability in alkaline solution. *Journal of Membrane Science* 244: 25–34.

118 Liu, L., Li, Q., Dai, J. et al. (2014). A facile strategy for the synthesis of guanidinium-functionalized polymer as alkaline anion exchange membrane with improved alkaline stability. *Journal of Membrane Science* 453: 52–60.

119 Li, X., Cheng, J., Liu, Y. et al. (2017). Improved conductivity and stability of anion exchange membrane modified with bi-phenylguanidinium bridged silsesquioxane. *International Journal of Hydrogen Energy* 42: 21016–21026.

120 Wang, J., Li, S., and Zhang, S. (2010). Novel hydroxide-conducting polyelectrolyte composed of an poly(arylene ether sulfone) containing pendant quaternary guanidinium groups for alkaline fuel cell applications. *Macromolecules* 43: 3890–3896.

121 Hollóczki, O., Terleczky, P., Szieberth, D. et al. (2011). Hydrolysis of Imidazole-2-ylidenes. *Journal of the American Chemical Society* 133: 780–789.

122 Li, W., Fang, J., Lv, M. et al. (2011). Novel anion exchange membranes based on polymerizable imidazolium salt for alkaline fuel cell applications. *Journal of Materials Chemistry* 21: 11340–11346.

123 Ran, J., Wu, L., Varcoe, J.R. et al. (2012). Development of imidazolium-type alkaline anion exchange membranes for fuel cell application. *Journal of Membrane Science* 415-416: 242–249.

124 Thomas, O.D., Soo, K.J.W.Y., Peckham, T.J. et al. (2011). Anion conducting poly(dialkyl benzimidazolium) salts. *Polymer Chemistry* 2: 1641–1643.

125 Price, S.C., Williams, K.S., and Beyer, F.L. (2014). Relationships between structure and alkaline stability of imidazolium cations for fuel cell membrane applications. *ACS Macro Letters* 3: 160–165.

126 Wang, J., Gu, S., Kaspar, R.B. et al. (2013). Stabilizing the imidazolium cation in hydroxide-exchange membranes for fuel cells. *ChemSusChem* 6: 2079–2082.

127 Hugar, K.M., Kostalik, H.A., and Coates, G.W. (2015). Imidazolium cations with exceptional alkaline stability: a systematic study of structure–stability relationships. *Journal of the American Chemical Society* 137: 8730–8737.

128 Gu, F., Dong, H., Li, Y. et al. (2014). Highly stable N3-substituted imidazolium-based alkaline anion exchange membranes: experimental studies and theoretical calculations. *Macromolecules* 47: 208–216.

129 Liu, L., He, S., Zhang, S. et al. (2016). 1,2,3-Triazolium-based poly(2,6-dimethyl phenylene oxide) copolymers as anion exchange membranes. *ACS Applied Materials and Interfaces* 8: 4651–4660.

130 Li, Y., Xu, T., and Gong, M. (2006). Fundamental studies of a new series of anion exchange membranes: membranes prepared from bromomethylated poly(2,6-dimethyl-1,4-phenylene oxide) (BPPO) and pyridine. *Journal of Membrane Science* 279: 200–208.

131 Kim, S., Yang, S., and Kim, D. (2017). Poly(arylene ether ketone) with pendant pyridinium groups for alkaline fuel cell membranes. *International Journal of Hydrogen Energy* 42: 12496–12506.

132 Gu, F., Dong, H., Li, Y. et al. (2014). Base stable pyrrolidinium cations for alkaline anion exchange membrane applications. *Macromolecules* 47: 6740–6747.

133 Strasser, D.J., Graziano, B.J., and Knauss, D.M. (2017). Base stable poly(diallylpiperidinium hydroxide) multiblock copolymers for anion exchange membranes. *Journal of Materials Chemistry A* 5: 9627–9640.

134 Ye, Y., Stokes, K.K., Beyer, F.L., and Elabd, Y.A. (2013). Development of phosphonium-based bicarbonate anion exchange polymer membranes. *Journal of Membrane Science* 443: 93–99.

135 Gu, S., Cai, R., Luo, T. et al. (2009). A soluble and highly conductive ionomer for high-performance hydroxide exchange membrane fuel cells. *Angewandte Chemie International Edition* 48: 6499–6502.

136 Zhang, B., Gu, S., Wang, J. et al. (2012). Tertiary sulfonium as a cationic functional group for hydroxide exchange membranes. *RSC Advances* 2: 12683–12685.

137 Zha, Y., Disabb-Miller, M.L., Johnson, Z.D. et al. (2012). Metal-cation-based anion exchange membranes. *Journal of the American Chemical Society* 134: 4493–4496.

138 Gu, S., Wang, J., Kaspar, R.B. et al. (2015). Permethyl cobaltocenium (Cp*2Co+) as an ultra-stable cation for polymer hydroxide-exchange membranes. *Scientific Reports* 5: 11668.

139 Wu, L., Zhou, G., Liu, X. et al. (2011). Environmentally friendly synthesis of alkaline anion exchange membrane for fuel cells via a solvent-free strategy. *Journal of Membrane Science* 371: 155–162.

140 Lin, X., Gong, M., Liu, Y. et al. (2013). A convenient, efficient and green route for preparing anion exchange membranes for potential application in alkaline fuel cells. *Journal of Membrane Science* 425-426: 190–199.

141 Wu, L., Xu, T., Wu, D., and Zheng, X. (2008). Preparation and characterization of CPPO/BPPO blend membranes for potential application in alkaline direct methanol fuel cell. *Journal of Membrane Science* 310: 577–585.

142 Wu, L. and Xu, T. (2008). Improving anion exchange membranes for DMAFCs by inter-crosslinking CPPO/BPPO blends. *Journal of Membrane Science* 322: 286–292.

143 Ran, J., Wu, L., Ge, Q. et al. (2014). High performance anion exchange membranes obtained through graft architecture and rational cross-linking. *Journal of Membrane Science* 470: 229–236.

144 Wu, L., Li, C., Tao, Z. et al. (2014). Anionic quaternary ammonium fluorous copolymers bearing thermo-responsive grafts for fuel cells. *International Journal of Hydrogen Energy* 39: 9387–9396.

145 Yang, Z., Hou, J., Wang, X. et al. (2015). Highly water resistant anion exchange membrane for fuel cells. *Macromolecular Rapid Communications* 36: 1362–1367.

146 Ge, Q., Liu, Y., Yang, Z. et al. (2016). Hyper-branched anion exchange membranes with high conductivity and chemical stability. *Chemical Communications* 52: 10141–10143.

147 Liu, Y., Zhou, J., Hou, J. et al. (2019). Hyperbranched polystyrene copolymer makes superior anion exchange membrane. *ACS Applied Polymer Materials* 1: 76–82.

148 Ge, Q., Liang, X., Ding, L. et al. (2019). Guiding the self-assembly of hyper-branched anion exchange membranes utilized in alkaline fuel cells. *Journal of Membrane Science* 573: 595–601.

149 Ge, X., He, Y., Liang, X. et al. (2018). Thermally triggered polyrotaxane translational motion helps proton transfer. *Nature Communications* 9: 2297.

150 Ge, X., He, Y., Guiver, M.D. et al. (2016). Alkaline anion-exchange membranes containing mobile ion shuttles. *Advanced Materials* 28: 3467–3472.

151 Ge, X., He, Y., Zhang, K. et al. (2021). Fast bulky anion conduction enabled by free shuttling phosphonium cations. *Research* 2021: 9762709.

152 Zhang, J., He, Y., Zhang, K. et al. (2021). Cation–dipole interaction that creates ordered ion channels in an anion exchange membrane for fast OH− conduction. *AICHE Journal* 67: e17133.

153 Zhang, J., Yu, W., Liang, X. et al. (2021). Flexible bis-piperidinium side chains construct highly conductive and robust anion-exchange membranes. *ACS Applied Energy Materials* 4: 9701–9711.

154 Zhang, J., Zhang, K., Liang, X. et al. (2021). Self-aggregating cationic-chains enable alkaline stable ion-conducting channels for anion-exchange membrane fuel cells. *Journal of Materials Chemistry A* 9: 327–337.

4

Polyacylation for Ion Exchange Membrane Preparation

Polymer electrolyte membranes (PEMs) have a critical role in electrochemical devices, particularly in fuel cells, where they act as the separator of anode and cathode electrodes, ion conductors, and provide water management [1–4]. However, there are some challenges associated with their application in fuel cells, particularly in alkaline fuel cells [5]. The synthesis of ion exchange membranes (IEMs) with high conductivity, mechanical strength, and stability at elevated temperatures, and high pH is necessary for the successful commercialization of fuel cells [4, 6]. The polyacylation of prequaternized monomers is a promising and versatile approach for synthesizing novel anion exchange polyelectrolytes for fuel cell applications [7–9]. In this chapter, we will discuss the principle of polyacylation and the polyacylation of perylene-based aliphatic polyimides (PIs), sulfonated poly (ether ketone)s (SPEKs), crowns, and other materials.

Membrane technology is a rapidly growing field with applications in various industries, including water treatment, food processing, pharmaceuticals, and energy production (Figure 4.1) [10–15]. In particular, polyelectrolyte membranes (PEMs) are essential components in many devices, including fuel cells, electrolyzers, batteries, and sensors [16–19]. PEMs act as ion conductors, separating the anode and cathode compartments, and preventing the mixing of reactants and products. They also help to maintain a neutral charge balance in the system and reduce the risk of short-circuiting or electrochemical degradation. PEMs are typically made of a polymeric matrix that contains functional groups capable of exchanging ions (Figure 4.1b). The most commonly used materials for PEMs are perfluorinated sulfonic acid membranes, such as Nafion (Figure 4.1c), due to their high proton conductivity and chemical stability. However, these materials are expensive, have limited availability, and pose environmental concerns due to the use of fluorinated compounds [20, 21]. Therefore, there is a need to develop alternative PEMs that are more sustainable, cost-effective, and environmentally friendly.

- Need for novel and sustainable membrane materials.

The development of new and sustainable PEMs is crucial for the advancement of many energy conversion technologies, such as fuel cells, and electrolyzers. The current state-of-the-art PEMs are based on perfluorinated ionomers, which suffer from

Ion Exchange Membranes: Design, Preparation, and Applications,
First Edition. Tongwen Xu and Yaoming Wang.

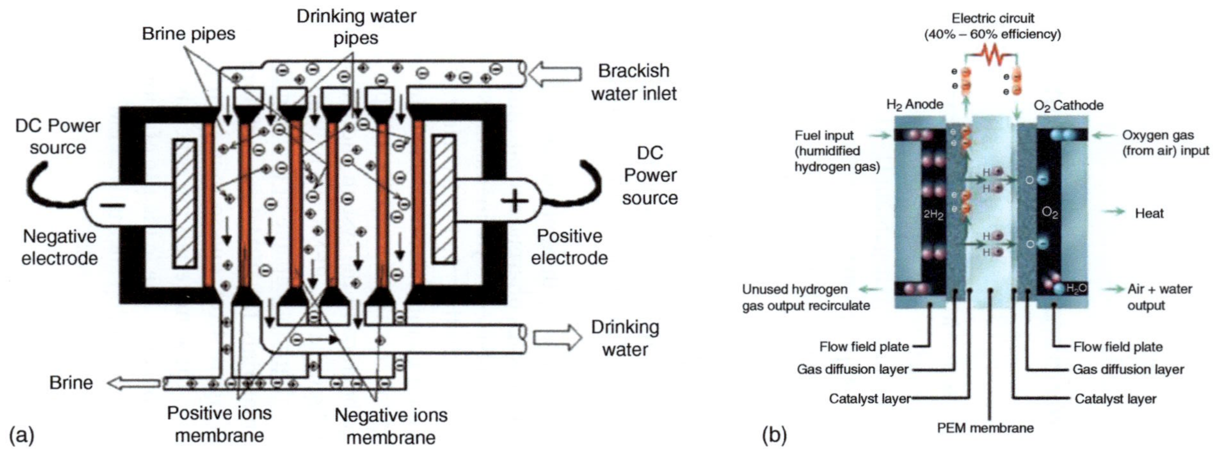

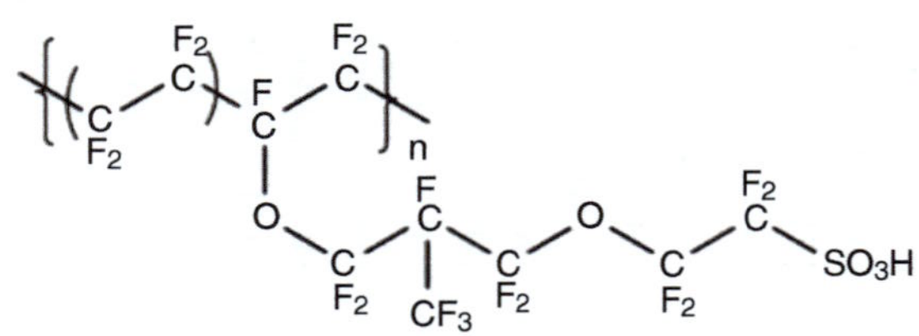

(c) Perfluorosulfonic acid (PFSA) polymer (Nafion)

Figure 4.1 (a) Water treatment membrane removing contaminants for safe drinking water/agricultural water production, (b) clean energy production (fuel cell), and (c) chemical structure of Nafion polymer.

several limitations, including high cost, limited availability, and environmental concerns. Additionally, these materials have poor alkaline stability, which limits their application in alkaline fuel cells, a promising alternative to acidic fuel cells due to their higher energy efficiency and the use of nonprecious metal catalysts [22–25].

To overcome these limitations, researchers have been exploring new approaches to design and synthesize PEMs with improved performance and sustainability. One promising strategy is to use aromatic polyelectrolytes that offer higher stability, mechanical strength, and thermal resistance compared to aliphatic polymers [7, 26–29]. Aromatic polymers also offer the possibility of tailoring the structure and properties of the material through different functional groups and architectures.

- Role of polyacylation in the synthesis of aromatic polyelectrolytes.

Polyacylation is a versatile and straightforward method for synthesizing aromatic polyelectrolytes with improved stability, conductivity, and mechanical strength (Figure 4.2).

In this approach, prequaternized monomers are reacted with acylating agents using Friedel–Crafts acylation, leading to the formation of long-chain pendent spacers [30]. This modification not only enhances the mechanical properties but also increases the accessibility of the functional groups and facilitates the transport of ions [31, 32]. Several studies have demonstrated the potential of polyacylation in the synthesis of aromatic polyelectrolytes for fuel cell applications [7–9]. For instance, Zilu et al. conducted a study in which they created sulfonated polyimide membranes with perylene-diimide units. These membranes were found to be both flexible and durable, with excellent proton conductivity and low swelling ratios. They were specifically designed for use in fuel cells. The materials used in the creation of these membranes exhibited self-assembly behavior, which was driven by both π–π and hydrophilic interactions. As a result, high power densities were achieved (Figure 4.3a, b). This research offers promising new materials that could be used in membrane-based fuel cell applications [33]. In another study, Zhang et al. proposed a new method of polyacylation using prequaternized monomers to create alkaline anion-exchange membranes (AAEMs) for fuel cells. This approach overcomes challenges with alkali resistance and synthetic difficulties, resulting in membranes with high conductivity, and exceptional alkaline stability. The polyacylation process creates aromatic anion-exchange polyelectrolytes with a long spacer, replacing the conventional benzyl-type spacer (Figure 4.3c, d). This innovative approach and the resulting exceptional membrane properties could pave the way for the practical use of these membranes in alkaline fuel cells [7]. Overall, polyacylation represents a promising strategy for the synthesis of sustainable

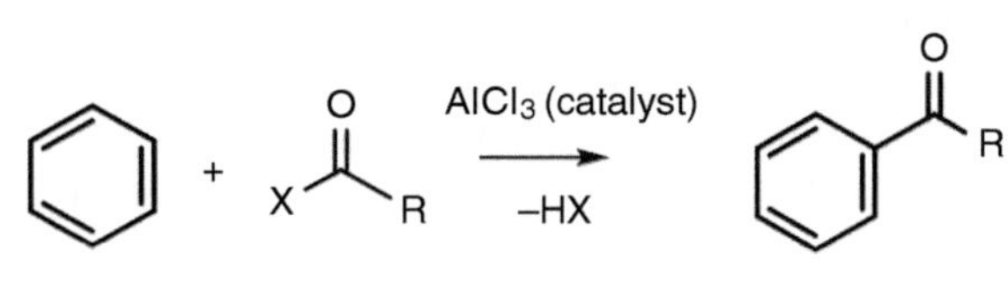

Figure 4.2 The process of introducing an acyl group onto an aromatic ring (arene), i.e. Friedel–Crafts acylation.

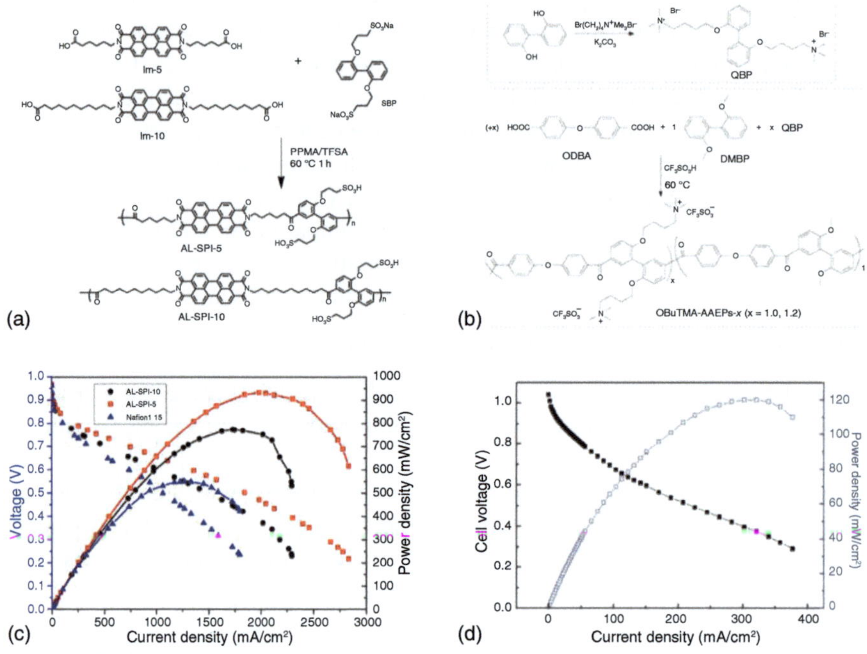

Figure 4.3 (a) Demonstration of the process of synthesizing diacid monomers that contain perylenediimide, (b) evaluation of fuel cell efficiency at 80 °C for AL-SPI-5,10 and Nafion 115, (c) depiction of the synthesis of (i) the QBP monomer and (ii) the OBuTMA-AAEPs-x polymer, and (d) graphical representation of the hydrogen–oxygen polarization and power density curves for a solid-state AFC utilizing an OBuTMA-AAEPs-1.0 membrane at 50 °C.

and high-performance PEMs. It enables the development of new and improved membrane materials with tailorable properties, which can contribute to the advancement of energy conversion technologies.

4.1 Principle of Polyacylation

Polyacylation is a chemical reaction in which an acyl group (RCO–) is added to a molecule, typically an organic compound containing a nucleophilic center (Figure 4.2). The resulting product is an acylated compound, which can have various applications in organic chemistry, biochemistry, and materials science [34–37]. In recent years, polyacylation has emerged as a promising strategy for synthesizing polyelectrolytes with tailored properties for applications in fuel cells, batteries, and other electrochemical devices [38–41]. The basic principle of polyacylation is the reaction of an acylating agent with a nucleophilic center on the target molecule, typically a polymer or monomer containing an amine or imine group. The acylating agent can be a carboxylic acid derivative, such as an acid chloride, anhydride, ester, or ketene. The reaction can be catalyzed by Lewis acids, such as aluminum chloride ($AlCl_3$), or tin tetrachloride, or by Brønsted acids, such as sulfuric acid, or trifluoroacetic acid. The polyacylation reaction can be carried out in solution, or the solid state, depending on the nature of the reactants and the desired product. In solution, the reaction can be performed in a variety of solvents, such as dichloromethane, chloroform, or dimethylformamide, and can be carried out under various conditions of temperature, time, and stoichiometry [42–46]. In the solid state, the reaction can be performed by heating the reactants in a sealed vessel, typically at high temperature and pressure, or by using mechanochemical methods, such as ball milling, or twin-screw extrusion (Figure 4.4a, b) [47–50].

The mechanism of acylation involves the attack of the nucleophilic center on the carbonyl carbon of the acylating agent, followed by the transfer of the acyl group to the nucleophile, and the elimination of a leaving group, typically a halide ion or an alcohol molecule. The reaction can proceed via an intermediate acylated species, which can be stabilized by neighboring groups or resonance effects and can lead to the formation of various products, depending on the nature of the reactants, and the reaction conditions [51]. Polyacylation can be used to modify the properties of polymers and monomers, such as their solubility, reactivity, and functionality, and to create new materials with specific properties, such as ion conductivity, mechanical strength, and chemical stability. In the context of fuel cells, polyacylation has been used to synthesize aromatic polyelectrolytes with high alkaline resistance and ion conductivity, which can replace the traditional perfluorinated ionomer membranes used in proton exchange membrane fuel cells (PEMFCs) [7, 33, 38]. These polyelectrolytes can be synthesized by reacting prequarternized monomers with acylating agents, such as acetic anhydride or benzoyl chloride, and can be tailored to exhibit specific properties, such as the length and flexibility of the pendent spacer, the degree of acylation, and the type of functional group [30, 39] (Figure 4.5).

4.2 Types of Acylation Reactions

Acylation reactions are classified based on the type of reactants and conditions used. This classification helps to categorize the reactions and provides a framework for understanding the mechanism and conditions required for each type of acylation reaction. By understanding the different types of acylation reactions, chemists can design and optimize reaction conditions for specific applications [52–54]. The importance of acylation reactions lies in their ability to modify the properties of organic compounds. By adding an acyl group, chemists can change the reactivity, solubility, and other physical and chemical properties of a molecule. For example, acylation reactions are commonly used in the synthesis of pharmaceuticals, where modification of the chemical structure can enhance bioactivity, and improve pharmacokinetics [55, 56]. Acylation reactions can be used to modify the properties of polymers, such as enhancing their solubility, thermal stability, and chemical resistance [57]. One application of acylation reactions in polymer science is in the preparation of IEMs, which are used in various electrochemical applications, such as fuel cells and electrodialysis.

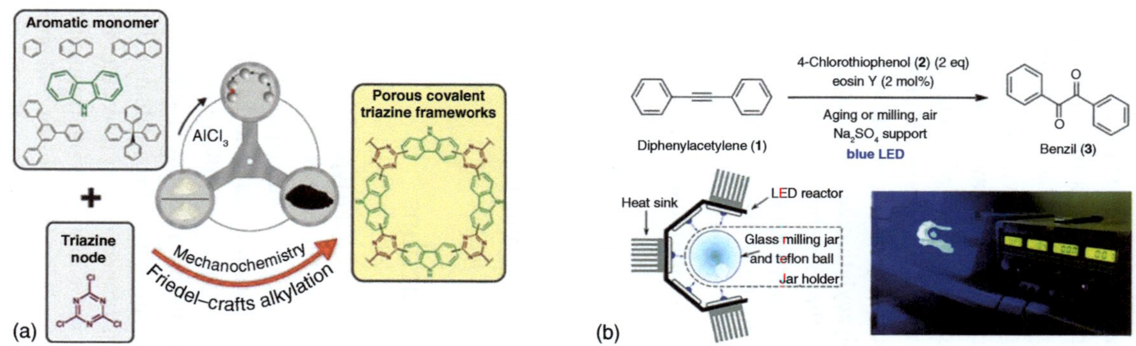

Figure 4.4 (a) Synthesizing porous covalent triazine frameworks by milling, (b) solid-state photocatalytic oxidation of diphenylacetylene using ball milling and visible light irradiation (solid-state polyacylation).

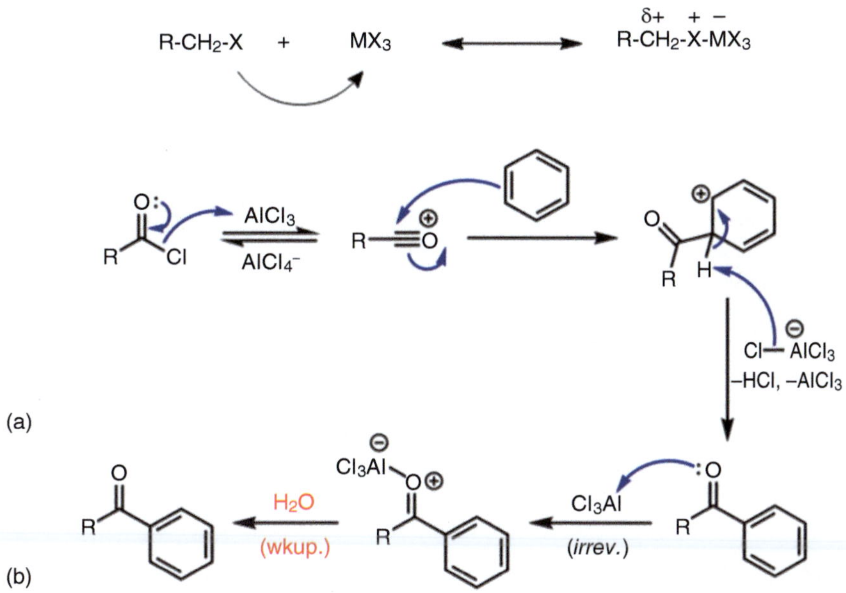

Figure 4.5 (a) Formation of carbocation: key to reactivity of haloalkanes, (b) mechanism of acylation reaction.

4.2.1 Acylation of Alcohols

Acylation of alcohols is a widely used method in organic chemistry that involves the transfer of an acyl group to an alcohol molecule. This reaction can introduce a carbonyl group into a molecule, which can then be modified further by subsequent reactions, including the synthesis of esters and other organic compounds (Figure 4.6a, b). In 2005, C.L. Padró studied the deactivation mechanism of the gas-phase synthesis of o-hydroxyacetophenone (o-HAP) via acylation of phenol with acetic acid. The study showed that coke formation occurred on Al-MCM-41 and zeolites HY and HBeta but not on HZSM-5. The study proposed that o-acetoxyacetophenone (o-AXAP) was the key intermediate responsible for the formation of carbonaceous deposits and activity decay [58]. In 2011, Murashige et al. reported on the control of reactions involving phenol derivatives and acyl chlorides for competitive O-acylations and C-acylations in acidic conditions. The study examined the use of trifluoromethanesulfonic acid (TfOH) as a catalyst and solvent and demonstrated the application of these reactions to the synthesis of homotyrosine derivatives [59]. Recently, two studies presented new methods for the selective acylation of phenols. Liao et al. developed a site-selective acylation method using a thioacid surrogate catalyzed by sodium thiosulfate [60] (Figure 4.6a). Tian et al. presented a metal-free, water-tolerant, and one-pot process for the ortho-acylation of phenols using an iodine source/hydrogen peroxide system (Figure 4.6b). These methods offer advantages such as high efficiency, low toxicity, and environmentally friendly conditions [61]. In addition, a recent study proposed a new method for the selective O-acylation of phenols using diethylaminosulfur trifluoride (DAST) as

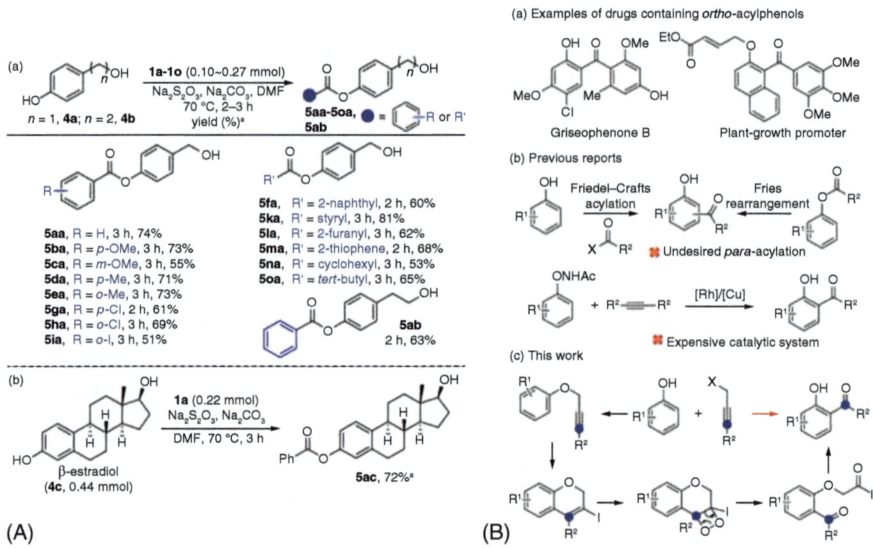

Figure 4.6 (A) Scheme 4 illustrates the selective O-acylation of phenol using diverse organic anhydrides and demonstrates the application of site-selective acylation in natural products, (B) the figure shows a strategy for a metal-free, one-pot process for the ortho-acylation of phenols.

both an activator for phenols and a deoxyfluorination reagent for organic salts. This method offers an efficient and convenient way to achieve selective O-acylation of phenols at room temperature in air [62].

Overall, these recent studies have contributed to the development of new methods for selective acylation reactions, which can be used to generate diverse acylated compounds with various applications. These methods offer advantages such as high efficiency, low toxicity, and environmentally friendly conditions, and represent valuable contributions to the field of organic chemistry.

4.2.2 Acylation of Amines

The study of the acylation of amines has been an important strategy in synthetic organic chemistry for many years. The process involves introducing an acyl group into an amine molecule, leading to the formation of amides, esters, and other organic compounds. In recent years, the selective acylation of amines has gained considerable attention due to its potential applications in the synthesis of natural products, pharmaceuticals, and other functional materials. Several approaches have been taken to achieve selective N-acylation, one of which involves the use of catalysts. Lewis acids, including ceria-yttria-based Lewis acids, have been studied for their effectiveness in achieving selective N-acylation [63] (Figure 4.7a, b). Boronic acids have also been explored as reusable catalysts in solvent-free direct condensations of amines and carboxylic acids, thanks to the use of molecular sieves as water scavengers.

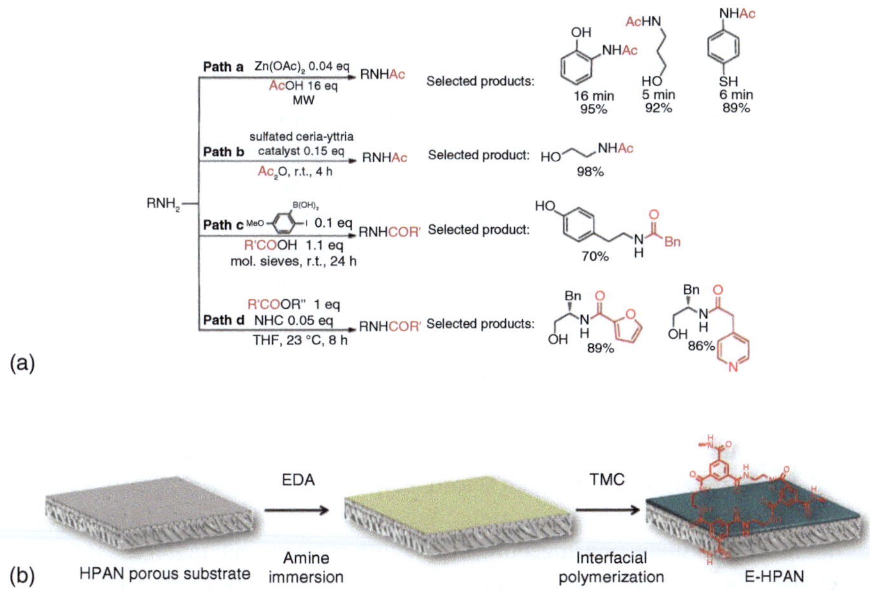

Figure 4.7 (a) Water-based N-acylation with high chemoselectivity, (b) illustration of how ion exchange groups can be produced through N-acylation of PEI in water.

A study by Nicolas et al. in [64] focused on the importance of amides as a component of biomolecules and synthetic products, which motivated the development of catalytic, direct amidation methods employing free carboxylic acids and amines. The study found that ortho-iodophenylboronic acid can catalyze direct amidation reactions at room temperature in the presence of 4A molecular sieves as a dehydrating agent. The study also optimized the arene core of ortho-iodoarylboronic acid catalysts with regard to the electronic effects of ring substitution. Contrary to expectations, electron-donating substituents were found to be preferable, in particular, an alkoxy substituent positioned para to the iodide. The optimal new catalyst, 5-methoxy-2-iodophenylboronic acid (MIBA, 4f), was demonstrated to be kinetically more active than the parent des-methoxy catalyst 4a, providing higher yields of amide products in shorter reaction times under mild conditions at ambient temperature. Catalyst 4f is recyclable and promotes the formation of amides from aliphatic carboxylic acids and amines and heteroaromatic carboxylic acids and other functionalized substrates containing moieties such as a free phenol, indole, and pyridine. Mechanistic studies demonstrated the essential role of molecular sieves in this complex amidation process. Dulla et al. reported another study in which the chemoselective acylation of amines was carried out in water under open air using acyclic or cyclic anhydrides or benzoyl chloride in the absence of any acid or base catalyst or a surfactant. Alkyl or aryl amines, including primary or secondary amines, amino acids, diamines, etc., afforded the corresponding N-acyl derivatives in good yields. Selective N-acylation was observed in the presence of –OH, –SH, –COOH, or –COOEt groups. As a practical alternative to the existing methods for the selective acylation of amines, this methodology may find applications. In contrast to older methods, this reaction takes place at room temperature [65]. N-Heterocyclic carbenes (NHCs) have also been used to activate carboxylic acid derivatives in N-chemoselective acylations of alkanolamines with esters. It has been suggested that NHC's Lewis base activity induces N-chemoselectivity [66]. In addition to catalysts, in-water activation of the acylating agent has been explored as a means of achieving selective N-acylation. Among in-water N-acylating reagents, di-tert-butyl dicarbonate (Boc_2O)has been extensively studied. Boc_2O is an efficient N-acylation reagent for primary and secondary amines under mild conditions. The reaction is carried out in an aqueous medium in the presence of an organic base and a surfactant, leading to the formation of the corresponding N-Boc-protected amides in good yields. This method is particularly useful for the synthesis of peptides and other biologically active compounds [67].

In conclusion, the selective acylation of amines is an important area of research in synthetic organic chemistry. The development of new catalysts and reagents has led to the development of efficient and selective methods for the synthesis of amides and other organic compounds. These methods have potential applications in the synthesis of natural products, pharmaceuticals, and other functional materials [68, 69]. The acylation of amines is important not only in synthetic organic chemistry but also in the preparation of materials with specific properties. In particular, polymers containing amine groups, such as polyethylenimine (PEI), can be acylated to form IEMs, which are used in various applications [70] (Figure 4.7b). The amine groups on the polymer backbone can be selectively acylated with an acylating agent to form amide linkages, resulting in the conversion of the amine group into a secondary or tertiary amide [71–73]. This modification introduces the desired properties, such as proton conductivity, mechanical stability, and chemical resistance. The degree of acylation can also be controlled to achieve the desired level of ion exchange capacity (IEC). For example, PEI can be acylated with an acylating agent to form IEMs with different IECs. These membranes have shown promising performance for ion separation [70].

4.2.3 Acylation of Enols

In recent years, there has been a growing interest in the field of acylation of enols, which involves the addition of an acyl group to the enol form of carbonyl compounds. This reaction is particularly important in the synthesis of pharmaceuticals, agrochemicals, and natural products [74, 75]. In 2022, Dr. Carmela reviewed the state of the art in oxidative NHC-catalyzed reactions that appeared in the literature from 2014 to the present, with a strong focus on crucial intermediates and their mechanistic implications [76]. The review discussed how oxidative NHC catalysis allows for oxidative and oxygenative transformations through specific manipulations of Breslow-type species depending upon the oxidant used (external oxidant or O_2/air). Hale et al. described a new method for ketone enolate C-acylation that utilizes alkyl pentafluorophenylcarbonates, thiocarbonates, and thionocarbonates as the reactive acylating agents and $MgBr_2 \cdot Et_2O$, DMAP, 4-(dimethylamino)pyridine, and i-Pr_2NEt as the reagents for enolization. A wide range of ketones were observed to undergo clean C-acylation via this protocol [74] (Figure 4.8a, b). Overall, these

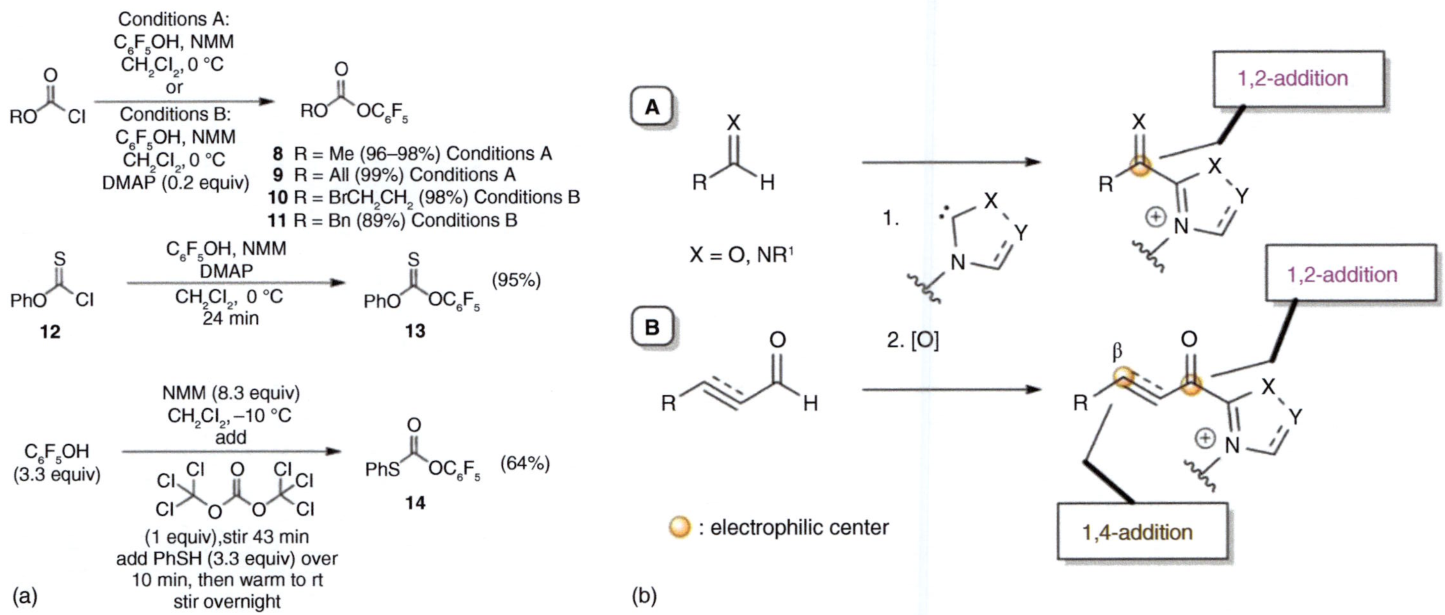

Figure 4.8 (a) Presents a new and easy method for C-acylation of ketone enolates to produce β-keto-esters, thionoesters, and thioesters in steps 8–11, 13, and 14, (b) illustrates possible techniques for straightforward acyl/imidoyl azoliums (A) and α,β-unsaturated (alkynyl) acyl azoliums (B).

works demonstrate the versatility of the acylation of enols in the synthesis of complex organic molecules. The development of enantioselective methods using chiral catalysts is particularly significant, as it allows for the synthesis of enantiopure products. Additionally, the use of earth-abundant metals such as copper and iron as catalysts is important from a sustainability perspective. Further research in this area is likely to lead to the development of even more efficient and selective methods for the acylation of enols, which will have important implications in the fields of pharmaceuticals, agrochemicals, and materials science (Figure 4.9).

4.2.4 Acylation of Carboxylic Acids

Acylation of carboxylic acids is a fundamental organic reaction that involves the conversion of a carboxylic acid into an acyl derivative through the substitution of the hydroxyl group (–OH) with an acyl group (–CO-R). This process is commonly used in organic synthesis for the preparation of esters, amides, anhydrides, and other important chemical compounds [77–80]. The acylation of carboxylic acids can be accomplished through different methods, such as reacting carboxylic acids with acyl chlorides, acid anhydrides, or esters. However, the general mechanism of acylation remains the same, which involves the activation of the acylating agent by the addition of a Lewis acid such as $AlCl_3$ [81, 82] or by using a coupling agent such as dicyclohexylcarbodiimide (DCC) [83]. This activation process makes the acylating agent more electrophilic and ready to react with the carboxylic acid. It then reacts with the carboxylic acid by a nucleophilic attack of the oxygen atom of the carboxylic acid on the electrophilic carbonyl carbon of the acylating agent. This step leads to the formation of a tetrahedral intermediate, which then eliminates the leaving group (usually a chloride ion or an alkoxide ion), regenerating the carbonyl group and releasing the desired acyl derivative. The acylation of carboxylic acids has numerous applications in organic synthesis, including the preparation of esters, amides, anhydrides, and other important chemical compounds [77–80]. Let us take a closer look at some of these applications. For example, the reaction between acetic acid and methanol can lead to the formation of methyl acetate, a common solvent and flavoring agent [84, 85]. Similarly, the acylation of carboxylic acids with amines can lead to the formation of amides [86, 87]. For example, the reaction between acetic acid and ammonia can lead to the formation of acetamide [88]. The acylation of amino acids with carboxylic acids can lead to the formation of peptides. This reaction is commonly used in solid-phase peptide synthesis, where a protected amino acid is attached to a solid support and then acylated with another protected amino acid to form a dipeptide. This process is repeated until the desired peptide sequence is obtained [89, 90]. The acylation of hydroxy acids with acid chlorides can lead to the formation of lactones [91, 92]. Cyclic esters are commonly found in natural products, such as antibiotics and pheromones [93, 94].

4.2.5 Acylation of Ketones

The acylation of ketones is a fundamental reaction in organic chemistry that involves the addition of an acyl group to a ketone compound [95, 96]. This reaction

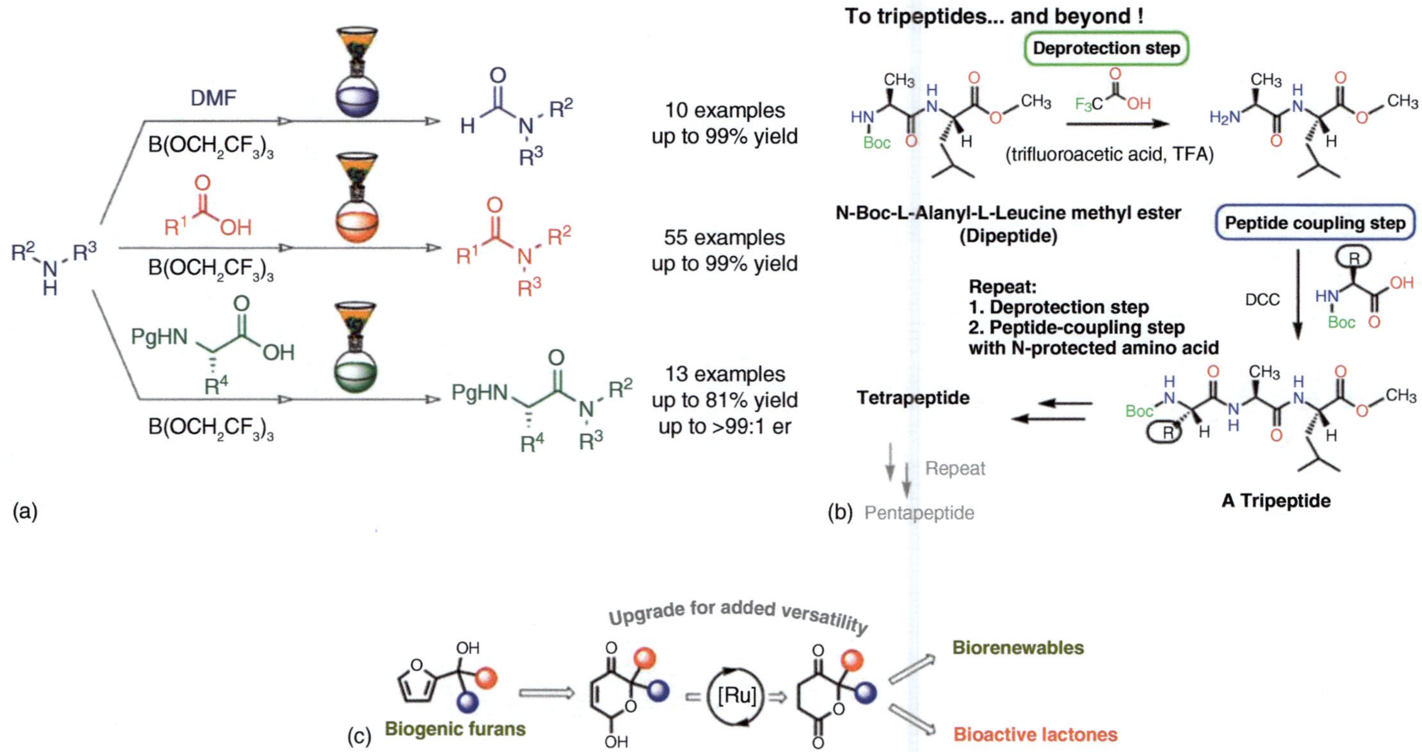

Figure 4.9 (a) B(OCH2CF3)3-Assisted solid phase extraction method for synthesizing amides from carboxylic acids and amines, (b) production of extended peptides – three-peptide and four-peptide chains, and (c) a viable pathway toward biorenewables and bioactive lactones that promotes sustainability.

is widely used in the synthesis of a variety of organic compounds, including pharmaceuticals, agrochemicals, and fragrances [56, 97, 98]. One of the main advantages of the acylation of ketones is its high selectivity and versatility. This reaction can be carried out using a wide range of acylating agents, such as acid chlorides, anhydrides, and esters, making it a highly adaptable process. Additionally, the acylation of ketones can be performed under mild conditions and with high yields, making it an efficient and cost-effective method for synthesizing a wide range of organic compounds. Moreover, this reaction can be easily controlled by selecting appropriate reaction conditions and catalysts, making it a highly tunable process. Overall, the acylation of ketones is a powerful tool for organic chemists, enabling the synthesis of a wide range of organic compounds with high selectivity, efficiency, and versatility [98]. For instance, the report by Charles in 2011 discusses the acylation of ketones under certain conditions to form β-ketones or β-keto aldehydes by replacing an alpha-hydrogen atom of the ketone with an acyl group. This reaction involves carbon–carbon condensation and can be carried out using various acylating agents, such as esters, acid anhydrides, acid chlorides, or formic esters. The acylation of ketones with esters, acid anhydrides, or acid chlorides has traditionally been carried out using basic or acidic reagents to facilitate the reaction. For example, the use of basic reagents such as sodium or sodium ethoxide is a common method for the acylation of ketones with esters [95]. Recently, Trisha described a method for synthesizing aromatic ketones using a heterogeneous catalyst composed of graphene oxide (GO)-supported PdO nanoparticles (PdO/GO). The use of a heterogeneous catalyst, as opposed to a traditional homogeneous catalyst, allows the reaction to take place under milder conditions, which can be more practical and cost-effective. The reaction involves coupling iodoarenes with aromatic aldehydes, facilitated by the PdO/GO catalyst. Notably, this method avoids the use of toxic CO gas as the source of the carbonyl group, which can be a safety hazard in some reactions. The practicality of this method was demonstrated by synthesizing 1,3-dihydroisobenzofuran, a more complex molecule than aryl ketones. Overall, Trisha's report highlights the importance of aromatic acylation in organic synthesis and presents a promising method for its synthesis using a PdO/GO catalyst [98].

4.2.6 Acylation of Amides

The acylation of amides is a unique acylation reaction that involves the transfer of an acyl group to an amide molecule. The reaction is typically carried out using an acid chloride or anhydride as the acylating agent and a base such as pyridine to catalyze the reaction. For instance, Meng et al. described an efficient method for the N-acylation of amides using a pyridine ring as an internal nucleophilic catalyst (Figure 4.10a). This method results in the formation of imides in moderate to excellent yields. The use of a pyridine ring as the catalyst provides several advantages, including air insensitivity, environmental friendliness, and ease of use. This methodology can be used to create a variety of imide scaffolds, which are commonly found in natural products and biologically active materials. Overall, this report presents a promising approach to synthesizing imides with high efficiency

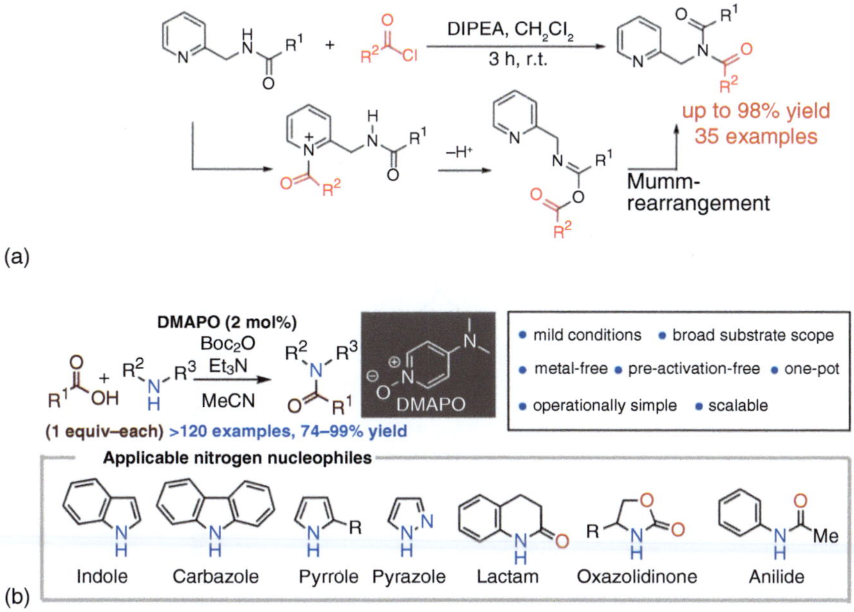

Figure 4.10 (a) Amides undergo N-acylation via catalysis by an internal nucleophile, (b) the direct N-acylation of less nucleophilic N-heterocycles with carboxylic acids is achieved in a one-pot reaction using DMAPO/Boc_2O as a mediator.

and environmental friendliness [99]. In another study in 2023, Atsushi reported that despite the significant progress made in the development of methodologies for amide bond formation in amines, direct N-acylation of less nucleophilic N-heterocycles and amides with carboxylic acids still poses a challenge. In this study, Atsushi and his team introduced a new one-pot method that enables the direct N-acylation of a wide range of nitrogen nucleophiles, such as indole, carbazole, pyrrole, pyrazole, lactam, oxazolidinone, and anilide, with carboxylic acids in high yield (Figure 4.10b). The method uses the 4-(N,N-dimethylamino) pyridine N-oxide (DMAPO)/Boc_2O system, which does not require preactivation of substrates. The new method shows excellent functional group tolerance and broad substrate scope. It is practical, operationally simple, and scalable, making it suitable for both academic and industrial laboratories. The direct N-acylation of less nucleophilic heterocycles and amides with carboxylic acids has many applications in the synthesis of natural products, pharmaceuticals, and agrochemicals [100]. This method has a broad substrate scope and functional group tolerance and is expected to have wide applications in both academic and industrial laboratories.

4.2.7 Acylation of Sulfonamides

The acylation of sulfonamides is a unique type of acylation reaction that involves the transfer of an acyl group to a sulfonamide molecule (Figure 4.11a,b) [101, 102]. The reaction is typically catalyzed by a base, such as a pyridine, and an acylating

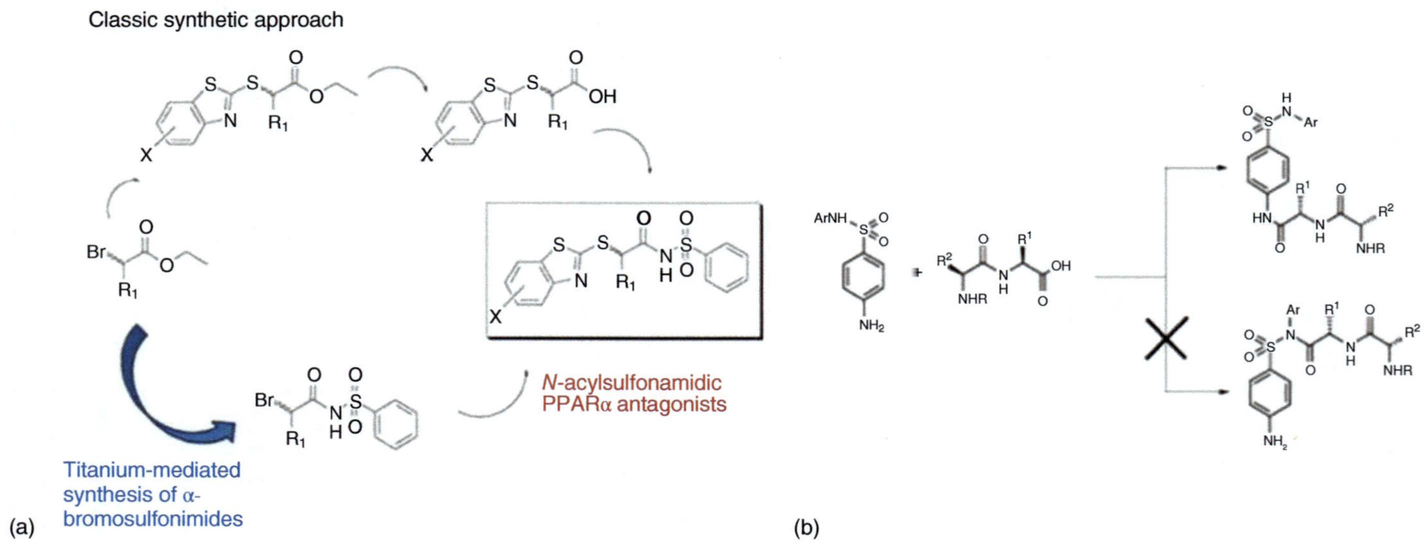

Figure 4.11 (a) Amino acids can act as agents that selectively acylate sulfonamides, (b) N-acylsu fonamide PPARα antagonists can be synthesized by promoting the acylation of sulfonamides with titanium.

agent, such as an acid chloride or anhydride [103, 104]. However, in 2018, Benali investigated the N-acylation of substituted sulfonamides using a Cesium salt of Wells–Dawson heteropoly acid as a catalyst in water, which resulted in high yields and easy recovery of the catalyst. The reaction yielded N-acylsulfonamides at room temperature and cyclic imides containing sulfonyl groups under refluxing conditions [105]. In 2019, Chen developed a convenient and fast method to prepare acyl sulfonamides from carboxylic acids and sulfonamides using the Mukaiyama reagent [103]. The reaction proceeded within one hour, yielding moderate to good yields under optimized conditions. Based on Mahmoud's research paper [106], aromatic sulfonamide-derived ethyl ester (p-TSAE) and its acyl hydrazide (p-TSAH) have been synthesized and tested as potential anticorrosive agents to protect mild steel in acidic conditions. The corrosion efficiency was measured through various electrochemical techniques, and both compounds showed excellent corrosion inhibition, with hydrazide demonstrating slightly better efficiency due to its strong interaction with the metal surface. The findings suggest that p-TSAE and p-TSAH have promising potential as effective anticorrosive agents in acidic environments.

In the synthesis of antimicrobial agents, the acylation of sulfonamides remains a widely used method. Recent research reported by Mahapatra et al. in 2023 demonstrated the synthesis of a series of N-heteroaryl-4-(1-[2-oxo-chromen-3-yl] ethylideneamino) benzenesulfonamide compounds through condensation reactions, which exhibited potent antimicrobial activity against terbinafine-resistant Trichophyton rubrum [107]. These findings, along with prior molecular docking optimization and in vitro anticancer activity screening against different cell lines, highlight the versatility and potential of sulfonamide acylation in the development of potent inhibitors and the creation of new chiral scaffolds for various chemical reactions. Overall, the acylation of sulfonamides has significant importance in the field of chemistry, with ongoing research aimed at improving the efficiency and effectiveness of this reaction for various applications.

4.2.8 Polyacylation of Polymers

In our previous discussion, we covered the different types of acylation reactions in detail, with a focus on small compounds. However, it is important to note that the acylation of polymers is of great significance because it allows the introduction of functional groups for ion exchange [108]. This is particularly relevant for membrane applications that require high selectivity and permeability [39]. Polymer acylation can be achieved using various methods, such as Friedel–Crafts acylation [109], Schotten–Baumann reaction (Figure 4.12b) [110], and acylation using acyl chlorides or anhydrides [111]. The choice of method depends on the type of polymer being used and the desired functional groups. Four polymers that have shown promise for membrane applications are perylene-based aliphatic polyimide, SPEK, and crown ether-based polymers. Perylene-based aliphatic polyimide is a highly selective membrane material that exhibits excellent stability under high temperatures and harsh conditions [33, 112, 113]. SPEKs, on the other hand, have good mechanical properties and are effective in separating gases and organic solvents [20, 114–116].

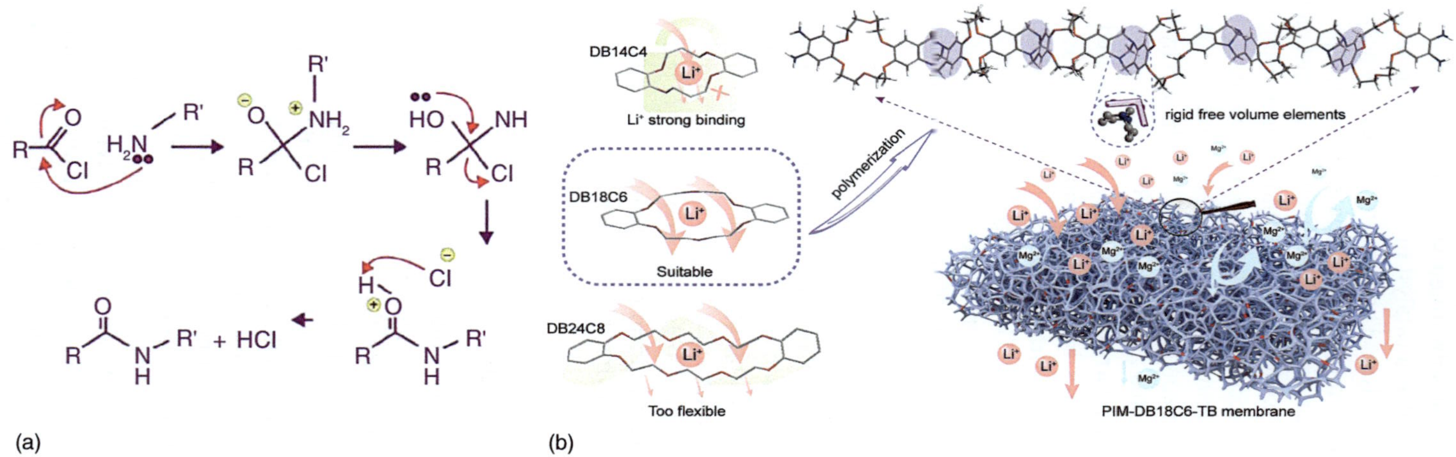

Figure 4.12 (a) The acylation reaction utilizing the Schotten–Baumann method, (b) Tröger's base membranes containing crown ether for effective separation of Li^+ and Mg^{2+} ions.

Crown ether-based polymers are known for their ability to selectively transport metal ions and have potential applications in ion separation and recovery (Figure 4.12b) [117]. Overall, the acylation of polymers is a critical tool in the development of advanced membranes for various applications. By introducing functional groups, it is possible to achieve high selectivity and permeability, making these materials invaluable in fields such as water treatment, energy production, and chemical processing [7, 32, 33, 39, 115, 117]. Polyacylation has several advantages over other methods for synthesizing polyelectrolytes, such as ease of synthesis, tunability of properties, and compatibility with a variety of substrates and conditions. In the context of fuel cell membranes, polyacylation has been used to synthesize a variety of polyelectrolytes with different properties and structures, such as linear or cross-linked, aromatic or aliphatic, and with various spacers and functional groups [39, 118].

One of the main applications of polyacylation in fuel cell membranes is the synthesis of AAEMs for use in alkaline fuel cells. AAEMs are a promising alternative to traditional acidic PEMFCs, as they can operate at high pH and do not require expensive platinum catalysts. However, the development of AAEMs has been hampered by several challenges, such as the lack of stability under alkaline conditions, low ion conductivity, and poor mechanical properties [119, 120]. Polyacylation offers a solution to these challenges by allowing the synthesis of polyelectrolytes with tailored properties, such as high ion conductivity, alkaline stability, and mechanical strength. In particular, the use of prequaternized monomers as a starting material for polyacylation has been shown to be a straightforward and versatile approach for synthesizing AAEMs with long pendent spacers instead of conventional benzyl-type spacers [7]. In the context of AAEMs, polyacylation offers a promising solution to the challenges of alkaline stability, ion conductivity, and mechanical strength, and has already led to the synthesis of novel polyelectrolytes with exceptional properties. Further research is needed to explore the full potential of polyacylation in the synthesis of fuel cell membranes and to optimize the properties of these materials for practical applications.

4.2.9 Advantages and Limitations of Polyacylation as a Synthetic Approach

The polyacylation of prequaternized monomers is a promising synthetic approach for producing polyelectrolytes with tailored properties for use in fuel cells and other electrochemical devices. However, like any synthetic approach, it has advantages and limitations.

Advantages:

i. Versatility: The polyacylation reaction can be carried out using a variety of acylating agents and prequaternized monomers, allowing for a wide range of polyelectrolytes to be synthesized with various properties.
ii. Straightforward synthesis: The polyacylation reaction is a simple and efficient way to synthesize polyelectrolytes with high yields and a low number of synthetic steps.

iii. Tunable properties: The properties of the resulting polyelectrolytes can be tailored by adjusting the length and flexibility of the pendent spacer, the degree of acylation, and the type of functional group.
iv. High alkaline stability: Polyacylated polyelectrolytes exhibit excellent alkaline stability, making them suitable for use in alkaline fuel cells.
v. Low cost: Compared to other synthetic approaches, such as the synthesis of perfluorinated ionomer membranes, the polyacylation approach is relatively inexpensive.

Limitations:

i. Limited monomer availability: The polyacylation approach requires the use of prequaternized monomers as starting materials, which can be difficult and expensive to obtain.
ii. Limited solubility: Some polyacylated polyelectrolytes can have limited solubility in common solvents, which can limit their processability and application.
iii. Potentially toxic reagents: The use of certain acylating agents, such as acetic anhydride, can pose health hazards and require proper safety precautions.
iv. Potential for side reactions: The polyacylation reaction can sometimes lead to side reactions, such as the formation of undesirable byproducts, which can affect the yield and purity of the resulting polyelectrolytes.
v. Narrow pH range: Polyacylated polyelectrolytes have been shown to exhibit high ion conductivity and alkaline stability, but they may have limited stability in acidic or neutral environments.

The advantages of the polyacylation approach, such as versatility, straightforward synthesis, and tunable properties, make it an attractive option for researchers working in this field. However, limitations, such as limited monomer availability, limited solubility, and potentially toxic reagents, should be carefully considered when selecting this approach for a specific application. Overall, the polyacylation approach shows great promise in advancing the development of high-performance polyelectrolytes for use in fuel cells and other electrochemical devices.

4.2.10 Polyacylation and Polymers

Polyacylation is an important polymerization reaction that offers several advantages over other types of polymerization reactions. One advantage is that it allows for the formation of high molecular weight polymers with good thermal and mechanical properties. Polyacylation reactions can also be carried out under mild conditions and with a wide range of monomers, making it a versatile method for polymer synthesis. The focus of this chapter is on perylene-based aliphatic PIs, SPEKs, and crown ethers, which are known for their unique properties and potential applications. These materials have high thermal stability, excellent mechanical properties, and good processability, making them attractive for various applications, including electronic and optoelectronic devices, membranes, and coatings. In addition to their inherent properties, the incorporation of crown ether moieties into the polyimide backbone can lead to enhanced ion transport properties, making these materials potentially useful in energy storage and conversion applications. Furthermore, this section explores the latest work being done with respect to polyacylation from a membrane point of view and provides insights into the design and synthesis of novel membranes for various applications. Overall, the advantages of polyacylation include the ability to form high molecular weight polymers with good thermal and mechanical properties and the versatility of the reaction with a wide range of monomers.

4.3 Perylene-based Polyimides

Perylene, a polycyclic aromatic hydrocarbon (PAH) with the chemical formula $C_{20}H_{12}$, has unique optical and electronic properties that make it highly useful in a range of applications. Despite its insolubility in water, it is soluble in organic solvents, making it highly versatile for use in a range of industries. Perylene-based aliphatic and aromatic PIs are two classes of polymers that incorporate perylene units into their backbone, resulting in materials with enhanced properties that are highly useful in various applications. Aliphatic PIs are characterized by the presence of nonaromatic (aliphatic) groups in their backbone. On the other hand, perylene-based aromatic PIs incorporate perylene units into their structure, alongside aromatic groups such as phenyl or naphthyl. The incorporation of perylene units into these PIs enhances their mechanical properties, thermal stability, and resistance to chemicals and radiation. PIs can be synthesized using various methods, including (a) traditional methods, and (b) polyacylation methods. Traditional methods involve the reaction of dianhydrides and diamines to form poly (amic acid) (PAA), which undergoes thermal cyclization to form PIs. On the other hand, polyacylation methods involve the reaction of dianhydrides with amines or alcohols to form poly (ester acid) (PEA), which undergoes thermal cyclization to form PIs. The synthesis of perylene-based aliphatic and aromatic PIs has been an area of active research, with many studies exploring various synthesis methods and their resulting properties. Perylene-based aliphatic and aromatic PIs are highly promising materials that offer enhanced properties compared to traditional PIs. The development of efficient and effective synthesis methods is crucial for the large-scale production of these materials, and ongoing research is expected to lead to even further improvements in their properties and potential applications.

4.3.1 Traditional Route

The traditional route for the synthesis of PIs, including perylene-based PIs, involves the reaction of dianhydrides with diamines to form polyamic acid, followed by imidization of the polyamic acid to form the final polyimide (Figure 4.13(a, b). For instance, in 2013, Masatoshi reported PIs based on alkyl-substituted cyclobutanetetracarboxylic dianhydrides (CBDAs). This research aimed to improve the solubility and mechanical properties of PIs. Copolymerization was found to be an

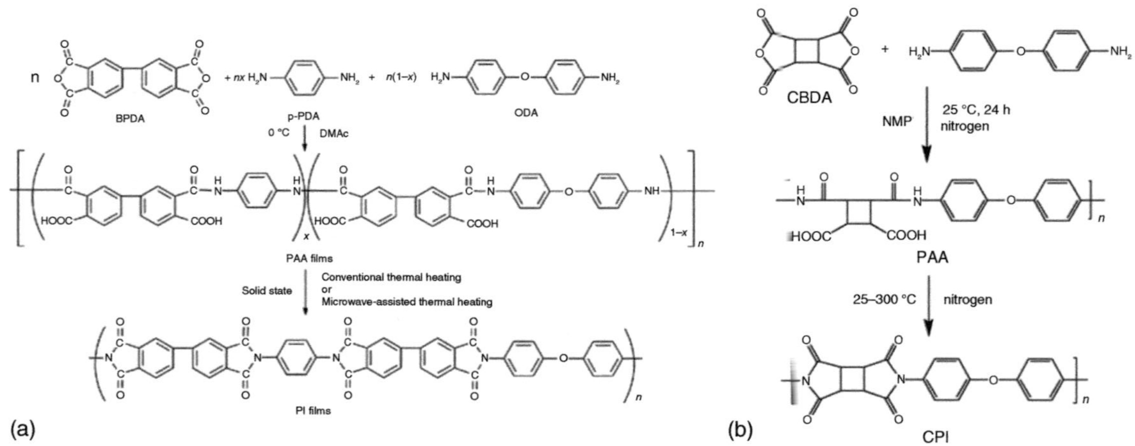

Figure 4.13 (a) PI films are prepared using BPDA, ODA, and p-PDA with $x = 0.7$, (b) a two-step procedure is employed, involving a thermal imidization process.

effective way to improve the solubility of DM-CBDA-based PIs. The copolyimide cast films prepared via chemical imidization displayed a further decrease in CTE without sacrificing other target properties. The research demonstrates the potential of DM-CBDA-based PIs as plastic substrates in display devices [121]. In 2015, Hongjiang published a review article and discussed the development of colorless and transparent polyimide (CPI) films. The review covers the latest research and development of CPI films, including synthesis chemistry, manufacturing processes, and engineering applications [122]. Later, in 2019, Lingren investigated the effects of microwave-assisted thermal imidization on the microstructures, mechanical properties, and thermal properties of PI films. The results showed that the PI films treated with microwave-assisted heating had a higher imidization degree and improved tensile strength and modulus compared to traditional thermal imidization. Moreover, the order degree of polymer chains was improved by the microwave-assisted imidization method, and the PI films showed excellent thermal stability [123]. This research demonstrates the potential of microwave-assisted thermal imidization as a rapid and efficient way to prepare high-performance polyimide materials. Michael's study in 2021 synthesized a perylene-based aromatic polyimide through a single condensation reaction step. The polyimide had multiple carbonyls, enabling high-capacity, and stable organic lithium and sodium ion batteries [124]. This research shows the potential of polyimide materials in energy storage applications. Li in 2023 reported that two series of PIs were synthesized from the one-step solution polycondensation of two diamines containing sulfonyl groups with aromatic and aliphatic dianhydrides. These PIs showed good organic solubility and were cast into films with high transparency of $T500 > 86\%$ [125]. The modification and advancement of polyimide synthesis methods have led to the development of high-performance polyimide materials with improved mechanical, thermal, and electrical properties (Figures 4.14–4.25).

The traditional route for synthesizing perylene-based PIs is widely used due to its simplicity, versatility, and ability to produce high-quality films with excellent thermal, mechanical, and electrical properties. However, this method has some drawbacks, such as long reaction times, harsh reaction conditions, and high costs. Additionally, perylene-based PIs synthesized using this route can exhibit poor film-forming ability, which can limit their practical applications. Therefore, ongoing research is focused on developing new synthesis methods and optimizing existing methods to address these limitations.

4.3.2 Polyacylation Route

The polyacylation route is a method for synthesizing PIs, including perylene-based PIs, in a one-step reaction. This route involves the direct reaction of aromatic diamines with dianhydrides in the presence of a polycondensation catalyst, which can be performed without the use of a solvent. The reaction proceeds through the formation of an intermediate polyamic acid, which undergoes imidization to form the final polyimide [126]. Compared to the traditional route, the polyacylation route can produce PIs with higher purity due to the absence of intermediate polyamic

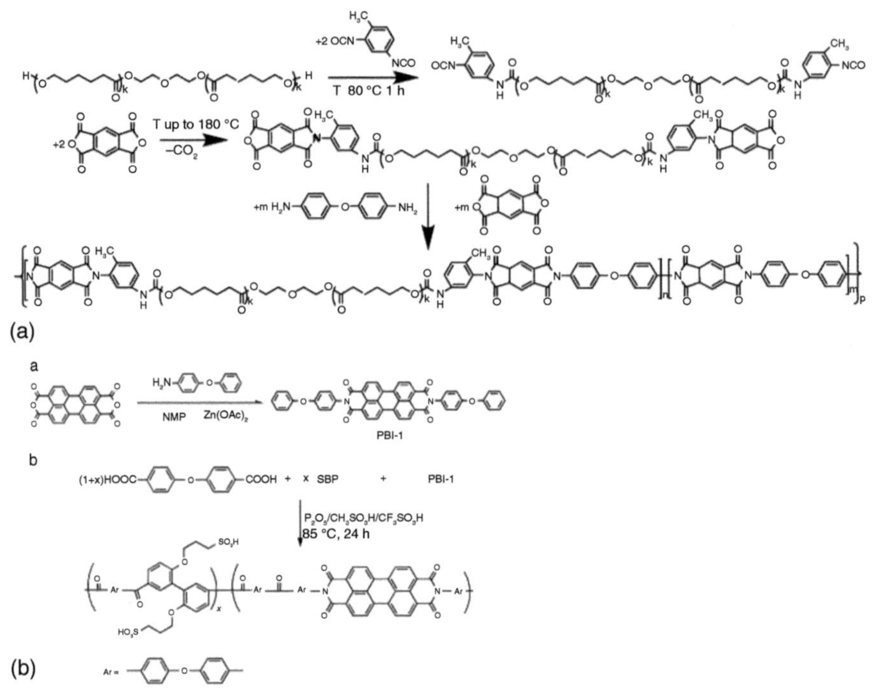

Figure 4.14 (a) The synthesis of a copolymer with a [(imide)n-(urethane)]m structure is achieved, (b) imide monomers undergo polyacylation using a straightforward method to create innovative sulfonated polyimide proton-exchange membranes, (c) a process for creating HFD-b-DFS block copolyamides is illustrated through a synthetic scheme.

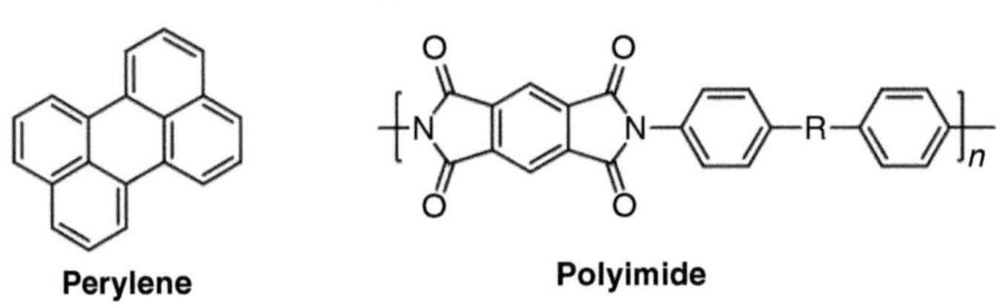

Figure 4.15 The molecular makeup of perylene and polyimide.

acid, which can sometimes lead to undesired side reactions and impurities. The polyacylation route offers an alternative method for synthesizing high-performance PIs with potential advantages over the traditional route [127]. In 2014, a study was conducted by a group to synthesize sulfonated polyimides (SPIs) using a novel molecular design. The SPIs consisted of alternating naphthalenediimide and sulfonated units separated by long aliphatic spacers. Unlike conventional aromatic-type imide units, the use of aliphatic-type naphthalenediimide units is expected to enhance the hydrolytic stability of proton exchange membranes. The segregation of naphthalenediimide units from sulfonic acid groups also helps to improve the hydrolytic stability of the SPIs. The aggregation of sulfonate groups increases proton conductivity, while the stacking of naphthalenediimide

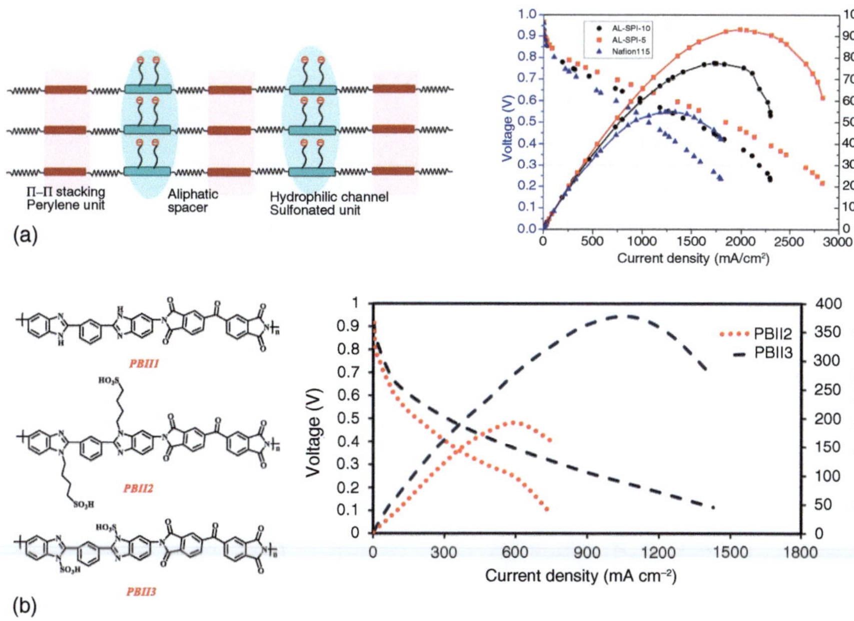

Figure 4.16 (a) A depiction is shown to illustrate how the polyimides self-assembled, (b) the fuel cell performance of AL-SPI-5,10 and Nafion 115 is compared at 80 °C, (c) the individual cell performance of PBII 2 and 3 is measured in a H_2/O_2 PEMFC under anhydrous conditions at 120 °C, and (d) PBII1-3′s structure is illustrated schematically.

moieties helps to restrain swelling. The synthesis of SPIs using this novel molecular design has shown great potential for enhancing the properties of PEMs, particularly their hydrolytic stability and proton conductivity [128]. We made a new monomer called polybenzimidazoles (PBI)-1 by combining 4-phenoxyaniline and 3,4,9,10-perylenetetracarboxylic dianhydride and used it to create SPIs through a polyacylation reaction with disodium 2,2-di (sulfopropyloxy)-biphenyl and 4,4-oxydibenzoic acid. These PIs showed good hydrophilic characteristics, high proton conductivity, and low linear swelling ratio due to their perylene structure. This suggests that they could be used as proton exchange membranes [129]. In 2018, our group synthesized sulfonated polyimide membranes composed of aliphatic-type perylenediimide units with varying aliphatic chain lengths. The synthesis involved mild polyacylation of a sulfonated diarene monomer and aliphatic perylenediimide dicarboxylic acid monomers. The resulting membranes were flexible and tough and exhibited microphase separation. They also showed exceptional proton conductivity while maintaining a low linear swelling ratio. This outstanding performance is due to the self-assembly behavior driven by the strong π–π interactions of the perylenediimide units and the hydrophilic interactions of the sulfonated groups [33]. Andrey and his coworkers developed a method for synthesizing a copolymer in 2022. The process involved copolycondensation of 4,4′-diaminodiphenyl ether in a DMAc solution with pyromellitic anhydride and macromonomer that had end anhydride groups. The macromonomer was produced using polycaprolactone,

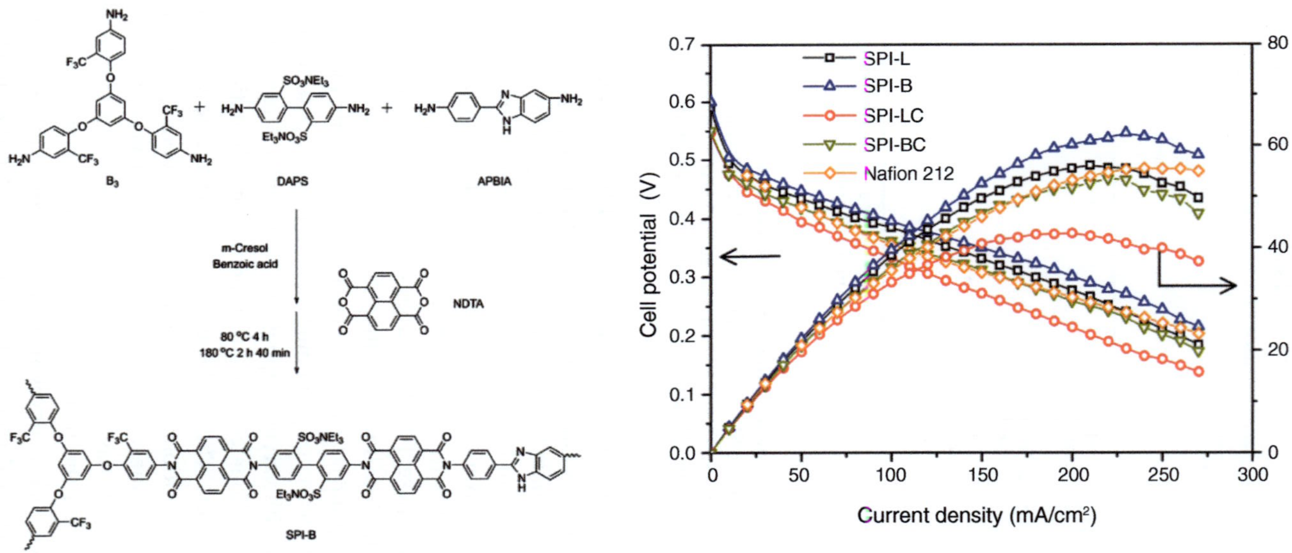

Figure 4.17 (a) The creation of SPI-B, (b) the individual cell efficiencies of SPI-L, SPI-LC, SPI-B, SPI-BC, and Nafion46 in MFCs using 2 M methanol/air as the anode/cathode.

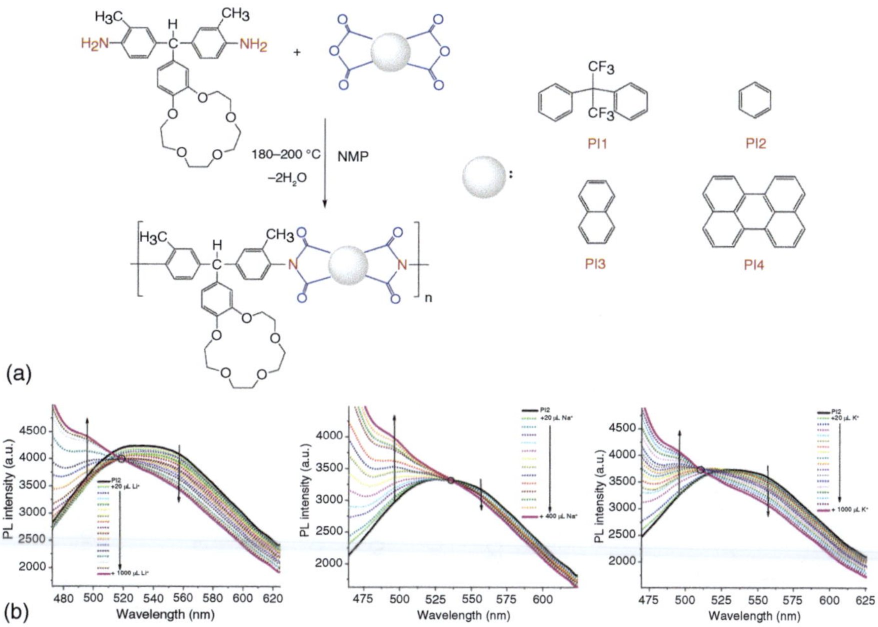

Figure 4.18 (a) A synthetic pathway for polyimides utilizing TPM-crown diamine, and (b) recording of PL spectra for PI2 while titrating with (a) Li^+, (b) Na^+, and (c) K^+.

2,4-toluene diisocyanate, and pyromellitic anhydride. The reaction mixture was then stirred under an argon current for four hours at room temperature to complete the diamine polyacylation reaction, after which the resulting solution was cast into films [130]. A new film fabrication method using insoluble polymers was also proposed in 2022. This involved a block copolymer solution containing both soluble and insoluble blocks, followed by the selective decomposition of soluble blocks. To demonstrate the concept, a copolymer ([imide]n-[polyurethane])m was synthesized with imide blocks and polyurethane blocks based on polycaprolactone. Through the selective hydrolysis of urethane blocks with acid, the first insoluble polyimide film was obtained, which closely resembled the homopolymer consisting of the same imide monomer, poly-(4,4′oxydiphenylene) pyromellitimide. This confirmed the feasibility of the proposed concept and its potential for creating organic solvent-resistant membranes [131].

4.3.3 Synthesis of Perylene-based Polyimide-based Ion Exchange Membranes

Polyimide-based IEMs are a preferred choice for electrochemical applications due to their remarkable features, such as high ionic conductivity, chemical, and thermal stability, and excellent mechanical strength. The preparation of these membranes involves two crucial steps. First, a polyimide precursor is synthesized using traditional and polyacylation methods, which we have previously discussed. Second, perylene is incorporated into the polyimide precursor to create

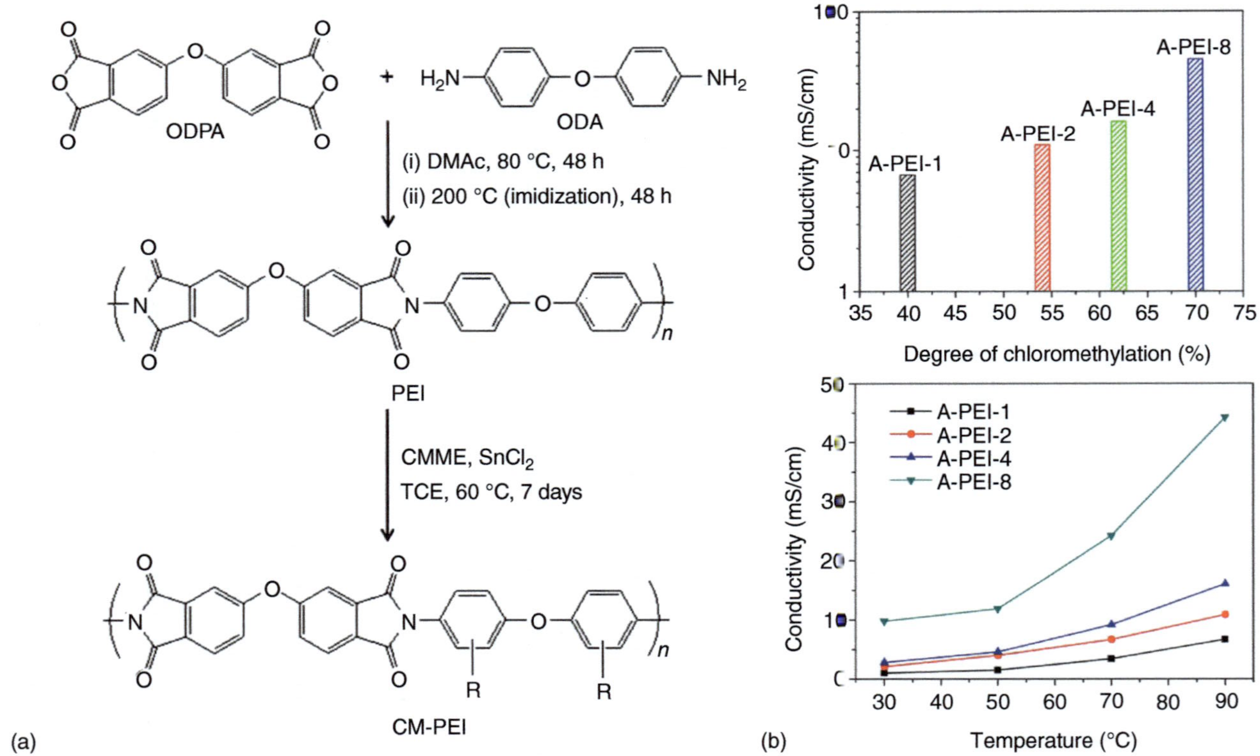

Figure 4.19 (a) The production of PEI and CM–PEI and their respective chemical compositions and (b) the measurement of hydroxide ion conductivity in A-PEI membranes at different temperatures and varying DC levels with a humidity level of 100%.

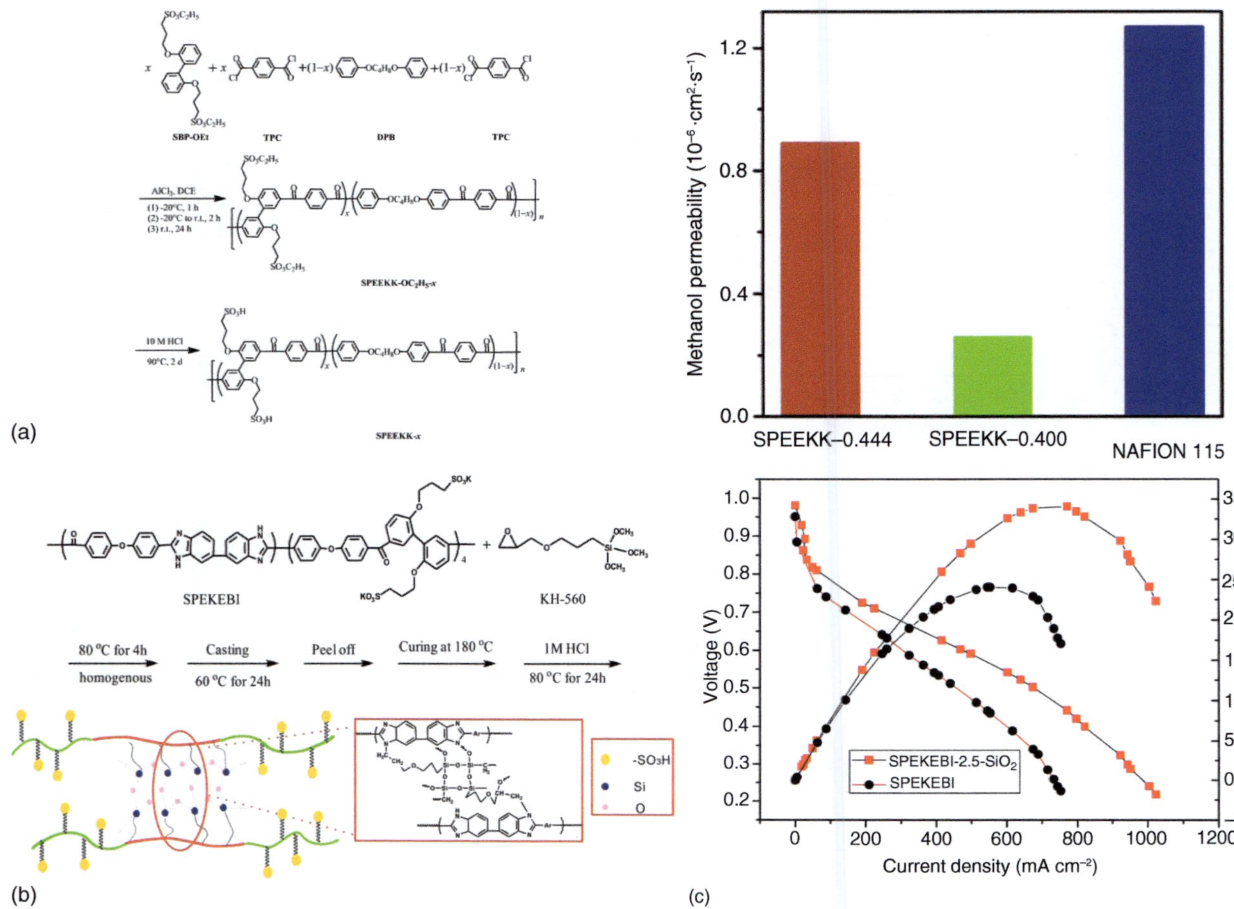

Figure 4.20 (a) Methanol permeability comparison of NAFION® 115 and SPEEKK-x, (b) synthesis of SPEKEBI-x-SiO_2 membranes, and (c) single fuel cell performance evaluation of SPEKEBI-2.5-SiO_2 and SPEKEBI membranes at 75 °C and 100% RH.

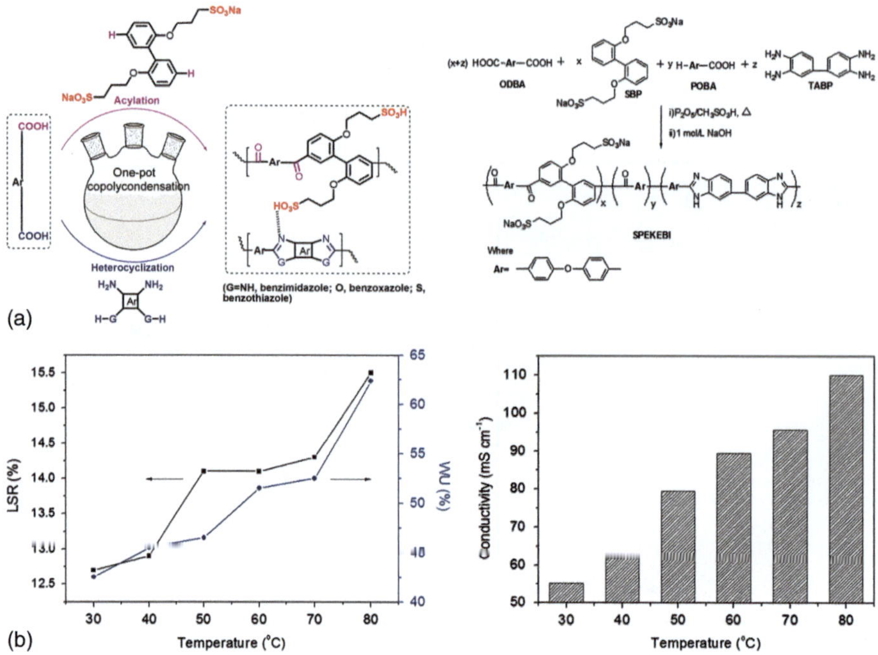

Figure 4.21 (a) Illustration of the "one-pot acylation/heterocyclization copolycondensation" concept in a schematic diagram and the synthetic pathway used for SPEKEBI, (b) changes in the linear swelling ratio (LSR), water uptake (WU), and proton conductivity characteristics of the SPEKEBI membrane at various temperatures.

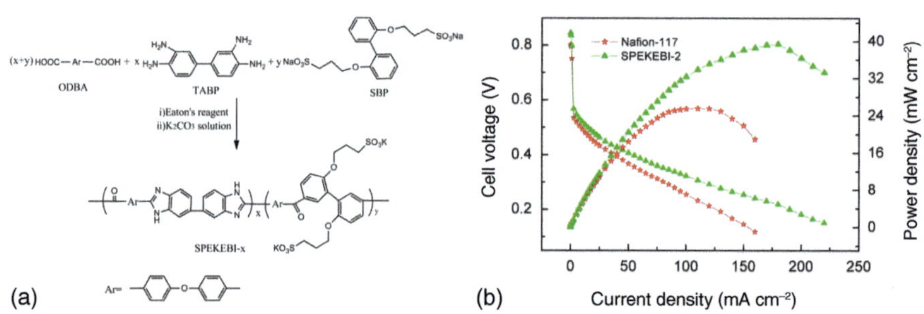

Figure 4.22 (a) The method for producing copolymers of SPEKBI-x, (b) comparison of cell voltage/power density with load current density for DMFCs utilizing SPEKEBI-2 membranes and Nafion® 117.

perylene-based-polyimide-based IEMs. This incorporation of perylene enhances the membrane's mechanical strength, chemical stability, and water retention capacity. These membranes can be further categorized into cation exchange membranes (CEMs) and anion exchange membranes (AEMs). Perylene-based-polyimide-based CEMs are applicable in desalination, fuel cells, and electrodialysis, while perylene-based-polyimide-based AEMs are ideal for alkaline fuel cells, electrochemical synthesis, and water electrolysis. The synthesis of perylene-based-

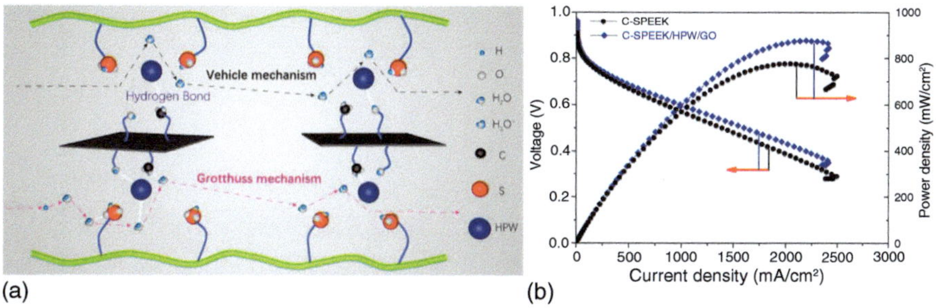

Figure 4.23 A composite proton exchange membrane made of sulfonated poly(etheretherketone), phosphotungstic acid, and carbon-based nanomaterials is crosslinked using the Friedel–Crafts method for fuel cell applications, (b) the performance of fuel cells using C-SPEEK and C-SPEEK/HPW/GO membranes is represented by their respective performance curves.

polyimide-based IEMs involves several steps, including thermal imidization, and ion exchange, to introduce ionic functionality into the polymer. Researchers can tailor the synthesis process to create perylene-based-polyimide-based IEMs with unique features suitable for various electrochemical applications.

4.3.4 Perylene and Polyimide-based CEMs

This section covers the topic of perylene-based-polyimide CEMs, which have various applications in electrochemistry, including fuel cells, electrodialysis, and desalination. These membranes possess a high degree of thermal and chemical stability, good mechanical strength, and excellent proton conductivity. Moreover, the incorporation of perylene into the polymer matrix enhances the membrane's stability, selectivity, and performance. The preparation of perylene-based-polyimide-based CEMs involves the synthesis of the polyimide precursor followed by the casting and subsequent imidization of the polymer.

The functionalization of the polymer with suitable cationic groups such as sulfonic acid or phosphonic acid enhances the cationic conductivity of the membrane. For example, our team conducted research in 2018 to develop and characterize sulfonated perylene polyimide membranes for PEMFC applications. The objective of the study was to investigate how perylene polyimide molecules with sulfonated groups could enhance the conductivity and fuel cell performance of the membranes. The study found that the proton conductivity of the AL-SPI-5 and AL-SPI-10 membranes increased with temperature and was higher than that of Nafion 115. This excellent proton conductivity performance was attributed to the regular stacking structure of the membranes, which was derived from the π–π interaction of perylene and the hydrophilic interaction of sulfonated groups. The stacking of the molecular structure enhanced the methanol permeability performance of the membrane without sacrificing proton conductivity. Additionally, the membranes showed improved resistance towards methanol and selectivity, demonstrating their potential for use in PEMFC applications. To evaluate the fuel cell performance of the membranes, the study compared two different membranes, AL-SPI-5

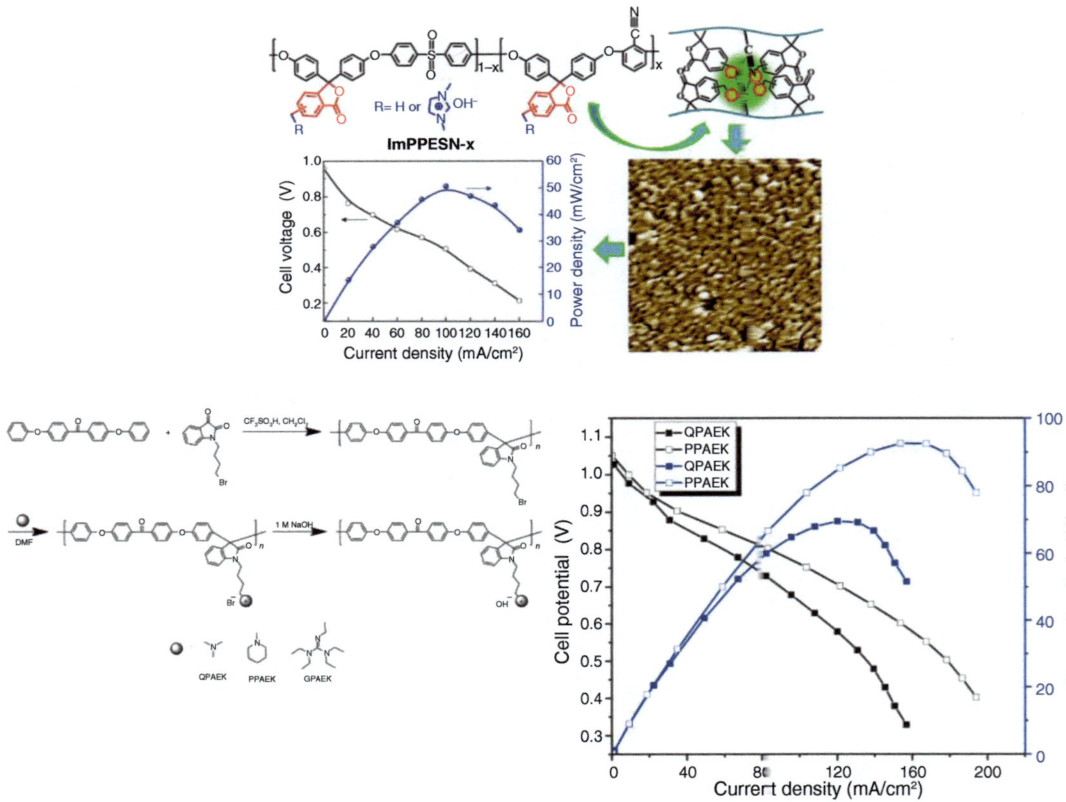

Figure 4.24 (a) The process used to synthesize the polymer BPAEK, (b) the performance of AEMs in H_2/O_2 AEMFC at a temperature of 60 °C, and (c) the use of a phenolphthalein-based poly (arylene ether sulfone nitrile) anion exchange membrane with side chains for fuel cells.

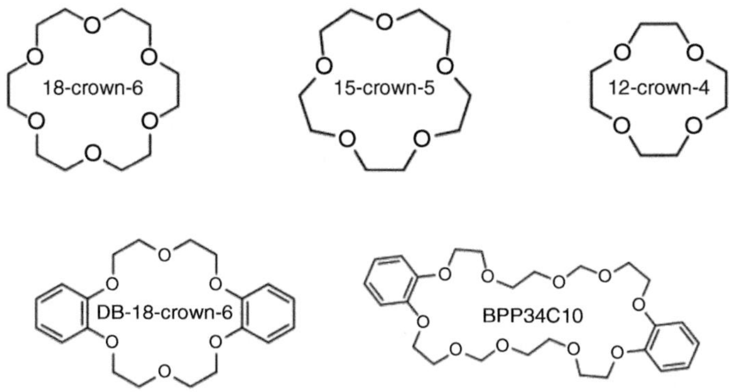

Figure 4.25 Examples of frequently used crown ethers include 18-crown-6, 15-crown-5, 12-crown-4, dibenzo-18-crown-6, and BPP34C10.

and AL-SPI-10, with Nafion 115 under fully humidified conditions. The results showed that the maximum power densities for the fuel cells based on AL-SPI-5 and AL-SPI-10 were 931.88 and 773.51 mW/cm^2, respectively, which were much higher than that of Nafion 115. These findings have important implications for the development of high-performance PEMFCs and can be applied to our team's work to improve energy efficiency and sustainability in various fields [33].

In 2019, a study was conducted to evaluate the performance of a cross-linked sulfonated polyimide (SPI) membrane that contained a highly branched polymer main chain as a PEM in a single direct methanol fuel cell. The study found that the cross-linked membrane with a branched structure demonstrated enhanced proton conductivity in comparison to a common cross-linked membrane, despite having a similar IEC value. Furthermore, the branched SPI membrane displayed improved water stability, as it retained 92.3% of its pristine weight after a 120-hour hydrolytic stability test. The study's most significant finding was that the branched SPI membrane's improved electrochemical properties resulted in a higher power density in a single direct methanol fuel cell. The membrane with a branched structure exhibited a power density of 53.4 mW/cm^2, while the common cross-linked membrane only had a power density of 43.0 mW/cm^2. These results indicate that the use of highly branched polymer main chains is a promising technique to enhance the electrochemical properties and fuel cell performance of cross-linked SPI membranes. This study's findings have significant implications for the development of more efficient and durable fuel cells, which could play a critical role in the advancement of sustainable energy systems [132]. In 2023, Tiantian and colleagues conducted a study on optimizing the performance of PEMs by controlling their morphology. While block copolymerization is an effective method, it is complicated, and simpler alternatives are needed. The researchers used a molecular engineering approach to modify the structure and volume of hydrophobic cross-linkable diamines and create a hydrophobic network structure to optimize the microphase morphology of SPIs. They synthesized three diamine monomers, each with tetrafluorostyrol pendant groups, and prepared three CSPI membranes with similar IECs but different main-chain structures using these monomers and cross-linking agents. CSPI membranes with a larger molecular volume and containing hydrophobic structural units in the main chain had a more pronounced phase separation morphology and higher proton conductivity. The CSPI-AP-10 membrane, prepared from the 6FATFVP monomer, had the best performance with high proton conductivity, and excellent methanol resistance. The direct methanol fuel cells (DMFCs) assembled from CSPI-AP-10 had a high maximum power density in both 2 and 10 M methanol, demonstrating its potential as a PEM in DMFC applications [133].

In 2022 and 2023, several studies were conducted to synthesize various types of PIs with high-performance properties and potential for use in various applications [134–137]. Two studies produced SPIs with high proton transfer and water retention potentials, while another study synthesized a novel polyimide with a broad-spectrum response, high photocatalytic activity, and strong π–π conjugation. Although the studies did not focus on fuel cell performance, based on the high proton transfer and water retention potentials, as well as the strong π–π conjugation of the synthesized polymers, it is likely that they could be useful as membranes in fuel cell applications. However, further studies are needed to confirm their performance in this specific application [138, 139]. In 2021, Jialing Wu provided an overview of the progress made in the development of potassium-ion batteries, noting the challenges posed by the lack of high-capacity, and cyclable electrode materials. However, Wu identified porous polyimide frameworks (PIFs) based on perylene and triazine as promising anode materials due to their high-density electroactive imide groups and robust molecular structure. The PIFs demonstrated a large reversible specific capacity, good rate performance, and long cycling stability, making them highly suitable for large-scale energy-storage applications [140]. Additionally, in 2021, Adriana–Petronela–Chiriac reported on the synthesis and investigation of imide-based materials for sensor applications. The study focused on novel PIs functionalized with benzo-15-crown-5 ether side chains, which were shown to exhibit electrochemical activity and be capable of detecting alkali metal ions. Chiriac also presented evidence for the occurrence of a ground-state intermolecular charge transfer complex responsible for the PIs strong light emission [141]. In 2023, Soujanya et al. discussed a strategy to synthesize hypercross-linked ionic polymers (HIPs) with strong visible-light absorption and abundant ionic sites. The PDI-HIPs demonstrated good surface areas, strong visible-light absorption, and good CO_2 uptake despite their relatively low surface area. Goudar's study highlighted the potential of PDI-HIPs as a dynamic network with good uptake of water up to seven times their weight [142]. In conclusion, although there is a vast body of literature regarding perylene and polyimide-based CEMs, this review specifically highlights the aforementioned studies and their significant contributions to the field.

4.3.5 Perylene and Polyimide-based AEMs

Recently, researchers have investigated the combination of perylene and PIs to create perylene-based polyimide AEMs. Perylene-based polyimide AEMs have been synthesized using various methods, including electrospinning, solution casting, and in situ polymerization. Electrospinning involves the use of an electric field to create nanofibers from a polymer solution containing perylene. The resulting AEMs have a high surface area and improved ion transport properties. In the solution casting method, a polymer solution containing perylene is cast into a thin film. This method can produce AEMs with different thicknesses and shapes, depending on the desired application. Insitu polymerization involves the polymerization of monomers in the presence of perylene to create a perylene-based polyimide membrane. Perylene-based polyimide AEMs have shown promising results in terms of their ion conductivity, stability, and mechanical properties. They exhibit high hydroxide ion conductivity, good thermal stability, and low gas permeability. These properties make them attractive for use in electrochemical devices such as fuel cells and water electrolyzers. For example, in 2009, a study was conducted to develop a novel polyimide-based alkaline AEM for potential fuel cell applications. This study aimed to prepare and characterize a novel derivative of poly (ether-imide) and evaluate its properties as an alkaline AEM. The researchers successfully prepared a poly (ether-imide) based membrane, which was characterized for its ionic conductivity in water, a critical metric of its applicability in a liquid-fed direct methanol fuel cell. The study found that the poly (ether-imide) based membrane was stable at all concentrations of KOH at room temperature but not at elevated temperatures. The measured conductivity of the formed membrane ranged from 2.28 to 3.51×10^{-3} S/cm at room temperature. The results indicate that functionalized poly (ether-imide) has suitable conductivity, suggesting that it can be used as an alkaline AEM in fuel cell applications [143]. Later, in 2019, advancements in the performance and properties of ionenes were reported through rational molecular design strategies. Kathryn reported that the use of imidazole-amine molecules with asymmetric reactivity enabled the synthesis of new bis (imidazole) diimide monomers, which were then polymerized via the Menshutkin reaction, followed by anion exchange to various molecular species well known in the literature. The study found that the functional groups present within the cationic backbone as well as the anion type strongly influence both the thermal and organizational properties of these new ionenes [144]. Another study conducted in 2019 by Byeong designed and synthesized a poly (ether imide) membrane that has good chemical and mechanical stabilities. The alkalized PEI (A-PEI) membrane was fabricated by solution casting of chloromethylated PEI (CM-PEI) followed by quaternization and alkalization. The study found that the membranes with a high degree of chloromethylation (DC) exhibited elevated hydroxide ion conductivity in the range of 6.7–44.2 mS/cm at 90 °C under 100% relative humidity (RH). The hydrophilic–hydrophobic phase separation was verified by atomic force microscopy (AFM) and small angle X-ray scattering (SAXS) measurements. Chemical stability was evaluated by measuring the durability of membranes while they were soaked in oxidative and alkaline solutions at 60 °C for 200 hours. Overall, these studies provide valuable insights into the development of alkaline AEMs for fuel cell applications, highlighting the importance of rational molecular design strategies and the need for membranes with high conductivity and long-term chemical stability. These findings have important implications for the future development of sustainable and efficient energy sources [145].

It should be noted that there are currently limited data available on the use of perylene-based polyimide AEMs in fuel cell applications. While some studies have reported promising results in terms of their ion conductivity, stability, and mechanical properties, there is still a need for further research to fully understand their potential for use in fuel cells [145]. One of the reasons why developing suitable alkaline AEMs for fuel cells is challenging is due to the harsh operating conditions within the fuel cell, such as high temperatures and exposure to corrosive chemicals. These conditions can lead to degradation of the membrane over time, resulting in decreased performance and shorter lifetimes of the fuel cell. Ongoing research in this field will be critical in advancing the development of fuel cells as a viable alternative energy source.

4.4 Polyacylation of SPEK-based IEMs

In our previous section, we highlighted the potential applications of polyethylene polyimide based IEMs in fuel cells and separation processes. However, to further improve the performance of IEMs, we suggest investigating the polyacylation of SPEK-based IEMs as a new avenue for their development. This can be categorized into polyacylation of SPEK-based CEMs and AEM. By introducing acyl groups into the backbone of SPEKs, the resulting materials can exhibit enhanced mechanical and chemical properties, making them even more suitable for various applications. These modified materials possess superior thermal stability, mechanical strength, and chemical resistance compared to unmodified SPEKs. Thus, the polyacylation of SPEK-based IEMs is a promising research area that could provide significant benefits for diverse applications. Further exploration of this area will enable us to discover more about the advancements in these materials and accelerate their development.

4.4.1 Polyacylation of SPEK-based CEMs

Polyacylation of SPEK-based CEMs is a promising research area that aims to enhance the performance of IEMs for various applications, including fuel cells. Recent studies have demonstrated the potential of polyacylation of SPEK-based CEMs in the development of high-performance membranes for fuel cells, exhibiting enhanced proton conductivity, thermal stability, and chemical resistance. For instance, an article published by Mingzhong in 2010 sheds light on the role of acylation in synthesizing and characterizing a novel type of soluble aromatic polymer known as poly(ether amide ether ketone ketone)s (PEAAEKKs). This paper details the electrophilic Friedel–Crafts copolycondensation reaction of N,N′-bis (4-phenoxybenzoyl)-m-phenylenediamine (BPBMPD) with a mixture of terephthaloyl chloride (TPC) and isophthaloyl chloride (IPC) using anhydrous $AlCl_3$ and N-methylpyrrolidone (NMP) in 1,2-dichloroethane (DCE). The results show that the synthesized PEAAEKKs are amorphous and possess high thermal stability and glass transition temperatures (Tgs) ranging from 220 to 231 °C. The incorporation of 1,3-dibenzoylaminobenzene moieties in the main chain enhances the solubility of the polymers. Furthermore, the synthesized polymers produce strong, transparent, and flexible films, with tensile strengths of 102.9–108.5 MPa, Young's moduli of 2.44–2.86 GPa, and elongations at break of 9.8–13.7%. The study offers an insightful exploration of PEAAEKK synthesis and characterization through acylation, highlighting their exceptional thermal and mechanical properties. Additionally, the paper underscores the potential applications of these materials in various fields, including coatings, membranes, and sensors. The authors did not examine the fuel cell performance of these materials; however, the research provides significant insights into the synthesis and characterization of PEAAEKKs, showcasing their outstanding thermal and mechanical properties. Overall, this study lays the groundwork for future research and advancements in the field of polymer synthesis and characterization, emphasizing the critical role of acylation in the development of new and promising materials [146].

In 2012, our group discovered the important role of polyacylation/acylation in the synthesis of a novel sulfonated aromatic diarene monomer, disodium 2,2′-di (sulfopropyloxy)-biphenyl (SBP), and its subsequent use in the synthesis of high-performance proton-exchange membranes for fuel cells. Our research showed that polyacylation directly from preionized monomers is a versatile approach to synthesizing side-chain-type sulfonated aromatic polyketones (SCT-SAPs) with higher proton conductivity and lower swelling ratio than Nafion®. This work represents a significant contribution to the field of fuel cell research and offers a new avenue for the synthesis of high-performance proton-exchange membranes. The synthesis of SBP was achieved via a simple and scalable one-step etherification reaction, and SBP was subsequently polymerized with 2,2′-dimethoxybiphenyl and 4,4′-oxydibenzoic acid using methanesulfonic acid (PPMA) as a catalyst and solvent. This is the first example of the synthesis of aromatic PEMs using sulfonated monomers via polyacylation [38]. A "one-pot acylation/heterocyclization copolycondensation" approach was proposed by our group in 2013, which allowed for orthogonal tuning of sulfonate groups and heterocyclic rings in the synthesis of side-chain-type sulfonated poly(ether ketone/ether benzimidazole) (SPEKEBI). This approach involved simultaneous acylation of sulfonated monomers and benzimidazolization in a single step, resulting in a molar ratio of 9/1 for side-chain sulfonate groups and main-chain benzimidazole rings. The corresponding membrane demonstrated a high proton conductivity of up to 55 mS/cm at 30 °C, surpassing that of conventional sulfonated benzimidazole-containing polymers. These findings highlight the potential of this approach for use in PEMFC [9].

In 2013, a study was conducted to synthesize sulfonated poly(aryl ether sulfone) copolymers (SPSFs) with phenyl pendant groups containing sulfonic acid groups on the backbone. The degree of sulfonation (DS) was controlled by altering the feed ratios of sulfonated to unsulfonated monomers. Postpolymerization reactions were used to produce a series of cross-linked SPSFs (CSPSFs) using 4,4′-thiodibenzoic acid (TDA) as a cross-linker. The carboxylic acid groups in TDA underwent Friedel–Craft acylation with the phenyl pendant rings in SPSFs to prepare PEMs for fuel cell applications. The study investigated the effect of DS and cross-linking density on the thermal, mechanical, and proton conductivity properties of the membranes.

The results showed that increasing the DS and temperature led to higher proton conductivity due to an increase in water uptake and enlarged ion channels. The cross-linking density had a slight negative effect on the proton conductivity of the membranes. The oxidative stability of the membranes improved with increasing cross-linking density, and the methanol permeability decreased as the water uptake decreased. The selectivity of the membranes increased with increasing cross-linking density and showed the highest value at 20%. However, excessive cross-linking density resulted in a decrease in proton conductivity and methanol permeability [147]. In 2013, we reported significant progress in the synthesis of poly (ether ketone)s that incorporated sulfonate groups, known as SPEK-x/y/z. Polyacylation and acylation techniques were utilized to copolymerize the presulfonated monomer SBP with two isomeric AB-type self-condensable comonomers, p-POBA and m-POBA. The resulting ionomers were then used in PEMs with high strength and precisely controlled IEC. Our study demonstrated that p-POBA exhibited better dimensional stability and proton conductivity at elevated temperatures above 60 °C than m-POBA, highlighting the isomeric effects of the polymer structural units. By optimizing IEC, we fine tuned the properties of the PEMs, with SPEK-1.0/2.2/0 showing acceptable swelling, high proton conductivity, and lower methanol permeability than commercial Nafion 115, making it a potential candidate for DMFC [148]. We explored further in different studies that the synthesis and properties of SPEK membranes, such as one of our studies published in 2014, focused on polyacylation and the synthesis of SPEKEBI membranes with side-chain structures using Eaton's reagent as a solvent and catalyst. To control the sulfonation content in relation to imidazole units, we varied the ratios of 3,3′,4,4′-tetraaminobiphenyl (TABP) and SBP. The resulting membranes showed excellent mechanical properties, thermal stability, high proton conductivity, and low methanol permeability. The study showed that the SCT-SPBI membrane exhibited better chemical and mechanical stability than previously studied materials. However, further refinement of its properties is necessary for specific applications. The oxidative stability of the membrane in Fenton's reagent (3% H_2O_2 containing $FeSO_4$ at pH 3) showed that the sulfonation content in the membrane influenced its weight retention. The SPEKEBI-4 membrane demonstrated a proton conductivity of 136 mS/cm at 30 °C, which was much higher than that of the Nafion® 115 membrane. Moreover, the methanol permeabilities of all the membranes were lower than that of the Nafion® 115 membrane. Passive DMFCs with a SPEKEBI-2 membrane exhibited a maximum power density of 39.3 mW/cm^2 at 25 °C, indicating its potential for use in PEM fuel cells [149].

Later, in 2017, we reported the successful incorporation of γ-(2,3-epoxypropoxy) propyltrimethoxysilane (KH-560) into SPEKEBI in various amounts as a cross-linker to create membranes with high ion exchange capacities and excellent performance for DMFCs. The study involved systematic measurements, such as Fourier transform infrared (FT-IR), scanning electron microscopy-energy-dispersive and X-ray photoelectron spectroscopy (XPS), to confirm the complete disappearance of epoxy groups in KH-560 and the existence of Si in the membranes. Compared to the unmodified SPEKEBI membrane, the resulting membranes showed increased mechanical strength and thermal stability, with a decrease in methanol permeability leading to an increase in the relative selectivity of SPEKEBI-x-SiO_2 membranes. Furthermore, the SPEKEBI-2.5-SiO_2 membrane showed a much higher peak power density than the pure SPEKEBI membrane. The SPEKEBI-2.5-SiO_2 membrane displayed the lowest Ea, indicating efficient enhancement of proton conductivity with the appropriate addition of KH-560 in the SPEKEBI membranes. However, the SPEKEBI-7.5-SiO_2 membrane showed higher proton conductivity than the pristine SPEKEBI membrane at high temperatures due to its low swelling effect. To control methanol permeability, the introduction of the SiO_2 network was found to be a better approach than acid-base interactions, as the SPEKEBI polymer had a very high sulfonation degree [150]. Similarly, we also successfully synthesized sulfonated/quaternized aromatic polyelectrolytes with precisely controlled ionized group content using polyacylation of sulfonated/quaternized monomers based on 2,2′-dihydroxy-1,1′-binaphthyl (DHBN). Both proton exchange membranes (PEMs) and AEMs made from these polyelectrolytes exhibit exceptional properties, with proton conductivity higher than Nafion 115 for PEMs and OH^- conductivity comparable to Tokuyama A901 for AEMs. The AEMs also demonstrated the ability to withstand 60 days of aging in 1 mol/l NaOH at 60 °C without degradation. We discovered that using optically active (S)-DHBN instead of racemic DHBN in the synthesis process resulted in an enhancement in proton conductivity for PEMs, leading to the possibility of optically active IEMs. The ionized monomers were made from commercially available racemic DHBN and optically active (S)-DHBN via a one-step etherification reaction in very high yields (>90%) and successfully polymerized with comonomers at different ratios (i.e. y/x = 2.0, 1.5, and 1.0) in Eaton's reagent. The resultant sulfonated and quaternized polyelectrolytes were then made into tough PEMs and AEMs, respectively. The PEM of S4PEKBN-1.0 exhibited higher proton conductivity than the Nafion 115 membrane while maintaining a low swelling ratio of 24.4%, even at 80 °C. Meanwhile, AEMs of Q4PEKBN-y/x achieved high OH– conductivity (28.5–53.7 mS/cm) at 30 °C and demonstrated outstanding alkaline stability through an aging test of being immersed in 1 mol/l NaOH at 60 °C for 60 days. A preliminary comparison between S4PEKBN-1.0 and (S)-S4PEKBN-1.0 also suggested that chirality might enhance the properties of PEMs, creating new possibilities for optically active IEMs [151]. Next, we proceeded to expand the polyacylation reaction to utilize sulfonated poly(ether ether ketone ketone) (SPEEKK) containing sulfopropyloxy pendants and butane-1,4-diylbis(oxy) spacers. The process involved Friedel–Crafts polyacylation of 1,4-diphenoxybutane (DPB), diethyl-3,3′-([1,1′-biphenyl]-2,2′-diylbis(oxy)bis(propane-1-sulfonate)) (SBP-OEt), and TPC under mild conditions, classifying it as polyacylation. The resulting SPEEKK membrane demonstrated high proton conductivity, moderate water uptake, and high IEC due to its flexible butane-1,4-diylbis(oxy) spacers and hydrophilic sulfopropyloxy sidechains. To make the membrane, SPEEKK-x was dissolved in NMP, filtered, and concentrated at 100 °C before being cast onto a glass plate and evaporated at 80 °C for 48 hours. The membrane displayed an increase in water uptake and linear swelling ratio with temperature. Although SPEEKK-0.444 had higher water uptake due to its higher IEC than SPEEKK-0.400, SPEEKK-0.400 exhibited superior oxidative stability at both 25 and 80 °C and lower methanol permeability than Nafion® 115. The complex hydrophobic structures formed contributed to its reduced methanol permeability by disrupting methanol permeation pathways [152].

Recently, Cetina-Mancilla et al. reported in 2023 that aromatic polysulfides and polysulfones, which are sulfur-containing polymers, are important materials with various applications, such as high-performance engineering plastics, gas separation, ultrafiltration, nanofiltration, ion-exchange membranes, and materials for biomedical applications. These polymers are typically produced through high-temperature nucleophilic substitution reactions involving dichlorobenzene with sodium sulfide (PPS) and dichlorodiphenyl sulfone with bis-phenols in the presence of basic catalysts (PAES). In this study, a novel method was developed to synthesize poly (arylene sulfide) in nearly quantitative yields through one-pot, room temperature, metal-free, highly regioselective, and nonstoichiometric step-polymerizations of diphenyl sulfide (DPS) with 1,1,1-trifluoroacetone and isatin using a mixture of trifluoromethanesulfonic acid (TFSA) with CH_2C_{12}. The resulting polymers were soluble in common solvents. The oxidation of the poly (arylene sulfide) with H_2O_2 at room temperature led to the production of poly(arylene sulfoxide) and poly(arylene sulfones) with high oxidation selectivity. Upon heating, the sulfoxide groups in the poly(arylene sulfoxide)s were converted into sulfide groups, and oxygen was released, resulting in the formation of nanofoams with a uniform pore distribution. The study also demonstrated the controlled incorporation of hydroxymethyl-, carboxy-, and "click"-able allyl- and propargyl-functionalities by room temperature postmodification reactions of polymers bearing reactive oxindole groups [153]. Polyacylation of SPEK-based CEMs has shown great potential in developing high-performance membranes for fuel cells. The studies we discussed have utilized acylation to synthesize new polymeric materials that exhibit remarkable thermal, mechanical, and chemical properties. By combining sulfonated aromatic diarene monomers with simultaneous acylation of sulfonated and benzimidazolized monomers, high-performance PEMs have been created, which demonstrate excellent proton conductivity and low swelling ratios. Nonetheless, further research is still needed to enhance the performance of these membranes and investigate their potential uses in other areas.

4.4.2 Polyacylation of SPEK-based AEMs

Polyacylation of SPEK-based AEMs is a promising approach for the development of high-performance alkaline exchange membranes with enhanced chemical, thermal, and mechanical stability [151, 154]. The use of acylation has been shown to improve the hydroxide conductivity and reduce the water uptake of AEMs, resulting in membranes with improved performance and durability in fuel cell applications [155]. Recent studies have explored the use of various acylation reagents and conditions to optimize the properties of the resulting AEMs. This discussion will focus on the recent advancements and challenges in developing polyacylated SPEK-based AEMs for fuel cell applications. Starting with a study published in 2016 by Lai, the researchers synthesized AEMs for alkaline fuel cells using phenolphthalein-based poly (arylene ether sulfone nitrile)s with imidazolium groups on the side chains. To enhance dimensional stability, strong polar nitrile groups were introduced into the polymer backbone, and varying their content helped understand their impact on AEM morphology and properties. All AEMs exhibited well-defined microphase-separated structures, with imidazolium groups aggregating with an increase in nitrile group content, resulting in larger and more interconnected ionic domains. The AEMs' ionic conductivity improved with increasing nitrile group content and temperature, displaying excellent conductivity to IEC or swelling ratio, good mechanical properties, and thermal and alkaline stability at 60 °C. The hydroxide conductivity of the ImPPESN-x membranes demonstrated high ionic conductivity across all temperatures. As the nitrile group content increased, the ImPPESN-x membranes had more significant and more interconnected ionic domains, creating efficient ion-conducting channels for higher conductivity and lower Ea value. Compared to other AEMs, ImPPESN-x membranes had higher hydroxide conductivity due to stronger conductive group aggregation and more efficient ion-conducting channels. The ImPPESN AEMs also exhibited excellent alkaline stability, retaining approximately 83% of ionic conductivity after 840 hours in a 2 M KOH solution at 60 °C [156]. In 2019, a study was conducted to develop a novel blend AEM for use in AEM fuel cells. The AEM contained a polymer of intrinsic microporosity (PIM), which had a loose packing structure and increased free volume, facilitating the formation of pathways for hydroxide ion transport. The AEM with 30% PIM showed a high conductivity of 16.9 mS/cm at 30 °C and a low IEC of 0.77 mmol/g, yielding an H_2/O_2 fuel cell power density of 163 mW/cm^2 at 60 °C. The membrane also demonstrated high mechanical robustness and alkaline stability compared to a membrane without PIM. The MmPSF/MmPIM-30% membrane exhibited the highest hydroxide conductivity and improved alkaline stability due to the interaction between the benzonitrile and morpholinium groups. This study provides a new and effective strategy to promote hydroxide ion conduction in AEMs, paving the way for better-balanced conductivity and stability. THSBI, a spiro-bisindane derivative, was synthesized via a condensation reaction between 3-methyl catechol and acetone under harsh acidic conditions, which formed a spirocyclic intermediate followed by polyacylation to form the final product. The yield of the pale solid product was 75%. This key reaction in the synthesis of THSBI provides insights into the polyacylation of SPEK-based AEMs and the potential for developing high-performance AEMs for fuel cell applications [157]. In 2020, our group successfully synthesized quaternized monomers containing two oxyhexyltrimethylammonium (OHeTMA) groups via "direct polyacylation of quaternized monomers" using inexpensive Eaton's reagent as the catalyst and solvent, replacing the costly CF_3SO_3H. By exchanging the quaternized monomer's counterion Br^- with $PF6^-$, we obtained high molecular weight Q6PEKBP-y/x with precisely controlled contents of side-chain OHeTMA groups. These polyelectrolytes were then used to prepare AEMs via solution casting. The AEMs exhibited high OH^- and HCO_3^- conductivity at 30°C and showed no sign of degradation in an accelerated aging test in 1 mol/l NaOH at 60 °C for 90 days, as confirmed by ^1H NMR and GPC. The use of Eaton's reagent and the exchange of the quaternized monomer's counterion resulted in easy synthesis and outstanding membrane properties, making the AEMs of Q6PEKBP-y/x promising candidates for applications in alkaline fuel cells and water electrolyzers [8].

In 2021, Ziqin reported a new polymerization process that involves reacting two monomers, 4,4′-bis(phenoxy)benzophenone and 1-(4-bromobutyl)indoline-2, 3-dione, in the presence of a superacid catalyst. This process, known as superacid-catalyzed polyhydroxyalkylation, results in the formation of a poly(aryl ether ketone) with a spatial cardo structure, called BPAEK. The resulting polymer has a high number average molecular weight and good reactivity, as evidenced by its polydispersity coefficient. The BPAEK polymer is then used to create three AEMs through a quaternization reaction using trimethylamine, N-methyl-piperidine, and N,N,N′,N′,N″-pentaethylguanidine. The resulting QPAEK and PPAEK ionomers are analyzed by ^{1}H NMR and FT-IR to confirm successful quaternization, and elemental analysis is conducted to ensure complete quaternization and exchange of Br^- for OH^-. These AEMs demonstrate excellent mechanical properties, good alkaline stability, and high OH^- conductivity. The PPAEK membrane is then tested in an H_2/O_2 fuel cell system and exhibits a high peak power density, indicating that the polyacylation reaction is essential in preparing the high-performance AEMs used in this study [154]. In summary, the successful commercialization of alkaline fuel cells relies heavily on the advancement of high-performance AEMs. To this end, utilizing polyacylation in SPEK-based AEMs has yielded positive outcomes, such as improved chemical, thermal, and mechanical stability, enhanced hydroxide conductivity, and reduced water absorption. Developing AEMs with desirable characteristics, such as low IEC, high conductivity, mechanical durability, and alkaline resistance, is crucial for achieving well-balanced conductivity and stability in fuel cell applications.

4.5 Polyacylation/Polyacylated Crown Ether IEMs

Crown ethers are macrocyclic polyethers capable of selectively binding metal ions through noncovalent interactions [158, 159]. They are named based on the number and type of atoms in their polyether ring, with larger rings being more selective. Crown ethers can be classified into two categories: aliphatic and aromatic. Polyacylation, a process of introducing multiple acyl groups onto the crown ether ring structure, can enhance their selectivity and stability for specific ions, making them ideal for ion separation and purification applications. The choice of crown ether backbone (aliphatic or aromatic) affects their properties, including ring size, flexibility, and metal ion binding ability. The type of crown ether used is a crucial factor in designing high-performance ion separation membranes. In this section, we discuss the synthesis of a series of acyl derivatives and the process of polyacylation of crown ethers, which have the potential to be used as membranes.

4.5.1 Acylation of Crown Ether

Acylation of crown ether refers to the process of adding an acyl group (RCO–) to a crown ether molecule. Acylation of crown ethers is a powerful tool for modifying

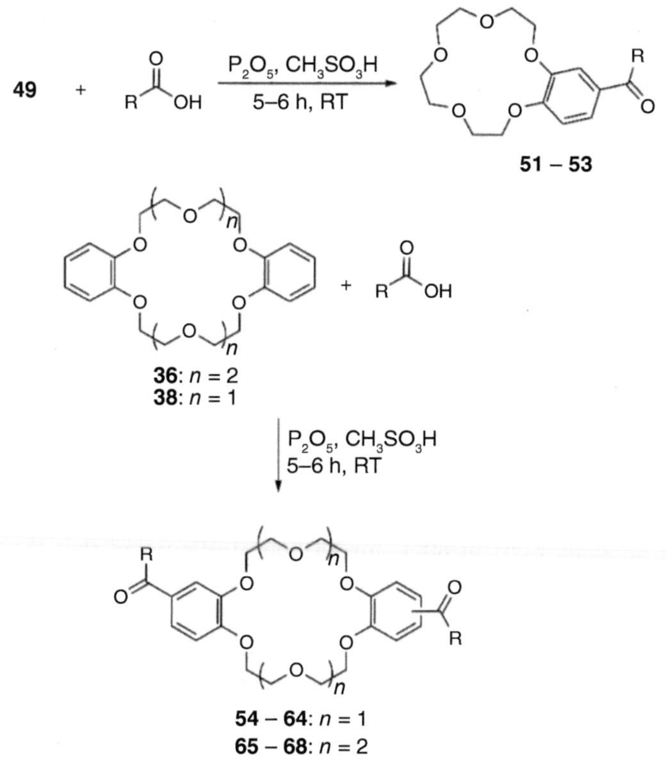

Figure 4.26 The process of using Eaton's reagent to acylate BCE and DBCE.

the properties of these important molecules and can be useful in a wide range of applications. For example, direct acylation can be used to modify benzene crown ether (BCE) and dibenzene crown ether (DBCE) by using carboxylic acids in the presence of Eaton's reagent or with carboxylic acids/anhydride and PPA. The first method was refined and displayed (Figure 4.26).

The process of acylation is widely used to modify the properties and functions of crown ethers, and polyacylation takes this technique to the next level. By attaching multiple acyl groups, polyacylation enables the creation of new materials with superior properties and specific functions. This technique is not only limited to crown ethers but also extends to membranes, offering the potential to develop advanced materials with customized functions for a broad range of applications. This section will focus on the fabrication of polymers via polyacylation and their use as membrane materials. Two categories of such membranes are poly-crown ether-based AEM and CEM. These membranes have shown extraordinary results.

4.5.2 Poly-Crown Ether-based AEM

In this section, we will discuss the significance, manufacturing process, and performance of poly-crown ether-based AEMs. These AEMs possess anionic functional groups that allow them to selectively conduct anions while blocking cations, which is a critical property for the operation of electrochemical devices that rely on ion separation [160]. They are highly regarded for their exceptional ion conductivity, robust mechanical properties, and remarkable chemical stability.

In this literature, we added some selected studies on the synthesis and characterization of novel polymers and copolymers containing crown ether units alternating with aliphatic units. We previously reported a new class of AEMs for use in AAEMFCs. The AEMs contain mobile ion shuttles, achieved by threading ionic linear guests into poly(crown ether) hosts via host-guest molecular interactions. This threading of guests into the hosts creates a comb-shaped polyrotaxane. The membrane was synthesized through polyacylation by reacting 1,4-bis (aminomethyl) benzene with methyl 4-formylbenzoate under reflux in an argon atmosphere for 24 hours, followed by reduction with lithium aluminum hydride ($LiAlH_4$). The resulting terminal hydroxyl and secondary amine groups were converted into bromomethyl and ammonium groups, respectively, through sequential reactions with hydrobromic acid (HBr) and ammonium hexafluorophosphate (NH_4PF_6). An accelerated alkaline stability test was performed under harsh conditions (4 M aqueous KOH at 80 °C for 98 hours). We noted that the sterically hindered headgroups might impede OH^- attack, but this effect might be weakened under such a harsh alkaline environment (4 m aqueous KOH at 80 °C). They suggest that one challenge for ion shuttle design will be to maintain high OH^- conductivity by preventing the diffusion loss of the shuttle in an alkali environment. More efforts will be directed toward improving the alkali stability by introducing more sterically hindered ammonium groups to impede OH– attack or replacing the ammonium cations with other more stable alkaline units [161]. The development of this new class of AAEMs is a significant advancement in the field of AAEM. Yang reported in 2019 the development of highly alkaline durable cross-linked AEMs using a sodium salt of 4,4′(5′)-di(hydroxybenzo)-18-crown-6 as the cross-linker and quaternary ammonium as the cation group. The AEMs showed promising potential for fuel cell applications, as demonstrated by their excellent ionic conductivity, ionic exchange capacity (IEC), mechanical properties, alkaline stability, and thermal stability. The best-performing membrane exhibited an ionic conductivity of 117 mS/cm at 80 °C and 100% RH, with only a 6.0% decrease after 480 h of testing in 2 M KOH at 80 °C. The addition of the crown ether cross-linker enhanced the ionic conductivity of the AEMs, with the highest ionic conductivity of 116.8 mS/cm achieved at 80 °C with a 10% crown ether content. The AEMs also showed good alkaline stability and mechanical properties, making them potentially useful for various applications, including fuel cells. The single-cell performance test indicated that the PPO-CE0.10-QA0.90 membrane has a strong potential for alkaline fuel cells. The durability test of the single cell using the PPO-QA and PPO-CE0.10-QA0.90 membranes demonstrated good durability of the single cell with PPO-CE0.10-QA0.90, owing to its decent alkaline stability under high pH conditions. Overall, the results suggest that the cross-linked AEMs have promising potential for fuel cell applications [162].

Our previous report, published in 2021, examined the difficulties associated with achieving high conductivity in AEMs. This is due to the limited mobility and rotation of tethered cationic groups in traditional AEMs. We introduced a new type of polyrotaxane AEM that employs free-shuttling phosphonium cations. This allowed for greater mobility and acceleration of the diffusion conduction of bulky anions. To create this new AEM, we dissolved dibenzo-24-crown-8 and compound 2 in $CH_3CN/CHCl_3$ and stirred the mixture at room temperature for 1 hour. We then added triphenylphosphine and allowed the reaction to continue for 24 hours. The resulting white precipitate was filtered and washed, and the remaining solid was dissolved in CH_2Cl_2. Et2O was added to this solution, resulting in a white precipitate that was filtered and washed to obtain rotaxane 6. Our polyrotaxane AEM displayed an exceptional HCO_3^- conductivity of 105 mS/cm at 90 °C, despite a lower IEC of 1.17 mmol/g. This highlights the potential of polyrotaxanes in developing highly conductive AEMs, which will stimulate further research into their applications in ion conduction processes [163] (Figures 4.27–4.29).

In 2022, Wang developed a new type of AEM for alkaline fuel cells and water electrolysis by incorporating hydrophilic crown ether units into poly(terphenyl piperidinium) (PDTP) through a superacid-catalyzed polycondensation process. The resulting PDTP membranes showed higher hydroxide conductivity and improved stability in alkaline conditions than the control poly(terphenyl piperidinium) (PTP) membranes. The highest hydroxide conductivity of 110 mS/cm was reached at 80 °C for the PDTP-10 membrane with an IEC of 2.57 mequiv./g. The introduction of hydrophilic crown ether alleviated the chemical degradation of PDTP membranes in 1 M NaOH at 80 °C compared to PTP membranes. The PDTP-10 membrane showed a peak power density of 621 mW/cm^2 at 60 °C in AEM fuel cells and a current density of 2000 mA/cm^2 at 2.1 V for water electrolysis. This study provides an effective strategy to design high-performance AEMs for alkaline fuel cells and water electrolysis by incorporating hydrophilic crown ether units into the polymer main chain [164]. Later, Chenshi in 2022 conducted research on AEMs and reported their synthesis and characterization. The new type of AEM, named B-CX%-P AEMs, was created by grafting dibenzo-18-crown-6 ether onto the polyvinyl alcohol matrix. The B-C30%-P AEM exhibited the highest conductivity among the tested membranes, with 235 mS/cm at 80 °C, and showed the highest power density, with 197 mW/cm^2. The synthesized membrane also demonstrated improved alkali stability and mechanical properties. The study demonstrated that grafting symmetrical bis-crown ether onto the polyvinyl alcohol matrix produced AEMs with excellent electrical conductivity and alkali stability. It also showed good alkali resistance stability and could maintain the stability of conductivity for a long time. The study highlights the importance of developing AEMs with improved properties to advance the field of fuel cell technology [165]. In 2023, Xu et al. reported on the synthesis of crown-ether block copolymer-based poly (isatin terphenyl) AEMs for electrochemical energy conversion devices. The polymers were synthesized using dibenzo-18-crown-6, terphenyl, and isatin via the acid-catalyzed polycondensation reaction and then quaternized by grafting (6-bromohexyl)-trimethylammonium bromides via the amide sites of isatin. The crown ethers in the polymer endowed the membrane with suitable water uptake, reasonable swelling, and interconnected ion-conducting channels, resulting in

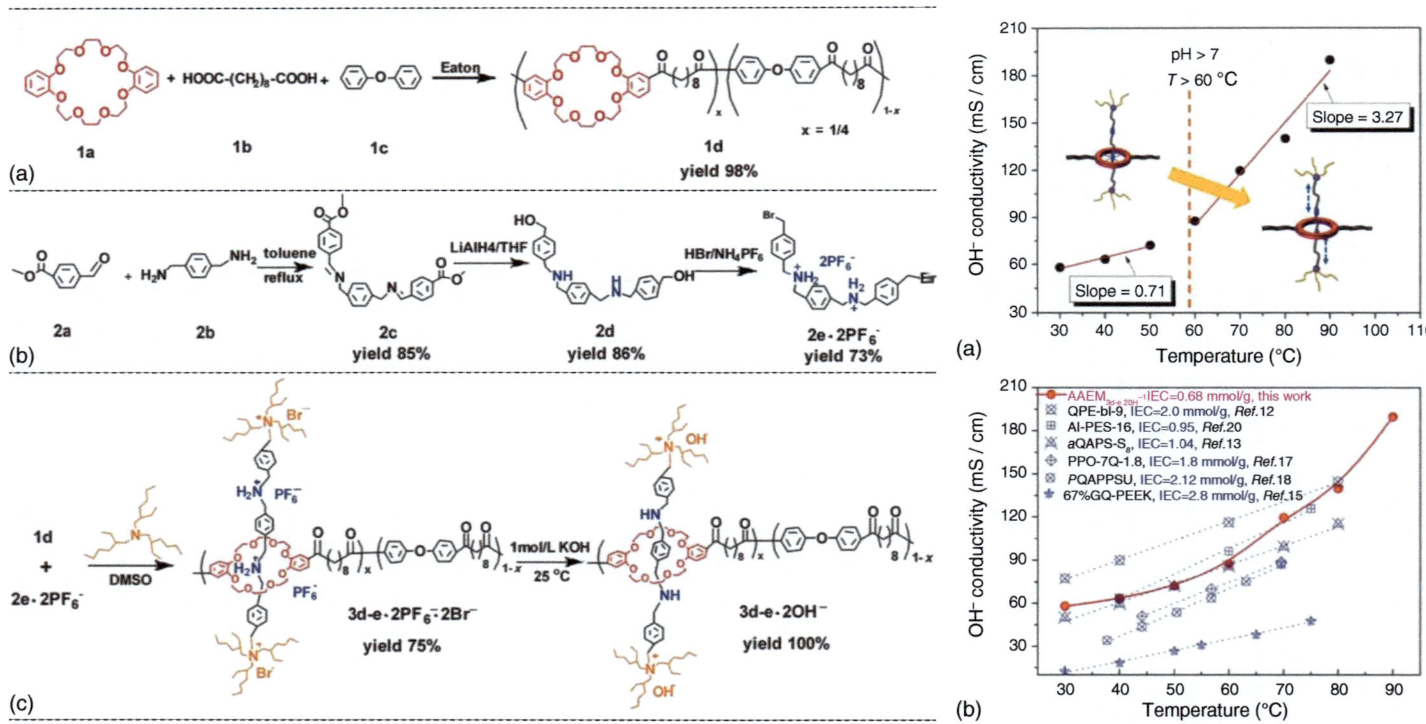

Figure 4.27 (a) Synthesizing AEMs based on poly(crown ether) with a precursor for guest side chains and a comb-shaped polyrotaxane that includes mobile ion shuttles, (b) investigating the OH^- conductivities of AAEM3d-e•2OH^- at different temperatures and comparing these results to the OH^- conductivities of reported comb-shaped AAEMs at varying temperatures.

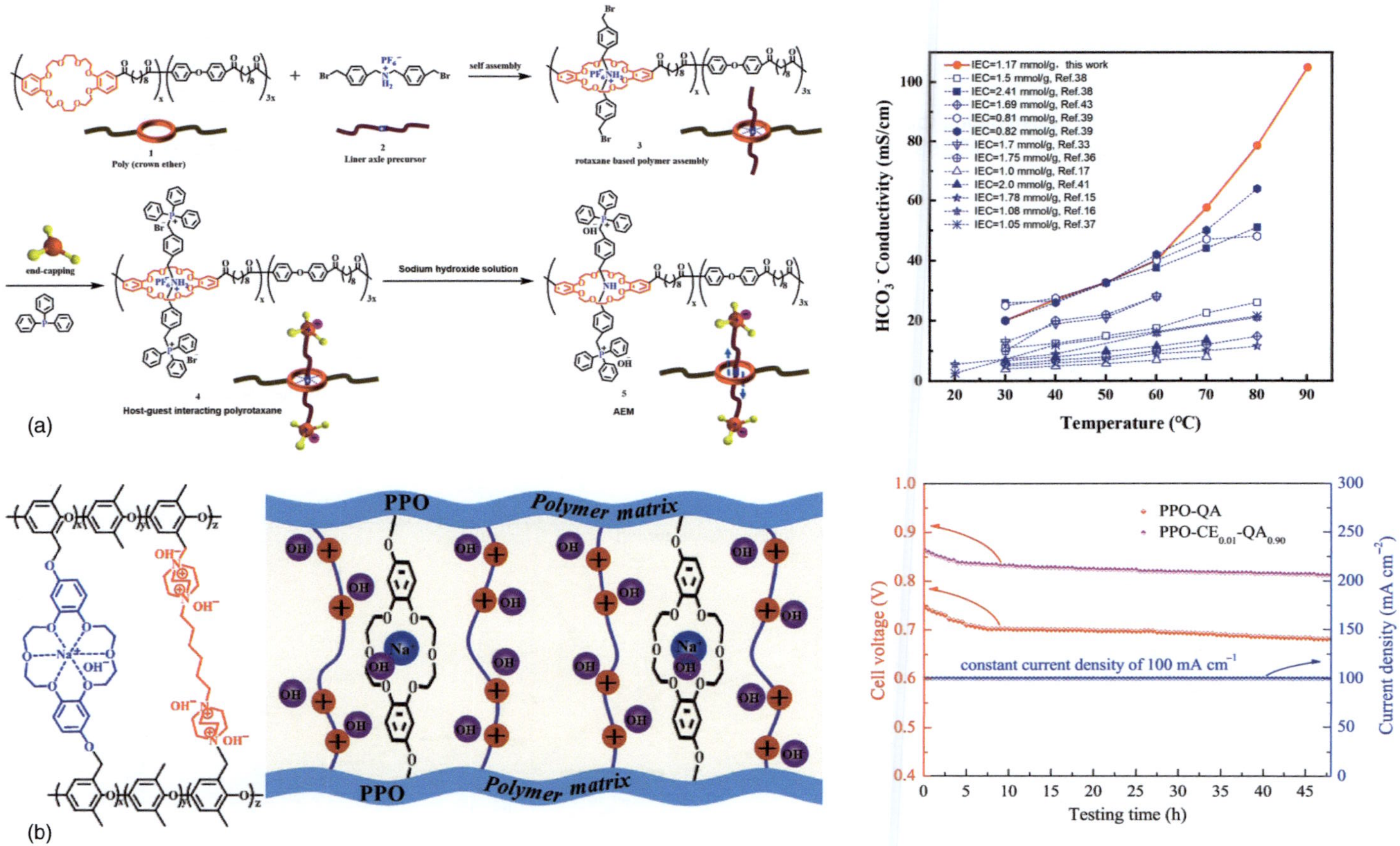

Figure 4.28 (a) Illustration of ion conduction in the conventional tethered AEM and polyrotaxane AEM, (b) comparison of temperature-dependent conductivity, (c) chemical structure and sketch of crosslinked AEMs, and (d) H_2/O_2 single-cell polarization curve.

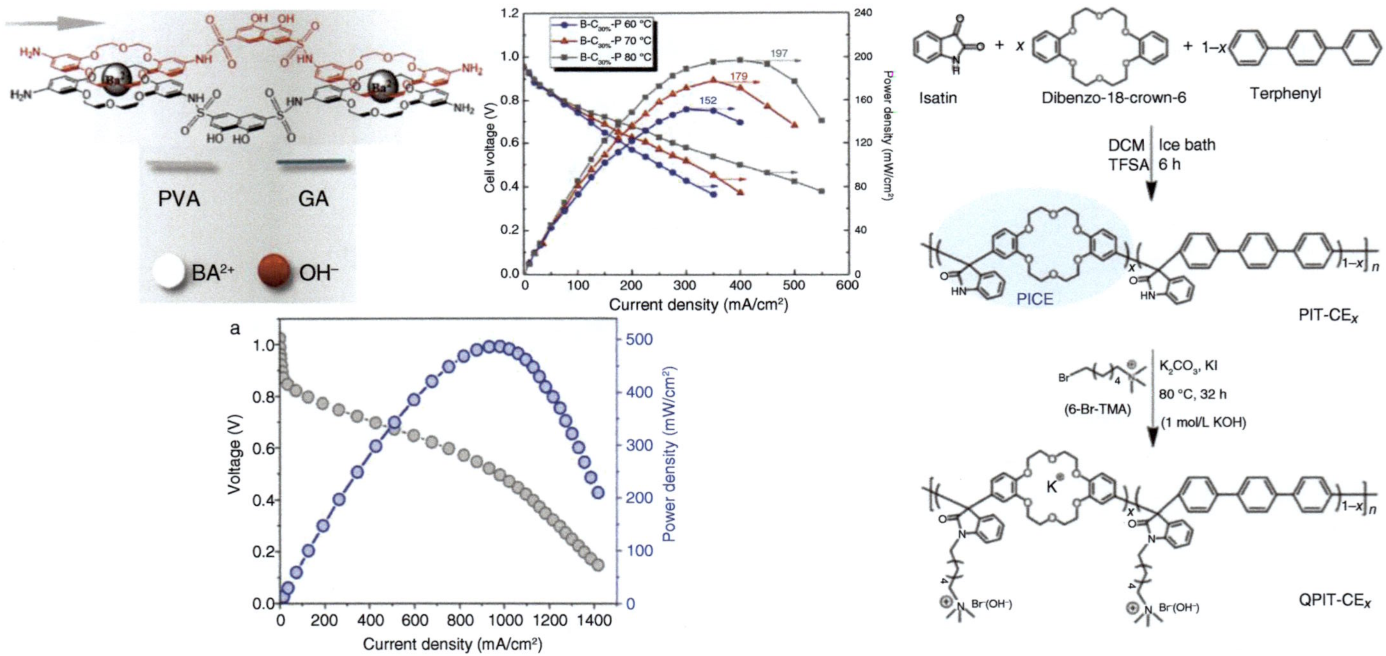

Figure 4.29 (a) A PVA-based anion exchange membrane (AEM) grafted with bis-crown ether and its single-cell performance at various temperatures, (b) preparation of QPIT-CEx ($0 \leq x \leq 0.5$) and PICE membranes, and (c) the fuel cell's performance with humidified H_2 and O_2 at a flow rate of 1 l/min at 60 °C without backpressure.

a hydroxide conductivity as high as 112 mS/cm and a bromide ion conductivity of 74 mS/cm at 80 °C. The membrane displayed acceptable stability, retaining 85% of the initial hydroxide conductivity after being immersed in 1 mol/l KOH at 80 °C for 1200 hours. The fuel cell fabricated with the prepared membrane demonstrated a peak power density of 487 mW/cm^2 at 60 °C with 71% of its initial voltage retained after operating at 150 mA/cm^2 for 40 hours. The presence of hydrophilic crown ethers was expected to benefit the formation of a suitable microphase morphology for anion conduction. This study introduced a new synthesis method for crown-ether block copolymer-based AEMs and evaluated their fuel cell performance, conductivity, stability, and the role of polyacylation [166].

4.5.3 Poly-crown Ether-based Noncharged Selective Membrane (PCENS-M)

Poly-crown Ether-based Noncharged Selective Membrane (PCENS-M) refers to a type of polymer material that contains crown ether units within its chemical structure and contains several ether groups, which can interact with metal ions through a process called complexation [167, 168] (Figure 4.30). In the case of PCENS-M, the material is designed to have a high affinity for specific metal ions, allowing it to selectively extract and separate them from a mixture. This property makes PCENS-M

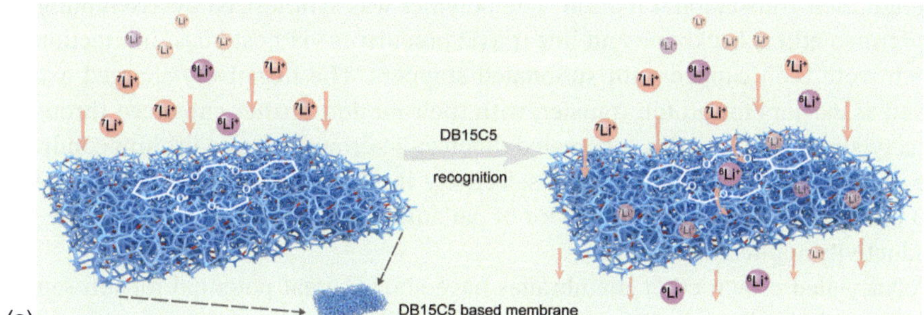

Figure 4.30 Preparation of dibenzo-15-crown-5-based lithium conductive membranes, namely, PIM-DB15C5-TB and Blend-DB15C5. (a) The design of a $^6Li^+/^7Li^+$ selective membrane having lithium-specific dibenzo-15-crown-5 units and abundant mass transport channels. The synthetic route and membrane formation procedure of PIM-DB15C5-TB (b) and Poly-acyl-DB15C5 (d). Molecular models of PIM-DB15C5-TB (c) and Poly-acyl-DB15C5 (e) [18]. Xinyue Huang et al. 2016/With permission of Elsevier.

useful in various applications, including metal ion sensing, catalysis, and wastewater treatment [169, 170]. In 2006, Zolotukhin et al. published a paper describing a one-step method for synthesizing these polymers using Friedel–Crafts chemistry in Eaton's reagent medium, followed by reduction of the carbonyl groups. The resulting polymers are soluble in common organic solvents and can be cast into flexible transparent films, making them useful for a variety of applications. They also discussed different methods for incorporating crown ethers into polymer matrixes, providing valuable information for researchers in the field. According to the authors, these polymers are suitable materials for investigating the effects of changes in size and electron-donating and electron-withdrawing capacity of crown ethers on their performance in membrane transport of ionic substances. This study presents a simple and reliable method for synthesizing polymers containing crown ether units and provides valuable insights into their potential applications in various fields [171] (Figure 4.31).

We recently presented a new type of polymer composed of rotaxanes that utilize thermally triggered translational motion to promote proton transfer. These rotaxanes consist of a ring threaded over an axle, featuring sulfonated stoppers that allow for movement relative to the axle at a temperature of approximately 60 °C. The resulting polymer exhibits a high proton conductivity of 260.2 mS/cm, with a relatively low ion-exchange capacity of 0.73 mmol/g. This method differs from conventional proton transfer polymers, as the rotaxanes promote an enhanced proton transfer rate through their translational motion. The polymer was synthesized by assembling a poly(crown ether) backbone and linear axle precursors via host-guest interactions, which were then capped with sulfonated stoppers. The negatively charged axles served as carriers for proton transfer, with their motion further enhanced through N-acetylation. Additionally, we created dibenzo-24-crown-8-based lithium conductive membranes and assessed their selectivity in electrodialysis experiments. We also investigated the protode behavior of palladium hydride for its potential use in conductivity applications [172].

Polyacylated crown ether membranes have shown great potential for ion separation and fuel cell applications due to their excellent ion transport properties and stability in harsh environments. The acylation of crown ethers provides a means to modify their chemical and physical properties, including their hydrophobicity and ion selectivity. These modifications result in membranes with improved selectivity and permeability for specific ions, making them ideal for use in separation processes. Additionally, the proton conductivity of these membranes can be enhanced by the introduction of acidic functional groups, making them attractive candidates for use as electrolytes in fuel cells. Overall, polyacylated crown ether membranes have shown promising results in both ion separation and fuel cell applications, and further research is needed to fully understand their potential in these fields.

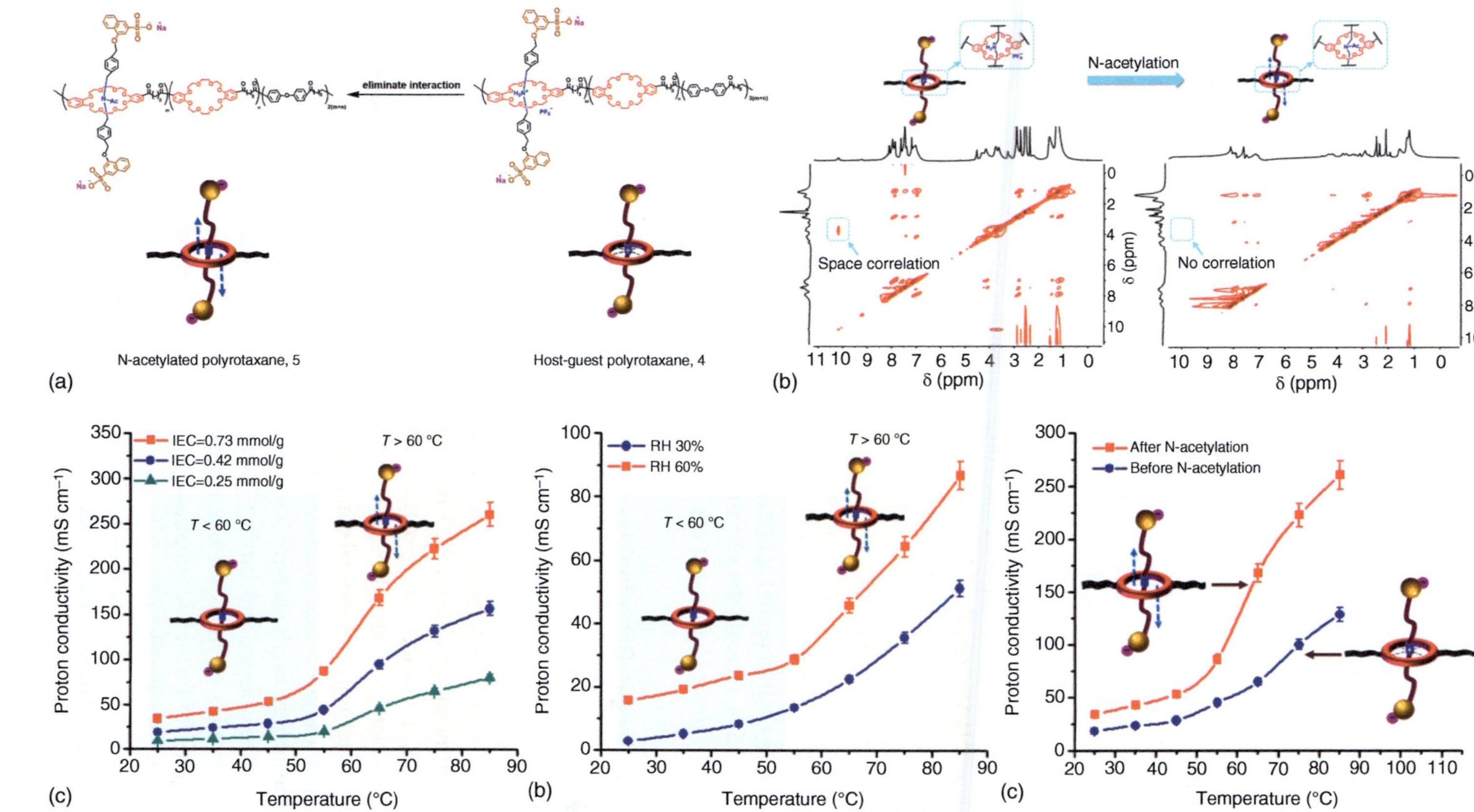

Figure 4.31 (a) Shows that the target polymer is obtained through the N-acetylation of 4. NOSEY spectra of polyrotaxane 4 and N-acetylated polyrotaxane 5, (b) NOSEY spectra of polyrotaxane 4 and N-acetylated polyrotaxane 5. (c) proton conductivity of polyrotaxane membranes before and after N-acylation.

4.6 Conclusion

The above sections provide a comprehensive review of polyacylation, a versatile method used to modify the properties of polymeric materials. Polyacylation involves introducing acyl groups onto polymer chains through a chemical reaction, which results in improved physical, chemical, and thermal properties. One class of polymeric materials that has been successfully modified through polyacylation is perylene-based aliphatic PIs. The introduction of acyl groups onto the polymer chains results in improved thermal stability, solubility, and mechanical properties, making them suitable for various applications, including cation and AEMs for separation and fuel cell applications. SPEK-based IEMs are another class of polymeric materials that have been modified through polyacylation. The introduction of acyl groups onto the sulfonic acid groups of SPEK improves their chemical stability and water uptake, resulting in improved IEC and selectivity. Similar to perylene-based aliphatic PIs, these membranes can be used as cation and AEMs for separation and fuel cell applications. Polyacylation has also been used to modify crown-based IEMs, resulting in improved thermal and chemical stability, as well as enhanced IEC and selectivity. In summary, polyacylation is a promising method for modifying the properties of polymeric materials. By improving their thermal stability, chemical resistance, solubility, and mechanical properties, these materials can be tailored for various applications. Future research in this field should focus on developing new polymeric materials with unique properties through polyacylation and exploring their potential applications in fields such as energy, the environment, and biomedicine. Overall, the versatility of polyacylation makes it a valuable tool in the field of polymer science and engineering.

4.6.1 Challenges/Opportunities for Further Development

Despite the excellent properties and performance of polyacylated PEM, challenges remain that need to be addressed for their further development and optimization. One challenge is the optimization of the degree of polyacylation, which affects the properties and performance of the membranes. The degree of polyacylation can be controlled by adjusting the reaction conditions, such as the reaction time, temperature, and concentration of the acylating agent. However, the optimal degree of polyacylation may vary depending on the application, and further studies are needed to determine the optimal conditions for each application. Another challenge is the optimization of the synthesis methods for the production of polyacylated PEM. The current methods have limitations in terms of scalability, reproducibility, and cost-effectiveness, and further research is needed to develop more efficient and cost-effective methods for the synthesis of these membranes. There are also opportunities for further development and optimization of polyacylated PEM. One opportunity is the development of new applications for these membranes, such as in sensors, actuators, and catalysis. The unique properties of these membranes make them suitable for these applications, and further research is needed to explore their potential. Another opportunity is the development of new materials and composites based on polyacylated PEM. These materials can have enhanced properties and performance, such as improved ion conductivity, mechanical strength, and thermal stability, and can be tailored for specific applications.

4.6.2 Outlook for the Future of Polyacylation in Membrane Research

Polyacylation has shown great potential in membrane research, and the development of polyacylated PEM has opened up new opportunities for the development of advanced materials for various applications. The unique properties and performance of these membranes make them suitable for a wide range of applications, and further research is needed to explore their full potential. In the future, it is expected that the optimization of the synthesis methods and the degree of polyacylation will lead to the development of more efficient and cost-effective polyacylated PEM. This will enable their wider use in various applications, including energy storage, water treatment, and drug delivery. Furthermore, the development of new materials and composites based on polyacylated PEM will lead to the development of advanced materials with enhanced properties and performance. These materials can be tailored for specific applications and can have a significant impact on various fields, such as energy storage, biomedical engineering, and environmental science.

References

1 Wang, Y.-J., Qiao, J., Baker, R., and Zhang, J. (2013). Alkaline polymer electrolyte membranes for fuel cell applications. *Chemical Society Reviews* 42 (13): 5768–5787.

2 Zhang, H. and Shen, P.K. (2012). Recent development of polymer electrolyte membranes for fuel cells. *Chemical Reviews* 112 (5): 2780–2832.

3 Hickner, M.A., Herring, A.M., and Coughlin, E.B. (2013). Anion exchange membranes: current status and moving forwards. *Journal of Polymer Science* 51 (24): 1727–1735.

4 Kim, Y.S. (2021). Polymer electrolytes with high ionic concentration for fuel cells and electrolyzers. *ACS Applied Polymer Materials* 3 (3): 1250–1270.

5 Song, C. (2002). Fuel processing for low-temperature and high-temperature fuel cells: challenges, and opportunities for sustainable development in the 21st century. *Catalysis Today* 77 (1–2): 17–49.

6 Merle, G., Wessling, M., and Nijmeijer, K. (2011). Anion exchange membranes for alkaline fuel cells: a review. *Journal of Membrane Science* 377 (1–2): 1–35.

7 Zhang, Z., Wu, L., Varcoe, J. et al. (2013). Aromatic polyelectrolytes via polyacylation of prequaternized monomers for alkaline fuel cells. *Journal of Materials Chemistry A* 1 (7): 2595–2601.

8 Zhang, Z. and Xu, T. (2020). Facile synthesis of poly (arylene ether ketone) s with pendent oxyhexyltrimethylammonium groups for robust anion exchange membranes. *Polymer* 210: 123035.

9 Zhang, Z. and Xu, T. (2013). One-pot acylation/benzimidazolization copolymerization approach to side-chain-type proton conductive membranes. *Journal of Membrane Science* 446: 121–124.

10 Lin, H., Gao, W., Meng, F. et al. (2012). Membrane bioreactors for industrial wastewater treatment: a critical review. *Critical Reviews in Environmental Science and Technology* 42 (7): 677–740.

11 Fane, A., Wang, R., and Jia, Y. (2011). Membrane technology: past, present and future. In: *Membrane and Desalination Technologies*, 1–45. Springer.

12 Singh, R. (2014). *Membrane Technology and Engineering for Water Purification: Application, Systems Design and Operation*. Butterworth-Heinemann.

13 Malaeb, L. and Ayoub, G.M. (2011). Reverse osmosis technology for water treatment: state of the art review. *Desalination* 267 (1): 1–8.

14 Razali, M., Kim, J.F., Attfield, M. et al. (2015). Sustainable wastewater treatment and recycling in membrane manufacturing. *Green Chemistry* 17 (12): 5196–5205.

15 Ang, W.L., Mohammad, A.W., Johnson, D., and Hilal, N. (2019). Forwards osmosis research trends in desalination and wastewater treatment: a review of research trends over the past decade. *Journal of Water Process Engineering* 31: 100886.

16 Lutkenhaus, J.L. and Hammond, P.T. (2007). Electrochemically enabled polyelectrolyte multilayer devices: from fuel cells to sensors. *Soft Matter* 3 (7): 804–816.

17 Hossain, S., Abdalla, A.M., Suhaili, S.B.H. et al. (2020). Nanostructured graphene materials utilization in fuel cells and batteries: a review. *Journal of Energy Storage* 29: 101386.

18 Huang, X., Li, X., Li, Y. et al. (2016). Trace dissolved ammonia sensor based on porous polyelectrolyte membrane-coated thin-core fibre modal interferometer. *Sensors and Actuators B: Chemical* 226: 7–13.

19 Yuan, W., Weng, G.M., Lipton, J. et al. (2020). Weak polyelectrolyte-based multilayers via layer-by-layer assembly: approaches, properties, and applications. *Advances in Colloid and Interface Science* 282: 102200.

20 Yee, R.S.-L., Rozendal, R.A., Zhang, K., and Ladewig, B.P. (2012). Cost effective cation exchange membranes: a review. *Chemical Engineering Research and Design* 90 (7): 950–959.

21 Wang, W., Luo, Q., Li, B. et al. (2013). Recent progress in redox flow battery research and development. *Advanced Functional Materials* 23 (8): 970–986.

22 Ye, Y. and Elabd, Y.A. (2012). Chemical stability of anion exchange membranes for alkaline fuel cells. In: *Polymers for Energy Storage and Delivery: Polyelectrolytes for Batteries and Fuel Cells*, 233–251. ACS Publications.

23 Chempath, S., Einsla, B.R., Pratt, L.R. et al. (2008). Mechanism of tetraalkylammonium headgroup degradation in alkaline fuel cell membranes. *Journal of Physical Chemistry C* 112 (9): 3179–3182.

24 Qiu, B., Lin, B., Si, Z. et al. (2012). Bis-imidazolium-based anion-exchange membranes for alkaline fuel cells. *Journal of Power Sources* 217: 329–335.

25 Lu, C., Long, C., Li, Y. et al. (2020). Chemically stable poly (meta-terphenyl piperidinium) with highly conductive side chain for alkaline fuel cell membranes. *Journal of Membrane Science* 598: 117797.

26 Kuroki, H., Miyanishi, S., Sakakibara, A. et al. (2019). Highly stable membrane–electrode assembly using ether-linkage-free spirobifluorene-based aromatic polyelectrolytes for direct formate solid alkaline fuel cells. *Journal of Power Sources* 438: 226997.

27 Rager, T., Schuster, M., Steininger, H., and Kreuer, K.-D. (2007). Poly (1, 3-phenylene-5-phosphonic acid), a fully aromatic polyelectrolyte with high ion exchange capacity. *Advanced Materials* 19 (20): 3317–3321.

28 Argun, A.A., Ashcraft, J.N., and Hammond, P.T. (2008). Highly conductive, methanol resistant polyelectrolyte multilayers. *Advanced Materials* 20 (8): 1539–1543.

29 Sailaja, G., Miyanishi, S., and Yamaguchi, T. (2015). A durable anion conducting membrane with packed anion-exchange sites and an aromatic backbone for solid-state alkaline fuel cells. *Polymer Chemistry* 6 (46): 7964–7973.

30 Ran, J., Wu, L., Ru, Y. et al. (2015). Anion exchange membranes (AEMs) based on poly(2,6-dimethyl-1,4-phenylene oxide) (PPO) and its derivatives. *Polymer Chemistry* 6 (32): 5809–5826.

31 Zhang, Q., Li, S., and Zhang, S. (2010). A novel guanidinium grafted poly(aryl ether sulfone) for high-performance hydroxide exchange membranes. *Chemical Communications* 46 (40): 7495–7497.

32 Wu, L., Xu, T., and Yang, W. (2006). Fundamental studies of a new series of anion exchange membranes: membranes prepared through chloroacetylation of poly(2,6-dimethyl-1,4-phenylene oxide) (PPO) followed by quaternary amination. *Journal of Membrane Science* 286 (1): 185–192.

33 Yao, Z., Zhang, Z., Hu, M. et al. (2018). Perylene-based sulfonated aliphatic polyimides for fuel cell applications: performance enhancement by stacking of polymer chains. *Journal of Membrane Science* 547: 43–50.

34 Velema, W.A. and Kool, E.T. (2020). The chemistry and applications of RNA 2′-OH acylation. *Nature Reviews Chemistry* 4 (1): 22–37.

35 Zhang, H., Liang, S., Wei, D. et al. (2022). Electrocatalytic generation of acyl radicals and their applications. *European Journal of Organic Chemistry* 2022 (39): e202200794.

36 Omae, I. (2004). Intramolecular five-membered ring compounds and their applications. *Coordination Chemistry Reviews* 248 (11–12): 995–1023.

37 Kumar, P., Sharma, U., and Ananthnag, G.S. (2022). The story of acyl phosphines: synthesis, reactivity, and catalytic applications. *Applied Organic Chemistry* 36 (6): e6672.

38 Zhang, Z., Wu, L., and Xu, T. (2012). Novel aromatic proton-exchange polyelectrolytes via polyacylation of presulfonated monomers. *Journal of Materials Chemistry* 22 (28): 13996–14000.

39 Ran, J., Wu, L., He, Y. et al. (2017). Ion exchange membranes: new developments and applications. *Journal of Membrane Science* 522: 267–291.

40 Li, X., Bu, F., Zhang, H., and Zhao, C. (2020). Facile synthesis of poly (arylene ether ketone) s containing flexible sulfoalkyl groups with enhanced oxidative stability for DMFCs. *International Journal of Hydrogen Chemistry* 45 (51): 27632–27643.

41 Deng, Y., Zhang, S., Lu, G., and Huang, X. (2013). Constructing well-defined star graft copolymers. *Polymer Chemistry* 4 (5): 1289–1299.

42 Dickerson, T.J., Reed, N.N., and Janda, K.D. (2002). Soluble polymers as scaffolds for recoverable catalysts and reagents. *Chemical Reviews* 102 (10): 3325–3344.

43 Kirschning, A., Monenschein, H., and Wittenberg, R. (2001). Functionalized polymers—emerging versatile tools for solution-phase chemistry and automated parallel synthesis. *Angewandte Chemie (International Ed. in English)* 40 (4): 650–679.

44 Ollivier, N., Behr, J.B., El-Mahadi, Q. et al. (2005). Fmoc solid-phase synthesis of peptide thioesters using an intramolecular N, S-acyl shift. 7 (13): 2647–2650.

45 Dryland, A. and Sheppard, R.C. (1986). Peptide synthesis. Part 8. A system for solid-phase synthesis under low pressure continuous flow conditions. *Journal of the Chemical Society, Perkin Transactions* 1: 125–137.

46 Mu, S., Guo, J., Yu, C. et al. (2015). A novel solid–solid phase change material based on poly (styrene-co-acrylonitrile) grafting with palmitic acid copolymers. *Journal of Macromolecular Science, Part A* 52 (8): 617–624.

47 Begum, S., Zehra, S.Q., and Siddiqui, B.S. (2006). Mild and efficient solvent-free Friedel–Crafts acylation of reserpine. *Synthetic Communications* 36 (21): 3203–3224.

48 Stolle, A. and Ranu, B. (2014). *Ball Milling Towards Green Synthesis: Applications, Projects, Challenges*. Royal Society of Chemistry.

49 Troschke, E., Grätz, S., Lübken, T., and Borchardt, L. (2017). Mechanochemical Friedel–Crafts alkylation—a sustainable pathway towards porous organic polymers. *Angewandte Chemie International Edition* 56 (24): 6859–6863.

50 Štrukil, V. and Sajko, I. (2017). Mechanochemically assisted solid-state photocatalysis (MASSPC). *Chemical Communications* 53 (65): 9101–9104.

51 Vollhardt, K.P.C. and Schore, N.E. (2003). *Organic Chemistry: Structure and Function*. MacMillan.

52 Katritzky, A.R., Meth-Cohn, O., and Rees, C.W. (1995). *Comprehensive Organic Functional Group Transformations: Synthesis: carbon with one heteroatom attached by a single bond*. Elsevier.

53 Wade, L.G. (1991). *Comprehensive Organic Transformations: A Guide to Functional Group Transformations*. ACS Publications.

54 Dubrovskiy, A.V., Kesharwani, T., Markina, N.A. et al. (2018). *Comprehensive Organic Transformations, 4 Volume Set: A Guide to Functional Group Preparations*, vol. 1. John Wiley & Sons.

55 Gotor-Fernández, V., Brieva, R., and Gotor, V. (2006). Lipases: useful biocatalysts for the preparation of pharmaceuticals. *Journal of Molecular Catalysis B: Enzymatic* 40 (3–4): 111–120.

56 Carey, J.S., Laffan, D., Thomson, C., and Williams, M.T. (2006). Analysis of the reactions used for the preparation of drug candidate molecules. *Organic & Biomolecular Chemistry* 4 (12): 2337–2347.

57 Ali, U., Karim, K.J.B.A., and Buang, N.A. (2015). A review of the properties and applications of poly (methyl methacrylate) (PMMA). *Polymer Reviews* 55 (4): 678–705.

58 Padró, C.L. and Apesteguía, C.R. (2005). Acylation of phenol on solid acids: study of the deactivation mechanism. *Catalysis Today* 107–108: 258–265.

59 Murashige, R., Hayashi, Y., Ohmori, S. et al. (2011). Comparisons of O-acylation and Friedel–Crafts acylation of phenols and acyl chlorides and Fries rearrangement of phenyl esters in trifluoromethanesulfonic acid: effective synthesis of optically active homotyrosines. *Tetrahedron* 67 (3): 641–649.

60 Liao, W. Jr.,, Lin, S.-Y., Kuo, Y.-S., and Liang, C.-F. (2022). Site-selective acylation of phenols mediated by a thioacid surrogate through sodium thiosulfate catalysis. *Organic Letters* 24 (23): 4207–4211.

61 Tian, J.-S., He, Y., Gao, Z.-Y. et al. (2021). Water-tolerant ortho-acylation of phenols. *Organic Letters* 23 (16): 6594–6598.

62 Zhang, Z., Zhao, Z., Liu, M. et al. (2022). Acylation of phenols to phenolic esters with organic salts. *Green Chemistry* 24 (24): 9763–9771.

63 Pandey, R.K., Dagade, S.P., Malase, K.M. et al. (2006). Synthesis of ceria-yttria based strong Lewis acid heterogeneous catalyst: application for chemoselective acylation and ene reaction. *Journal of Molecular Catalysis A: Chemical* 245 (1–2): 255–259.

64 Gernigon, N., Al-Zoubi, R.M., and Hall, D.G. (2012). Direct amidation of carboxylic acids catalyzed by ortho-iodo arylboronic acids: catalyst optimization, scope, and preliminary mechanistic study supporting a peculiar halogen acceleration effect. *The Journal of Organic Chemistry* 77 (19): 8386–8400.

65 Dulla, B., Vijayavardhini, S., Rambau, D. et al. (2014). Catalyst/surfactant free chemoselective acylation of amines in water. *Current Green Chemistry* 1 (1): 73–79.

66 Movassaghi, M. and Schmidt, M.A. (2005). N-heterocyclic carbene-catalyzed amidation of unactivated esters with amino alcohols. *Organic Letters* 7 (12): 2453–2456.

67 Chankeshwara, S.V. and Chakraborti, A.K. (2006). Catalyst-free chemoselective N-tert-butyloxycarbonylation of amines in water. *Organic Letters* 8 (15): 3259–3262.

68 Ibrahim, T.S., I.A. Seliem, S.S. Panda, et al. An efficient greener approach for N-acylation of amines in water using benzotriazole chemistry. *Molecules*, 2020. 25, DOI: https://doi.org/10.3390/molecules25112501.

69 Heller, S.T., Schultz, E.E., and Sarpong, R. (2012). Chemoselective N-acylation of indoles and oxazolidinones with carbonylazoles. *Angewandte Chemie (International Ed. in English)* 124 (33): 8429–8433.

70 Hou, L., Wu, B., Yu, D. et al. (2018). Asymmetric porous monovalent cation perm-selective membranes with an ultrathin polyamide selective layer for cations separation. *Journal of Membrane Science* 557: 49–57.

71 Mourya, V.K. and Inamdar, N.N. (2008). Chitosan-modifications and applications: opportunities galore. *Reactive and Functional Polymers* 68 (6): 1013–1051.

72 Pattabiraman, V.R. and Bode, J.W. (2011). Rethinking amide bond synthesis. *Nature* 480 (7378): 471–479.

73 Boas, U., Brask, J., and Jensen, K.J. (2009). Backbone amide linker in solid-phase synthesis. *Chemical Reviews* 109 (5): 2092–2118.

74 Fukuyama, T., Picard, B., and Ru, I. (2023). Cross-coupling reactions catalyzed by RuHCl $(CO)(PPh_3)_3$. *Catalysis Science and Technology* 13: 600–610.

75 Zhang, Y., Zhang, Y., Lin, J. et al. (2023). Radical acylation: concepts, synthetic applications and directions. *Organic Chemistry Frontiers* 10: 1056–1085.

76 De Risi, C., Carmine, G.D., Ragno, D. et al. (2023). Oxidative N-heterocyclic carbene catalysis. *Chemistry - A European Journal* 29 (4): e202202467.

77 Taussat, A., R.M. deFigueiredo, and J.M. Campagne,Direct catalytic amidations from carboxylic acid and ester derivatives: a review. *Catalysts*, 2023. 13, DOI: https://doi.org/10.3390/catal13020366.

78 Yang, J., Huang, H., and Zhao, J. (2023). Active ester-based peptide bond formation and its application in peptide synthesis. *Organic Chemistry Frontiers* 10: 1817–1846.

79 Anishchenko, V.M., Redko, A.M., Dykun, O.M., and Rybachenko, V.I. (2023). Efficient synthesis of mixed aliphatic benzoic anhydrides in a two-phase system. *Journal of Molecular Structure* 1281: 135051.

80 Gabdullin, A.M. et al. (2023). TaCl5-catalyzed amidation of carboxylic acids with amines. *ChemistrySelect* 8 (7): e202204298.

81 Chandra, K.L., Kadikova, R.N., Mozgovoj, O.S., and Ramazanov, I.R. (2002). Lewis acid catalyzed acylation reactions: scope and limitations. *Organic Chemistry* 58 (7): 1369–1374.

82 Donahue, M.G. and Rustin, G. (2023). Synthesis of multi-substituted pyrroles by lewis acid promoted cyclization of N-sulfonyl vinylogous carbamates and amides.

83 Neises, B. and Steglich, W. (2003). Esterification of carboxylic acids with dicyclohexylcarbodiimide/4-Dimethylaminopyridine: tert-butyl ethyl fumarate: (E)-2-Butenedioic acid, ethyl 1, 1-dimethylethyl ester. *Organic Syntheses* 63: 183–183.

84 Gomez-Castro, F.I. et al. (2013). Alternatives for the production of biodiesel by supercritical technologies: a comparative study. In: *Computer Aided Chemical Engineering* (ed. A. Kraslawski and I. Turunen), 7–12. Elsevier.

85 Gueven, A. (2013). Chemical fingerprints of Raki: a traditional distilled alcoholic beverage. *Journal of the Institute of Brewing* 119 (3): 126–132.

86 Guo, Y. et al. (2021). Efficient synthesis of primary and secondary amides by reacting esters with alkali metal amidoboranes. *Nature Communications* 12 (1): 5964.

87 Lanigan, R.M., Starkov, P., and Sheppard, T.D. (2013). Direct synthesis of amides from carboxylic acids and amines using B(OCH2CF3)3. *The Journal of Organic Chemistry* 78 (9): 4512–4523.

88 Yadav, G.D. and Borkar, I.V. (2009). Synthesis of n-butyl acetamide over immobilized lipase. *Journal of Chemical Technology and Biotechnology* 84 (3): 420–426.

89 Gutte, B. (1995). *Peptides: Synthesis, Structures, and Applications*. Elsevier.

90 Perez Espitia, P.J., Soares, N., Coimbra, J. et al. (2012). Bioactive peptides: synthesis, properties, and applications in the packaging and preservation of food. *Comprehensive Reviews in Food Science and Food Safety* 11 (2): 187–204.

91 Pawar, G.G., Kale, A., Sah, P., and Kapur, M. (2023). Total synthesis of resorcylic acid lactone (−)-neocosmosin a using palladium-catalyzed α-arylation of enones as the key step. *European Journal of Organic Chemistry* 26: e202201277.

92 Dangalov, M. et al. (2023). Ru-catalyzed isomerization of achmatowicz derivatives: a sustainable route to biorenewables and bioactive lactones. *ACS Catalysis* 13 (3): 1916–1925.

93 Liu, X. and He, L.-N. (2018). Synthesis of lactones and other heterocycles. In: *Chemical Transformations of Carbon Dioxide* (ed. X.-F. Wu and M. Beller), 145–176. Cham: Springer International Publishing.

94 Latos, P., Brzęczek-Szafran, A., Krzywiecki, M. et al. (2023). Highly active and stable catalysts for Baeyer–Villiger oxidation-traditional postloading vs in situ synthesis of Fe/N/C nanoparticles. *Applied Catalysis A: General* 651: 119027.

95 Hauser, C.R., Swamer, F.W., and Adams, J.T. (2004). The acylation of ketones to form β-diketones or β-keto aldehydes. In: *Organic Reactions*, vol. 8, 59–196. Wiley.

96 Katritzky, A.R., Meher, N.K., and Singh, S.K. (2005). γ, δ-unsaturated β-diketones by acylation of ketones. *The Journal of Organic Chemistry* 70 (19): 7792–7794.

97 Ren, S.-C., Yang, X., Mondal, B. et al. (2022). Carbene and photocatalyst-catalyzed decarboxylative radical coupling of carboxylic acids and acyl imidazoles to form ketones. *Nature Communications* 13 (1): 2846.

98 Bhattacharya, T., Suchand, B., Sreenivasulu, C. et al. (2020). Heterogeneous direct acylation strategy to diaryl ketones and their application to 1, 3-dihydroisobenzofurans. *ChemistrySelect* 5 (4): 1349–1352.

99 Yao, M., Chen, X., Lu, Y. et al. (2021). N-acylation of amides through internal nucleophilic catalysis. *Journal of Chemical Research* 45 (1–2): 141–146.

100 Umehara, A., Shimizu, S., and Sasaki, M. (2023). DMAPO/Boc2O-mediated one-pot direct N-acylation of less nucleophilic N-heterocycles with carboxylic acids. *ChemCatChem* e202201596.

101 Ammazzalorso, A., Tricca, M., Bruno, I. et al. (2015). Titanium-promoted acylation of sulfonamides to N-acylsulfonamide PPARα antagonists. *Synthetic Communications* 45 (22): 2546–2554.

102 Gomes, P., Gomes, J.R.B., Rodrigues, M., and Moreira, R. (2003). Amino acids as selective sulfonamide acylating agents. *Tetrahedron* 59 (38): 7473–7480.

103 Chen, L. and Luo, G. (2019). Facile synthesis of acyl sulfonamides from carboxyic acids using the Mukaiyama reagent. *Tetrahedron Letters* 60 (3): 268–271.

104 Boddula, R. and Asiri, A.M. (2020). *Green Sustainable Process for Chemical and Environmental Engineering and Science: Microwaves in Organic Synthesis*. Elsevier Science.

105 Benali, N., Bougheloum, C., Alioua, S. et al. (2018). Efficient N-acylation of sulfonamides using cesium salt of Wells–Dawson heteropolyacid as catalyst: synthesis of new N-acyl sulfonamides and cyclic imides. *Synthetic Communications* 48 (24): 3099–3112.

106 Bedair, M.A., Abuelela, A.M., Alshareef, M. et al. (2023). Ethyl ester/acyl hydrazide-based aromatic sulfonamides: facile synthesis, structural characterization, electrochemical measurements and theoretical studies as effective corrosion inhibitors for mild steel in 1.0 M HCl. *RSC Advances* 13 (1): 186–211.

107 Mahapatra, M., Mahapatra, P., Sahoo, S. et al. (2023). Design, synthesis, and in-silico study of chromen-sulfonamide congeners as potent anticancer and antimicrobial agents. *Journal of Molecular Structure* 135190.

108 Morizur, V., Hector, D., Olivero, S. et al. (2016). Metal sulfonate polymers as catalysts for the heterogeneous acylation of aromatic derivatives. *European Journal of Organic Chemistry* 2016 (18): 3126–3129.

109 Matsumoto, K., Ogawa, T., and Jikei, M. (2017). Nonstoichiometric polycondensation based on Friedel–Crafts acylation in superacids for the syntheses of aromatic polyketones. *Polymer Chemistry* 8 (47): 7297–7300.

110 Baumann, E., and Schotten, C. (1884). Ueber die Oxydation des Piperidins. *Berichte der deutschen chemischen Gesellschaft* 17 (2): 2544–2547.

111 Bejblová, M., Procházková, D., and Vlk, J. (2009). Acylations with long-chain acid anhydrides and acyl chlorides over zeolite beta. *Topics in Catalysis* 52: 178–184.

112 Yin, Y., Yamada, O., Tanaka, K., and Okamoto, K.-I. (2006). On the development of naphthalene-based sulfonated polyimide membranes for fuel cell applications. *Polymer Journal* 38 (3): 197–219.

113 Yamazaki, K. and Kawakami, H.J.M. (2010). High proton conductive and low gas permeable sulfonated graft copolyimide membrane. *Macromolecules* 43 (17): 7185–7191.

114 Chen, J.H., Liu, Q., Zhu, A. et al. (2008). Dehydration of acetic acid using sulfonation cardo polyetherketone (SPEK-C) membranes. *Journal of Membrane Science* 308 (1–2): 171–179.

115 Chen, J.H., Liu, Q., Zhu, A. et al. (2008). Dehydration of acetic acid by pervaporation using SPEK-C/PVA blend membranes. *Journal of Membrane Science* 320 (1–2): 416–422.

116 Aziz, M.A. and Shanmugam, S. (2018). Sulfonated graphene oxide-decorated block copolymer as a proton-exchange membrane: improving the ion selectivity for all-vanadium redox flow batteries. *Journal of Materials Chemistry A* 6 (36): 17740–17750.

117 Dong, Y., Liu, Y., Li, H. et al. (2023). Crown ether-based Tröger's base membranes for efficient Li+/Mg2+ separation. *Journal of Membrane Science* 665: 121113.

118 Liu, M., Wu, D., Yu, S., and Gao, C. (2009). Influence of the polyacyl chloride structure on the reverse osmosis performance, surface properties and chlorine stability of the thin-film composite polyamide membranes. *Journal of Membrane Science* 326 (1): 205–214.

119 Ran, J., Wu, L., Varcoe, J. et al. (2012). Development of imidazolium-type alkaline anion exchange membranes for fuel cell application. *Journal of Membrane Science* 415: 242–249.

120 Zhao, Y., Yu, H., and Xie, F. (2014). High durability and hydroxide ion conducting pore-filled anion exchange membranes for alkaline fuel cell applications. *Journal of Power Sources* 269: 1–6.

121 Hasegawa, M., Horiuchi, M., Kumakura, K., and Koyama, J. (2014). Colorless polyimides with low coefficient of thermal expansion derived from alkyl-substituted cyclobutanetetracarboxylic dianhydrides. *Polymer International* 63 (3): 486–500.

122 Ni, H.J., Liu, J.-G., Wang, Z.-H., and Yang, S.-Y. (2015). A review on colorless and optically transparent polyimide films: chemistry, process and engineering applications. *Journal of Industrial and Engineering Chemistry* 28: 16–27.

123 Zhou, L., Li, Y., Wang, Z. et al. (2019). Preparation of polyimide films via microwave-assisted thermal imidization. *RSC Advances* 9 (13): 7314–7320.

124 Raj, M.R., Kim, N., and Lee, G. (2021). A perylene-based aromatic polyimide with multiple carbonyls enabling high-capacity and stable organic lithium and sodium ion batteries. *Sustainable Energy & Fuels* 5 (1): 175–187.

125 Li, X., Wang, M., Mushtaq, N. et al. (2023). Colorless polyimide films with low birefringence and retardation: synthesis and characterization. *Polymer* 265: 125579.

126 Ohya, H., Kudryavsev, V., and Semenova, S.I. (1997). *Polyimide Membranes: Applications, Fabrications and Properties.* CRC Press.

127 Kuznetsov Alexander, A. and Tsegelskaya Anna, Y. (2019). Synthesis of polyimides in the melt of benzoic acid. In: *Solvents, Ionic Liquids and Solvent Effects* (ed. G.-M. Daniel and M. Magdalena). IntechOpen: Rijeka Chapter 3.

128 Zhang, Z. and Xu, T. (2014). Proton-conductive polyimides consisting of naphthalenediimide and sulfonated units alternately segmented by long aliphatic spacers. *Journal of Materials Chemistry* 2 (30): 11583–11585.

129 Yao, Z., Zhang, Z., Wu, L., and Xu, T. (2014). Novel sulfonated polyimides proton-exchange membranes via a facile polyacylation approach of imide monomers. *Journal of Membrane Science* 455: 1–6.

130 Didenko, A.L., A.K. Ivanov, V.E. Smirnova, et al. Selective destruction of soluble polyurethaneimide as novel approach for fabrication of insoluble polyimide films. *Polymers*, 2022. 14, DOI: https://doi.org/10.3390/polym14194130.

131 Sulub-Sulub, R., Loría-Bastarrachea, M.I., González-Díaz, M.O., and Aguilar-Vega, M. (2023). Synthesis and characterization of block sulfonated amphiphilic aromatic copolyamides for cation conductive membranes. *Polymer Bulletin* 80 (1): 429–444.

132 Hu, M., Zhang, B., Chen, J. et al. (2019). Cross-linked polymer electrolyte membrane based on a highly branched sulfonated polyimide with improved electrochemical properties for fuel cell applications. *International Journal of Energy Research* 43 (14): 8753–8764.

133 Xie, T., Pang, Y., Fan, H. et al. (2023). Controlling the microphase morphology and performance of cross-linked highly sulfonated polyimide membranes by varying the molecular structure and volume of the hydrophobic cross-linkable diamine monomers. *Journal of Membrane Science* 666: 121177.

134 Wang, G., Yu, Y., Li, Y. et al. (2023). Sulfonated polyimide membrane containing poly [bis (4-aminodiphenyl bissulfonate) phosphoronitrile] flexible chains for vanadium redox flow battery. *Polymer* 267: 125667.

135 Yu, L., Wang, L., Yu, L. et al. (2019). Aliphatic/aromatic sulfonated polyimide membranes with cross-linked structures for vanadium flow batteries. *Journal of Membrane Science* 572: 119–127.

136 He, Z., Wang, G., Wei, S. et al. (2021). A novel fluorinated acid-base sulfonated polyimide membrane with sulfoalkyl side-chain for vanadium redox flow battery. *Electrochimica Acta* 399: 139434.

137 Xu, W., Long, J., Liu, J. et al. (2022). A novel porous polyimide membrane with ultrahigh chemical stability for application in vanadium redox flow battery. *Chemical Engineering Journal* 428: 131203.

138 Tanriverdi, A.A., Yildiko, U., Tekes, A. et al. (2022). Synthesis, characterization and affinity detection of sulfonated polyimides: confirmation of proton transfer in quantum theory simulations. *Polymer Bulletin* 80: 9853–9880.

139 Yildiko, U. and Tanriverdi, A.A. (2022). A novel sulfonated aromatic polyimide synthesis and characterization: energy calculations, QTAIM simulation study of the hydrated structure of one unit. *Bulletin of the Korean Chemical Society* 43 (6): 822–835.

140 Wu, J., Di, S., Huang, W. et al. (2021). Porous polyimide framework based on perylene and triazine for reversible potassium-ion storage. *Materials Chemistry Frontiers* 5 (19): 7184–7190.

141 Chiriac, A.-P. and Damaceanu, M.-D. (2021). A novel approach towards crown-ether modified polyimides with affinity for alkali metal ions recognition. *Journal of Molecular Liquids* 322: 114929.

142 Goudar, S.H., Ingle, D., Sahu, R. et al. (2023). Perylene diimide-containing dynamic hypercrosslinked ionic porous organic polymers: modulation of assembly and gas storage. *ACS Applied Polymer Materials* 5: 2097–2104.

143 Wang, G., Weng, Y., Chu, D. et al. (2009). Preparation of alkaline anion exchange membranes based on functional poly(ether-imide) polymers for potential fuel cell applications. *Journal of Membrane Science* 326 (1): 4–8.

144 O'Harra, K.E., Kammakakam, I., Bara, J.E., and Jackson, E.M. (2019). Understanding the effects of backbone chemistry and anion type on the structure and thermal behaviors of imidazolium polyimide-ionenes. 68 (9): 1547–1556.

145 Oh, B.H., Kim, A.R., and Yoo, D.J. (2019). Profile of extended chemical stability and mechanical integrity and high hydroxide ion conductivity of poly(ether imide) based membranes for anion exchange membrane fuel cells. *International Journal of Hydrogen Energy* 44 (8): 4281–4292.

146 Cai, M., Zhu, M., Sun, Y., and Qian, J. (2010). Synthesis and characterization of soluble aromatic poly(ether amide amide ether ketone ketone)s by electrophilic Friedel–Crafts solution polycondensation. *Reactive and Functional Polymers* 70 (3): 182–188.

147 Li, D., Li, Z., Hu, F. et al. (2014). Pendant crosslinked poly(aryl ether sulfone) copolymers sulfonated on backbone for proton exchange membranes. *Polymer Engineering & Science* 54 (9): 2013–2022.

148 Zhang, Z. and Xu, T. (2014). Poly(ether ketone)s bearing pendent sulfonate groups via copolyacylation of a sulfonated monomer and isomeric AB-type comonomers. *Journal of Polymer Science Part A: Polymer Chemistry* 52 (2): 200–207.

149 Cui, M., Zhang, Z., Yuan, T. et al. (2014). Proton-conducting membranes based on side-chain-type sulfonated poly(ether ketone/ether benzimidazole)s via one-pot condensation. *Journal of Membrane Science* 465: 100–106.

150 Yao, Z., Cui, M., Zhang, Z., et al. Silane cross-linked sulfonted poly(ether ketone/ether benzimidazole)s for fuel cell applications. *Polymers*, 2017. 9, DOI: https://doi.org/10.3390/polym9120631.

151 Zhang, Z. and Xu, T. (2019). High Performance ion exchange membranes prepared via direct polyacylation of racemic and (S)-1,1′-binaphthyl-based cationic/anionic monomers. *Macromolecular Materials and Engineering* 304 (3): 1800547.

152 Zeng, G., Yan, L., Yue, B. et al. (2021). Synthesis of the sulfonated poly(ether ether ketone ketone) proton exchange membrane with butane-1,4-diylbis(oxy) spacers and sulfopropyloxy pendants through Friedel-Crafts polyacylation. *International Journal of Hydrogen Energy* 46 (54): 27770–27783.

153 Cetina-Mancilla, E., Reyes-Garcia, G.A., Rodriguez-Molina, M. et al. (2023). Room temperature, simple and efficient synthesis and functionalization of aromatic poly(arylene sulfide)s, poly(arylene sulfoxide)s and poly(arylene sulfone)s. *European Polymer Journal* 184: 111800.

154 Li, Z., Yu, R., Liu, C. et al. (2021). Preparation and characterization of side-chain poly(aryl ether ketone) anion exchange membranes by superacid-catalyzed reaction. *Polymer* 222: 123639.

155 Li, L., Wang, J., Hussain, M. et al. (2021). Side-chain manipulation of poly (phenylene oxide) based anion exchange membrane: alkoxyl extender integrated with flexible spacer. *Journal of Membrane Science* 624: 119088.

156 Lai, A.N., Zhuo, Y., Lin, C. et al. (2016). Side-chain-type phenolphthalein-based poly(arylene ether sulfone nitrile)s anion exchange membrane for fuel cells. *Journal of Membrane Science* 502: 94–105.

157 Gong, S., Li, L., Ma, L. et al. (2020). Blend anion exchange membranes containing polymer of intrinsic microporosity for fuel cell application. *Journal of Membrane Science* 595: 117541.

158 Pedersen, C. (1967). Cyclic polyethers and their complexes with metal salts. *Journal of the American Chemical Society* 89 (26): 7017–7036.

159 Pedersen, C. (1970). Crystalline salt complexes of macrocyclic polyethers. *Journal of the American Chemical Society* 92 (2): 386–391.

160 Lin, C., Gao, Y., Li, N. et al. (2020). Quaternized Tröger's base polymer with crown ether unit for alkaline stable anion exchange membranes. *Electrochimica Acta* 354: 136693.

161 Ge, X., He, Y., Guiver, M.D. et al. (2016). Alkaline anion-exchange membranes containing mobile ion shuttles. *Advanced Materials* 28 (18): 3467–3472.

162 Yang, Q., Li, L., Gao, X.L. et al. (2019). Crown ether bridged anion exchange membranes with robust alkaline durability. *Journal of Membrane Science* 578: 230–238.

163 Ge, X., He, Y., Zhang, K. et al. (2021). Fast bulky anion conduction enabled by free shuttling phosphonium cations. *Research Journal* 2021: 9762709.

164 Wang, X., Huang, Y., He, M. et al. (2022). Poly(terphenyl piperidinium) containing hydrophilic crown ether units in main chains as anion exchange membranes for alkaline fuel cells and water electrolysers. *Journal of Membrane Science* 653: 120558.

165 Shang, C., Wang, Z., Wang, L., and Wang, J. (2020). Preparation and characterization of a polyvinyl alcohol grafted bis-crown ether anion exchange membrane with high conductivity and strong alkali stability. *International Journal of Hydrogen Energy* 45 (33): 16738–16750.

166 Xu, S., Wei, W., Su, X., and He, R. (2023). Crown-ether block copolymer based poly(isatin terphenyl) anion exchange membranes for electrochemical energy conversion devices. *Chemical Engineering Journal* 455: 140776.

167 Kwon, S.-M., Lee, K.P., Tanaka, K., and Ohta, K. (1999). Simultaneous determination of anions and cations by ion-exclusion chromatography–cation-exchange chromatography with tartaric acid/18-crown-6 as eluent. *Journal of Chromatography A* 850 (1): 79–84.

168 Dong, Y., Zhu, Q., Zhou, W. et al. (2023). Dibenzo-15-crown-5-based Tröger's base membrane for 6Li+/7Li+ separation. *Separation and Purification Technology* 309: 122990.

169 Wang, X., Zheng, X., Liu, X. et al. (2022). K+-Responsive crown ether-based amphiphilic copolymer: synthesis and application in the release of drugs and Au nanoparticles. *Polymers* 14 (3): 406.

170 Kimura, K. and Shono, T. (1992). Applications of crown compounds to analytical and separation chemistry: ion sensor and liquid chromatography. In: *Studies in Organic Chemistry*, 198–264. Elsevier.

171 Zolotukhin, M.G., Hernández, M.C.G., Lopez, A.M. et al. (2006). Film-forming polymers containing in the main-chain dibenzo crown ethers with aliphatic (C10–C16), aliphatic–aromatic, or oxyindole spacers. *Macromolecules* 39 (14): 4696–4703.

172 Ge, X., He, Y., Liang, X. et al. (2018). Thermally triggered polyrotaxane translational motion helps proton transfer. *Nature Communications* 9 (1): 2297.

5

Superacid–Catalyst Polymerization for IEMs Preparation

5.1 Definition and Types of Superacid

Superacid is an acid system that is more acidic than 100% sulfuric acid. It is several orders of magnitude more acidic than ordinary strong acids such as sulfuric acid and hydrochloric acid. For example, by mixing hydrofluoric acid and antimony pentafluoride in a molar ratio of 1 : 0.3, the acidity of the mixture is approximately 100 million times stronger than that of anhydrous sulfuric acid (100%). Superacids are of great value in chemistry and chemical industry applications, both as inorganic and organic protonation reagents and as highly active catalysts. Many chemical reactions that used to be extremely difficult or impossible to achieve under normal conditions can be accomplished exceptionally well in the presence of superacids. Simple superacids include trifluoromethyl sulfonic acid and fluorosulfonic acid.

The term "superacid" first appeared in 1927 in a chemical literature article published by Hall and Conant in the *Journal of the American Chemical Society*. In an investigation of the hydrogen-ion activity of nonaqueous solutions, these authors noted that sulfuric acid and perchloric acid in glacial acetic acid solutions are able to form salts with various weak bases, such as ketones and other carbonyl compounds. These weak bases cannot form salts with the same acid in aqueous solutions. The authors attribute this high acidity to the ionization of these acids in glacial acetic acid, thereby increasing the concentration of $CH_3COOH_2^+$, which is a species less solvated than H_3O^+ in aqueous acids. They proposed to call these solutions "superacid solutions."

Until 1960, Olah's studies of obtaining stable solutions of highly electron-deficient ions focused on nonaqueous systems with very high acidity. Gillespie then proposed a broader and widely accepted definition of superacid: any acid system stronger than 100% sulfuric acid, i.e. $H_0 < -12$. Fluorinated sulfuric acid and trifluoromethanesulfonic acid (TfOH) are examples of Brønsted acids that are more acidic than concentrated sulfuric acid, with H_0 values of approximately −15.1 and −14.1, respectively.

Ion Exchange Membranes: Design, Preparation, and Applications,
First Edition. Tongwen Xu and Yaoming Wang.

To make the acidity exceed this limit, one must start with an already strong acid ($H_0 \approx -10$) and then add a stronger acid to it to increase ionization. This can be achieved by adding a strong Brønsted acid (HB) capable of ionizing in the medium or by adding a strong Lewis acid (L), which shifts the autoprotonation equilibrium by forming a conjugate acid. The equation is displayed in Figure 5.1. In both cases, a significant increase in acidity can be observed from the H_0 value of pure HA, as shown in Figure 5.2. This large acidity jump elevates the strong acid into the superacid region.

$$HA + B \rightleftharpoons H_2A^+ + B^-$$

$$2HA + L \rightleftharpoons H_2A^+ + LA^-$$

Figure 5.1 Dissociation equilibrium of Brønsted and Lewis acids in strong acids.

Gillespie's definition of superacid is based on the Brønsted acid system. As Lewis acids also cover a broader range of acidity beyond the strength of systems conventionally used, Olah recommended using anhydrous aluminum chloride ($AlCl_3$) as a reference. We identify Lewis superacids as Lewis acids stronger than anhydrous $AlCl_3$.

Although superacids are all more acidic than 100% sulfuric acid, the difference in acidity between the various superacid systems may be as or greater than that between pure sulfuric acid and its 0.1 M aqueous solution (Figure 5.3).

As mentioned before, superacids are similar to conventional acid systems, including Brønsted and Lewis acids and their conjugate acid systems. Protic (Brønsted type) superacids include strong parent acids and their mixtures, whose acidity

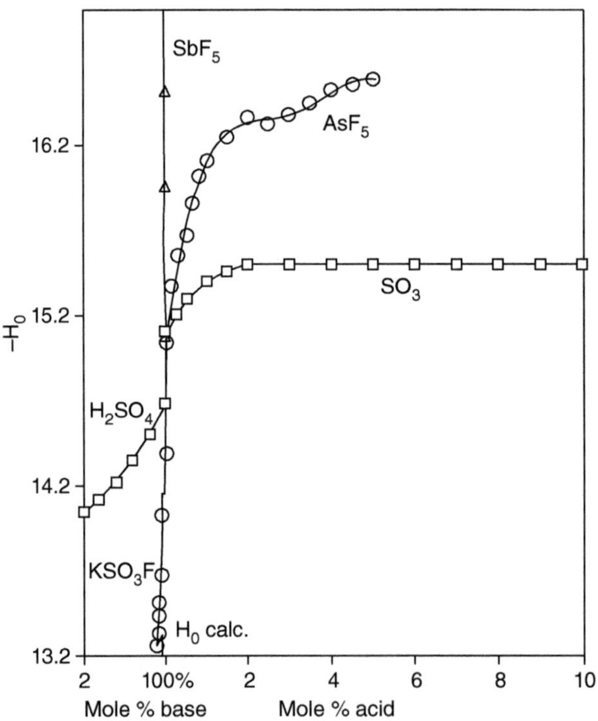

Figure 5.2 The change in acidity of HSO_3F [1].

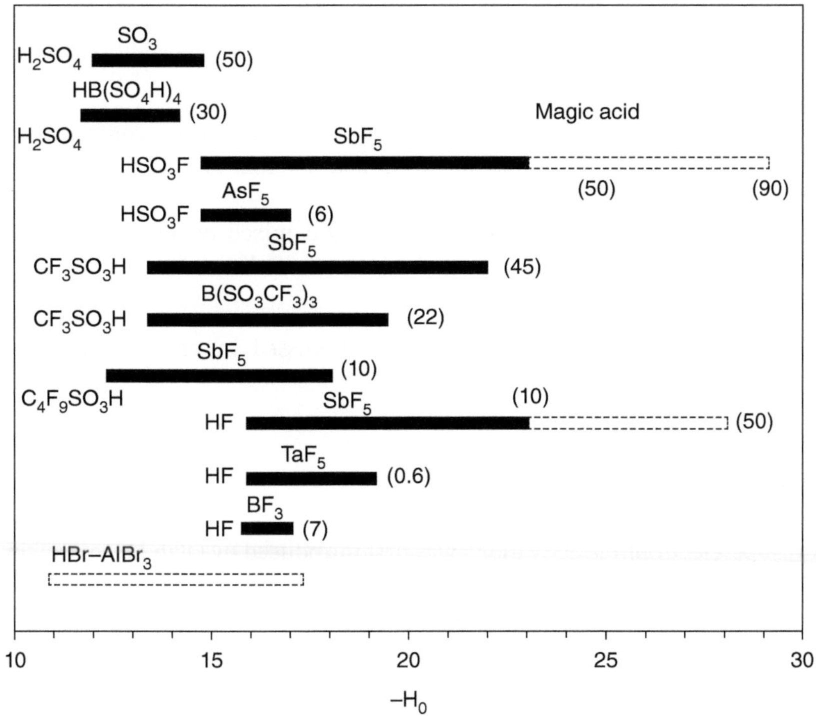

Figure 5.3 Acidity range of different superacids . Source: Olah et al. [2]/American Chemical Society.

can be further enhanced by various combinations with Lewis acids (conjugated acids). The most frequently used superacid systems include (i) Brønsted superacids, such as perchloric acid ($HClO_4$), halogenosulfonic acid (HSO_3Cl, HSO_3F), perfluoroalkanesulfonic acid (CF_3SO_3H, $RFSO_3H$), hydrogen fluoride, and carborane superacids ($H[CB_{11}HR_5X_6]$). (ii) Lewis superacids, such as SbF_5, AsF_5, PF_5, TaF_5, NbF_5, BF_3, tris (pentafluorophenyl) borane, and tris (trifluoromethanesulfonate) boron.

Binary superacid: (i) Binary Brønsted superacids, such as HF–HSO_3F, HF–CF_3SO_3F, and $HB(HSO_4)_4$. (ii) Conjugate Brønsted–Lewis superacid: (a) The combination of Brønsted oxacid (H_2SO_4, HSO_3F, CF_3SO_3H, $RFSO_3H$) and Lewis acid (SO_3, SbF_5, AsF_5, TaF_5, and NbF_5); (b) the combination of HF and Fluorinated Lewis acid (SbF_5, PF_5, TaF_5, NbF_5, and BF_3); (c) conjugate Friedel–Crafts acids, such as HBr–$AlBr_3$ and HCl–$AlCl_3$.

Ternary superacids include HSO_3F–HF–SbF_5, HSO_3F–HF–CF_3SO_3H, and HSO_3F–SbF_5–SO_3.

Solid superacid: Tanabe's group has long studied the acid-base properties of solids. They described them in detail for the first time. According to the properties of the acid sites, solid superacids can be further classified into different types. The acidity of solid acids depends on the properties of their chemical structure. As a part of the chemical structure, it usually has Lewis or Brønsted sites. The acidity of the latter can be further enhanced by complexing with Lewis acids. Solid superacids can also be obtained by depositing or inserting strong acids into inert or low-acidity carriers.

1. Zeolite acid.
2. Polymeric resin with sulfonic acid, including sulfonic acid resin compounded with Lewis acid and perfluoropolymer resin acid (Nafion-H and Nafion-Silica nanocomposites).
3. Enhanced acid solids, including metal oxides and mixed oxides modified by Brønsted acids and Lewis acids and metal salts combined with Lewis acids.
4. Graphite intercalated superacid.

These categories are relative and should be discussed in practical fields of application.

5.2 Principle of Superacid Catalyst

The term "superelectrophilic activation" was first introduced by Olah [3] to explain electrophilic reactivity in superacids. That is, when the electrophilic reagent reacts with the superacid to form the dicationic intermediate product, its reactivity will be significantly improved. Based on this activation phenomenon, superacids can be used as catalysts in various reaction systems.

Olah et al. [3] observed that the electrophilic reactivity of onium salts was enhanced in superacidic media. For example, nitronium salt showed higher reactivity in strong acids, which was interpreted by Olah as further protonation of nitronium cations in superacid solutions. Although the nitro ion (1 in Figure 5.4) itself is considered a strong electrophilic reagent, the increased degree of protonation makes the ion more positively charged (2 and 3 in Figure 5.4) and therefore has higher electrophilic reactivity. For example, the protonation of acetyl cations in the superacid system HF–BF_3 leads to the formation of superelectrophilic cations with high reactivity (Figure 5.4).

This type of activator is called a superelectrophilic reagent, which can react with unusually weak nucleophiles. For example, in the presence of nitronium superelectrophiles (2 or 3 in Figure 5.4), these substances have been proven to react with less active aromatic hydrocarbons and alkanes (including methane). That is, electrophiles capable of further interaction (coordination, solvation) with strong Brønsted or Lewis acids can be activated in this way. The resulting enhancement of their reactivity is greatly improved compared to that of their parent compounds under conventional conditions and indicates the formation of superelectrophiles (i.e. electrophiles with doubly electrodeficient [dipositive] nature whose reactivity substantially exceeds that of their parent electrophiles). This phenomenon is called superelectrophilic activation of an electrophilic reagent in a superacid medium and is depicted in Figure 5.5. This activation can develop superelectrophilic reagents (electrophiles with doubly electrodeficient). Compared with parent compounds under conventional conditions, their reactivity was greatly improved. The improved reactivity leads to an easier reaction, and it speeds up the reaction to some extent.

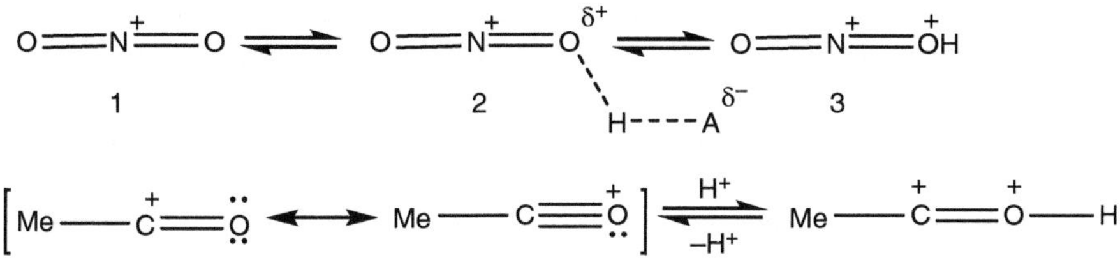

Figure 5.4 Further protonation of the nitrocation and acetyl cation in superacid. Source: Cruz et al. [4].

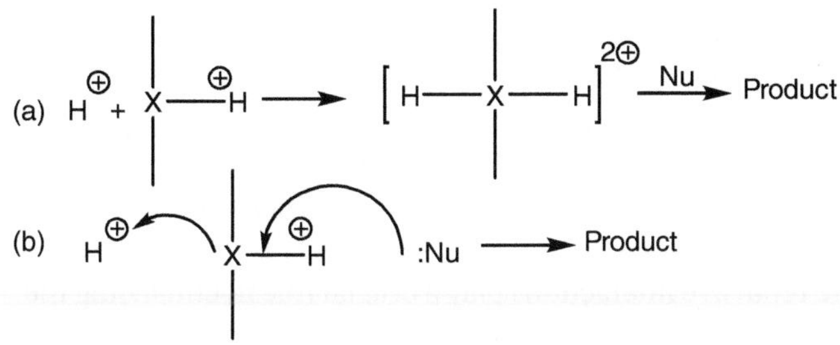

Figure 5.5 Activation by superdocks. (a) formation of a dication; (b) electrophilic assistance [3].

Since Olah formalized the concept of superelectrophilic activation, numerous scholars have carried out a great deal of work in this area, including synthetic methods by superelectrophiles, direct spectral characterization of these ions, and studies through computational and kinetic methods. An increasing number of experimental observations and related theoretical calculations have demonstrated that a variety of electrophiles can further interact with Brønsted or Lewis superacids to be greatly activated. Many subsequent studies have proven that superacid-catalyzed reactions can be widely used. With the application of Olah theory to the hydroxyalkylation reaction of carbonyl compounds with aromatic hydrocarbons in superacid systems, great success has been achieved. At present, superacid-catalyzed reactions have been widely used in the synthesis of aromatic condensed monomers and their precursors and corresponding polymers. Among them, the reactions using TfOH as the reaction medium are most concerning.

Due to the strong acidity of superacids, they are very capable of ionizing hydrogen ions, so superacids have very strong proton-donating ability. It can be combined with compounds having lone pair electrons, such as aldehydes, ketones, alcohols, etc. At the same time, it can be combined with compounds containing π electrons, such as alkenes and alkynes. It can also be combined with compounds having σ electrons. As the proton-donating ability of superacids is much stronger than that of general inorganic acids, they can be used as powerful catalysts for some organic reactions, such as the catalytic isomerization of alkanes, polymerization of alkanes, alkylation of aromatic hydrocarbons, oxidation, esterification, polymerization, and other reactions. The catalytic effect of superacids is superior to that of traditional catalysts.

In organic reactions, superacids are used as catalysts to form intermediates called carbocations. In 1901, Norris described carbocation for the first time when he discovered triphenyl carbocation. Due to the existence of a triphenyl ring and its stereoscopic effect, the triphenyl carbocation is in a stable state. Triphenyl carbocations are formed by ionization of triphenyl carbinol or conversion of halogenated hydrocarbons under weakly acidic conditions. After that, Olah used the weak nucleophilic properties of solvents in strong acid conditions to enhance the lifetime of carbocations. Unstable carbocation was observed by NMR. These studies on carbocations laid a foundation for superelectrophilic reactions catalyzed by superacids, which are especially suitable for studying reactions in which unstable intermediate carbocations are formed.

The key to successfully performing chemical reactions involving carbocations is the formation of stable carbocations in low-nucleophilicity media. The superelectrophile is a reaction intermediate produced by further protonation. The second protonation increases the electron deficiency, resulting in an increase in the electrophilicity of the intermediate, which leads to a significant increase in reactivity. In particular, charge–charge repulsive interactions play a crucial role in enhancing the reactivity of dicationic and tricationic superelectrophilic intermediates.

The acidity level also plays an important role in catalysis, and in many catalytic reaction systems, without an appropriate acidity level, conversion may occur at a much lower rate or not at all. For example, n-butane, in the presence of superacid, can break the C—H bond to form hydrogen. It can also break carbon–carbon bonds and produce methane, and it can be isomerized to produce isobutane. There are many reports about the application of superacids in organic synthesis. Brouwer and Kiffin combined acetaldehyde or acetone with isobutane under the catalysis of HF:SbF_5 or HF:BF3 to form tert-butyl cations. Under superacid conditions, the acetaldehyde or acetone cationic groups could be completely protonated. Their work overturned the long-standing belief that hydrogen ions cannot be transferred from isobutane to carbocation. Later, Olah explained the reaction mechanism of this reaction, and he pointed out that under the condition of superacid, carbocation (a in Figure 5.6) was protonated to form superelectrophilic material (b in Figure 5.6).

In summary, due to the characteristics of strong acidity and high activity, superacid can be used as a catalyst with less dosage, high yield, mild reaction

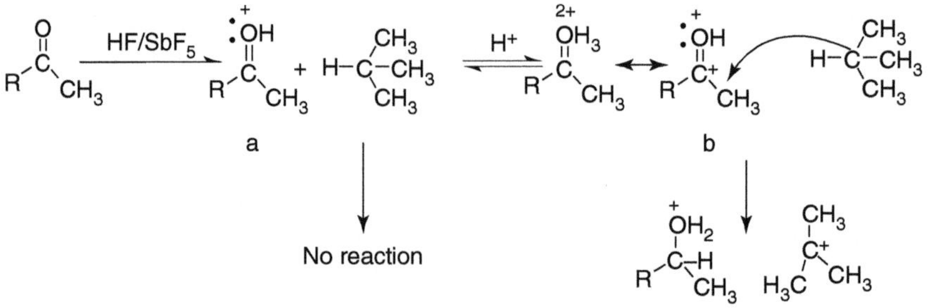

Figure 5.6 Mechanism of hydrogen ion transfer catalyzed by superacid. Source: Klumpp and Anokhin [5].

conditions, and less time consumption. Therefore, superacid has opened up a very broad path both from theoretical research and the application of organic synthesis in just a few decades.

5.3 Superacid-catalyzed Reaction for Polymer Synthesis

The catalytic reaction of superacids is widely used in the condensation of various aromatic monomers and polymers. As shown below (Figure 5.7), the acid-catalyzed condensation reaction of monomers with ketone or aldehyde functional groups and aromatic compounds is called the hydroxyalkylation reaction, which is a kind of Friedel–Crafts aromatic electrophilic substitution reaction.

Depending on the monomer structure and reaction conditions, alcohols, diaryl compounds, or mixtures of these compounds may be obtained. The acid-catalyzed addition reaction of aldehydes and ketones has long been known. For example, in 1872, A. Bayer reported the condensation reaction of chloroacetaldehyde and benzene in sulfuric acid. However, due to low yield and the formation of low molecular weight polymers, these Friedel–Crafts reactions of carbonyl did not receive the same degree of attention compared with corresponding reactions involving alcohol, alkyl, or halide for a long time. It was not until Olah et al. proposed the concept of superelectrophilic activation to explain the high reactivity of electrophilic substances in superacids that the hydroxyalkylation reaction achieved remarkable results. Later, the hydroxyalkylation reaction catalyzed by superacid, as a highly effective Friedel–Crafts reaction, was widely used for the synthesis of many monomers, attracting many scholars to research and report. The reactions using TfOH as the reaction medium are of most concern. TfOH with an acidity of 714 (two orders of magnitude higher than 100% sulfuric acid) is considered to be the strongest of all known organic monoacids. This is an ideal nonoxidizing medium to produce highly active substances in many electrophilic reactions.

In addition to the synthesis of small molecules, hydroxyalkylation reactions catalyzed by superacids have been widely used in the synthesis of some polymers. Mainly, ketones or aldehydes and aromatic hydrocarbons containing electron-withdrawing substituents are polymerized under the catalysis of superacid to form a series of high polymers, such as the polycondensation of various fluorinated carbonyl compounds with aromatic hydrocarbons. In these reactions, fluorine-substituted groups with electron-withdrawing effects activate carbon–oxygen ions, thus boosting the electrophilic reactivity. The resulting intermediate alcohol reacts with another aromatic compound to obtain a diaromatic

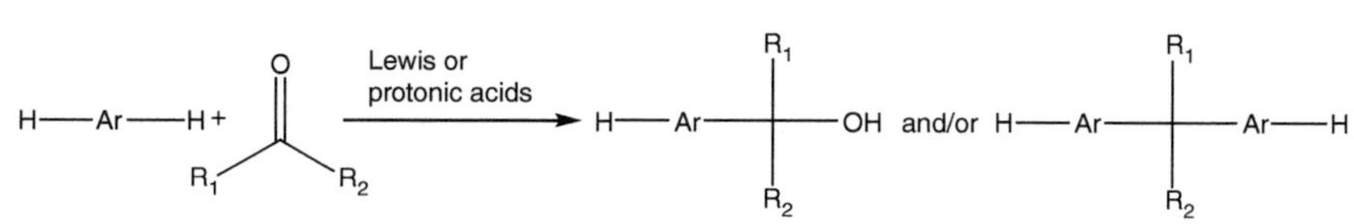

Figure 5.7 Hydroxyalkylation reaction. Source: Romero-Hernández et al. [6].

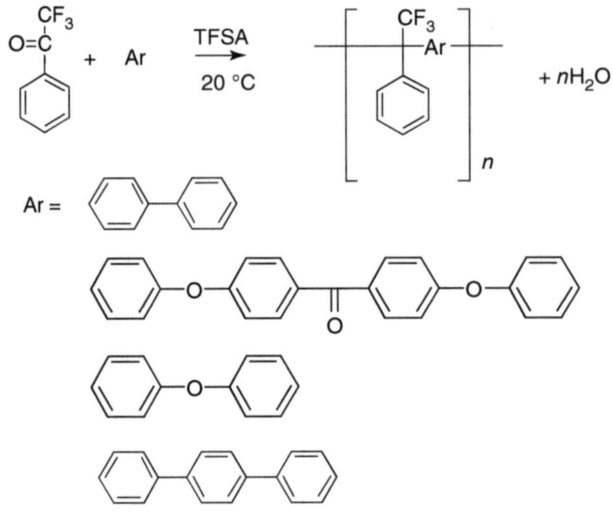

Figure 5.8 Superacid-catalyzed condensation of different aromatic hydrocarbons with trifluoroacetophenone. Source: Lira et al. [7]/American Chemical Society.

hydrocarbon product, which, as an aromatic compound, can react further with other intermediate alcohols. It is worth mentioning that the fluorinated substituents in the carbonyl compounds in this polymer eventually become substituents in the polymer chain and provide new polymer structures and properties. For example, the polycondensation reaction of trifluoroacetophenone with various aromatic hydrocarbons (biphenyl, diphenyl ether, and 4,4′-diphenyl oxy benzophenone) (Figure 5.8).

In fact, the polycondensation reaction catalyzed by superacid does not conform to the classical step polymerization theory. In A_2+B_2-type step polymerization [6], two bifunctional monomers with different functional groups react with each other. For this polymerization, specifying only the chain length "n" is not sufficient since the number of A_2 groups (i) and B_2 groups (j) may be different, giving different end groups. Type A and type B polymer molecules have functional groups A and B as end groups, respectively, while type M polymer molecules have functional group A at one end of the molecule and functional group B at the other end. Polymer molecules are represented by P_{ij}, where i represents the number of A_2 groups and j represents the number of B_2 groups. Since nonstoichiometric conditions may exist in the step polymerization of $A_2 + B_2$, an unbalanced stoichiometric ratio defined by Eq. (5.1) is needed. In Eq. (5.1), functional group A is the limiting reagent.

$$r = \frac{[A_2]_0}{[B_2]_0} \leq 1 \tag{5.1}$$

There are four possible reactions in step polymerization. In the traditional $A_2 + B_2$ step polymerization theory, these four reactions occur at the same reaction rate, that is, a single kinetic rate constant k is used in all four reactions.

$$A \sim A + B \sim B \xrightarrow{k} A \sim B \tag{5.2}$$

$$A \sim A + A \sim B \xrightarrow{k} A \sim A \tag{5.3}$$

$$B \sim B + A \sim B \xrightarrow{k} B \sim B \tag{5.4}$$

$$A \sim B + A \sim B \xrightarrow{k} A \sim B \tag{5.5}$$

In this framework, the classical theory of step polymerization of $A_2 + B_2$ is based on four assumptions: (a) functional groups A and B react at the same rate, independent of the type of molecule, (b) the k value is independent of chain length (diffusion control effects are absent or negligible), (c) cyclization is absent, and (d) the chain length (molecular weight) decreases as the stoichiometric ratio r decreases, implying that the maximum chain length (molecular weight) is reached when $r = 1$. However, the polyhydroxyalkylation reaction catalyzed by superacid significantly deviates from the classical behavior of $A_2 + B_2$ step polymerization, and a significant acceleration of the polymerization rate is observed when excess carbonyl compounds are added. This is because in the polyhydroxyalkylation reaction catalyzed by superacid, the assumed equal reactivity of functional groups is invalid, and the activity of the monomer and the intermediate is not the same, so the reaction rates of the first step and the second step are not the same. Therefore, higher molecular weight polymers are usually obtained by using excessive carbonyl compounds in such polymerization reactions. In addition, electron-withdrawing substituents that do not bind directly to the carbon cation center can also improve the electrophilicity of the substrate. Therefore, compared with unsubstituted fluorenone, 2,7-dinitrofluorenone is easier to polymerize with aromatic hydrocarbons under the catalysis of superacid to form high molecular weight polymers (Figure 5.9).

Acenaphthenequinone, isatin, and their derivatives have also been studied for superacid-catalyzed polyhydroxyalkylation to obtain high molecular weight polymers due to the influence of two adjacent electron-withdrawing carbonyls (Figure 5.10).

In addition, a similar electrophilic reagent activation mechanism may exist in superacid-catalyzed reactions involving carbonyl compounds that contain nitrogen heterocycles (Figure 5.11), such as the reaction of piperidone with benzene in the presence of TFSA, which can produce polymers with good yields (80%–99%). Under the same conditions, cyclohexanone only produces a small amount of aldol condensation products. This is because the protonation of the amino group activates the carbonyl carbon, and this activation is similar to that induced by electron-withdrawing

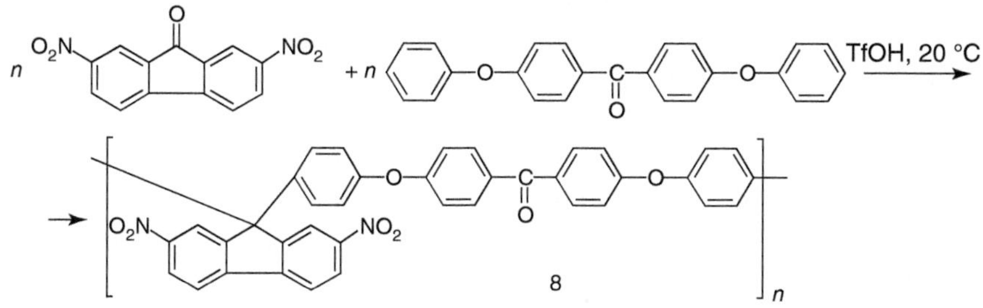

Figure 5.9 Superacid-catalyzed condensation of nitrofluorenone. Source: Rusanov et al. [8]/IOP Publishing.

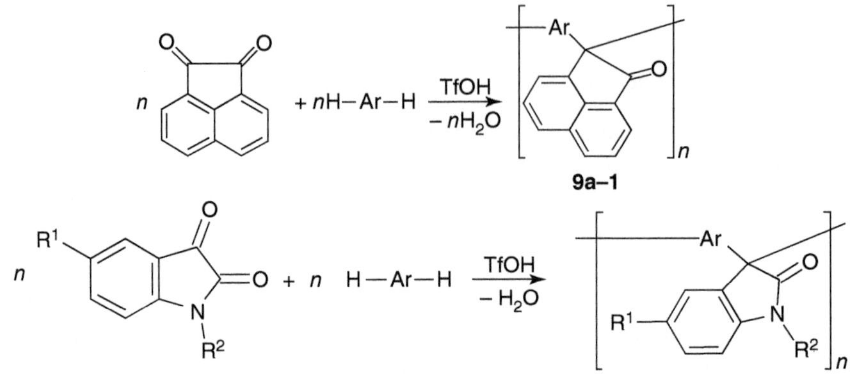

Figure 5.10 Superacid-catalyzed condensation reaction of acenaphthene quinone and isatin. Source: Rusanov et al. [8]/IOP Publishing.

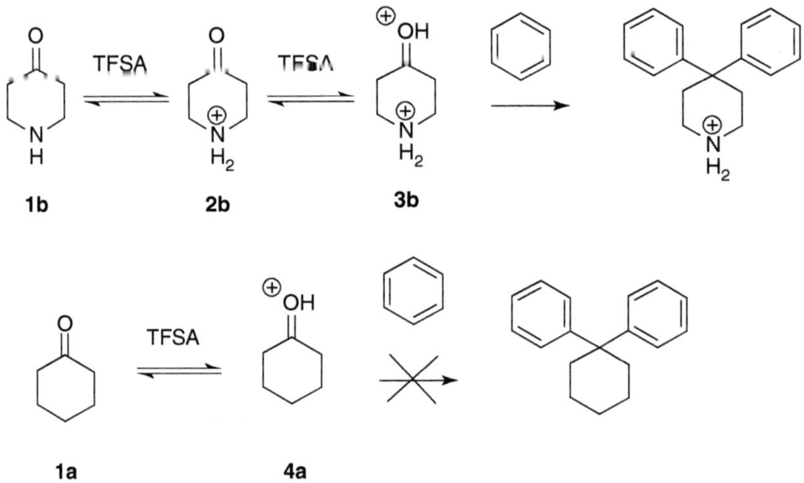

Figure 5.11 The superacid-catalyzed reaction of carbonyl compounds containing heterocyclic nitrogen. Source: Lira et al. [7]/American Chemical Society.

substituents in carbonyl compounds. Therefore, piperidone can be regarded as a promising monomer for polymer preparation in superacid catalytic systems.

The preparation of polymers by superacid-catalyzed polyhydroxyalkylation has the following advantages: (i) the reaction can be performed at room temperature and atmospheric pressure, (ii) the structure of the monomer is various and easy to obtain, (iii) high reaction rate and production rate, (iv) high regioselectivity, (v) acceptable reaction time, (vi) it is convenient to introduce functional groups that can chemically modify the polymers, (vii) easy to polymer purification, and (viii) the resulting polymer has good physical, chemical, and thermal properties, and high molecular weight. Therefore, in recent years, it has attracted a large number of scholars to study and show broad application prospects, such as in gas separation membranes, ion exchange membranes, and other fields. In the following chapter, the application of superacid catalysis in the preparation of ion exchange membranes will be introduced in detail.

5.4 Superacid-catalyst Polymerization for IEM Preparation

There are many methods for the preparation of ion exchange membranes, such as nucleophilic aromatic substitution, Diels–Alder polymerization, metal-catalyzed coupling reactions, Friedel–Crafts polyhydroxyalkylation reactions catalyzed by superacids, and cyclo-polycondensation reactions. The preparation of ionic membranes by superacid catalysis is a new method developed in recent years. Compared with other methods, the preparation of ionic membranes by superacid catalysis has many advantages, such as high catalytic efficiency and fast reaction speed; reaction conditions are simple and mild; polymers containing only C–C bonds in the backbone can be prepared; the degree and position of quaternary ammonium (QA) groups can be controlled more precisely; the reaction conditions do not contain metal-based catalysts; the resulting polymer can have high or ultrahigh molecular weights; the ionic membrane varies and has great solubility in polar solvents, good mechanical strength, and other excellent properties; initial materials and reagents are readily available; and the reaction kinetics are controllable. Therefore, it has been recognized and widely used.

First, the preparation of ionic membranes by superacid catalysis has the advantages of high catalytic efficiency and fast reaction speed. Strong Brønsted acids or Lewis acids can enhance the reactivity of electrophiles by superelectrophilic activation. Superelectrophilic activation plays an important role in the Friedel–Crafts polycondensation of ketones or aldehydes with aromatic hydrocarbons to form C—C bonds. George A. Olah et al. proved through many experimental observations and relevant theoretical calculations that various electrophilic molecules capable of further interaction (coordination) with Brønsted or Lewis superacids could be greatly activated by them, and such activation could produce superelectrophiles. Examples are onium and carboxonium ions, acyl cations, halonium, azonium, carbazonium ions, and even certain substituted carbocations. This activation produces superelectrophiles, that is, electrophiles of doubly electron-deficient (dipositive) nature whose reactivity significantly exceeds that of their parent monocations under conventional reaction conditions. The improved reactivity leads to an easier reaction, and it speeds up the reaction to some extent. Of course, the time required for the reaction also depends on the activity of the monomer itself. Once the molecular weight of the polymer increases, it begins to form a reactive gel-like precipitate, and the initial homogeneous solution phases separate.

Superelectrophiles are highly reactive intermediates. They are the de facto reactive intermediates of many electrophilic reactions in superacidic systems (including those involving solid superacids) and should be differentiated from energetically low-lying, much more stable intermediates, which are frequently observable, and even isolable but not necessarily reactive without further activation. George A. Olah's group observed that in some cases, additional activation of electrophilic cells can be achieved through further primary hydrolysis (or electrophilic) interactions. Such interaction of these ions with superacid systems can in the limiting case lead to extremely reactive de facto dications, although electrophilic assistance by the strong acids through polarized donor–acceptor interactions should have a similar effect. Electrophilic activation can play a significant role in Friedel–Crafts and superacid-catalyzed reactions, solid-acid-catalyzed systems, and even enzymatic reactions. In superacidic systems, anions are generally associated with fluorine bridges to form oligomeric anions. This helps to delocalize the negative charge and thus to enhance the proton donor ability of the systems. This also takes place in solid, polymeric superacids, such as Nafion-H.

Second, the preparation of ionic membranes catalyzed by superacids has the advantage of mild and simple reaction conditions. Because of their high acid strength, superacids are ideal acid catalysts, which can make some otherwise difficult reactions proceed smoothly at lower temperatures. This reaction condition is simple and mild and has been widely used in the petrochemical industry and organic synthesis. In addition, we can use superacid catalysis to prepare a polymer backbone containing only C—C bonds. Traditional anion exchange membranes (AEM) are usually prepared by attaching QA groups to the backbone of preformed polymers. In AEM, the main role of the polymer backbone is to provide mechanical stability through the entanglement of the polymer chains. Since most of the previous alkaline polymers applied to alkaline membranes were prepared by nucleophilic aromatic substitution reactions of aromatic diols and dihalide monomers, they inevitably contained aryl ether bonds along the backbone. Usually, an aromatic polymer backbone containing an aryl ether bond is first synthesized and then functionalized by chloromethylation or benzyl bromide followed by a QA reaction with trimethylamine. Although the synthesis process is also simple, chloromethylation reactions often require toxic reagents and long reaction times to achieve the desired degree of functionalization.

On the other hand, although aryl ether bonds allow freer rotation along the polymer backbone and improve the solubility of the polymer, the presence of such ether bonds also makes the alkaline membrane unstable under high pH conditions. As shown in Figure 5.12 [9], although poly (arylene ether sulfone) is quite stable in an alkaline environment, when an ionic group, such as benzyltrimethylammonium (BTMA), is attached to the polymer, it allows the polymer to swell and hydroxide ions to attack the polymer backbone. This will lead to a break in the main chain and loss of hydroxide ions, resulting in membrane brittleness, and a drop in ion conductivity, respectively. When the adjacent benzene contains electron-withdrawing substituents, the effect will be more serious, which will lead to the loss of electrons in the benzene, thus increasing the fragility of the backbone. Moreover, Miyanishi and Yamaguchi suggested that the decomposition of cation groups is not derived from the instability of the ionic group itself but rather triggered by the cleavage of nearby ether bonds [10]. Even when the backbone is not functionalized with ionic groups, the presence of C—O bonds in the main chain may cause hydroxyl radical-initiated degradation of the membrane under fuel cell operating conditions [11, 12]. Thus, aryl ether-containing polyaromatics are not suitable for the long-term operation of alkaline membrane fuel cells (AMFCs), and polymer main chain degradation needs to be resolved to improve the overall durability of AMFCs. One of the most promising approaches to resolve the chemical stability of polymeric materials used in AMFCs is to prepare aryl ether-free polyaromatics. This approach allows polymer electrolytes

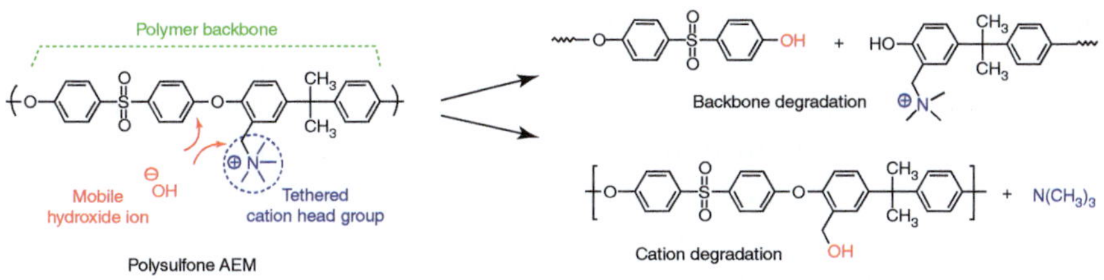

Figure 5.12 Structure of poly(arylene ether sulfone) AEM and its degradation mechanism of the polymer backbone and cation headgroup under alkaline conditions. Source: Noh et al. [9]/American Chemical Society.

to possess desirable characteristics of the polyaromatic backbone, such as high glass transition temperature, high impact strength, and toughness, good thermal, chemical, and mechanical stability, and low water uptake (WU) compared to polyolefinic electrolytes [13].

The grafting degree and position of the QA group can be controlled more precisely by using superacid catalysis to prepare an ion exchange membrane. Zolotukhin et al. first proposed preparing linear polycyclic aromatic hydrocarbons by a superacid-catalyzed polyhydroxyalkylation reaction of fluorinated carbonyl-containing compounds [14]. Because the QA precursor is acid resistant, it can be grafted into the polymer structure at the polymerization stage, and then the alkyl bromide can realize QA by a simple tertiary amine Menshutkin reaction. A comparison between the synthesis of typical AEM and superacid-catalyzed membranes is shown in Figure 5.13.

Another advantage of preparing ion-exchange membranes by superacid catalysis is that the reaction conditions do not contain metal-based catalysts, which can

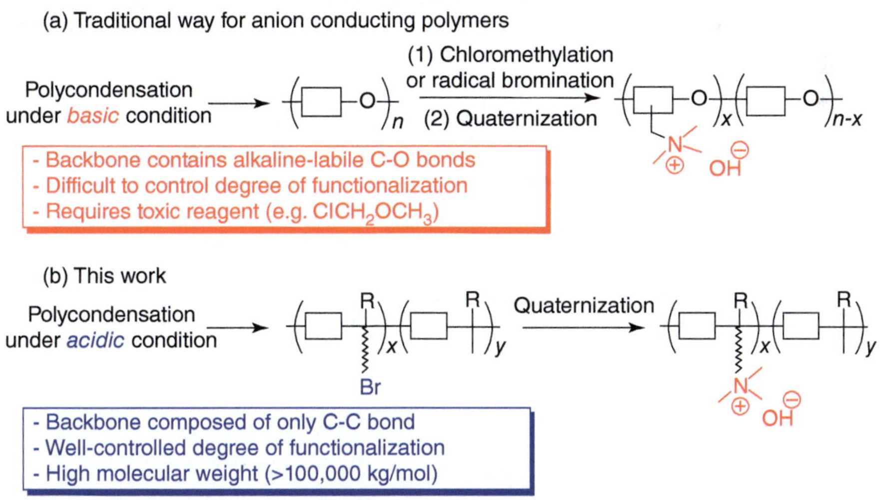

Figure 5.13 Graphical illustrations of the synthesis of (a) typical AEM membranes and (b) superacid-catalyzed QA-aromatic polymers. Source: Lee et al. [15]/American Chemical Society.

save costs. Polymers prepared by superacid catalysis can have high or ultrahigh molecular weights. The polymers are obtained by superacid-catalyzed polyhydroxyalkylation involving two consecutive steps: slow and fast. Polymer syntheses with a decrease in the reactivity of intermediates lead to narrow polydispersity polymers, while nonstoichiometric reactions result in polymers of high and ultrahigh molecular weight [16]. In addition, the polymerization has a high regional selectivity for the 4 and 4′ positions of the terminal aryl of the aromatic monomer, so the polymer can have a high molecular weight. However, to achieve successful superelectrophilic activation and growth of the long polymer chain, we must choose a ketone monomer with electron-withdrawing groups [9]. As reported by Lee et al., although the polymers' repeating units consisted of rigid biphenyl and methyl groups, they had high molecular weights. These polymers could also dissolve in organic solvents, such as $CHCl_3$ and tetrahydrofuran (THF). The selection of the monomer is the key to preparing high molecular weight polymers. Because of the rigid aromatic group in the backbone of the polymer, QA aryl-free alkaline membranes require a higher molecular weight to provide sufficient toughness. According to the selection of carbonyl-containing electrophilic monomers and phenyl monomers, a variety of polyaromatic compounds with different structures and properties can be prepared by superacid-catalyzed polymerization.

Polycyclic aromatic hydrocarbons with flexible skeleton structures of sp^3 carbon hybrids can be formed in the main chain prepared by superacid catalysis, which enables the ion exchange membrane to have good solubility in polar solvents and will not make the obtained membrane brittle and affect its mechanical strength. High molecular weight, QA-tethered poly (biphenyl alkylene) without alkaline labile C–O bonds were synthesized via acid-catalyzed polycondensation reactions by Lee et al. for the first time (Figure 5.14). To add a cationic functional group to the polyaromatic main chain, they prepared a bromopentyl side chain for QA.

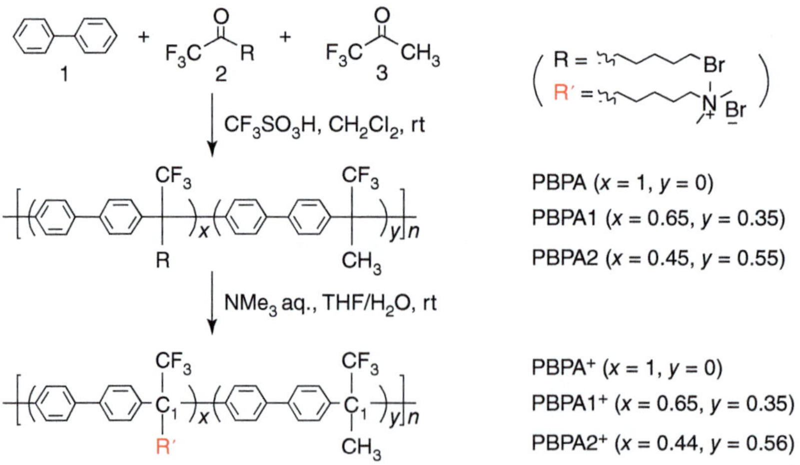

Figure 5.14 Synthetic route for quaternary ammonium polymers. Source: Lee et al. [15]/American Chemical Society.

The IEC of the obtained membrane was controlled by different ratios of trifluoromethyl alkyl ketones. The alkyl bromide was subsequently converted to trimethylpentyl ammonium (TMPA) via the Menshutkin reaction. Although the main structure was a rigid aromatic nucleus, it had good solubility in polar organic solvents due to its more flexible backbone structure with the presence of a quaternary sp^3 carbon [15].

As the monomers, solvents, and catalysts used in the preparation of ion exchange membranes by superacid catalysis are not very complex, another advantage is the easy availability of initial materials and reagents. Moreover, changes in structural factors and acidity of the reaction medium make it possible to adjust the reactivity of the carbonyl and aromatic components and thus control the reaction kinetics. In contrast, other approaches to synthesize aryl ether-free polyaromatics have some disadvantages. Park and Kim [17] have reported that there exist many kinds of approaches, such as Diels–Alder polymerization, metal-catalyzed coupling reactions, acid-catalyzed Friedel–Crafts polyhydroxyalkylation, and cyclopolycondensation.

The Diels–Alder reaction is a concerted [4+2] cycloaddition between a conjugated diene and a dienophile into a cyclohexene. The key monomers to synthesize the backbone of Diels–Alder polyphenylene (DAPP), bis(cyclopentadienone) and its derivatives, are not commercially available, so they need to be synthesized via multiple-step reactions. Diels–Alder polymerization does not require a metal catalyst and can produce polymers consisting of completely rigid phenyl backbones. Due to the rigid and stiff nature of the polymer backbone, which is entirely composed of phenyl rings, Diels–Alder polymerization requires a higher molecular weight than other polyaromatics to obtain good mechanical properties. Additionally, the bulky backbone structure may lead to high free volume between the polymer chains, which may increase the reactant gas permeability compared to linear polymers. In addition, the synthetic limitation of incorporating cation precursor moieties into the monomer for easy quaternization is that it also requires multiple-step postpolymerization functionalization for the AEM application of these materials [17].

Transition-metal-catalyzed-cross-coupling reactions are another class of reactions that have been utilized in polymer chemistry to build a polymer backbone composed of aromatic C—C bonds. Choosing appropriate aryl monomers is very important, as they greatly affect the properties of polymer electrolytes. The polyaromatics made by the synthetic method commonly feature their p-conjugated backbone structure as well as good chemical and thermal stabilities and are used widely for optical and electronic applications. The disadvantage of the utilization of Suzuki coupling reactions is the requirement of expensive palladium metal catalysts and boron-containing monomers, which would economically limit the approach, thus scaling up the synthesis of the materials. While nickel-catalyzed polymerization is cost-effective due to less expensive nickel catalysts and aryl chloride monomers, it should be noted that a large quantity of nickel catalyst is required for polymerization (2.5 equivalents molar amount of aryl chloride monomers) [17].

The main difference between the polymeric ionenes and the previously introduced polymers prepared by the different synthetic routes is that the former has ionic groups as part of the polymer main chain, while the latter contains ionic groups as the pendant group on the side chain. The nature of cyclo-polycondensation directly leads to the formation of cationic groups and the polymer backbone, which is simple, however, the resulting polymers have high IECs that reduce their mechanical stability and limit their practical applicability without physical modification [17].

The above are some advantages of the preparation of ion exchange membranes prepared by superacid catalysis. More advantages of this preparation method need to be explored by researchers in practical operation. However, there is no denying that using superacid catalysis to prepare ion exchange membranes has many advantages that other preparation methods cannot compare, so it has attracted the interest of a large number of researchers in recent years and has been widely studied and applied in various application fields. Alkaline AEM are usually polymers with free OH^- anions. The overall stability of alkaline AEM is limited by the polymeric backbone and organic cations. Lee et al. [15] synthesized high molecular weight, QA-tethered poly (biphenyl alkylene) without alkaline unstable C—O bonds via acid-catalyzed polycondensation reactions for the first time (Figure 5.14). The IECs of the membranes could be easily controlled by adjusting the feed ratio of the two ketone monomers (7-bromo-1,1,1-trifluoroheptan-2-one and methyl trifluoromethyl ketone) in the polymerization. The polymer prepared by superacid catalysis not only exhibited high molecular weight but was also soluble in common organic solvents, which was convenient for the next QA reaction. The QA polymer was obtained by the substitution reaction with trimethylamine to convert the bromopentyl group on the side chain of the precursor polymer to TMPA. The prepared anion exchange membrane showed a high hydroxide ion conductivity of up to 122 mS/cm. After 30 days of immersion in 1 mol/l NaOH solution at 80 °C, the ion exchange capacity of the membrane did not change, showing remarkable chemical stability.

Zolotukhin and coworkers [18] first reported in 2004 that the polyhydroxyalkylation reaction between piperidone and electron-rich phenyl monomers produced polymers containing aromatic rings and piperidine rings in the backbone structure. Marino and Kreuer [19] investigated the alkaline stability of 26 kinds of different QA groups for temperatures up to 160 °C and NaOH concentrations up to 10 mol/l. Most QAs exhibit unexpectedly high alkaline stability with the exception of aromatic cations. β-Protons were found to be far less sensitive to nucleophilic attack than previously suggested, whereas the presence of benzyl groups, nearby heteroatoms, or other electron-withdrawing species significantly promoted degradation reactions. Cyclic QAs proved to be exceptionally stable, with the piperidine-based 6-azonia-spiro [5.5]undecane featuring the highest half-life under the chosen conditions.

Jannasch and coworkers [20] first thought of adding piperidine to the superacid-catalyzed anion exchange membrane backbone structure in 2018 (Figure 5.15). A series of poly(arylene piperidinium)s (PAPipQs) without any alkali-sensitive aryl ether bonds or benzylic sites were prepared and studied as AEMs for alkaline fuel cells. Medium molecular weight PAPipQs were then synthesized in polycondensations of *N*-methyl-4-piperidone and either bi- or terphenyl via superelectrophilic

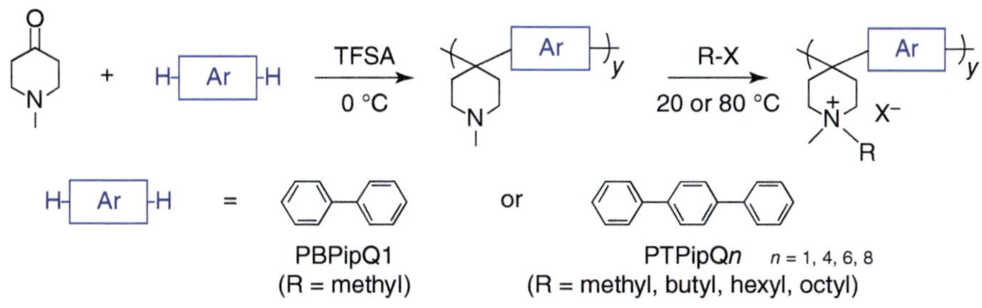

Figure 5.15 Synthetic pathways to PAPipQs based on biphenyl or *p*-terphenyl and carrying different alkyl extender chains [R]. Source: Olsson et al. [20]/American Chemical Society.

activation in TfOH. Film-forming PAPipQs were subsequently prepared in Menshutkin reactions with methyl, butyl, hexyl, and octyl halides. AEMs based on poly (terphenyl dimethylpiperidinium) showed the best performance with no structural degradation after storage in 2 M aq. NaOH at 60 °C after 15 days and a mere 5% ionic loss at 90 °C. The authors introduced longer alkyl chains into the piperidine cation to reduce water absorption to obtain higher conductivity. However, the presence of longer pendant *N*-alkyl chains (butyl to octyl) was found to significantly promote Hofmann ring-opening elimination reactions, and the degradation rate increased with increasing alkyl chain length.

On this basis, Jannasch and coworkers [21] continued to introduce cyclic QA to the piperidine rings. First, they prepared poly(biphenyl piperidine) by superacid-catalyzed polycondensation. They introduced different *N*-spirocyclic QA cations via cyclo-quaternization of the piperidine rings (Figure 5.16). The resulting polymers and AEMs were free of diaryl ether linkages and benzylic C—H bonds and showed very high thermal stability and hydroxide ion conductivity.

Jannasch and coworkers [22] systematically varied the structures of the cation and the backbone polymer. Two monomers were initially prepared by attaching 4-benzylpiperidine groups to trifluoroacetophenone and m-terphenyl through Suzuki coupling reactions. Polymerizations followed by quaternizations were then carried out to obtain poly(terphenyl alkylene) with approximately the same ionic

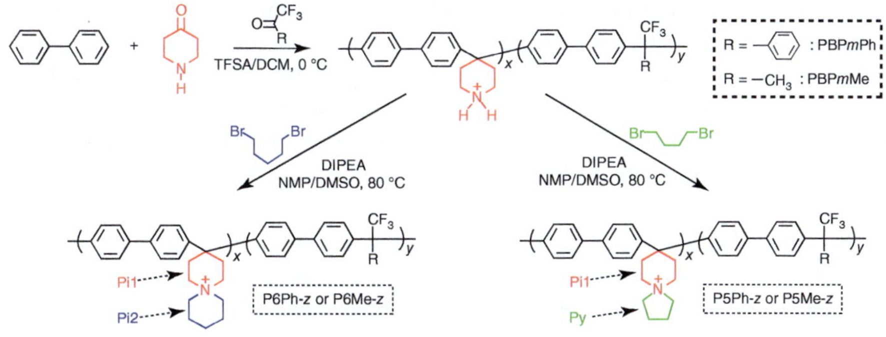

Figure 5.16 Synthesis pathway to poly(arylene alkylene)s functionalized with *N*-spirocyclic QAs via copolycondensation and cyclo-quaternization. Source: Pham et al. [21]/Royal Society of Chemistry.

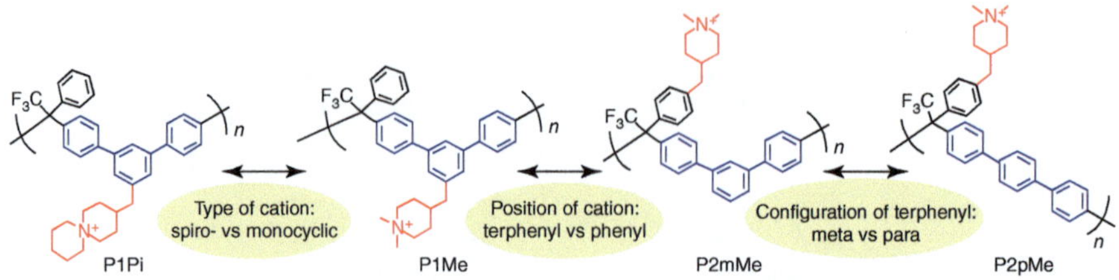

Figure 5.17 Overview of the variations in the molecular design of the cationic polymers synthesized and investigated as HEMs. Source: Huongpham et al. [22]/Royal Society of Chemistry.

contents (Figure 5.17). The author found that the flexibility of polymer chains and the functionalization of cations promoted ionic aggregation and conductivity. Functional synthesis of biphenyl and ketone monomers with piperidone could systematically change the type and location of cations and rigidity of main chain polymers. Although all the designed AEM showed high alkaline stability, those AEMs functionalized by N,N-dimethylpiperidinium also showed high thermal and alkaline stability, high hydroxide conductivity, low swelling ratio (SR), and WU. In particular, placing N,N-dimethylpiperidinium cations on the side phenyl of ketone monomer residues, rather than on the more rigid skeleton terphenyl units, led to a further increase in the alkaline stability and hydroxide conductivity of the anion exchange membrane. The large spirocyclic cation provided very high thermal stability for the anion exchange membrane, but the ion aggregation and conductivity were relatively low. In addition, when compared at the same skeleton and position, spirocyclic cations exhibited lower alkaline stability than N,N-dimethylpiperidinium cations, possibly due to the greater degree of bond angle deformation of the former cations when attached to the rigid polymer backbone, which might promote degradation.

Jannasch [23] further reported aryl ether-free 2,7-diphenylflfluorene-based copolymers, as shown in Figure 5.18. They synthesized 2,7-diphenylfluorene monomers through Suzuki coupling to increase polymer backbone stiffness. Then, they obtained polymers through superacid-catalyzed Friedel–Crafts polycondensation of the coupled fluorenyl monomer and trifluoroacetophenone, followed by quaternization reactions. The membrane's OH^- conductivity reached 100 mS/cm at 80 °C and showed excellent thermal stability and alkaline stability. Immersed in 2 M NaOH aqueous solution at 90 °C for more than 672 hours, the NMR result showed that the double-ring cage quinuclidine structure was completely free of any open-ring β-elimination. In addition, the β-elimination in the hexyl spacer group was less than 2%.

Although the anion exchange membrane prepared by superacid catalysis has achieved excellent performance, Maurya et al. [24] believed that anion exchange ionomers should have minimum adsorption to the catalyst and high oxidation stability, but most aryl anion exchange ionomers showed high phenyl adsorption, such as poly (aryl piperidine) homopolymer ionomers. Young Moo Lee [25]presented a poly (diphenyl-terphenyl piperidinium) copolymer containing an aliphatic chain to

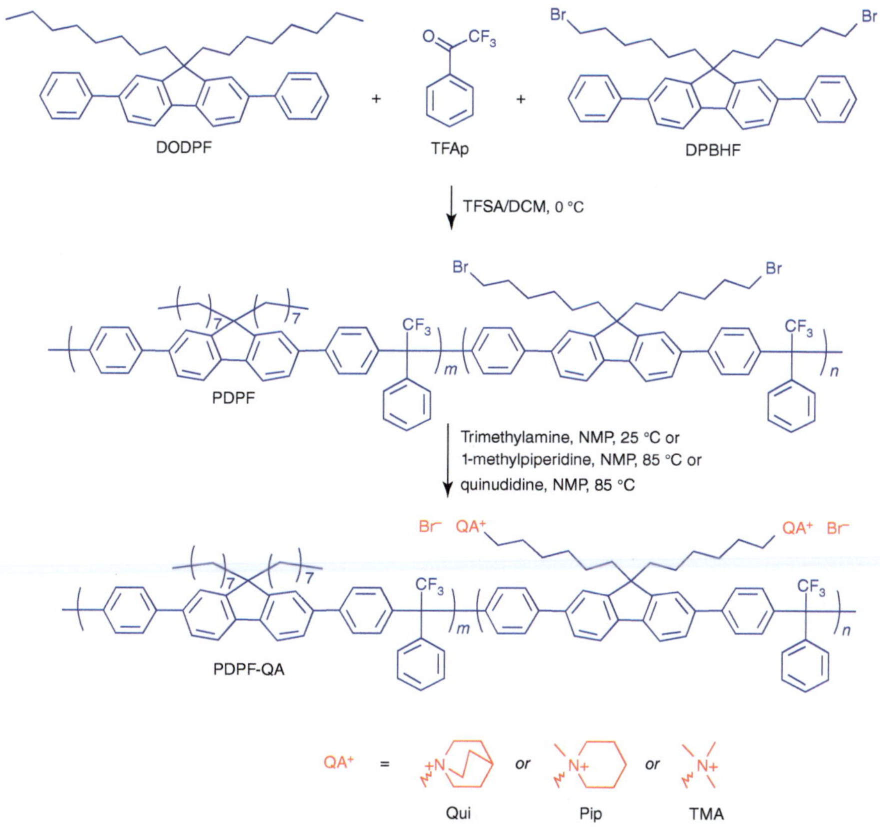

Figure 5.18 Preparation of the PDPF copolymer by superacid-catalyzed polycondensation of DODPF, DPBHF, and TFAp, followed by quaternization reactions to attach Qui, Pip, and TMA cations, respectively. Source: Andrit et al. [23]/Royal Society of Chemistry.

reduce the phenyl content and adsorption of AEIs and to increase the mechanical properties of AEMs. They obtained poly (diphenyl-terphenyl piperidinium) from a Friedel–Crafts condensation reaction of p-terphenyl, 1,2-diphenylethane, and N-methyl-4-piperidone (Figure 5.19). The ion conductivity can be as high as 166 mS/cm at 80°C. The anion exchange membrane also showed excellent peak power density, reaching 2.58 W/cm^2 in alkaline fuel cells.

On this basis, Young Moo Lee [26] presented poly(fluorenyl aryl piperidinium) ionomers for AEM (Figure 5.20). The fluorene (FLN) segment in the copolymers improved the rigidity and phase-separated morphology of AEMs to increase the dimensional stability and ion conductivity. Moreover, AEIs with rigid FLN groups are expected to improve the water vapor permeability (or water diffusivity) and decrease the phenyl adsorption effect. They obtained poly (fluorenyl aryl piperidinium) from the Friedel–Crafts condensation reaction of diphenyl/p-terphenyl, FLN, and N-methyl-4-piperidone. The obtained anion exchange membrane had an extremely high OH$^-$ conductivity of 208 mS/cm at 98 °C. The membrane also exhibited excellent mechanical properties and alkaline stability, and its tensile

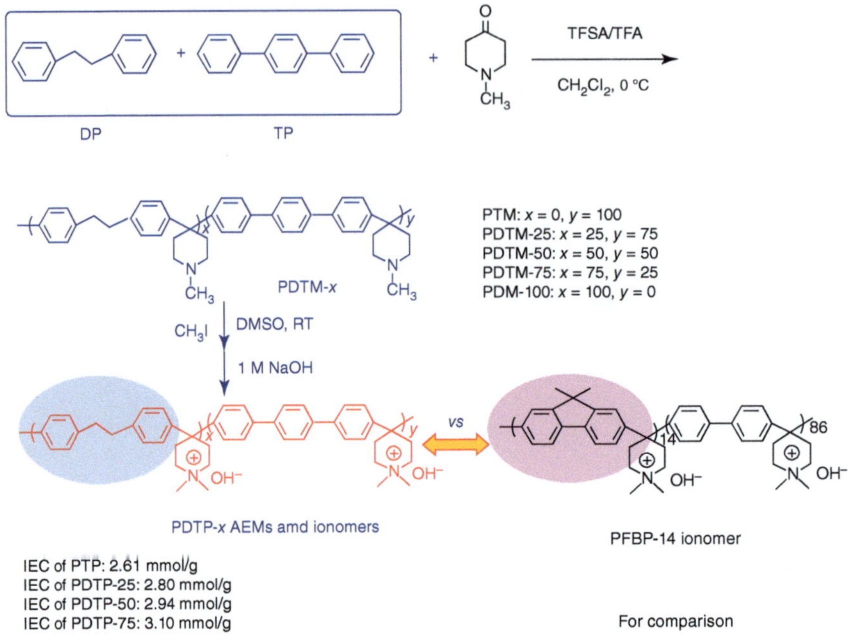

Figure 5.19 The synthesis routes of aliphatic chain- and fluorine-containing poly(aryl piperidinium)s. Source: Chen et al. [25]/John Wiley & Sons.

strength reached 84.5 MPa. Less than 3% of the ion conductivity was lost after immersion in 1 mol/l NaOH solution at 80 °C for 2000 hours.

Many domestic scholars have also begun to use superacid catalysis to prepare high-performance AEM. Nanwen Li [27] found that a membrane with a QA cation group and poly([fluorene alkylene]-co-[biphenyl alkylene]) (PFBA) backbone exhibited the same long-term alkaline stability as membranes with a bulky imidazolium cation group in combination with the same polymer backbone. Herein, they attached QA and bulky imidazolium to long-side-chain type PPO, PFBA, and PFA backbones and studied the alkaline stability of AEMs prepared using these polymers. The anion exchange copolymer was obtained by the Friedel–Crafts condensation reaction of biphenyl, long-chain fluorene alkyl, and trifluoroacetone, followed by a QA reaction with trimethylamine/imidazole. The prepared anion exchange membrane reached a high conductivity of 80 mS/cm at room temperature. In addition, the membrane also showed excellent alkaline stability, with almost no ion conductivity loss after 30 days of immersion in 2 mol/l NaOH at 80 °C. Hydrogen/oxygen fuel cells were tested at 60 °C. The results showed that the polymer backbone and cations had significant effects on the performance of fuel cells. The prepared anion exchange membrane PFBA-QA-0.4 peaked power density of 610 mW/cm^2 in alkaline fuel cells. The durability test showed that the voltage of the PFBA-QA-0.4 membrane can be maintained above 0.7 V for 74 hours at 200 mA/cm^2. In the subsequent study, Li and coworkers [28] prepared highly conductive poly ([terphenyl piperidinium]-co-[oxindole terphenylylene]) (PTP-x)

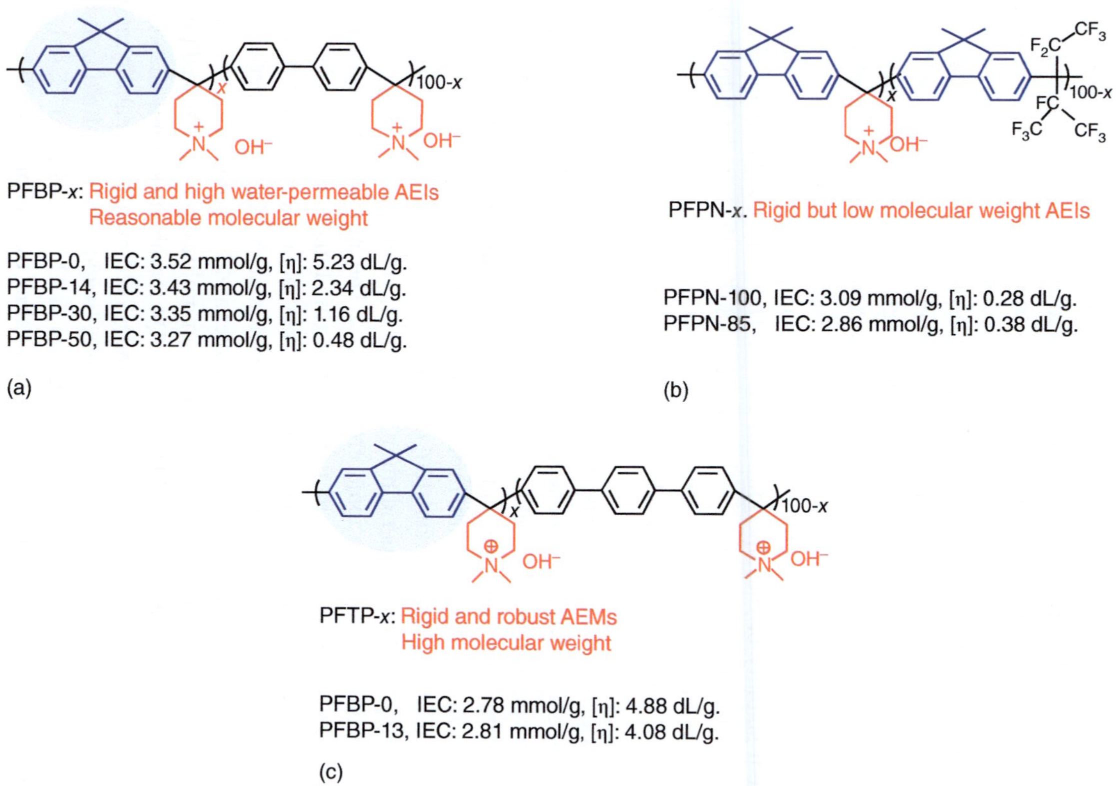

Figure 5.20 Synthesis route of poly(aryl piperidinium) of the fluorenyl backbone. Source: Chen et al. [26]/Springer Nature.

AEMs with aryl ether-free backbones by a superacid-catalyzed polycondensation reaction of p-terphenyl, isatin and N-methyl-4-piperidone and subsequent quaternization with methyl iodide. The highest OH^- conductivity of the prepared anion exchange membrane reached 64.4 mS/cm at 20 °C and 128.9 mS/cm at 80 °C. The ion conductivity loss was less than 10% after immersion in 1 mol/l NaOH solution at 80 °C for more than 300 hours. The membrane showed good initial performance in an alkaline electrolyzer. The current density was 910 mA/cm^2 at 55 °C and increased to 1000 mA/cm^2 at 75 °C. Recently, Li and coworkers [29] found that when the alkyl side chain was directly placed on the backbones, comb-shaped AEMs that possessed high conductivity and durability led to improved performance in alkaline fuel cells. Herein, they proposed a copolymerization strategy to prepare a comb-shaped poly(arylene indole piperidinium) copolymer (PITP-CnQx) with the aim of increasing the conductivity and maintaining the alkaline stability, in which the hydrophobic alkyl side chain was directly fixed on the aryl–ether free backbone. The hydrophobic alkyl side chain directly attached to the backbone was not only responsible for inducing microphase separation but also expected to maintain the excellent stability of the piperidinium cation. The OH^- conductivity of this membrane reached 134.5 mS/cm at 80 °C. More than 80% of the initial conductivity could be maintained after immersion in 1 mol/l NaOH solution for 1200 hours. The membrane could achieve a peak power density of 445 mW/cm^2 in alkaline fuel cells.

Researchers also continuously solved the existing problems in the preparation of AEM by superacid catalysis through new designs and made innovations in structure to optimize the performance of ion exchange membranes. It is generally believed that cross-linking can effectively optimize the performance of AEM. However, common cross-linking agents with hydrophobic structures will cause embrittlement of AEM. Chen and co-workers [30] designed a series of tunable multication-cross-linked poly (biphenyl piperidinium) (PBP-xQ4) membranes and prepared them by embedding flexible cationic cross-linkers. The number of alkyl spacers between the cationic groups in the flexible cross-linkers was optimized to five and six to achieve optimum performance. They obtained polymers from a Friedel–Crafts condensation reaction of diphenyl and N-Methyl-4-piperidone. Then, they performed quaternizations of this polymer and 1,16-dibromo-5,11-(N,N-dimethylammonium) hexadecane bromide to obtain multication-cross-linked poly (biphenyl piperidinium) copolymers. These PBP-xQ4 membranes simultaneously had high ion conductivity (155 mS/cm at 80 °C) and good dimensional stability, and these membranes could be folded into many shapes. In addition, cross-linking also promoted the alkaline stability of these membranes in 2 M NaOH at 80 °C, only a slight degradation was detected by ^{1}HNMR after alkaline treatment for 1800 hours. This might be because the good microphase structure of the membrane improved the ion transport efficiency and durability. The open circuit voltage of PBP-20Q4 reached 1.05 V, which was close to the theoretical value of 1.23 V, indicating almost no feed-gas penetration in the membrane. In addition, the PBP-20Q4 membrane realized a maximum power density (P_{max}) of 234 mW/cm^2 at a current density of 450 mA/cm^2 (Figure 5.24b). The results showed that the multication cross-linking strategy is promising for the application of poly (biphenyl piperidinium)-based alkaline membranes in fuel cells. The maximum power density of these cross-linked membranes can be further improved by increasing dimensional stability, which will be the goal of future studies.

The alkaline stability of the spiral ring structure membrane is generally excellent, but as the spiral ring rigidity is too strong and the dimensional stability is poor, the molecular weight is difficult to increase, and the prepared membrane is too brittle. To solve this problem, Zhu et al. [32] introduced a profitably cross-linked poly (biphenyl piperidinium) segment into the polymer backbone to improve the membrane formation performance and optimize the microphase separation of the membrane, which significantly improved the ion conductivity and dimensional stability of the membrane. The ionic conductivity of the cross-linked membrane reached 116.1 mS/cm at 80 °C, and the membrane exhibited alkaline stability for more than 2000 hours at 80 °C in a 3 mol/l NaOH aqueous solution. Zhu and coworkers [33] found that m-terphenyl could effectively regulate the morphology and properties of AEM due to its spatial twisted structure as a polymer monomer. They obtained polymers from a Friedel–Crafts condensation reaction of m-terphenyl, trifluoroacetophenoneand N-methyl-4-piperidone. Then, they performed quaternization of this polymer and 5-bromo-N,N,N-trimethylpentan-1-ammonium bromide to obtain poly (meta terphenyl piperidinium) copolymers. Because of the self-assembly of copolymers, these multication side chains with folded and spatially twisted structural backbones were bound by multication side chains, resulting in a good microphase separation morphology. The ion conductivity of the membrane reached 164 mS/cm at 80 °C. In addition, stable piperidine cations and long alkyl spacer chains helped to improve the alkaline stability of the membrane. After being immersed in 2 mol/l NaOH aqueous solution at 80 °C for 1500 hours, the ion conductivity and ion exchange capacity of the membrane only decreased by 11.67 and 12.73%, respectively. The peak power density of the anion exchange membrane in alkaline fuel cells could also reach 269 mW/cm^2.

Azaspirocyclic cations are among the most alkali-stable cations. Zhu and coworkers [34] synthesized and studied poly (biphenyl piperidine) AEM grafted with different proportions of azaspirocyclic cations on the side chain. The WU and SR of the membrane were limited by the increase in molecular weight. It was found that increasing the grafting ratio of the side chain cation increased the SR and WU of membranes, and the ion conductivity first increased and then decreased. Additionally, proper grafting of the side chain cation could effectively improve the hydrophilic–hydrophobic microphase separation structure. The membrane reached the highest hydroxide ion conductivity of 117.43 mS/cm at 80 °C. After being immersed in 2 mol/l NaOH aqueous solution at 80 °C for 1400 hours, the degradation rate was less than 14%. Overall, this study provides a method for reducing the WU and SR of membranes and for increasing the ion conductivity of alkaline stable AEMs.

To prepare AEMs with high conductivity and alkaline stable ion channels, Zhang et al. [35] of our research group reported a self-aggregating side chain design strategy with the best alkaline stability. They inserted dipolar ethylene oxide (EO)

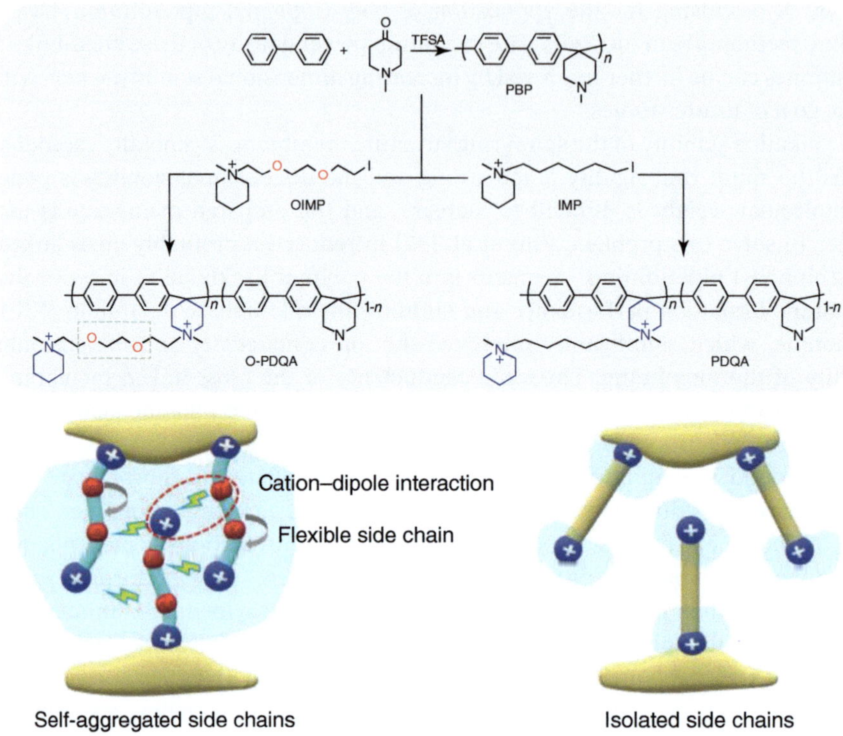

Figure 5.21 Synthetic routes of O-PDQA (with the self-aggregated side chain due to the cation–dipole interaction) and PDQA (isolated side chain). The schematic diagram shows the function of the self-aggregated side chain, which can promote the formation of the aggregated conducting region. Source: Zhang et al. [35].

spacers in the cationic side chain. They precisely manipulated the polyelectrolyte self-assembly process to form the required microstructures with ion conduction channels (Figure 5.21). This concept of constructing ion-conduction highways using self-aggregating cationic chains will benefit many fields, which involve anion-conductive electrolytes (fuel cells, redox flow batteries, electrolyzers, etc.). Simulation and nanoscale microscopy analyses verify the self-assembly process of the flexible side chain with cation–dipole interactions to construct interconnected ionic highways for fast water and ion transportation. The prepared O-PDQA AEM exhibited higher hydroxide conductivity (106 mS/cm at 80 °C).

TEM micrographs and atomic force microscopy (AFM) phase images of PDQA provide visible evidence for the presence of ionic clusters(Figure 5.22c, d, e, f). In the TEM image of O-PDQA, the aggregated ionic clusters (dark regions) penetrated the hydrophobic phase of the polymer backbone (light regions). The tapping mode AFM phase image also shows a similar phase-separated morphology with interconnected ionic clusters (ca. $d = 7$ nm). In contrast, there is no apparent microphase separation for PDQA due to its poor self-assembly ability. The ionic clusters are dispersed in the polymer matrix and isolated from each other. Additionally, in the AFM image of PDQA, no distinct phase difference was found. These results highlight the role of

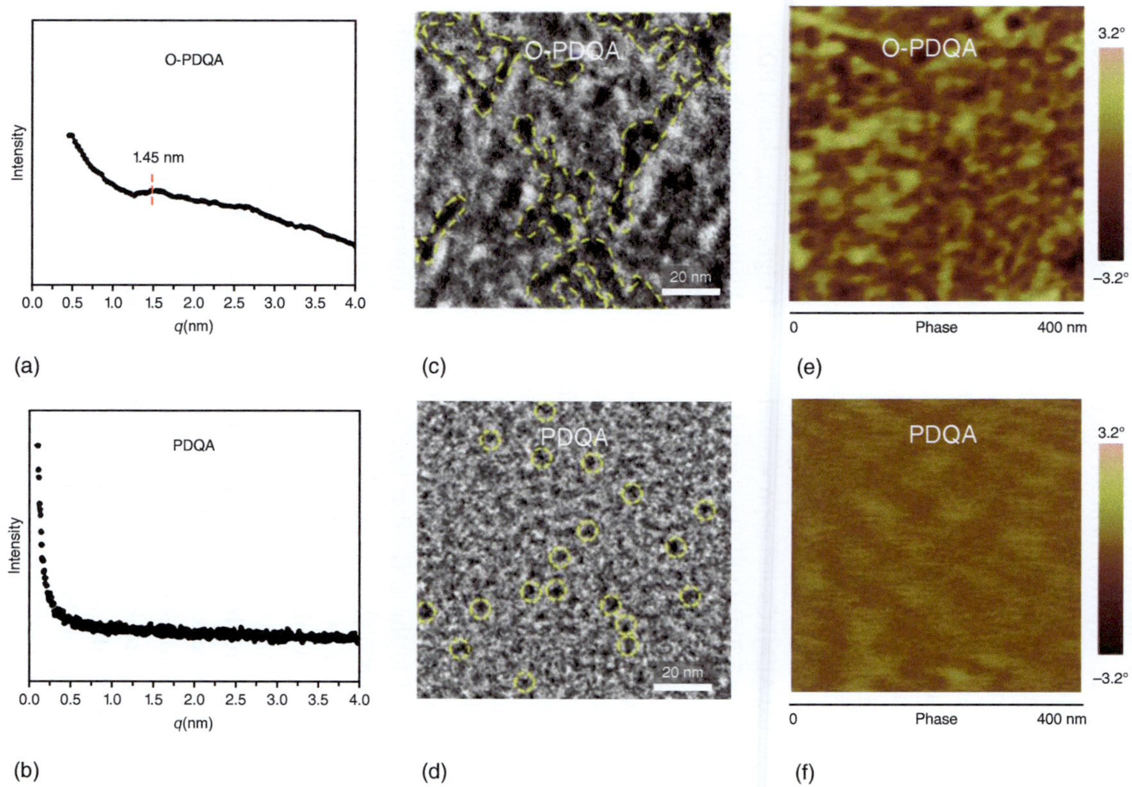

Figure 5.22 Microstructural characterization of exemplar AEMs. SAXS spectra of (a) O-PDQA (IEC = 1.93 mmol/g) and (b) PDQA (IEC = 1.98 mmol/g) AEMs. TEM images of (c) O-PDQA and (d) PDQA, where the bright and dark regions correspond to the hydrophobic backbone and hydrophilic QA ionic clusters, respectively. AFM phase images of (e) O-PDQA and (f) PDQA membrane. Source: Zhang et al. [35]/With permission of Royal Society of Chemistry.

EO segments in facilitating phase separation by providing cation–dipole interactions and improving side-chain mobility.

To investigate the benefits of the self-aggregated side chain for ion transport, the hydroxide conductivity of O-PDQA was compared with that of PDQA, as shown in Figure 5.23a, b. At similar IECs, O-PDQA exhibits higher ion conductivity than PDQA over the whole temperature range (30–80 °C). The introduction of EO spacers optimized the hydration environment for ion transport in the channel. EO spacers could serve as additional hydrated OH hopping sites by forming a hydrogen-bonded network with water. To evaluate the long-term alkaline stability of the as-prepared AEMs, O-PDQA, and PDQA membranes were treated in 2 M aq. NaOH at 80 °C for 1080 hours (Figure 5.30c). Both O-PDQA and PDQA maintain high hydroxide conductivity after 1080 hours, highlighting the excellent alkaline stability of piperidinium groups and the aryl–ether-free backbone. Specifically, conductivity retention >96% was observed for the O-PDQA membrane, which is slightly higher than that of PDQA (93.2%). The AEMFC containing O-PDQA produces a peak power density of 1.18 W/cm^2 at 70 °C without back pressure, which is competitive with other previously reported AEMFCs and approximately twofold higher than that of PDQA (0.60 W/cm^2) (Figure 5.30d). O-PDQA-3 with higher conductivity reduced the internal resistance of the cell. O-PDQA-3 shows reduced charge-transfer resistance, thus accelerating the electrochemical reaction in the fuel cell and giving a higher power density. This is due to the fast ion transport and water back-diffusion provided by the well-connected hydrophilic channel. The results further highlight the benefit of the self-aggregation of the EO-containing side chain in O-PDQA AEMs.

Our research group [31] also prepared a series of AEMs with aryl–ether-free polymer backbones bearing long flexible alkoxy-containing bis-piperidinium cationic side chains and additional hydrophobic alkyl chains. The rotatable hydrophilic cation strings (bis-piperidinium linked by a flexible ether oxygen spacer) were grafted to the aryl–ether-free poly (isatin biphenylene) main chain via a long hydrophobic 6-alkyl spacer (Figure 5.24). The benefits of using the long hydrophobic spacer are twofold. First, the spacers can distance the cation strings from the main chain to mitigate the restriction from the main chain. Second, the unreacted hydrophobic spacer away from the cationic chains can further drive the aggregation of the hydrophilic ionic clusters. The synergistic effect can accelerate the self-aggregation process to form the desired ionic channels for fast ion transfer. Meanwhile, piperidinium cations and aryl–ether-free main chains were adopted to guarantee alkaline stability. Various morphology analyses confirmed that this structure can promote the self-assembly of cationic groups. Due to the enhanced flexibility and interaction between functional side chains, the prepared AEMs have excellent self-assembly capability and ion conductivity (Figure 5.25a, b). The hydroxide ions in AEMs can be chemically converted to bicarbonate or carbonate in the air. Thus, the ion conductivity of AEMs in bicarbonate and chloride forms was also measured. A high HCO_3^- conductivity of 62.17 mS/cm and Cl^- conductivity of 55.63 mS/cm were observed at 80 °C, which were competitive compared with those of previous reports (Figure 5.32c). Therefore, the AEMs have sufficient CO_2 tolerance, which has potential application in a harsh environment. To assess the

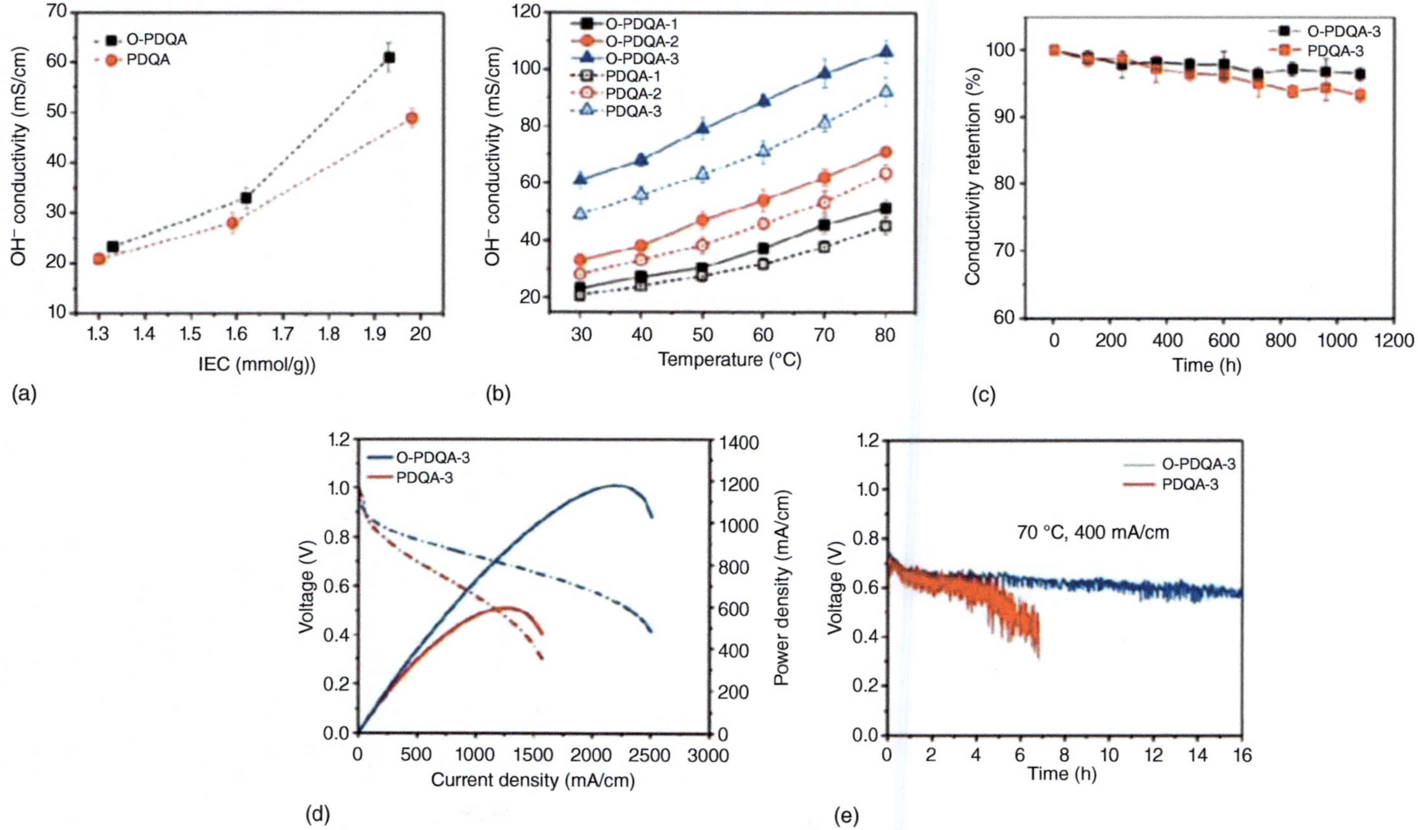

Figure 5.23 (a) The OH conductivities as a function of IEC of the O-PDQA and PDQA membranes at 30 °C, (b) the OH^- conductivities of O-PDQA and PDQA at increasing temperatures, (c) conductivity retention of the O-PDQA and PDQA membranes when immersed at 80 °C in 2 M aq. NaOH for different time intervals, (d) *I*–*V* polarization curves (dash) and power density curves (solid) of O-PDQA-3 and PDQA in H_2/O_2 AEMFC tests at 70 °C, and (e) AEMFC durability of O-PDQA and PDQA operating at 400 mA/cm^2, 70 °C, 0.5 l/min gas flow at 100% RH. Source: Zhang et al. [35]/Royal Society of Chemistry.

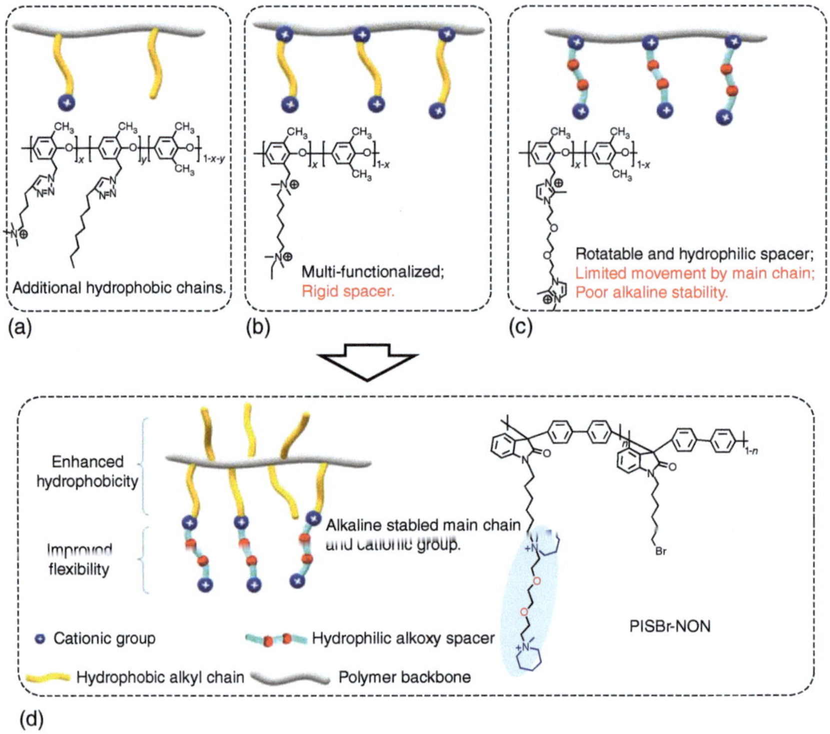

Figure 5.24 Structures and schematic illustrations of the (a) additional hydrophobic chain-grafted AEM, (b) multifunctional side-chain-type AEM, (c) rotatable side-chain-type AEM, and (d) the targeted PISBr-NON AEM structure in this work. Source: Zhang et al. [31]/American Chemical Society.

alkaline stability, the AEM samples were immersed in a 2 M NaOH solution at 80 °C for accelerated aging, and the ion conductivity of the AEMs decreased to 93.45% of the initial value after 1080 hours, indicating its excellent tolerance in a high pH environment for long periods. The open-circuit voltage of the AEM is 1.01 V, implying its excellent properties to prevent gas penetration. The AEM achieved a peak power density of 489.8 mW/cm^2 under a current density of 867.2 mA/cm^2 at 70 °C without back pressure and 514.4 mW/cm^2 under a current density of 815.2 mA/cm^2 with a 0.1 MPa back pressure (Figure 5.25d). Compared with reported AEMs, the PISBr-NON-4 AEM shows high ion conductivity but a very general peak power density. This implies that the fuel cell performance of AEMs is not always positively correlated to the conductivity. Additionally, the same problem appeared for the stability issues. It reminds us that a high ex situ performance (including the ion conductivity and alkaline stability) may not give a corresponding in situ performance in the device. The high-frequency resistance (HFR) of the PISBr-NON AEM is approximately 40–45 mΩ (Figure 5.25e), which is relatively higher compared to the advanced AEM in the literature. The unsatisfactory performance mainly stems from the ohmic resistance of the various layers (catalyst layer and gas diffusion layers, GDLs) and contact resistance.

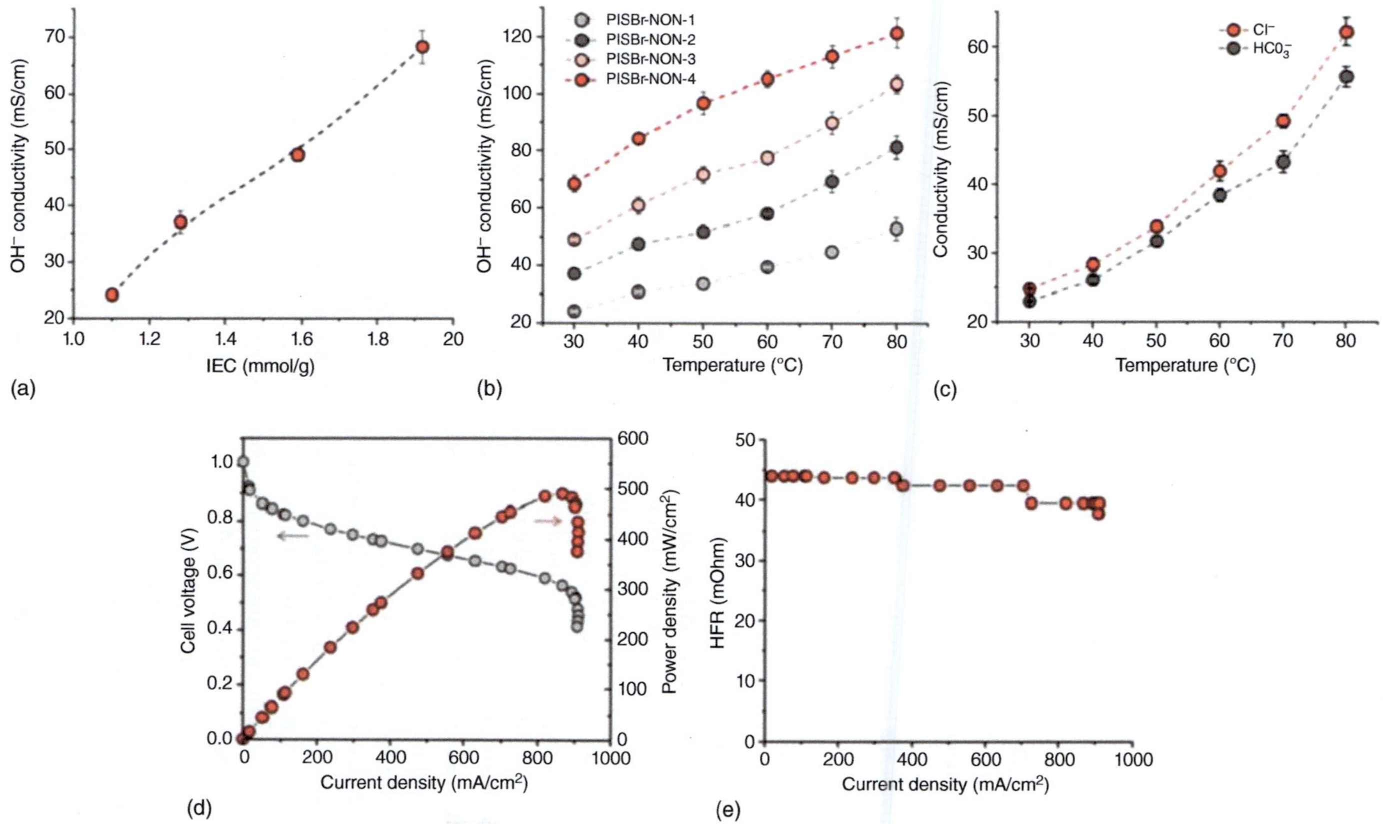

Figure 5.25 (a) Hydroxide conductivity of PISBr-NON AEMs as a function of IEC at 30 °C, (b) hydroxide conductivity of the PISBr-NON (IEC = 1.88 mmol/g) AEM as a function of temperature, (c) HCO_3^- and Cl^- conductivities of the PISBr-NON (IEC = 1.88 mmol/g) AEM as a function of temperature, (d) H_2/O_2 single-cell AEMFC performance, and (e) HFR test of PISBr-NON (IEC = 1.88 mmol/g) at 70 °C and 100% RH. Source: Zhang et al. [31]/American Chemical Society.

Our research group [37] designed an "ether-free" main chain with twisted structures by introducing spirobifluorene units with noncoplanar structures. The side chains are linked by C—C bonds at the twisted sites via the Suzuki coupling reaction. Eventually, novel all-carbon twisted backbone-based AEMs are obtained. The all-carbon backbone without any heteroatoms guarantees the excellent chemical stability of the membranes. Moreover, due to the noncoplanar structure of the spirobifluorene units, the backbone of the main chain is distorted to reduce the steric hindrance, and the side chain gains more removable space (Figure 5.26a). All of these factors facilitate the aggregation of functional groups and the formation of ion channels, endowing the membrane with excellent conductivity at a low IEC level.

In general, the high OH^- conductivity of AEMs depends on a sufficient amount of ion exchange group content to form continuous hydrophilic ion channels. However, with increasing IEC, the WU and SR of the membrane usually increase significantly. Excessive swelling will lead to a decrease in the dimensional stability and mechanical properties of the membrane, which will seriously affect its application. Therefore, it is of great significance to obtain an AEM with high conductivity at a lower IEC level. The AEMs prepared in this work were maintained at a lower IEC level of less than 1.5 mmol/g. Accordingly, the prepared membrane has low WU and excellent dimensional stability. The WU rate of the QPS-PB-4 membrane at 30°C was 50.2%, and the SR of all membranes was not more than 10% (Figure 5.26b). On the other hand, even at low IEC levels, the prepared membranes still showed superior hydroxide conductivity. The conductivity of the QPS-PB-4 membrane was 50.1 mS/cm at 30 °C and increased gradually with temperature, reaching 94.4 mS/cm at 80 °C (Figure 5.26c). This property might benefit from the unique twisted backbone structure that reduces the steric hindrance of the main chain to the side chain, which provides more space for the side chain to swing and enhances the aggregation of functional groups. Furthermore, the swing of the side chain promotes the diffusion and continuous interconversion between the different hydrated states of hydrated hydroxide ions. Therefore, even at a low IEC level, a continuous and efficient ion channel can still be formed.

To verify this conjecture, simulated calculations of the swing ability of side chains grafted onto twisted and conventional backbones were performed. As shown in Figure 5.26d, e, the color from blue to red reflects the increased probability density distribution of the angle distance between cationic groups and the main chain plane. The ranges of d-value changes for both the conventional structure and the twisted grafting structure are approximately 2 Å, suggesting similar folding behavior of the side chains. However, the θ values of the twisted grafting structure show a large angular variation of approximately 70°, which is in sharp contrast to the conventional structure of approximately 30°. This reflects the enhanced swing ability of the side chain due to the low steric hindrance brought by the twisted structure.

The microphase separation morphology of the prepared AEMs was observed via AFM (Figure 5.27a, b). The dark areas represent the hydrophilic domains on the membrane surface, and the light-colored areas represent the hydrophobic domains. Consistent with the calculated results, although the QPSPB-4 membrane only has a low IEC of 1.48 mmol/g, it still exhibits a rather pronounced

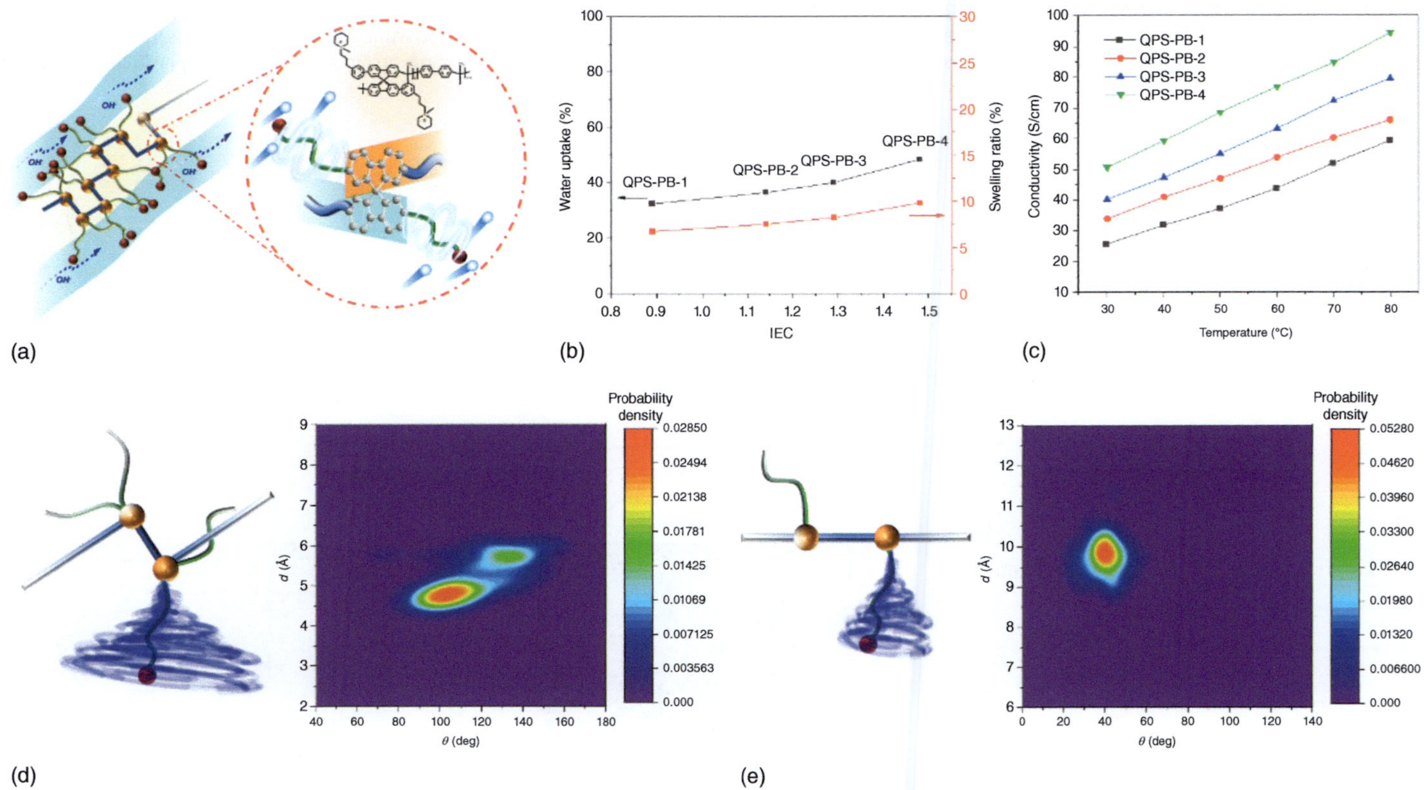

Figure 5.26 (a) Conduction mechanism of QPS-PB membrane, (b) water uptake and swelling ratio of QPS-PB of different IEC, (c) OH^- conductivity of QPS-PB at different IEC as a function of temperature, (d), (e) simulated calculations of the swing ability of side chains on different backbones. Source: Song et al. [37]/with permission of Elsevier.

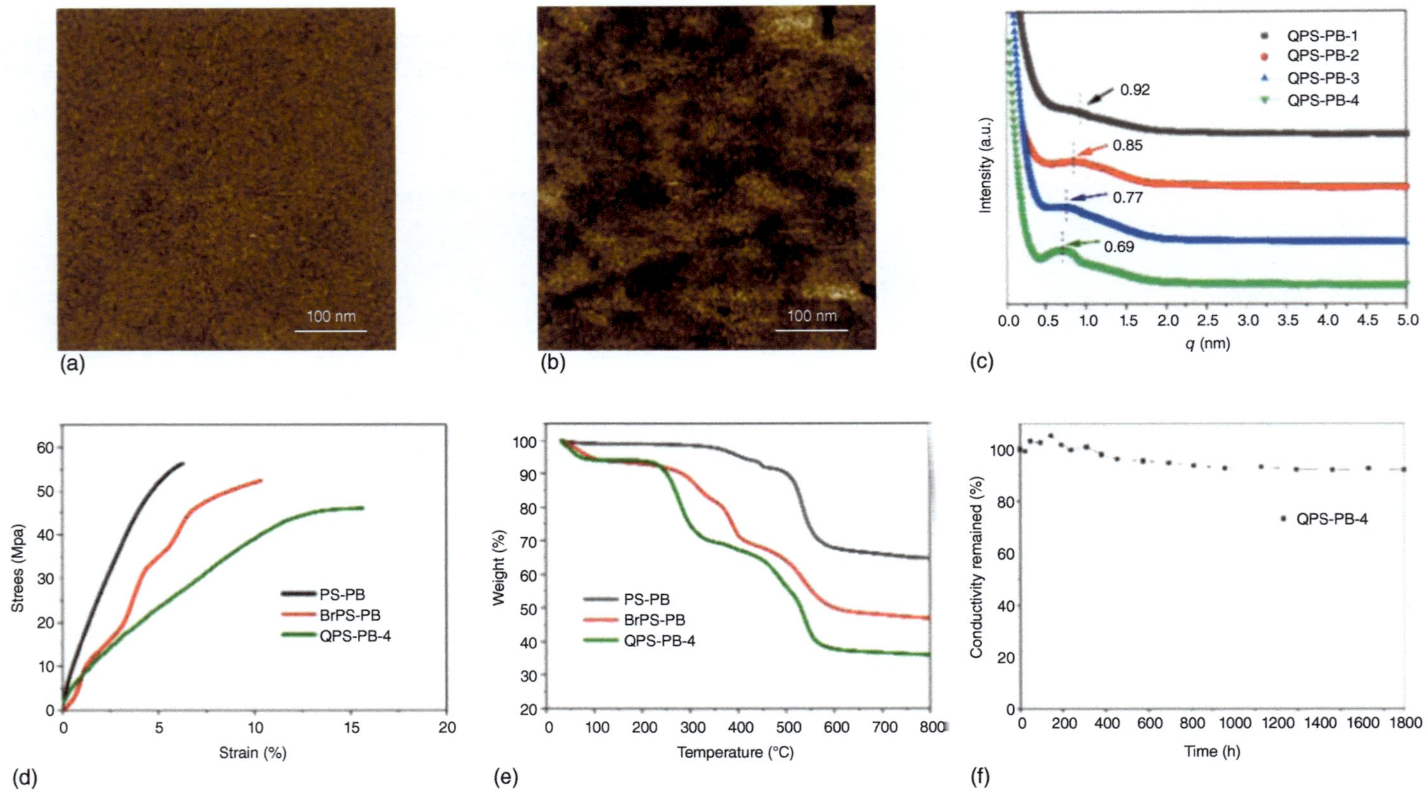

Figure 5.27 (a, b) AFM phase images of the BrPS-PB and QPS-PB-4 membranes, (c) SAXS spectra of different membranes, (d) mechanical properties of the different membranes, (e) thermal stability of the different membranes, and (f) conductivity change of the QPS-PB-4 membrane treated in 1 M NaOH solution at 80 °C for different time durations. Source: Song et al. [37]/With permission of Elsevier/CC BY-NC-ND 4.0.

hydrophilic–hydrophobic microphase separation structure. This result is because the twisted backbone reduces the steric hindrance of side chain swing and makes it easier for ion exchange groups to aggregate. Moreover, the same result is obtained by further characterization using small-angle X-ray scattering (SAXS) spectroscopy (Figure 5.27c).

Stabilities are important indicators for evaluating AEMs, including mechanical strength, thermal stability, and most importantly, chemical stability under a strongly alkaline environment. The twisted main chain PS-PB has an outstanding tensile strength of up to 56.3 MPa (Figure 5.27d). After grafting and functionalization, although the strength decreases slightly, the QPS-PB-4 membrane still maintains an excellent mechanical strength higher than 45 MPa. The TGA curves of different materials are shown in Figure 5.27e. For BrPS-PB and QPS-PB-4, the weight loss below 100 °C is caused by the evaporation of water. The apparent weight loss between 250 and 350 °C corresponds to the degradation of side chains and ion exchange groups. The PS-PB polymer degrades significantly after 500 °C, which proves the excellent thermal stability of the prepared twisted backbone.

As shown in Figure 5.27f, the OH^- conductivity of the QPS-PB-4 membrane decreased by less than 8% after immersion in 1 M NaOH at 80 °C for 1800 hours, and the FTIR spectra of the samples treated for different times did not change significantly, which proves the good chemical stability of the AEMs. This is due to the AEMs designed in this work adopting the "ether free" polymer without a sensitive aryl–ether bond. In addition, the Suzuki coupling grafting method used in this work makes the main chain and the side chain directly connected by C—C bonds, avoiding sensitive sites such as benzyl and carbonyl possibly arising from the traditional grafting reaction. This all-carbon structure without any miscellaneous atoms gives the membrane excellent alkaline resistance.

Our research group fabricated a freestanding ultrathin AEM by a superacid catalyst reaction (Figure 5.28) [36]. After years of effort, the conductivity and stability of current AEMs have been greatly improved, and they have basically met the demand of fuel cell applications. To further improve the performance of AEMFCs, researchers have noted that reducing the thickness of AEMs is considered a promising development direction for AEMs. Based on this, our research group developed an ultrathin AEM. First, a polymeric backbone with a grafting site was prepared. The whole polymer did not contain heteroatoms or degradation groups, which could ensure the chemical stability of the membrane. Tertiary amine ionic liquid was used for grafting. Three kinds of alkaline membrane materials with different functional group contents were prepared: QABP-1 (IEC = 1.68 mmol/g), QABP-2 (IEC = 2.01 mmol/g), and QABP-3 (IEC = 2.2 mmol/g).

Then, the morphology of the prepared alkaline membrane was characterized. All of the AFM phase images of QABP-*x* display well-defined phase separation (Figure 5.29a, b, c). This apparent phase separation structure is due to the polarity difference between the hydrophobic backbone and hydrophilic side chain, which leads to ion cluster aggregation, forming a regular hydrophilic channel. The author deeply visualized the internal ion cluster aggregation behavior by TEM (Figure 5.29d, e, f). In the TEM image of QABP-1, the ionic clusters were small

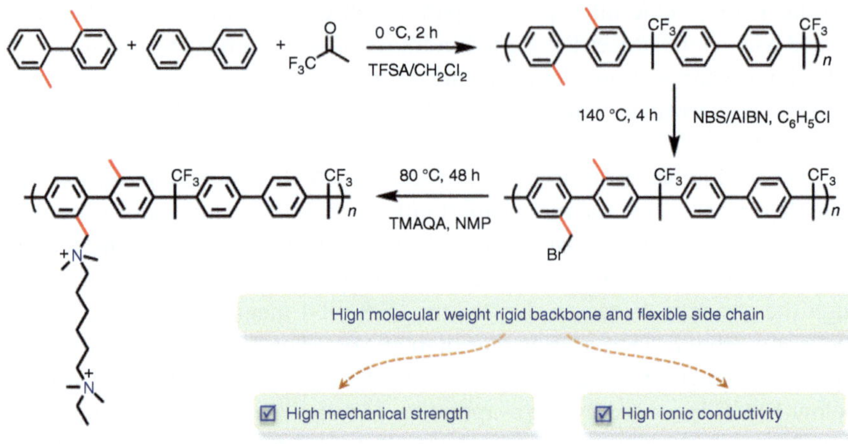

Figure 5.28 Synthetic procedure of QABP-*x* (*x* = 1, 2, 3). Source: Song et al. [36]/Royal Society of Chemistry.

and isolated. For QABP-2, more ion clusters were detected and interconnected to form penetrating ion domains. With a further increase in the IEC value, a large and wider ionic cluster almost dominated the entire field of vision in the QABP-3 TEM image. The excessive ionic cluster led to a high WU and SR, so the properties of the membrane would be influenced. The test of OH^- conductivity also confirmed this. SAXS profiles were used to further quantitatively monitor the length scale of ion channels (Figure 5.29 g, h, i). The expanded hydrophilic domains of *d*-spacing from 6.9 to 9.3 nm confirmed the results of AFM and TEM. In particular, the sharp scattering peaks of QABP-2 suggested a well-defined interconnected network between the hydrophilic channels.

Using the prepared polymer material, a 10 μm freestanding AEM could be prepared. This membrane had excellent mechanical strength and dimensional stability. SEM images presented a uniform and flat morphology (Figure 5.30a), which indicates that the AEMs are free of defects. The author performed ionic conduction tests on membranes of different IECs prepared. The results showed that the membranes had excellent OH^- conductivity (Figure 5.30b). The conductivity of QABP-3 first increased and then decreased with increasing temperature. This was because the ionic clusters of the membrane were too large. After the temperature rose, the WU and SR of the membrane increased rapidly, leading to a decrease in conductivity. Although the membranes' thickness was only 10 μm, they could absolutely meet the fuel cell test, showing good performance. Durability testing was performed on the QABP-2 membrane with the best performance; as a result, the QABP-2-based AEMFC can operate for more than 140 hours (Figure 5.30d).

Recently, our research group conducted an in-depth exploration of scale and batch preparation. In addition, we realized the mass production of AEMs. Our products had excellent OH^- conductivity of 70 ± 10 mS/cm @ 30 °C and 150 ± 10 mS/cm @ 80 °C (Figure 5.31a) and a high tensile strength of 40 ± 5 MPa @ R.T. In addition, after immersion in 1 mol/l NaOH solution at 80 °C for 5000 hours, the membrane

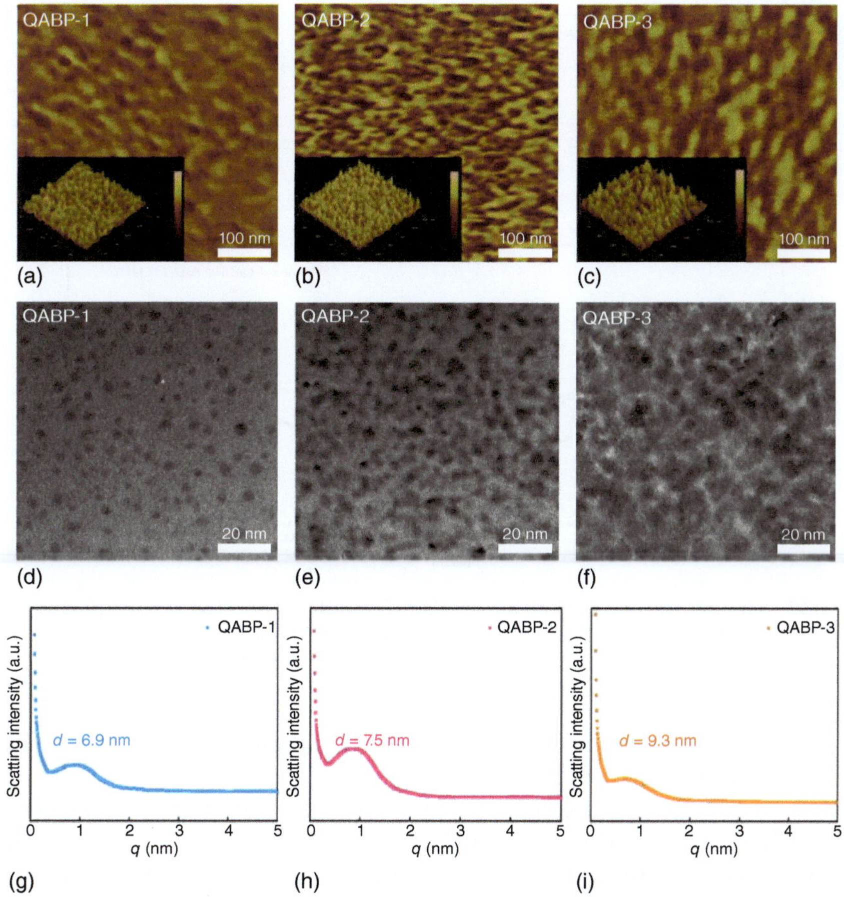

Figure 5.29 Microstructure characterization of the fabricated QABP-*x* (*x* = 1, 2, 3). (a–c) The AFM phase images (3D phase image is shown in inner illustration) and (d–f) TEM images of QABP-*x* in the Br^- form. (g–i) SAXS profiles of QABP-*x* in the OH^- form. Source: Song et al. [36]/Royal Society of Chemistry.

conductivity decreased by less than 4.5% (Figure 5.31b). This result indicated that the membrane was suitable for all kinds of alkaline application environments. Membrane samples have been provided to more than 20 scientific research institutes at home and abroad, involving fields such as fuel cells, water electrolysis for hydrogen production, electrochemical synthesis of ammonia, and carbon dioxide reduction.

Our group's industrially produced AEMs (MTCP-50) showed excellent performance in water electrolysis for hydrogen production. We tested the performance in a 1 mol/l KOH aqueous solution and pure water. The results showed that in a 1 mol/l KOH solution, at 1.8 V@90 °C, the current could reach 5.1 A/cm^2, which could be matched with those of most proton membranes (Figure 5.32b). At the same time, the alkaline membrane had good performance even when electrolyzed in pure water (Figure 5.32c). Finally, the operational stability of the membrane was

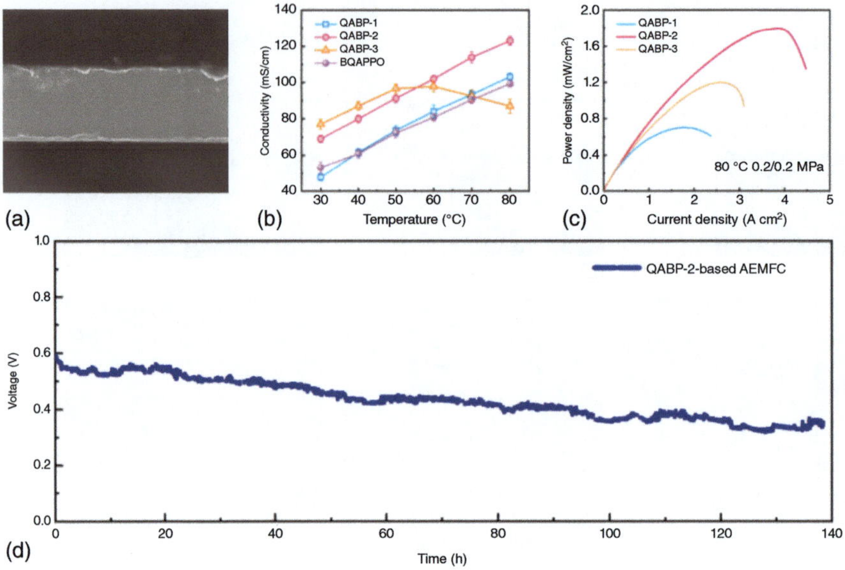

Figure 5.30 (a) Cross-section SEM image of QABP-2, (b) OH^- conductivity of the prepared AEM, (c) AEMFC performance of the prepared AEM, (d) in situ durability test of the QABP-2-based AEMFC. Source: Song et al. [36]/Royal Society of Chemistry.

tested in 1 mol/l KOH solution at 60 °C and 0.5 A/cm^2. It can operate for more than 2000 hours now (Figure 5.32d).

In 2019, Yushan Yan and coworkers [38] published the work of poly (aryl piperidinium) membranes for alkaline fuel cells. The anion exchange membrane showed similar WU and ion conductivity to the Nafion membrane. The PAP-based and air (CO_2 free)-fed membrane electrode assembly (MEA) achieved a high peak power density of 920 mW/cm^2(Figure 5.33a) and good durability with low Pt electrocatalyst loadings at 95 °C, which showed great promise for next-generation HEMFCs (Figure 5.33b).

In 2021, Nanwen Li and coworkers [39] partially grafted naphthalene-based polybenzimidazole (NPBI) with cationic side chains to synthesize an NPBI-QAx membrane, and then it reacted with bromobutane to achieve a fully grafted polymer named NPBI-QAx-By. The cell with NPBI-QA55-B45 AEM achieved a peak power density of 260.1 mW/cm^2 at 440.5 mA/cm^2. Moreover, the durability of the MEA with the NPBI-QA55-B45 AEM was investigated with a current density of 200 mA/cm^2. The decay rate of the voltage exhibited was approximately 1.98 mV/h with increasing testing time. After 110 hours of testing, the voltage decreased from 0.61 to 0.33 V. After careful analysis, it was found that there was no obvious change in the chemical structure of the NPBI-QA55-B45 membrane before and after the durability test. The performance decline might possibly be attributed to the interfacial delamination between the catalyst layer and membrane, the aggregation of catalyst particles, and the degradation of the binder. Additionally, to explore anion exchange membrane-based alkaline water electrolyzers, Park et al. [40] synthesized

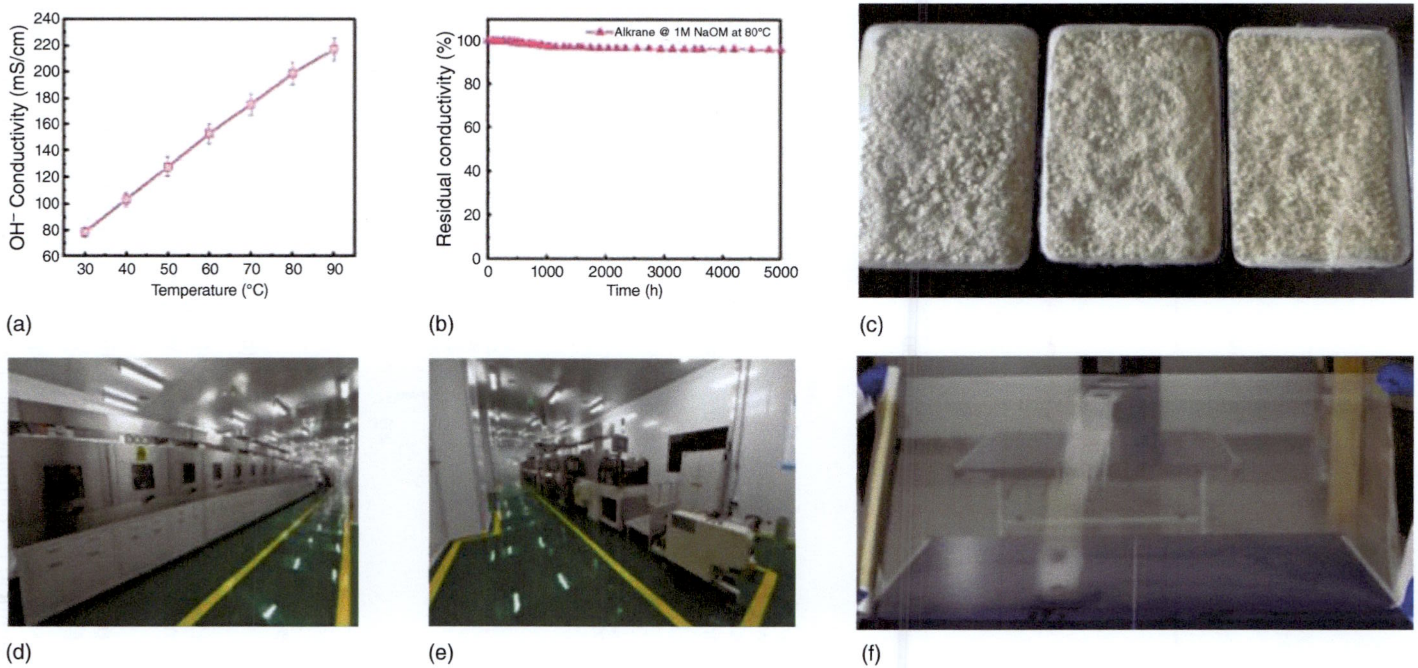

Figure 5.31 Basic properties and production lines of alkaline membrane continuously prepared products.

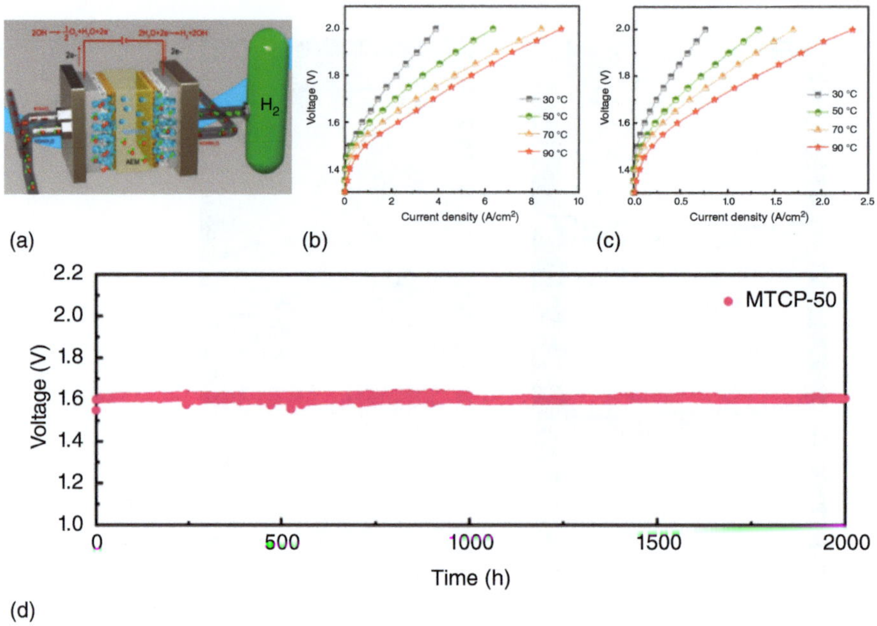

Figure 5.32 (a) MEA diagram of water electrolysis for hydrogen production, (b) performance of water electrolysis for hydrogen production of the MTCP-50 membrane in 1 mol/l KOH solution, (c) performance of water electrolysis for hydrogen production of the MTCP-50 membrane in pure water, and (d) stability testing of the membrane in 1 mol/l KOH solution, 60 °C @ 0.5 A/cm^2.

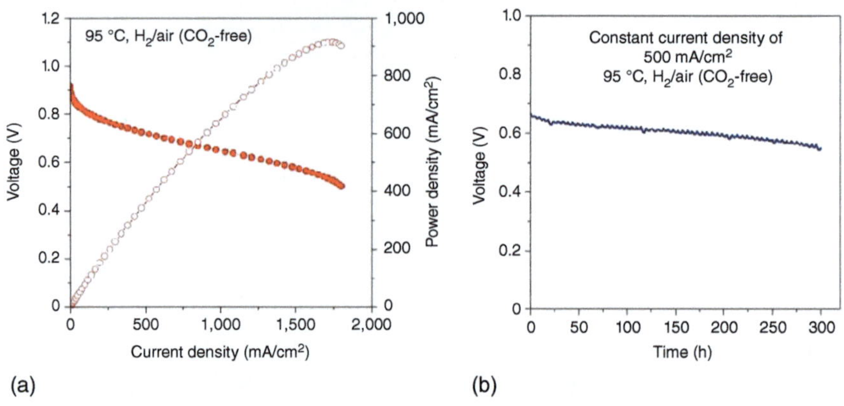

Figure 5.33 (a) Performance of PAP HEMFC at 95 °C, (b) HEMFC durability test at 95 °C at a constant current density of 500 mA/cm^2. Source: Wang et al. [38]/Springer Nature.

two QA aromatic polymers by superacid catalysis. Under the condition of a constant current density of 200 mA/cm^2, both membranes showed excellent battery voltage stability in steady-state operation. MEA using BPN1-100 and TPN1-100 was used to test the electrochemical performance of the BPN1-100 and TPN1-100 membranes for alkaline water electrolysis. A baseline membrane marked as A201 (Tokuyama

Corporation) was used for comparison. The electrochemical performance was compared over the range of the current densities applied. The MEA using BPN1-100 exhibited a current density of 400 mA/cm^2 at 2.1 V, which performed better than TPN1-100 with a difference of 0.25 V, which can be attributed to its high IEC and ion conductivity. Meanwhile, BPN1-100 and TPN1-100 also showed good voltage stability. Although steady-state testing for a longer period of time would be required for a better evaluation of the materials, the stabilization of the cell voltage at a constant current density showed the great potential of biphenyl- and terphenyl-based polymer membrane electrolytes for alkaline water electrolysis. The alkaline stability test of the bulk membranes showed excellent chemical stability even at 95 °C. After the water electrolysis operation, however, mechanical failure of membranes was observed for both BPN1-100 and TPN1-100 during the post-operation visual inspection of the AEMs. This result suggested that although these membranes were chemically stable, they were not mechanically stable enough under fully hydrated conditions; additional treatment of the AEMs, for example, cross-linking, or making composite membranes, was necessary to reduce swelling in water and ensure better mechanical durability of the AEMs for use in alkaline water electrolysis.

Jang Yong Lee et al. [41] highlighted an anion-exchange membrane and ionomer based on quaternized poly-carbazole (QPC-TMA) with a rigid ether-free and curved backbone structure composed of carbazole monomers (Figure 5.34). The developed ACP exhibited excellent ionic conductivity, as well as chemical and mechanical stability. The QPC-TMA membrane showed better performance than the commercial membrane FAA-3 under all test conditions. The AEMWE using QPC-TMA demonstrated outstanding stability and state-of-the-art performance (3.5 A/cm^2 at 1.9 V) (Figure 5.34d), which was the first report of an AEMWE (Figure 5.35) that outperformed the best-performing proton-exchange membrane water electrolyzers.

Sulfonated polybenzene has very high intrinsic chemical and mechanical stability because of its full aromatic backbone, so it has very good application prospects as a proton exchange membrane. However, sulfonated aryl polymers are generally prepared by chemical coupling reactions. Compared with AEM, the use of TfOH and other superacids as catalysts to prepare proton exchange membranes is very rare. Because the chemical coupling method requires a high reaction temperature and gas protection, which greatly limits the commercial application of proton exchange membranes, the use of superacid catalysis to prepare proton exchange membranes has excellent application prospects in the future.

Jannasch [42] synthesized main chain polymers by efficient superacid-catalyzed Friedel–Crafts polycondensations of commercially available perfluoroacetophenone and p-terphenyl (Figure 5.36). Thermogravimetric analysis (TGA) revealed high thermal stability with the onset of thermal decomposition at $T_{d,onset}$ = 418 °C. No glass transition was detected by differential scanning calorimetry (DSC) up to this temperature, confirming the very high chain rigidity of the polymer backbone. Next, the sulfonic acid groups were introduced in the p-position of the PFAp residues via a straightforward thiolation–oxidation route. The main chain polymer was dissolved in N,N-dimethylacetamide before adding an excess of NaSH at 25 °C.

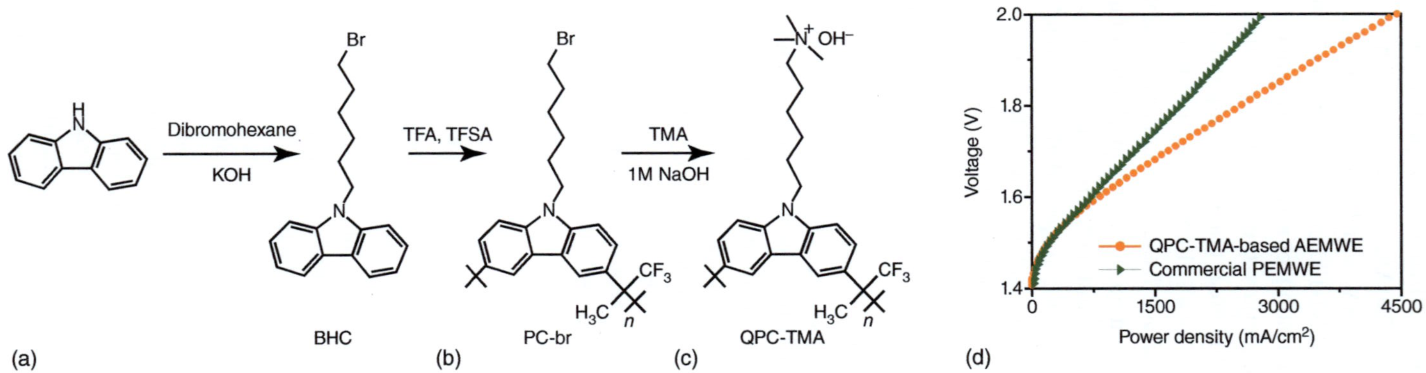

Figure 5.34 Synthetic scheme of (a) BHC, (b) Pc-br, (c) QPC-TMA Source: (a-c) Min et al. [41]/Royal Society of Chemistry. (d) performance comparison of the AEMWE and commercial PEMWE based on the QPC−TMA (polarization curve). Source: (d) Adapted from Min et al. [41].

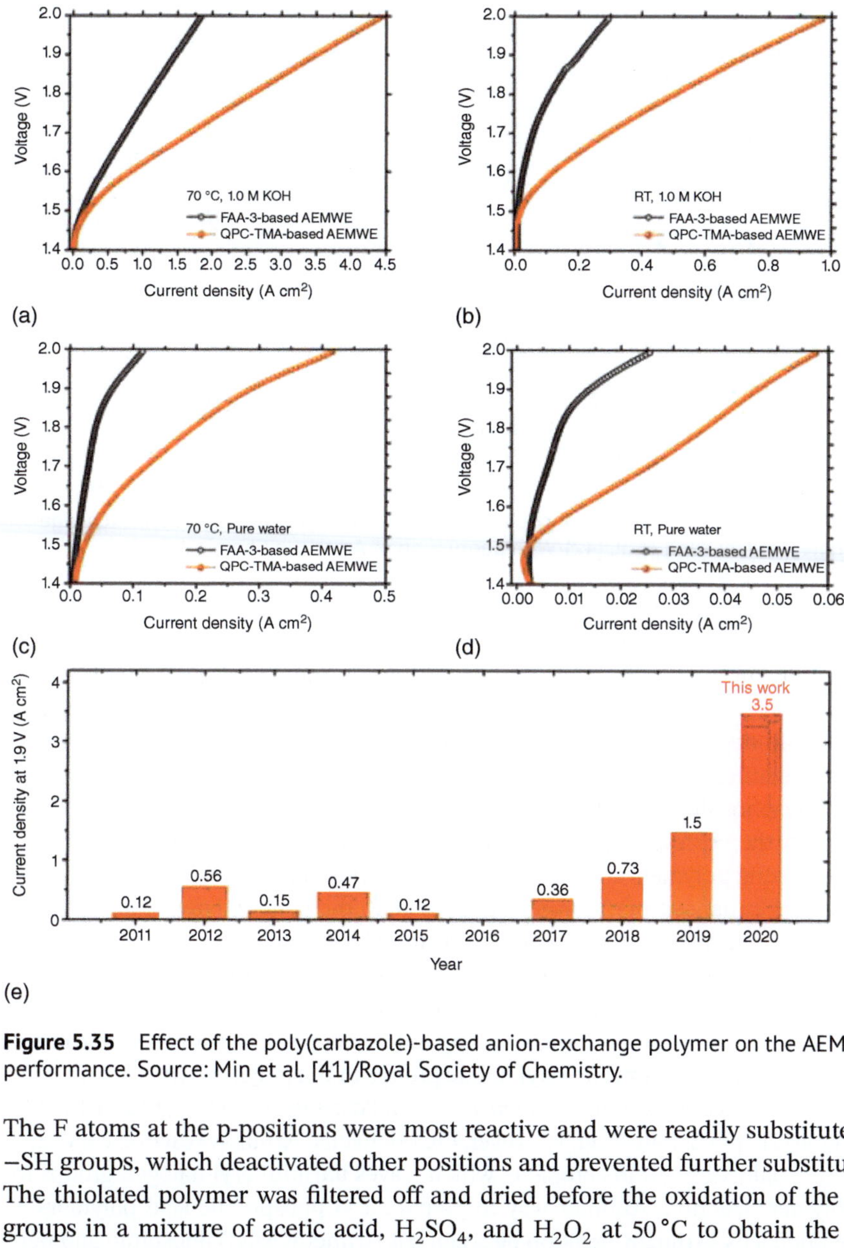

Figure 5.35 Effect of the poly(carbazole)-based anion-exchange polymer on the AEMWE performance. Source: Min et al. [41]/Royal Society of Chemistry.

The F atoms at the p-positions were most reactive and were readily substituted by −SH groups, which deactivated other positions and prevented further substitution. The thiolated polymer was filtered off and dried before the oxidation of the thiol groups in a mixture of acetic acid, H_2SO_4, and H_2O_2 at 50 °C to obtain the final poly (p-terphenyl perfluorophenylsulfonic acid) sample. This route requires neither elaborate monomer synthesis nor metal catalysis. Moreover, the resulting polymer SP100 has no heteroatom linkages and is structurally well defined with precisely sequenced and strongly acidic groups along the backbone. Despite the high T_g, the polymer shows excellent solubility in, e.g. DMSO and DMAc, while retaining complete insolubility in hot water.

In addition to sulfonated aromatic polymers, phosphonic proton exchange membranes have also received considerable attention. Compared with sulfonated proton

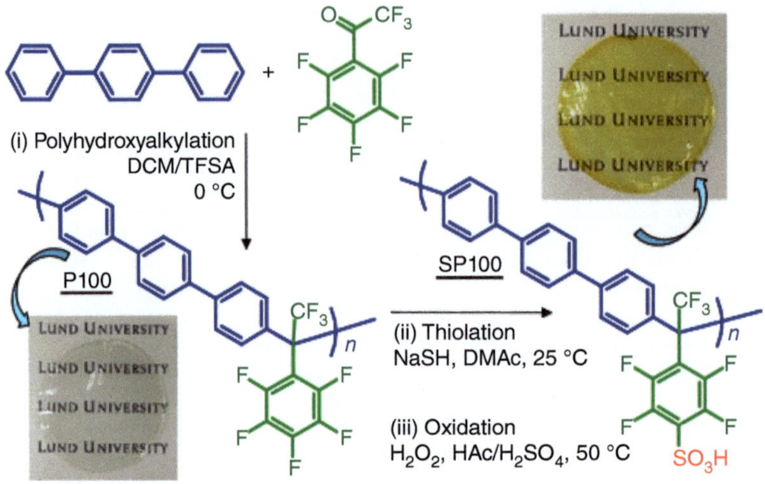

Figure 5.36 Synthesis route of polyaromatic perfluorophenylsulfonate proton exchange membrane. Source: Kang et al. [42]/With permission of American Chemical Society.

exchange membranes, the corresponding phosphonic proton exchange membranes usually have better thermal and chemical stability, lower water SR, and lower permeability in fuel cells. Aryl phosphonic acids and alkyl phosphonic acids are generally amphoteric, having strong hydrogen bonding and exhibiting a considerable degree of self-dissociation, resulting in anhydrous conductance in high-temperature-high-concentration systems. Nonetheless, because of its relatively low acidity, phosphonic acid results in lower WU and electrical conductivity compared with the corresponding sulfonated proton exchange membrane. However, by synthesizing polymers containing high concentrations of phosphonic acid or by using acidic enhanced perfluorophosphonic acid, the conductivity of phosphonic proton exchange membranes can be significantly improved.

In general, phosphonates are more challenging to treat with aromatic polymers than sulfonates. Most phosphonic aromatic polymers are synthesized by the classical Michaelis–Arbuzov reaction involving the nucleophilic substitution of trialkyl phosphite esters and benzyl or alkyl halide groups in the polymer structure. In contrast, aryl halides are generally less reactive, so a Pd catalyst is often used to directly transfer aryl to phosphate ester. However, the Michaelis–Arbuzov reaction catalyzed by Pd is usually difficult to complete, which leaves harmful aryl halogen groups in the polymer structure. Another way to prepare aryl phosphonic acid polymers is by the Michaelis–Arbuzov reaction between tris (trimethylsilyl) phosphite and perfluorinated aryl polymer. The fluorine atom of the perfluorophenyl group is highly activated by the electron-withdrawing fluorine atom nearby and is likely to undergo aromatic nucleophilic substitution.

On the basis of a sulfonated proton exchange membrane, the Patric Jannasch group [43] further used superacid catalysis to synthesize a series of poly (p-terphenyl perfluoroalkylene)s. They first carried out TfOH -catalyzed polyhydroxylations involving p-terphenyl, 2,2,2-trifluoroacetophenone and perfluoroacetophenone. Subsequently, the resulting polymers were quantitatively and selectively

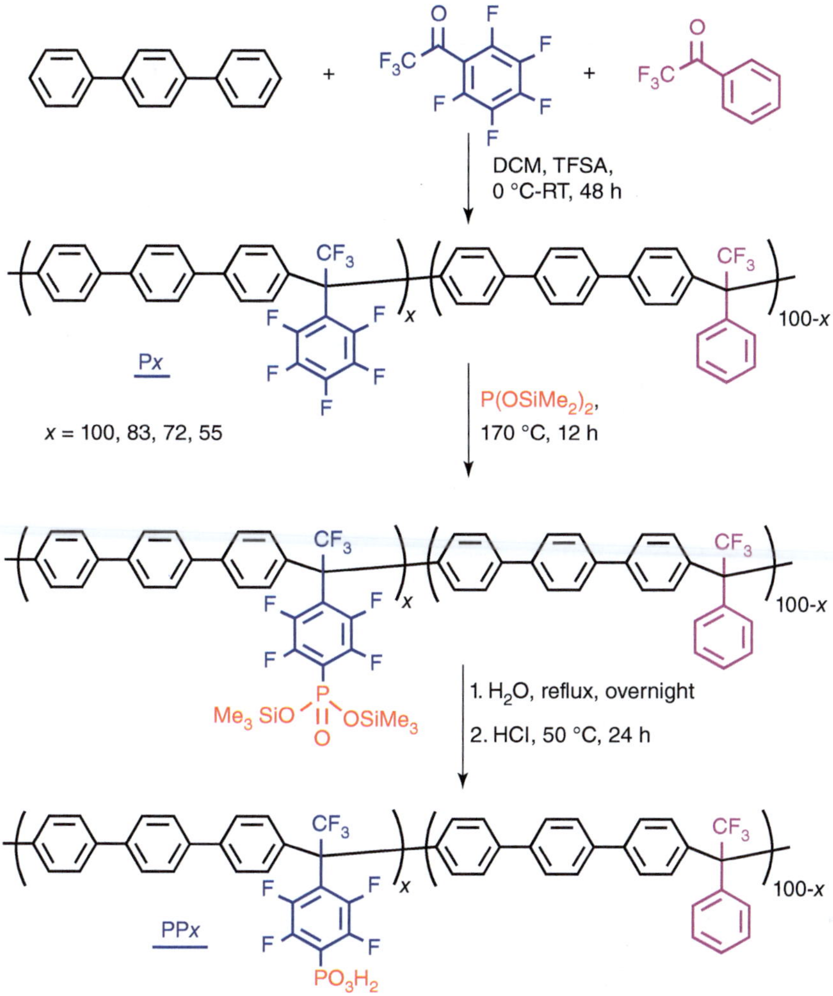

Figure 5.37 Synthetic pathway to poly(arylene perfluorophenylphosphonic acid)s in the PPx series by superacid-mediated polycondensation, phosphonation via a Michaelis–Abuzov reaction, and hydrolysis. Source: Kang et al. [43]/with permission of Elsevier.

phosphonated in the para positions of the pendant perfluorophenyl units by employing an efficient Michaelis–Arbuzov reaction (Figure 5.37). As in the case of the sulfonated polymers, the substitution of the F atoms in the para position was complete and selective. Hence, the two series of polymers were functionalized with the same concentration of acid groups in the same position of the identical polymer structure. This allows for direct comparison and benchmarking of phosphonic acid versus sulfonic acid in proton exchange membranes (PEMs) in regard to, e.g. morphology, stability, and conductivity.

Based on the main-chain polymer synthesized by a superacid catalyst, our research group developed a kind of proton exchange membrane MBTF using the Menshutkin reaction (Figure 5.38a). This membrane had not only high mechanical

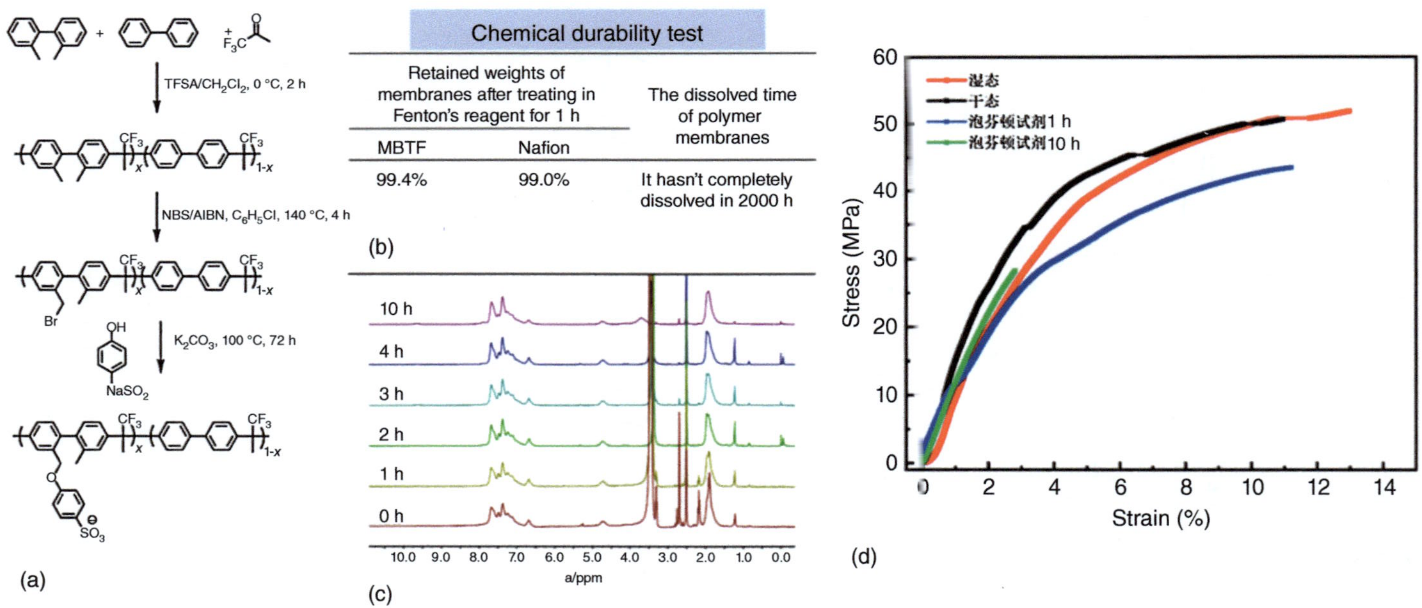

Figure 5.38 (a) Synthetic route of MBTF proton membrane, (b) weight retention rate of MBTF proton membrane after immersion in Fenton reagent, (c) NMR testing of MBTF proton membrane after immersion in Fenton reagent, (d) tensile strength of MBTF proton membrane after immersion in Fenton reagent.

strength but also excellent chemical stability. After immersion in the Fenton reagent at 80°C for an hour, the weight retention rate was higher than that of the Nafion membrane. In addition, the membrane did not completely dissolve after immersion for 2000 hours. These performances are better than those of most proton membranes reported in the literature. To further confirm the changes in proton membrane structure and strength after Fenton reagent immersion, we used NMR and tensile testing to test the change after immersion for 10 hours (Figure 5.38b, c). The NMR showed that the membrane structure did not have an obvious degradation peak after immersion for 10 hours. Tensile testing showed that after immersion for 10 hours, the proton membrane still had a certain mechanical strength (Figure 5.38d). This membrane had excellent conductivity with an IEC of 1.0 mmol/g (Figure 5.39a). The fuel cell test showed that the membrane performance could reach 1.7 W/cm^2(Figure 5.39b). At the same time, the membrane's fuel cell testing has been in continuous operation for more than 2000 hours, demonstrating excellent stability.

5.5 Others

Ion exchange membranes are widely used in energy generation and conversion technologies to solve the problems of global pollution and energy shortages. Recent concerns about sustainability and alternative energy sources have led to increasing attention to the research and development of new energy devices, such as fuel cells, water electrolysis for hydrogen production, flow batteries, carbon dioxide reduction, and electrochemical ammonia synthesis.

These technologies usually need to work in an electrochemical environment with strong acids and bases. To ensure the long-term stable operation of the equipment, the chemical stability of ion exchange membranes is the key index to evaluate their properties. As these technologies continue to be industrialized, they put forward increasingly higher requirements for the service life of ion exchange membranes. Improving the chemical stability of ion exchange membranes has become a great challenge for researchers. Previous studies have focused on the stability of ion-exchange groups and partially fluorinated polymers, but little attention has been given to the chemical stability of the polyphenyl ether polymer backbone used in traditional ion-exchange membranes. With further research on the polymer degradation mechanism and understanding of aryl–ether bond breaking in acid-base environments, synthesizing ether-free polymer backbones by superacid catalysts has attracted much attention in the field of ion exchange membranes.

Synthesizing ether-free polymer backbones by superacid catalysts has advantages that traditional condensation polymers cannot compare with, which are listed as follows: It has high catalytic efficiency, fast reaction speed, and gentle reaction conditions. The initial materials are easy to obtain. The reaction condition does not contain a metal-based catalyst. The resulting polymers can have high or ultrahigh molecular weights. The polymers have excellent mechanical strength and other properties. Most importantly, we can construct a polymer backbone structure with

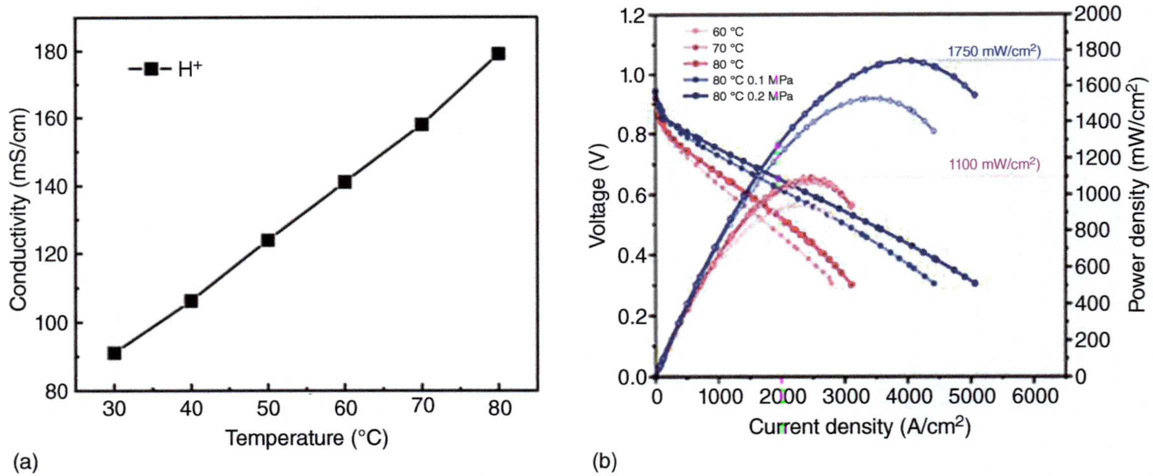

Figure 5.39 (a) The ionic conductivity of the MBTF proton membrane varying with temperature, (b) H_2/O_2 single-cell AEMFC performance of the MBTF proton membrane.

only C—C bonds in the main chain by this method, thus avoiding the introduction of aryl–ether bonds in traditional condensation reactions. Recent studies have proven that the breakage of sensitive aryl–ether bonds under attack in various harsh environments is an important reason for the chemical degradation of ion exchange membranes.

Ion exchange membranes prepared by ether-free polymers synthesized by the superacid catalysis method were first applied in the field of alkaline membrane fuel cells to solve the problem of alkaline stability, which has puzzled the field for many years, showing excellent performance. After that, many research groups at home and abroad carried out studies of ether-free main-chain polymer ion exchange membranes. Various structures of ether-free main chain ion exchange membranes catalyzed by superacids have been reported. Based on the ether-free main chain catalyzed by superacid, our group further improves the properties of ion-exchange membranes by introducing secondary structures and functions such as twisted structures, dipole interactions, and π–π stacking interactions. The research results have fully demonstrated that this ether-free main chain ion exchange membrane has obvious advantages in conductivity and alkaline stability compared with traditional poly(aryl ether) polymer ion exchange membranes.

The broad application prospects of ether-free polymers prepared by superacid catalysis in ion exchange membranes have gradually become a concern in other application fields, such as water electrolysis for hydrogen production, which also needs to operate in alkaline environments, and proton membrane fuel cells, and vanadium redox flow batteries, which need to operate in acidic environments. In recent years, the preparation of ion exchange membranes catalyzed by superacids has been reported in related fields, showing excellent performance and application prospects. Overall, with increasing attention and interest from researchers, the preparation of ion-exchange membranes by superacid catalysis is on the rise. At present, the main research direction is mainly in the synthesis of a new structure ether-free main chain ion exchange membrane to give the membrane corresponding characteristics and explore the application of ether-free main chain ion exchange membrane in new fields, such as electrochemical ammonia synthesis and CO_2 reduction, which has attracted much attention recently.

References

1 Gillespie, R. and Peel, T. (1973). Hammett acidity function for some superacid systems. II. Systems sulfuric acid-[fsa], potassium fluorosulfate-[fsa], [fsa]-sulfur trioxide, [fsa]-arsenic pentafluoride, [sfa]-antimony pentafluoride and [fsa]-antimony pentafluoride-sulfur trioxide. *Journal of the American Chemical Society* 95 (16): 5173–5178.

2 Olah, G.A. and Klumpp, D.A. Aue D.D.(2007). *Superelectrophiles and Their Chemistry*. Hoboken, NJ: John Wiley & Sons, Inc. viii+ 302 pp. $125.00. ISBN 978-0-470-04961-7 [M]. ACS Publications. 2008.

3 Olah, G.A. (1993). Superelectrophiles [M]. *Angewandte Chemie International Edition* 32: 767.

4 Cruz, A.R., Zolotukhin, M.G., Morales, S.L. et al. (2009). Use of 4-piperidones in one-pot syntheses of novel, high-molecular-weight linear and virtually 100%-hyperbranched polymers. *Chemical Communications* 29): 4408–4410.

5 Klumpp, D.A. and Anokhin, M.V. (2020). Superelectrophiles: recent advances. *Molecules* 25 (14): 3281.

6 Romero-Hernández, J.E., Cruz-Rosado, A., Zolotukhin, M.G. et al. (2017). Modelling of superacid catalyzed step-growth polymerization of isatin and biphenyl or terphenyl monomers. *Macromolecular Theory and Simulations* 26 (5): 1700031.

7 Lira, A.L., Zolotukhin, M., Fomina, L. et al. (2007). Superelectrophilic activation of 4-heterocyclohexanones. implications for polymer synthesis. A theoretical study. *The Journal of Physical Chemistry A* 111 (51): 13606–13610.

8 Rusanov, A.L.V., Chebotarev, V.P., and Lovkov, S.S. (2008). Superelectrophiles in the synthesis of condensation monomers and polymers. *Russian Chemical Reviews* 77 (6): 547.

9 Noh, S., Jeon, J.Y., Adhikari, S. et al. (2019). Molecular engineering of hydroxide conducting polymers for anion exchange membranes in electrochemical energy conversion technology. *Accounts of Chemical Research* 52 (9): 2745–2755.

10 Miyanishi, S. and Yamaguchi, T. (2016). Ether cleavage-triggered degradation of benzyl alkylammonium cations for polyethersulfone anion exchange membranes. *Physical Chemistry Chemical Physics* https://doi.org/10.1039/C6CP00579A.

11 Mohanty, A.D., Tignor, S.E., Krause, J.A. et al. (2016). Systematic alkaline stability study of polymer backbones for anion exchange membrane applications. *Macromolecules* 49 (9): 3361–3372.

12 Kreuer, K.D. (2001). On the development of proton conducting polymer membranes for hydrogen and methanol fuel cells. *Journal of Membrane Science* 185 (1): 29–39.

13 Joo, P.E. and Seung, K.Y. (2018). Quaternized aryl ether-free polyaromatics for alkaline membrane fuel cells: synthesis, properties, and performance – a topical review. *Journal of Materials Chemistry A* 6: 15456–15477.

14 Diaz, A.M., Zolotukhin, M.G., Fomine, S. et al. (2007). A novel, one-pot synthesis of novel 3F, 5F, and 8F aromatic polymers. *Macromolecular Rapid Communications* 28 (2): 183–187.

15 Lee, W.H., Kim, Y.S., and Bae, C. (2015). Robust hydroxide ion conducting poly(biphenyl alkylene)s for alkaline fuel cell membranes. *ACS Macro Letters* 4 (8): 814–818.

16 Cruz, A.R., Hernandez, M., Guzmán-Gutiérrez, M. et al. (2012). Precision synthesis of narrow polydispersity, ultrahigh molecular weight linear aromatic polymers by A 2 + B 2 nonstoichiometric step-selective polymerization. *Macromolecules* 45 (17): 6774–6780.

17 Park, E.J. and Kim, Y.S. (2018). Quaternized aryl ether-free polyaromatics for alkaline membrane fuel cells: synthesis, properties, and performance—a topical review. *Journal of Materials Chemistry A* 6 (32): 15456–15477.

18 Peña, E.R., Zolotukhin, M., and Fomine, S. (2004). Factors enhancing the reactivity of carbonyl compounds for polycondensations with aromatic hydrocarbons. A computational study. *Macromolecules* 37 (16): 6227–6235.

19 Marino, M. and Kreuer, K. (2015). Alkaline stability of quaternary ammonium cations for alkaline fuel cell membranes and ionic liquids. *ChemSusChem* 8 (3): 513–523.

20 Olsson, J.S., Pham, T.H., and Jannasch, P. (2018). Poly(arylene piperidinium) hydroxide ion exchange membranes: synthesis, alkaline stability, and conductivity. *Advanced Functional Materials* 28 (2): 1702758.

21 Pham, T.H., Olsson, J.S., and Jannasch, P. (2018). Poly(arylene alkylene)s with pendant n-spirocyclic quaternary ammonium cations for anion exchange membranes. *Journal of Materials Chemistry A* 6: 16537–16547.

22 Huongpham, T., Olsson, J.S., and Jannasch, P. (2019). Effects of the N-alicyclic cation and backbone structures on the performance of poly(terphenyl)-based hydroxide exchange membranes. *Journal of Materials Chemistry A* 7: 15895–15906.

23 Andrit, A., Pham, T.H., Olsson, J.S. et al. (2019). Ether-free polyfluorenes tethered with quinuclidinium cations as hydroxide exchange membranes. *Journal of Materials Chemistry A* 7: 27164–27174.

24 Maurya, S., Noh, S., Matanovic, I. et al. (2018). Rational design of polyaromatic ionomers for alkaline membrane fuel cells with >1 W cm^{-2} power density. *Energy & Environmental Science* 11: 3283–3291.

25 Chen, N., Hu, C., Wang, H.H. et al. (2021). Poly(alkyl-terphenyl piperidinium) ionomers and membranes with an outstanding alkaline-membrane fuel-cell performance of 2.58 W cm^{-2}. *Angewandte Chemie International Edition* 60 (14): 7710–7718.

26 Chen, N., Wang, H.H., Kim, S.P. et al. (2021). Poly(fluorenyl aryl piperidinium) membranes and ionomers for anion exchange membrane fuel cells. *Nature Communications* 12: 2367.

27 Yang, K., Chu, X., Zhang, X. et al. (2020). The effect of polymer backbones and cation functional groups on properties of anion exchange membranes for fuel cells. *Journal of Membrane Science* 603: 118025.

28 Xu, H., Yh, A., Lei, L.C. et al. (2020). Piperidinium functionalized aryl ether-free polyaromatics as anion exchange membrane for water electrolysers: performance and durability. *Journal of Membrane Science* 621: 118964.

29 Zhou, X., Wu, L., Zhang, G. et al. (2021). Rational design of comb-shaped poly(arylene indole piperidinium) to enhance hydroxide ion transport for H_2/O_2 fuel cell. *Journal of Membrane Science* 631: 119335.

30 Chen, N., Lu, C., Li, Y. et al. (2019). Tunable multications-crosslinked poly(arylene piperidinium)-based alkaline membranes with high ion conductivity and durability. *Journal of Membrane Science* 588: 117120.

31 Zhang, J., Yu, W., Liang, X. et al. (2021). Flexible bis-piperidinium side chains construct highly conductive and robust anion-exchange membranes. *ACS Applied Energy Materials* 4 (9): 9701–9711.

32 Zhu, H., Li, Y., Chen, N. et al. (2019). Controllable physical-crosslinking poly(arylene 6-azaspiro[5.5]undecanium) for long-lifetime anion exchange membrane applications. *Journal of Membrane Science* 590: 117307.

33 Lu, C., Long, C., Li, Y. et al. (2020). Chemically stable poly(meta-terphenyl piperidinium) with highly conductive side chain for alkaline fuel cell membranes. *Journal of Membrane Science* 598: 117797.

34 Wang, F., Li, Y., Li, C.H. et al. (2020). Preparation and study of spirocyclic cationic side chain functionalized polybiphenyl piperidine anion exchange membrane. *Journal of Membrane Science* 620 (1–2): 118919.

35 Zhang, J., Zhang, K., Liang, X. et al. (2021). Self-aggregating cationic-chains enable alkaline stable ion-conducting channels for anion-exchange membrane fuel cells. *Journal of Materials Chemistry A* 9: 327–337.

36 Song, W., Liang, X., Zhang, H. et al. (2022). Ultrathin anion exchange membranes with an improved OH^- transfer rate for high-performance AEMFCs. *Journal of Materials Chemistry A* 10 (40): 21503–21511.

37 Zhang, H., Zhang, Y., Zhang, F. et al. (2021). Enhancing side chain swing ability by novel all-carbon twisted backbone for high performance anion exchange membrane at relatively low IEC level. *Journal of Membrane Science Letters* 1 (2): 100007.

38 Wang, J., Zhao, Y., Setzler, B.P. et al. (2019). Poly(aryl piperidinium) membranes and ionomers for hydroxide exchange membrane fuel cells. *Nature Energy* 4: 392–398.

39 Wang, Y., Qiao, X., Liu, M. et al. (2021). The effect of –NH on quaternized polybenzimidazole anion exchange membranes for alkaline fuel cells. *Journal of Membrane Science* 626: 119178.

40 Park, E.J., Capuano, C.B., Ayers, K.E. et al. (2018). Chemically durable polymer electrolytes for solid-state alkaline water electrolysis. *Journal of Power Sources* 375: 367–372.

41 Min, S.C., Ji, E.P., Kim, S. et al. (2020). Poly(carbazole)-based anion-conducting materials with high performance and durability for energy conversion devices. *Energy & Environmental Science* 13: 3633–3645.

42 Kang, N.R., Pham, T.H., and Jannasch, P. (2019). Polyaromatic perfluorophenylsulfonic acids with high radical resistance and proton conductivity. *ACS Macro Letters* 8 (10): 1247–1251.

43 Kang, N.R., Pham, T.H., Nederstedt, H. et al. (2021). Durable and highly proton conducting poly(arylene perfluorophenylphosphonic acid) membranes. *Journal of Membrane Science* 623: 119074.

6

Microporous Polymers for IEM Preparation

6.1 Ion Transport Behavior in Nanospace-confined Membranes

When ions enter channels with sizes approaching the size range of ions and then diffuse in channels, they will face a total energy barrier coming from steric hindrance imposed by the channel shape and size and strong interaction with their closest vicinity (e.g. the electrostatic interaction with charged groups fixed on the channel walls) [1, 2]. Thus, the efficiency of ion transport through membranes is determined by the channel structure and channel chemistry. We conclude that the factors affecting ion transport within nano space-confined membranes are physical factors (shape and size effects) and chemical factors (electrical effect, dielectric effect, and wettability) [3].

The steric effect refers to the steric hindrance of channels and is usually manifested in their sieving capacity, which prevents the entrance of ions with a larger size than the channel size into channels. For smaller ions that have already entered channels, their physical collisions with channel walls and viscous effects deriving from the breaking and formation of chemical bonds with the channel wall become essential to precise ion separation [1]. The perspective of controlling the channel size and channel shape of membranes helps design membranes with precise ion separation.

The chemical constitution of channel walls is also vital to ion transport. As the bare ions are in a very unstable state, they always absorb water molecules to become fully or partially hydrated. This means that the easy entrance of water into membranes plays an important role in helping ion transport. To this end, improving the wettability of membranes to water is needed. The wettability depends on the chemical constituents on the channel walls, and hydrophilic modification is needed to enhance the wettability. For example, membranes made from hydrophobic polyxanthene absorbed nearly no water, but those from sulfonated polyxanthene absorbed water and effectively conducted cations [4].

The interaction between ions and fixed groups on the channel wall also dominates the transport behavior of ions. As most hydrated ions enter the rigid subnanometer channel from the bulk solution, they face partial dehydration to rearrange the water shell. However, this dehydration makes the resulting ions less stable than those in the fully hydrated state. Ions within membrane channels have to develop

Ion Exchange Membranes: Design, Preparation, and Applications,
First Edition. Tongwen Xu and Yaoming Wang.

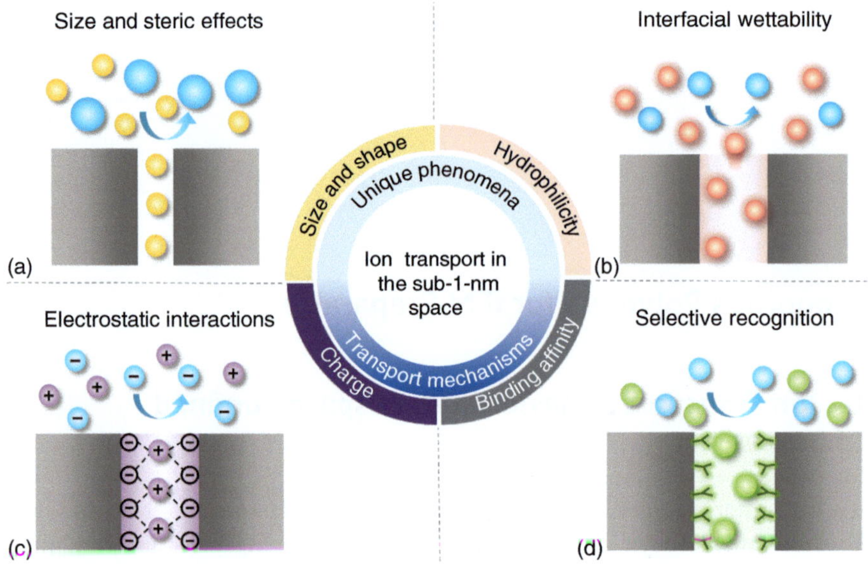

Figure 6.1 Schematics show the critical effects for ion/molecule transportation across sub-1-nm channels, including (a) size/steric effect, (b) interfacial wettability, (c) electrostatic interactions, and (d) selective recognition.

interactions between channels and dehydrated ions to overcome dehydration-based energy barriers. These successive interactions within the confined channels help ions diffuse forward from one site to the next. To offer an interaction site, functional groups should be anchored to the channel surface. These groups include charged groups that form electrostatic interactions and polar moieties that induce dielectric effects [5]. Therefore, positively or negatively charged groups (such as sulfonate groups and quaternary ammonium) are commonly used because they not only form strong electrostatic interactions to attract counter ions but also easily dissociate when another counter ion approaches. How to fix groups to channels is the dominant aspect when designing ion-selective membranes.

Ion transport within nanospace-confined membranes is a very complicated behavior. Both physical factors and chemical factors must be considered when studying the behavior and designing high-performing membranes. Microporous polymers containing high free volumes for mass transport are emerging as excellent materials for constructing ion-conducting membranes. In our Section 6.2, we discuss the principles of microporous polymers and their hydrophilic modification for membranes (Figure 6.1).

6.2 Principle of Microporous Polymers

Polymers used in engineering plastics or rubbers, such as poly(phenyl ether), poly(ether sulfone), polystyrene, polyethylene, and polybutadiene, consist of chains that are flexible enough to pack and intertwine efficiently. Thus, membranes from these polymers contain nearly no free volume. Increasing the rigidity and contortion of polymer backbones may frustrate their packing, thus leading to a large amount of intermolecular free volume with a width of ≤2 nm. Soluble polymers featuring this characteristic are termed polymers of intrinsic microporosity (PIMs), which were first demonstrated by Neil B. McKeown et al. [6]. They employed nucleophilic substitution of commercial monomers (2,3,5,6-tetrafluoroterephthalonitrile and 5,5′,6,6′-tetrahydroxy-3,3,3′,3′-tetramethyl-1,1′-spirobisindane) to synthesize the typical PIM (PIM-1, Figure 6.2a). PIM-1 has rigid and contorted units of the spiro-center, which are greatly conducive to ineffective chain packing, and thus results in a high surface area of 850 m^2/g. Diverse dibenzodioxane-based PIMs were subsequently reported, and their free volume could be tuned via variations in the chemical structure of either the tetrahalide monomers or the aromatic tetrol monomers.

Tröger's base (TB), a V-shaped bicyclic amine unit, which is formed via the acid-catalyzed condensation of dimethoxymethane and aromatic amine, is also commonly used to yield these backbones, and the resulting polymers are called TB-based PIMs [7, 8]. Carta et al. [7] successfully introduced the ethanoanthracene unit (a bridged bicyclic ring system) into TB-based polymers and obtained a shape-persistent ladder polymer (PIM-EA-TB, Figure 6.2b) with a surface area of 1024 m^2/g, which is higher than those for all other reported PIMs.

As seen from the above-mentioned examples, the key characteristic of PIMs is their contorted and rigid backbones. Any reactions that can form these kinds of polymer backbones have the potential to create PIMs. After approximately 20 years of development, much progress in diversifying PIMs has been achieved. In addition to dibenzodioxane-based or TB-based polymers, polyxanthene (Figure 6.2d) is another typical kind of PIM material that is synthesized via superacid-catalyzed polycondensation between bisphenol monomers and carbonyl compounds with strong electron-withdrawing substituents linked to the ketone [9]. This polycondensation is established by the electrophilic aromatic substitution of phenol fragments with ketone linkages, and the subsequent cyclodehydration of hydroxyl-carrying intermediates. Carbinol formation is the rate-determining step, as seen from the calculation of the reaction pathways [10]. Different from the classical step-growth polycondensation requiring a 1 : 1 mole ratio of monomers to ensure high molecular weight, polymerizations starting with nonstoichiometric amounts of monomers lead to high molecular weight polymers. Most polyxanthenes are soluble in common organic solvents (such as methylene chloride, chloroform, and *N,N*-dimethylformamide, and can be processed into self-supporting films. The nonplanar xanthene units would frustrate the chain packing and thus lead to microporosity. The pore volume is mainly determined by the tortuosity and rigidity of the polymer backbone, which can be tuned by the selection of monomers with variations in chemical structure. For example, a ladder polymer synthesized from 3,6-dihydroxy-9-trifluoromethyl-9-phenylxanthene (a monomer presenting an overall butterfly-like conformation) exhibited a fully fused-ring main chain and had a high BET surface area of 587 m^2/g [11]. Other typical types of PIMs include polyimide (Figure 6.2c) [12–15] and CANAL polymers (Figure 6.2e) [16–19].

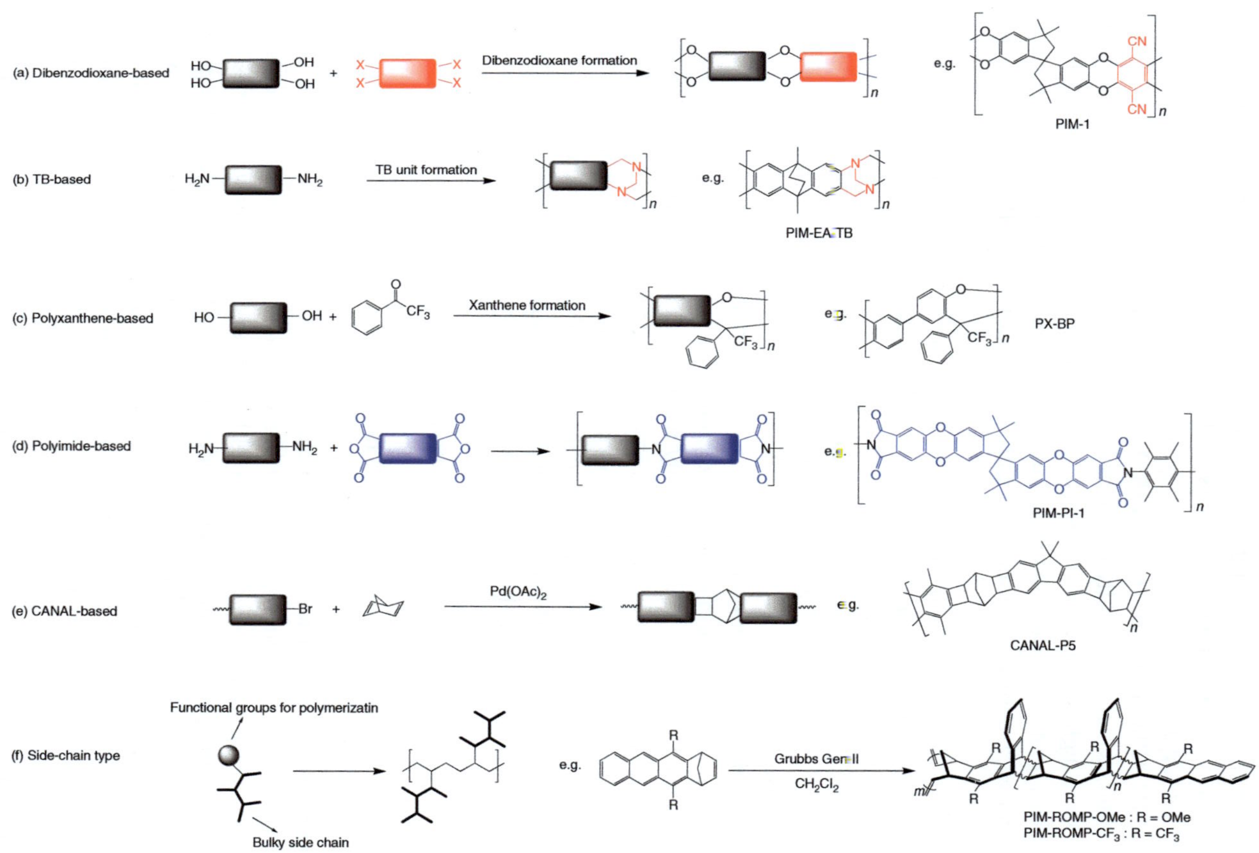

Figure 6.2 Typical types of polymers of intrinsic microporosity.

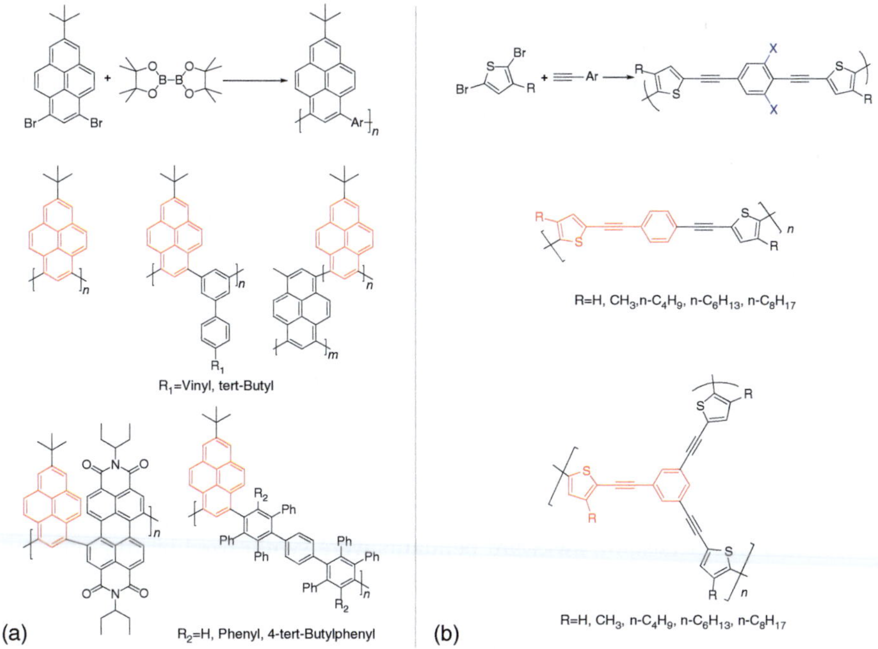

Figure 6.3 The synthesis of soluble conjugated microporous polymers containing pyrene units (a) or thiophene units (b).

Conjugated macromolecules are other typical microporous polymers and are always synthesized by Suzuki coupling, Yamamoto coupling, or metal-free Diels-Alder reactions. Cheng et al. reported the first example of soluble conjugated microporous polymers through a two-step $(A_4 + B_2)$-type Suzuki coupling polymerization (Figure 6.3a) [20]. The obtained material, a statistical hyperbranched copolymer, had bulky tert-butyl groups that could inhibit the packing of the backbone and promote the solubility of the polymer in common organic solvents, such as dichloromethane, tetrahydrofuran, and toluene. This polymer is microporous, and its microporosity was largely determined by the processing conditions. The solids obtained from the rapid precipitation of polymers exhibited a surface area of 505 m^2/g, while the polymer film prepared from the slow evaporation of solvent had a very low surface area of 12 m^2/g. This result confirmed that the microporosity was derived from the combination of backbone rigidity and chain packing. The authors then widened the variety of linear conjugated microporous polymers and found that the surface area of polymers exhibited a roughly linear relationship with their chain length [21]. This was because of the exacerbating chain packing of polymers with high molecular weights.

The rigidity and contortion of the polymer backbone, molecular weight, and processing conditions are key factors for the microporosity of conjugated polymers. For microporous conjugated polymers that are processable, the introduction of bulky pendant substituents should also be considered, as they are prone to imparting both microporosity and solubility. Ren et al. [22] synthesized a series of thiophene-substituted conjugated microporous polymers containing varied alkyl chains through Sonogashira–Hagihara cross-coupling polymerization between ethynylbenzene monomers and dibromothiophene with carbon-branched chains (Figure 6.3b) and investigated the influence of side chains on polymer properties. Improving the length of the alkyl chains improved the solubility of the corresponding polymers in common organic solvents (such as toluene, dichloromethane, and butylene oxide but decreased the BET surface area, presumably because the bulky pendants would occupy a certain amount of free voids. The highest BET surface area for the soluble polymer is 182 m^2/g. These soluble polymers can be processed into free-supporting membranes.

All the above-mentioned microporous polymers are constituted by linear and structurally rigid ladder-type main chains. What would happen if the noncompliant and rigid units were introduced to side chains instead of the main chain? Zhao et al. [23] reckoned that the awkward side chains would prevent the compact folding of the polymers, thus creating a large number of voids (Figure 6.4a). To test this prediction, a series of one-dimensional linear polymers having flexible backbones and shape-persistent, awkward pendants were synthesized by ring-opening metathesis polymerization (ROMP) of a three-dimensional stereorandom molecule with a benzonorbornene moiety (Figure 6.4c). It should be noted that the extreme steric hindrance of the large monomeric units may inhibit ROMP. With optimization, a number-average molecular weight of approximately 340,000 could be achieved, and the resulting polymers were soluble in common organic solvents, such as dichloromethane, chloroform, and tetrahydrofuran.

ROMP polymers are porous, and their pore size is distributed in a wide range from 1 nm to >50 nm. Different from conjugated polymers, the degree of polymerization contributed little to the porosity of ROMP polymers. The topology of the polymer and the structure of side chains mainly determined the porosity. For example, replacing the $-CF_3$ groups with –OMe groups would improve the rotational freedom, resulting in a significantly decreased BET-specific surface area from 700 to 146 m^2/g (Figure 6.4c) [23]. Manipulating the side-chain length and dispersity of ROMP polymers is an effective method to tune their porosity [25]. ROMP polymers

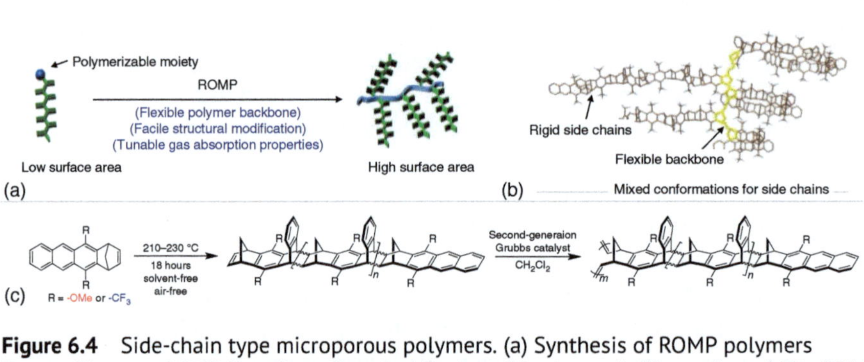

Figure 6.4 Side-chain type microporous polymers. (a) Synthesis of ROMP polymers Source: Zhao et al. [23]/American Chemical Society, (b) schematic illustration of CF_3-ROMP with five repeating units Source: He et al. [24]/John Wiley & Sons, and (c) the synthesis of example ROMP polymers, CF_3-ROMP, and OMe-ROMP.

could absorb large amounts of organic vapors. These materials were also highly permeable and plasticization-resistant for CO_2 separation [24].

These ROMP polymers can be further modified through the Diels-Alder reaction of anthracenyl end groups, diversifying the variety of side-chain-type microporous polymers. As ROMP using Grubbs' catalysts shows good tolerance to many groups, some functionalities can be introduced by copolymerization to endow the resulting materials with both microporosity and specific functions. This related research remains largely unexplored.

Microporous polymers discussed above (i.e. PIMs) are soluble in common organic solvents, which endows them with good processability. The solution-casting approach can be employed to make self-supporting membranes from these materials. However, membranes from linear polymers may suffer from the aging problem, that is, the obvious decrease in microporosity. Cross-linking the polymer backbones may improve the anti-aging properties. The major difficulty with cross-linked polymers is their insolubility and poor processability. To overcome this problem, the organic sol–gel approach has been developed. Zhu et al. [26] presented the preparation of triazine-framework-based membranes through aromatic nitrile trimerization catalyzed by trifluoromethanesulfonic acid (Figure 6.5a). This membrane was no longer soluble in any common solvents and was microporous, with a high BET surface area of $\sim$738 m^2/g. This membrane was demonstrated to be effective in CO_2/N_2 separation. With a similar synthetic procedure, Yang et al. [27] prepared a series of fluorinated triazine framework membranes (Figure 6.5b). These membranes showed enhanced CO_2-philic properties, and their CO_2 separation performance significantly improved as the fluorine content within the membranes increased.

Membranes constructed by microporous polymers have a large amount of free volume on the scale of $\leq$2 nm. These voids may act as ion transport channels, and membranes from PIMs can provide materials to probe into the ion transport behavior within nano space-confined channels. First, hydrophilic PIMs are needed. For this purpose, attaching charged groups to PIMs to produce IEMs has received increasing attention. The synthetic strategies for hydrophilic PIMs are summarized in the Section 6.3.

6.3 IEMs Derived from Microporous Polymers

PIMs were originally designed for gas separation [28] and organic nanofiltration [29], and were made up of hydrophobic chains. These PIMs are not suitable for membranes used in aqueous environments. Direct postmodification of these PIMs seems to be a simple method for hydrophilic PIMs, including post-quaternization, post-sulfonation, and post-hydrolysis.

6.3.1 Positively Charged Microporous Polymers

Positively charged groups that use nitrogen as the center are synthesized via amine chemistry. TB-based PIMs contain bicyclic amine units that can be easily made into

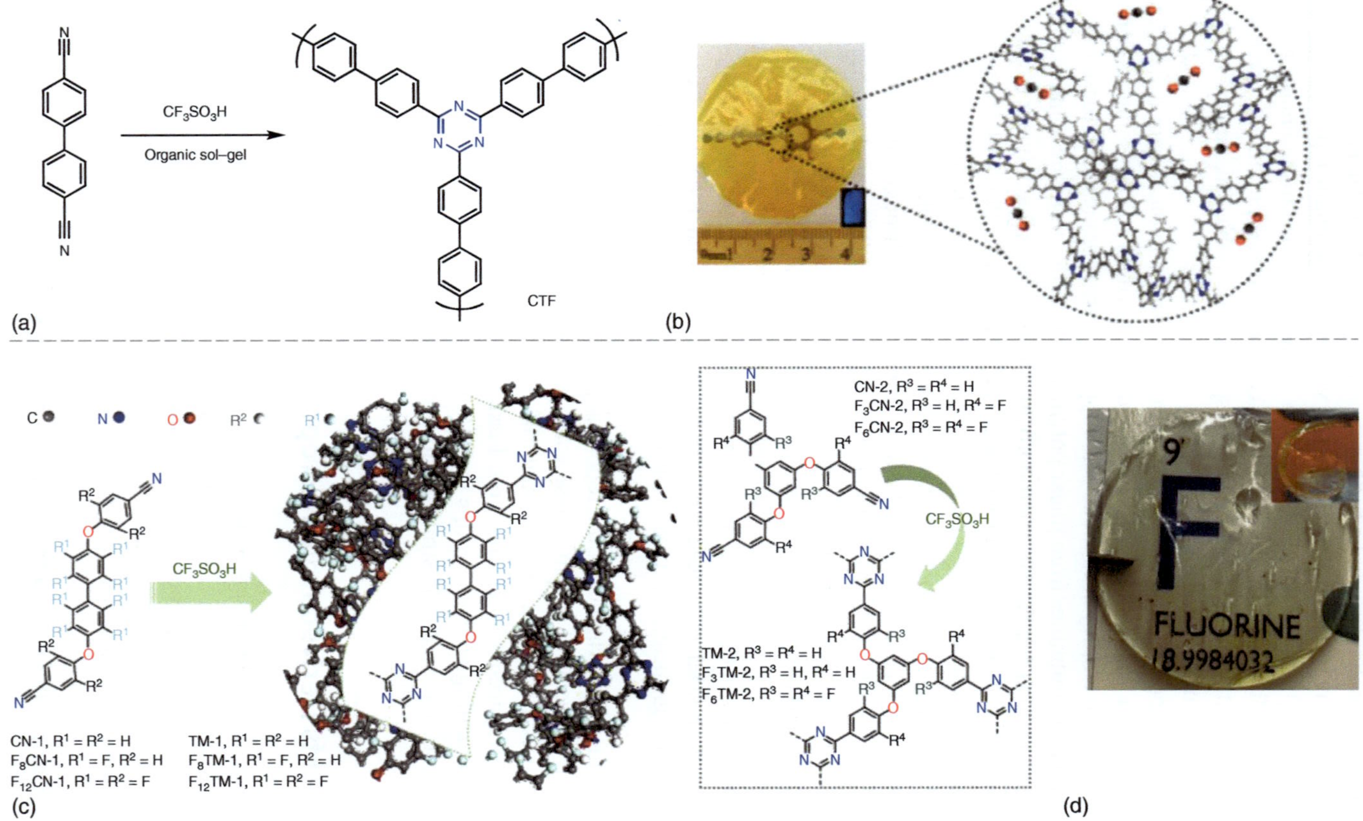

Figure 6.5 Self-standing covalent triazine framework membranes fabricated via organic sol–gel method. Membranes constructed by rigid (a)Source: (a) Zhu et al. [26] or flexible, fluorinated (b) frameworks. Source: (b) Zhu et al. [26]/American Chemical Society (d) [27]. Yang et al., 2020/With permission from Elsevier.

charged groups. For example, the bicyclic amine can react with acid and easily convert into a protonated compound. As this reaction is very reactive, the protonation process of membranes made by TB polymers can be successfully achieved just through a heterogeneous reaction of membranes that are immersed in an aqueous acid solution [30]. The resultant membranes (Figure 6.6b R-TB^+) became increasingly hydrophilic as the reaction degree increased and may allow water or ions to pass through. Marken et al. confirmed the reversible protonation and deprotonation of TB-based PIM membranes and used these membranes to construct ionic diodes [31]. The bistable and switchable states of the ionic diodes were contingent on the pH of the electrolytes. When an electric field was applied to restrict proton diffusion into the membranes, the corresponding ionic diode was in a closed state, as the membrane in this situation was hydrophobic. With the release of an electric field, protons could permeate into the membrane and react with amine groups, resulting in a hydrophilic open state.

Zhou et al. also performed the protonation of TB-based polymer membranes and employed them as ion-selective membranes [30]. Ion-diffusion experiments on these membranes revealed their excellent size-sieving performance. The DMBP-TB^+ membrane well balanced the ion permeance and ion selectivity, with a high Li^+ flux of 190.8 mmol/m/h and a Li^+/Mg^{2+} selectivity of 21. This selectivity is mainly attributed to the difference in the size of different ions. When proton/vanadium ions permeated through the DMBP-TB^+ membrane, a much higher selectivity of above 6300 was realized. These results encouraged the authors to explore the potential of protonated TB-based membranes in vanadium redox flow batteries [32]. The battery assembled with the DMBP-TB^+ membrane showed a very low area-specific resistance of $0.57\,\Omega\,cm^2$ and a high power density of $710.9\,mW/cm^2$ at ~100% state of charge. In addition, this battery exhibited a much more stable long-term cycling performance than that assembled with the commercial Nafion 117 due to the extremely high proton/vanadium selectivity of the DMBP-TB^+ membrane.

The large number of micropores in R-TB^+ membranes not only contributes to fast and selective ion transport but also induces the syphoning effect that helps membranes retain solvents at very high temperatures. Tang et al. took advantage of these characteristics and employed phosphoric acid-doped TB-based membranes in fuel cells with a wide range of temperatures (−20 to 200 °C) [33]. They found the syphoning effect of the membranes with an average diameter of 0.66 nm, which enabled very high retention of phosphoric acid at both low and high humidity and thus a very high proton conductivity retention (a value more than three orders of magnitude than that for the conventional control sample, that is, phosphoric acid-doped polybenzimidazole membrane). The corresponding phosphoric acid-doped proton exchange membrane fuel cells operated at 15 °C retained a 95% peak power density after 150 start-up/shut-down cycles and could realize over 100 cycles even at a very low temperature of −20 °C. This work suggested the great potential of charged microporous polymers as membrane separators in fuel cells with a very wide operational range of temperatures.

It seems to be a very convenient method to produce hydrophilic PIMs through the protonation of TB-based PIMs. The deprotonation of protonated tertiary amine

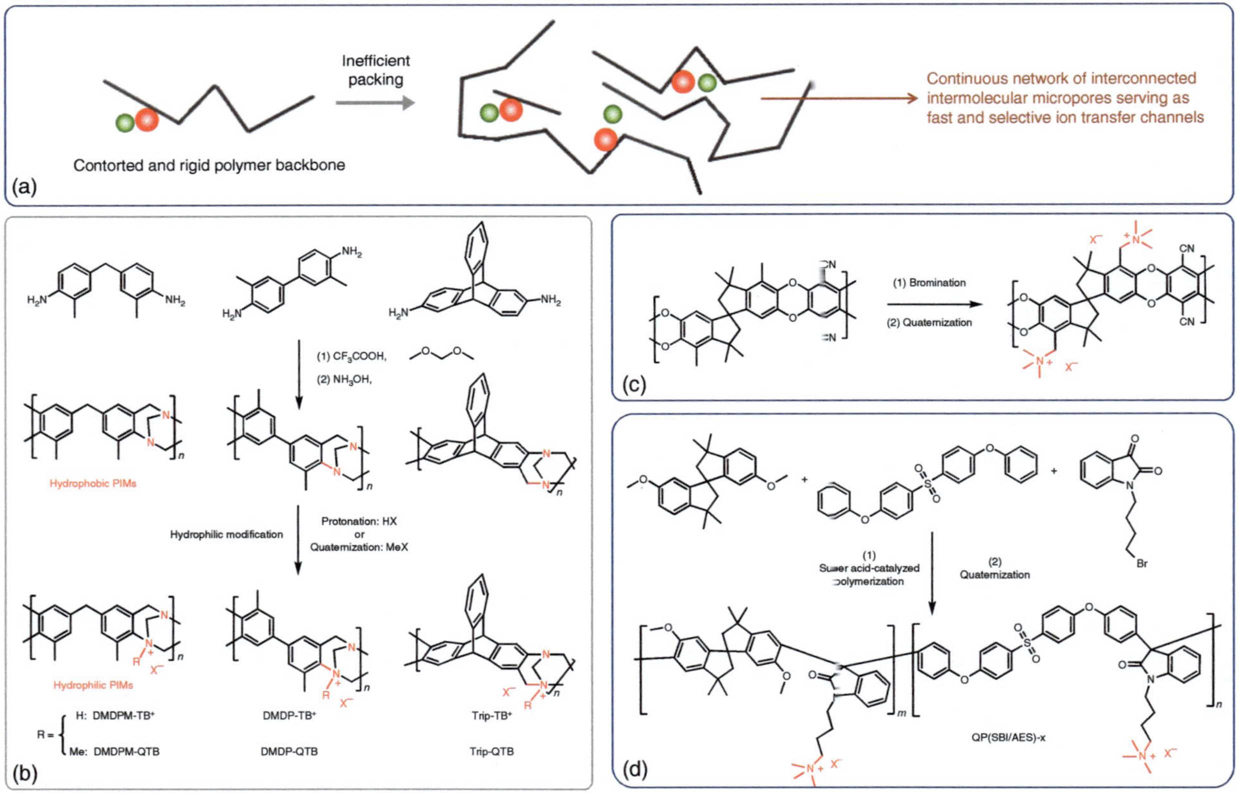

Figure 6.6 Representative AEMs from positively charged PIMs. (a) Schematic of the contorted and rigid backbone of charged PIMs creating microporosity. Synthetic route for AEMs from TB polymers (b), dibenzodioxane-based ladder polymers (c), and spirobisindane-containing poly (ether sulfone) (d).

could easily occur in an aqueous alkaline solution, and the R-TB^+ membranes become hydrophobic at high pH. This reversibility endows R-TB^+ membranes with the potential to act as an ionic diode [34] but causes loss of ion-conducting performance of the membranes in an alkaline solution. This problem can be avoided via quaternization of the tertiary amine with a haloalkane. The resulting quaternary ammoniums are dissociated in an aqueous solution at any pH. In 2016, Yang et al. [35] first performed quaternization by directly immersing TB-based polymer membranes in a methyl iodide solution, obtaining positively charged polymer membranes (R-QTB, Figure 7.6b). Among them, the DMBP-QTB membrane, with a low ion exchange capacity of 0.82 mmol/g, presented the highest hydroxide ion conductivity of 164.4 mS/cm. This efficient ion transport was mainly ascribed to intrinsic microporosity. This work demonstrated the great potential of R-QTB as an ion-conducting membrane and inspired a series of studies relating to positively charged TB-based PIMs [36, 37].

Similar to protonated polymers, R-QTB could also sieve ions according to the ion size, as proven by ion-diffusion experiments [30]. The DMBP-QTB membrane retarded the permeation of ions >8.2 Å in size but allowed smaller cations to pass effectively. A high Li^+/Mg^{2+} selectivity of 152.2 could be realized for this membrane, indicating the possibility of R-QTB membranes for extracting Li^+.

Even with excellent performance in selective conducting ions, R-QTB membranes suffered from a problem, that is, easy degradation in aqueous alkaline solutions, which prevented their application in anion exchange membrane fuel cells. Ishiwari et al. first studied the chemical degradation of positively charged TB units [38, 39]. The degradation started with the N-methylation of the tertiary amine, followed by the ring-opening of the quaternized TB units in an alkaline solution (Figure 6.7a). With this transformation, the methylene bridge was removed, resulting in diazacyclooctane polymer backbones that were flexible enough to pack effectively (Figure 6.7b). Therefore, the resulting polymers showed a decrease in both ion exchange capacity and microporosity. The diazacyclooctane polymers are promising for constructing separation membrane materials. The presence of secondary amine groups endowed these membranes with improved resistance to CO_2-induced plasticization and an H_2/CH_4 selectivity of 186.7 (Figure 6.7c) [43].

Diazacyclooctane polymers have secondary amine groups that can be further functionalized. This means that the rigid-to-flexible transformation offers an opportunity to expand the variety of ladder polymer derivatives. The amine groups also enable the pH-responsive capacity of membranes due to the reversibility between protonation and deprotonation (Figure 6.7d) [41]. Notably, diazacyclooctane polymers can undergo coordination or elimination of a Lewis acid, enabling reversible interconversion between flexible and rigid conformations (Figure 6.7e) [42]. This suggested the opportunity to design stimuli-responsive polymers.

As R-QTB membranes are subjected to chemical degradation in aqueous alkaline solutions, preparing positively charged membranes from other types of PIMs is needed. Ishiwari et al. [44] synthesized a dibenzodioxane-based PIM containing benzylic methyl groups and modified this polymer via bromination and subsequent quaternization, affording trimethylbenzyl ammonium-appended

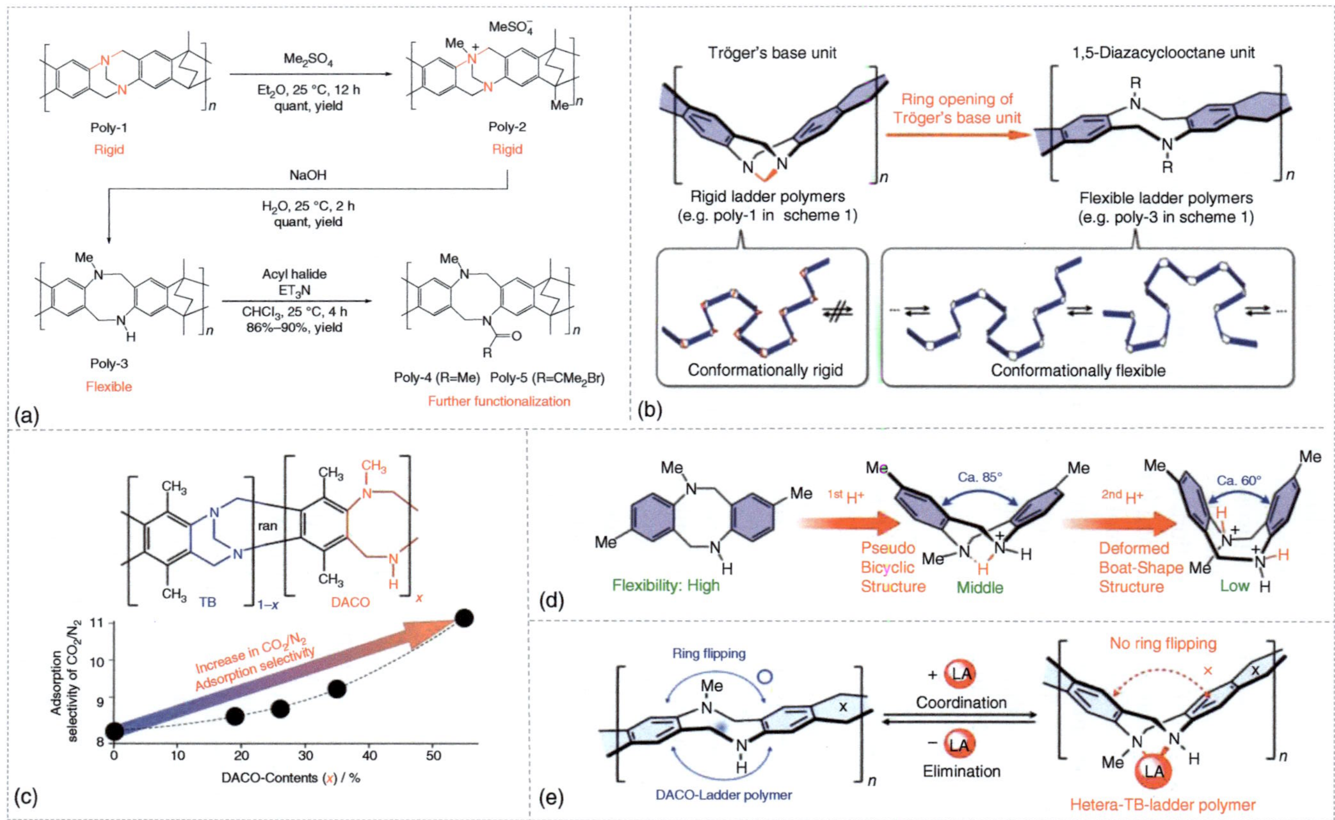

Figure 6.7 TB-based polymers undergoing rigid-to-flexible conformation changes. (a) Converting quaternized TB units into conformationally flexible diazacyclooctane units via the ring-opening reaction, (b) schematic illustration presenting the structural change of TB-based polymers after ring opening. Source: (a, b) Ishiwari et al. [38], (c) increased adsorption selectivity of CO_2/N_2 by the polymers with both TB and diazacyclooctane. Source: Inoue et al. [40]/John Wiley & Sons, (d) the conformation difference between the mono-protonated and diprotonated DACO-containing ladder polymers. Source: Ishiwari et al. [41], and (e) the coordination or elimination of a Lewis acid enables reversible interconversion between flexible and rigid conformations. Source: Ishiwari et al. [42].

PIMs (Figure 6.6c). The quaternization process was carried out by immersing the membrane in a trimethylamine solution. The resultant membrane had a BET surface area of 381 m^2/g and presented a good hydroxide conductivity of 65 mS/cm at 80 °C.

Huang et al. reported another positively charged dibenzodioxane-based PIM (QPIM-1) that was synthesized by modifying the typical PIM (PIM-1) [45]. The modification involved two steps: the animation of nitrile groups and the following quaternization. QPIM-1 was soluble in some polar organic solvents and could be processed into robust membranes from a solution-casting method. The CO_2 adsorption confirmed the high microporosity of the QPIM-1 membranes, with a BET value of 337 m^2/g. The QPIM-1 membrane with an ion exchange capacity of 2.12 mmol/g exhibited a hydroxide ion conductivity of 57 mS/cm at 20 °C. Importantly, this membrane, even with high microporosity, could prevent the crossover of hydrogen gas, and thus, a high open-circuit voltage of 1.038 V could be realized for the fuel cell with this membrane as the separator. This was presumed due to the occupancy of free volumes by water molecules.

Directly modifying existing PIMs seems to be a shortcut for synthesizing charged PIMs but highly depends on their reactivity. Li et al. [46] employed a super acid-catalyzed polymerization to synthesize new kinds of microporous poly(spirobiindane-aryl ether sulfone) copolymers and then carried out quaternization of the brominated alkane with trimethylamine to afford polymers with positively charged side chains (P(SBI/AES)-X, Figure 6.6d). Notably, the flexible side chains ensured good self-assembly ability of the charged moieties, and the P(SBI/AES)-X membranes thus had a phase-separation microstructure. Benefiting from both the microstructure and microporosity, P(SBI/AES)-X membranes, with an ion exchange capacity of 1.59 mmol/g presented a high hydroxide ion conductivity of 110 mS/cm at 80 °C. These membranes were assembled into single-cell fuel cells, which displayed a peak power density of 593 mW/cm^2. The major problem with P(SBI/AES)-X is its poor chemical stability in alkaline aqueous solutions, impeding its wide application.

The current membranes made from negatively charged PIMs are mainly based on post-modification, which cannot precisely control the degree of functionality. In addition, they suffer from poor alkaline stability. Zhou et al. [47] proposed the synthesis of ionic-conjugated microporous polymer membranes. To overcome the problem of poor film-forming properties of most conjugated polymers because of their brittleness, the authors employed the electropolymerization strategy to construct thin films directly onto a plate electrode (Figure 6.8). The membranes contained both very rigid monomers (i.e. 1,3,5-tris (*N*-carbazolyl) benzene, for maintaining structural uniformity) and flexible, charged monomers (i.e. 6-(9*H*-carbazol-9-yl)-*N*,*N*,*N*-trimethylhexan-1-aminium bromide, for enhancing mechanical flexibility and introducing functional groups). The membranes had a thickness of only ~40 nm and exhibited a high Young's modulus of 4 GPa on average. The membranes were found to be microporous, with a BET-specific surface area >600 m^2/g and a sharp pore-size distribution peak at 0.74 nm. Compared with the commercial Nafion 117, these conjugated membranes presented similar ion

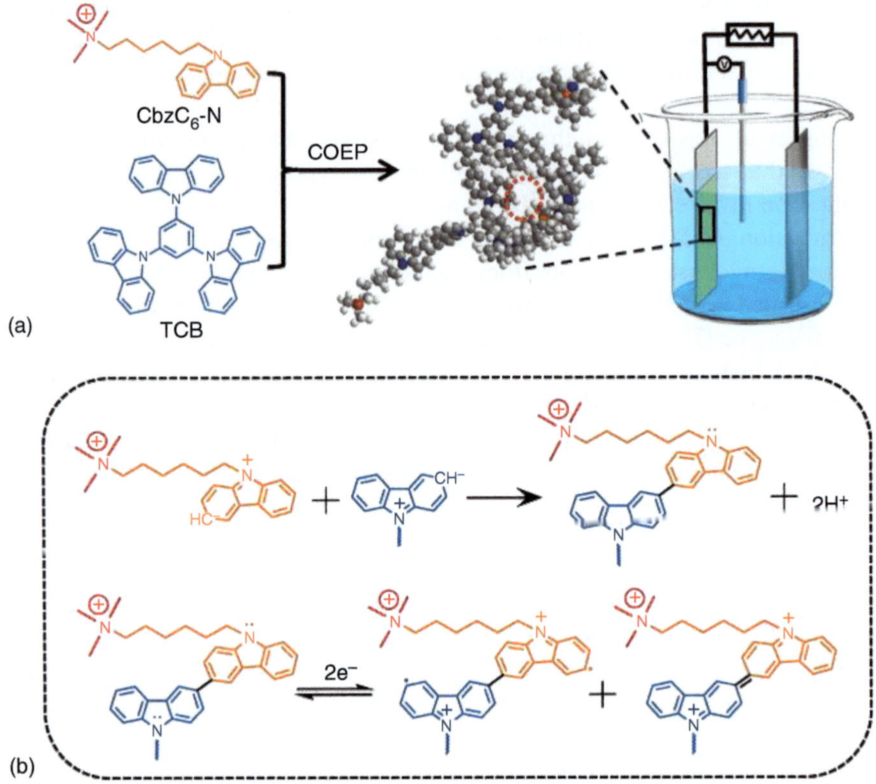

Figure 6.8 Ionic conjugated microporous polymer membranes made by electro-polymerization. These membranes had a thickness of low to approximately 40 nm. Source: Zhou et al. [47].

conductivity but showed a dominant advantage in ion selectivity, having an order of magnitude higher ion selectivity of K^+/Mg^{2+} and Li^+/Mg^{2+}.

As discussed previously, the development of positively charged PIMs is still in the early stage and is facing the challenge of extending the synthetic approach and material variety. In particular, the bottom-up approach, that is, the polymerization of charged monomers, is highly expected. In addition, the above-reported PIMs are subjected to chemical degradation, and thus, appending chemically stable cations to PIMs is very desired to promote the development of IEMs derived from PIMs. The electro-polymerization strategy is emerging as an important method to construct charged microporous polymer membranes but may still show difficulty in self-standing, homogeneous ion exchange membranes.

6.3.2 Negatively Charged Microporous Polymers

IEMs made from positively charged microporous polymers conduct anions and repel cations, while IEMs from negatively charged microporous polymers are required to conduct cations and reject anions. Current negatively charged microporous polymers are mainly synthesized from hydrolysis, amidoxime, and sulfonation, as discussed in detail below.

6.3.2.1 Hydrolysis of Dibenzodioxin-based Microporous Polymers

Most dibenzodioxin-based PIMs are synthesized using 2,3,5,6-tetrafluoro-1,4-benzenedicarbonitrile as one of the starting materials and have a large number of nitrile groups. Nitrile chemistry is widely employed to modify these PIMs. For example, the hydrolysis of nitriles yields carboxylated groups that are ionizable in an aqueous alkaline solution. As shown in Figure 6.8a, Du et al. first used this reaction to modify PIM-1 through immersion of membranes in a strong alkaline solution, producing hydrophilic membranes with mechanical robustness [48]. The presence of carboxylated groups was confirmed by proton nuclear magnetic resonance and Fourier transform infrared spectroscopy. The degree of functionality was tuned by controlling the reaction time and temperature, as confirmed via carbon elemental analysis. The conversion of nitriles into carboxylated groups greatly changed the solubility of the resulting polymers. Primitive PIM-1 can dissolve in chloroform and tetrahydrofuran but is insoluble in polar aprotic solvents (methylpyrrolidone, dimethyl sulfoxide, dimethylformamide, dimethylacetamide, etc.). In contrast, the hydrolyzed polymer membranes were readily soluble in polar aprotic solvents but insoluble in dichloromethane and chloroform.

Budd et al. explored the base-catalyzed hydrolysis mechanism and found that the products are a mixture of ammonium carboxylate, amide, sodium carboxylate, and carboxylic acid [49]. This conclusion was further confirmed by Santoso et al. in a subsequent work [50]. They observed that the nitriles mainly converted into intermediate amide ($-CONH_2$) at the initial stage and then into carboxylic groups. A much harsher hydrolysis condition or longer hydrolysis time was needed for the second conversion because $-NH_2$ was a poor leaving moiety. Jeon et al. prepared a highly carboxylate-functionalized polymer membrane (approximately 92 mol% carboxylated moieties) by prolonging the time of the alkaline hydrolysis process to approximately 360 hours [51]. The introduction of carboxylic groups resulted in smaller interchain distances and a higher CO_2 affinity. The enhanced packing of polymer chains was ascribed to the strong intermolecular hydrogen bonding between adjacent carboxylic acid groups, leading to a significant decrease in surface area from 735 to 284 m^2/g. The hydrolyzed polymer membranes were effective in separating CO_2 from the gas mixture, with a very high selectivity of 53.6 for CO_2/N_2.

The carboxylic acid groups within these polymer membranes are negatively charged in alkaline conditions and thus provide great potential for hydrolyzed polymer materials to serve as ion-conducting membranes. Relative research deserves more attention.

6.3.2.2 Amidoxime of Dibenzodioxin-based PIMs

As early as 2012, Patel et al. first reported the amidoxime of dibenzodioxin-based PIMs [52]. They found that PIM-1 reacted with hydroxylamine to convert into amidoxime groups, and the conversion was completed under reflux in tetrahydrofuran for 20 consecutive hours. With the introduction of amidoxime, the resulting polymers (AO-PIM-1, Figure 6.8b) showed reduced microporosity, with a pore volume decreasing from 0.57 to 0.28 cm^3/g. This was possible because the interchain packing was enhanced by hydrogen bonding and the bulkiness of the functional groups. The amidoxime groups are negatively charged in aqueous alkaline solutions with pH values above 13 [53] and thus can serve as cation-conducting groups in strongly alkaline solutions. Baran et al. found that membranes made from amidoxime-containing PIMs could balance ion conductivity and selectivity well, and thus show great potential in aqueous flow batteries [54].

Tan et al. [55] further explored the influence of pore architecture and ionic conductive functionality on the performance of size-exclusion selectivity and ion conductivity. With an increase in microporosity or content of functionality, the membranes from amidoxime-modified PIMs exhibited enhanced ion transfer efficiency. By tuning the topology of the polymer main chains, Ye et al. extended the variety of amidoxime-modified PIMs [56]. All the above work demonstrated the great potential of these materials as separators in anthraquinone | $K_4[Fe(CN)_6]$ alkaline aqueous flow batteries. However, amidoxime moieties are no longer charged in a neutral aqueous solution, which may cause obvious performance degradation of membranes used in neutral media.

Wang et al. blended carboxylate-modified PIM-1 with amidoxime-functionalized PIM-1 to fabricate high-performing ion-selective membranes for aqueous organic redox flow batteries [57]. The blend membranes were microscopically miscible, which was attributed to the establishment of cohesive interchain interactions (salt bridges and hydrogen bonds). The micropore channels fixed with ionizable functional moieties promoted ion conduction, and the steric effect of micropores helped to reject the permeation of redox-active molecules with relatively large sizes. It is noteworthy that the blend membranes surpassed the trade-off (encountered by most aqueous flow battery membranes) between ion conductivity and selectivity.

6.3.2.3 Post-sulfonation of PIMs or Bottom-up Approach

The ion-conducting performance of both carboxylated- and amidoxime-containing PIMs relies heavily on pH. These membranes may lose their high ionic conductivity in acidic solutions. Applying sulfonate groups to PIMs is free of this problem. In 2014, Kim et al. first tried to directly sulfonate PIM-1 using chlorosulfonic acid or concentrated sulfuric acid as a sulfonating agent but only detected degradation or crosslinked products [58]. The brittleness of the obtained sulfonated polymers limited their preparation for self-standing robust membranes.

In 2020, for this problem, Zuo et al. reported the first example of sulfonated PIMs (SPX-BP and SPX-HFP in Figure 6.9a) [59]. They used fluorinated polymers of polyxanthene that were stable in strongly acidic conditions as the raw materials for post-sulfonation. Polyxanthene was produced via trifluoromethanesulfonic acid-catalyzed step-growth polymerization of a,a,a-trifluoroacetophenone and bisphenol monomers (4,4-biphenol, bisphenol AF), and chlorosulfonic acid was used as a sulfonating agent in the second step. The resulting polymers were soluble in some polar solvents (such as dimethyl sulfoxide, *N*,*N*-dimethylformamide, and dimethylacetamide) and could be processed into mechanically robust membranes.

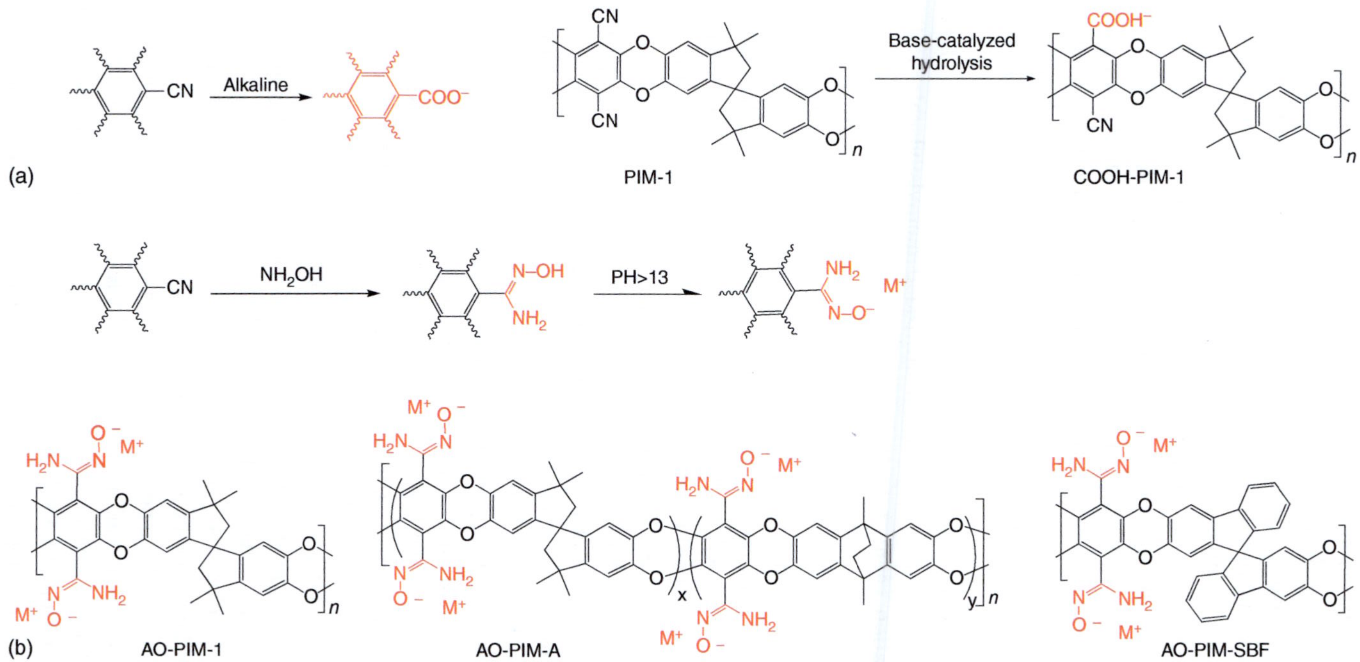

Figure 6.9 Negatively charged PIMs made from hydrolysis (**a**) or amidoxime (**b**) of dibenzodioxin-based PIMs.

Similar to carboxylated PIMs and amidoxime-containing PIMs, sulfonated polymers exhibited a decrease in microporosity. Even so, SPX-BP constructed by the rigid main chains retained a large number of interconnected voids, as proven by the carbon dioxide adsorption isotherm. With a similar ion exchange capacity, SPX-BP had a high proton conductivity that surpassed that of most microphase-separated membranes. Compared with the commercial Nafion 117, the SPX-BP membrane exhibited much higher selectivity of small ions (e.g. potassium ions) to redox anions (e.g. 2,6-dihydroxyanthraquinone). Based on these advantages, the SPX-BP membrane showed great effectiveness in anthraquinone | $K_4[Fe(CN)_6]$ alkaline aqueous flow batteries. The cell with the SPX-BP membrane presented a very low area-specific resistance of $0.70\,\Omega\,cm^2$ and could run stably at a current density of $60\,mA/cm^2$ for 1000 consecutive cycles. A low capacity loss rate of 0.05% per hour was found, and only a 7.7% contribution was ascribed to the crossover of iron ions.

To avoid side reactions of conventional dibenzodioxin-based PIMs during the post-sulfonation process, Ye et al. utilized trimethylsilyl chlorosulfonate, a weaker sulfonation agent, to sulfonate PIM-1 [60]. They tuned the sulfonation degree by adjusting the amount of sulfonation agent, producing sulfonated PIMs (sPIM-SBF, Figure 6.9b) with ion exchange capacities ranging from 0.53 to 1.86 mmol/g. These membranes surpassed the trade-off between ferri/ferrocyanide permeability and apparent ionic conductivity. A potassium ferrocyanide/2,6-di(3-phosphenopropyloxy) anthraquinone aqueous flow cell assembled with the sPIM-SBF membrane could run stably for 2100 continuous cycles at $100\,mAcm^{-2}$.

To avoid side reactions, chemically stable PIMs or weaker sulfonation agents are required. All the sulfonated PIMs discussed previously were prepared via homogeneous reaction. Dong et al. carried out heterogeneous sulfonation of a PIM-1 membrane by immersion of the membrane in a trioxide/dichloromethane solution and successfully introduced sulfonate groups to PIM-1 [61]. Notably, the reaction time had to be kept short to prevent the chemical degradation of ether linkages.

All the above-mentioned sulfonated PIMs were synthesized from postmodification, which poorly controlled the sulfonation degree and reactive sites. The bottom-up approach, that is, direct polymerization of charged monomers, is comparatively controllable and thus is receiving increasing attention. Zhou et al. [62] reported the first example of sulfonated PIMs that were synthesized via a one-step synthetic approach based on Friedel–Crafts hydroxyalkylation between 6,6′-dimethoxy-3,3,3′,3′-tetramethyl-1,1′-spirobiindane (as the contorted and rigid unit for inducing microporosity) and commercial benzaldehyde derivatives (including both ionic and nonionic units), as shown in Figure 6.9c. The degree of sulfonation of the resulting polymers could be easily controlled by tuning the ratio of ionic monomers to nonsulfonated benzaldehyde monomers. The carbon dioxide adsorption isotherms confirmed the microporosity of the sulfonated polymers. For example, the highest BET surface area of SPIM-3Fa-x–y reached $122\,m^2/g$. Combining fast ion transport and high cation permselectivity, these sulfonated poly(phenyl-alkane)s of intrinsic microporosity membrane proved to be excellent in advanced osmotic energy conversion, enabling a maximum power density of up to $9.58\,W/m^2$ under a 50-fold salinity gradient.

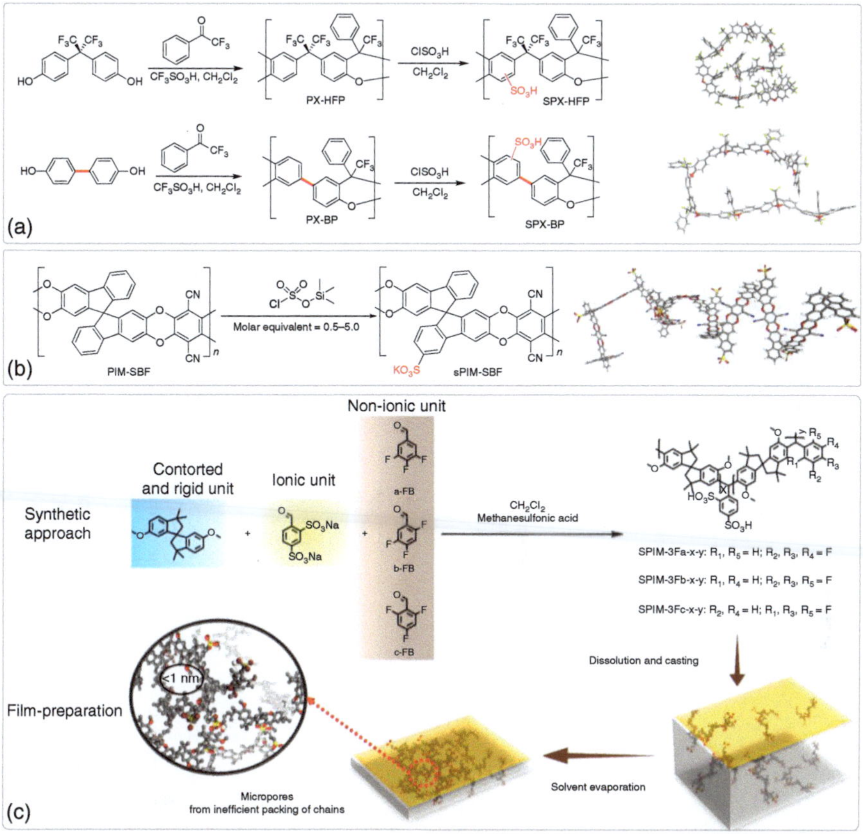

Figure 6.10 CEMs are derived from (a) sulfonated polyxanthene, (b) a spirobifluorene-based ladder polymer, and (c) sulfonated poly(phenyl-alkane)s of intrinsic microporosity. The molecular models of the polymer chains are given. The first two sulfonated PIMs are made from post-sulfonation, while the last is from bottom-up synthesis. All the polymers are solution-processed into self-standing membranes.

Compared with IEMs made from conventional polymer materials (represented by perfluorocarbon Nafion® and the developed hydrocarbon-based polyelectrolytes) whose ion-conducting performance is based on the microphase-separated morphology (Figure 6.10a), the recently emerging microporous polymer membranes discussed above having semirigid ion micropore channels exhibit both size-exclusion-induced selectivity and free-volume-induced permeability (Figure 6.10b), and thus have the potential to break the selectivity-permeability trade-off encountered by microphase-separated membranes [63]. However, IEMs made from semirigid, nonnetwork PIMs tend to swell to some extent after absorbing a certain amount of water [60], which may deteriorate mechanical robustness and ion selectivity. In addition, these PIMs may age over long-term service, which may lead to a decrease in ion permeability [64].

The research team led by T.W. Xu and Z. J Yang, along with their collaborators, proposed a new type of ion-transport membrane – triazine framework membranes (Figure 6.11) [65]. The design concept combines (i) grafting the charged groups to

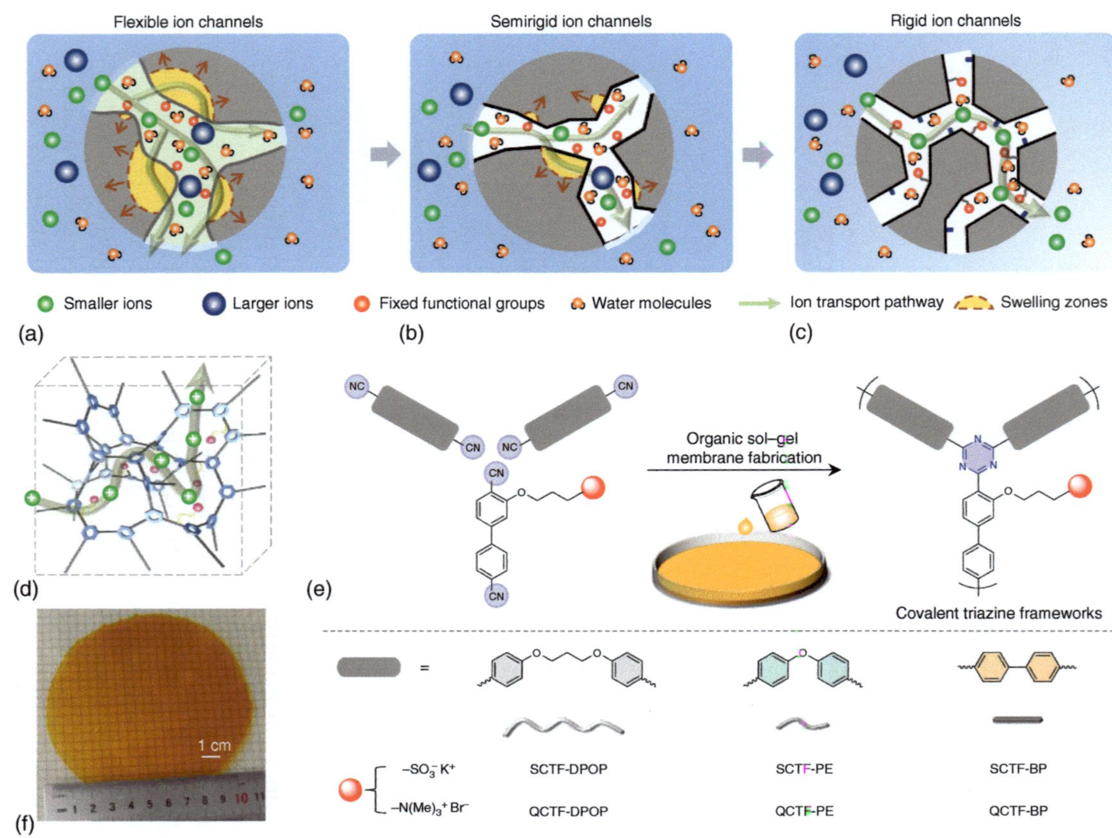

Figure 6.11 Schematic illustrations showing IEMs with flexible (a), semirigid (b), and rigid (c) ion channels. Source: (a-c) Zuo et al. [65]. IEMs with flexible ion channels refer to membranes with microphase-separated morphology that are derived from the self-assembly of charged groups into the non-ionic matrix formed by hydrophobic flexible polymer backbones. IEMs with semirigid ion channels are constructed by PIMs. The last type of IEMs is expected to be constructed by swelling-resistant 3D polymer frameworks. Preparation of self-standing covalent triazine framework membranes via an organo-sol–gel approach (e), and a photo of a membrane with a diameter of >10 cm (f).

the backbone framework to isolate the absorbed water within micropore channels from the hydrophobic skeletons, which apply robust subnanometer confinement to facilitate selective ion transport, and (ii) engineering charged, polar groups onto pore walls, which endow multiple interactions of ions and membrane to synergistically promote the ion flow on account of coordination mechanisms. The self-standing covalent triazine framework membranes were constructed from trifluoromethanesulfonic acid-catalyzed organic sol–gel reactions of charged or noncharged aromatic nitrile monomers (Figure 6.11e,f). This bottom-up synthesis enabled the easy regulation of both pore structure and pore chemistry just through the rational selection of monomers.

Compared to traditional materials, these polymer framework membranes exhibited much-enhanced capacity in both anti-swelling and anti-aging, with an extremely low swelling ratio of 3.1% on water absorption. Their rigid channels ensure high selectivity from size-sieving, thus enabling extremely low permeability of active materials with relatively large sizes. With proper control over the chemistry of rigid pore channels by engineering charged and polar moieties into polymer frameworks, the researchers observed near-frictionless ion flow within the all-rigid triazine framework membrane, with an ion diffusion coefficient close to that in water. This ultrahigh ion diffusivity is confirmed to be achieved by multiple interactions under robust micropore confinement. In addition, the SCTF-BP membrane could efficiently block active materials (e.g. 2,6-dihydroxyanthraquinone), showing a very low permeability of 1.27×10^{-11} cm^2/s, which is much lower than that for the commercial Nafion 212 membrane (6.14×10^{-10} cm^2/s, [55]).

The great effectiveness of framework polymer materials as ion-conducting membranes was demonstrated using 2,6-dihydroxy anthraquinone/K4[$Fe(CN)_6$] aqueous organic redox flow batteries. The neat membrane presented a very low area-specific resistance of 0.17 $\Omega\,cm^2$ and thus enabled a fast-charging flow battery. This battery could run stably at extreme current densities, from 200 to 500 mA/cm^2, with both exceptionally high energy efficiency and high-capacity utilization. These data related to energy efficiency and capacity utilization far surpass those for otherwise identical cells assembled with commercial membranes and state-of-the-art ion-sieving membranes.

This work highlights the importance of secondary interactions in the development of high-performing ion-transport membranes. The design strategy proposed in this study is believed to be broadly applicable, considering the numerous options of organic reactions and functional monomers that can be utilized to construct polymer frameworks, and directs the fit-for-purpose design of membranes according to practical application demand.

6.4 Conclusion and Outlook

Solution-processable charged PIMs may combine high performance in both ion conductivity and ion selectivity with excellent processability and have been used as ion-conducting membranes in a wide range of applications, especially aqueous organic flow batteries (an emerging large-scale energy storage technology [66, 67]).

Membrane micropore channels formed by inefficient chain packing largely determine the performance in ion conduction and ion selection. Control over pore architecture and pore chemistry is crucial to designing charged PIMs. All factors that influence chain packing behavior should be considered, including the rigidity, contortion, and chemical constitution of polymer backbones.

Post-modification is still the main synthetic strategy for charged PIMs. The addition of charged groups always results in an obvious decrease in microporosity, presumably due to the bulkiness of charged groups and enhanced chain packing. However, the charged groups alter the pore chemistry and create a hydrophilic environment, which is vital to the performance related to ion transfer. In general, a higher content of charged elements leads to higher ion conductivity, but its rational regulation must be taken into consideration to circumvent the swelling issue caused by too much water uptake. In addition, an increase in the microporosity may also benefit ion conduction but is sometimes accompanied by a decrease in ion selectivity.

Based on the discussion so far, the development of charged PIMs is in its infancy until 2023. Their synthetic approach is almost limited to post-modification. New synthetic strategies that provide easy control over pore architecture and pore chemistry are required to diversify the material variety and explore the complicated relationship between membrane performance and microstructure. A bottom-up approach, that is, polymerization of charged monomers into charged PIMs, may be an excellent choice. Another important issue for PIMs is that they may age over time, which results in a decrease in microporosity [64]. Cross-linked polymer membranes may offer assistance to restrict chain movement and alleviate the aging problem; however, show poor processability due to their insolubility.

Very recently reported triazine framework membranes may provide an excellent solution to the above-mentioned problems. These membranes are readily constructed during the bottom-up polymerization reaction with the concomitant formation of microporosity, and their micropore size and chemistry are well-tuned. Considering the great variety of organic reactions and functional monomers in building organic frameworks, it is anticipated that more diversified polymer frameworks will be employed to build IEMs, and these membranes will play an increasing role in a wide range of applications.

References

1 Epsztein, R., DuChanois, R.M., Ritt, C.L. et al. (2020). Towards single-species selectivity of membranes with subnanometre pores. *Nature Nanotechnology* 15: 426–436.

2 Zhou, X., Wang, Z., Epsztein, R. et al. (2020). Intrapore energy barriers govern ion transport and selectivity of desalination membranes. *Science Advances* 6: eabd9045.

3 Zhu, Q., Li, H., Wu, W. et al. (2023). Solution-processable amorphous microporous polymers for membrane applications. *Progress in Polymer Science* 137.

4 Zuo, P., Li, Y., Wang, A. et al. (2020). Sulfonated microporous polymer membranes with fast and selective ion transport for electrochemical energy conversion and storage. *Angewandte Chemie, International Edition* 59: 9564–9573.

5 Shen, J., Liu, G., Han, Y., and Jin, W. (2021). Artificial channels for confined mass transport at the sub-nanometre scale. *Nature Reviews Materials* 6: 294–312.

6 Budd, P.M., Ghanem, B.S., Makhseed, S. et al. (2004). Polymers of intrinsic microporosity (PIMs): robust, solution-processable, organic nanoporous materials. *Chemical Communications* 230–231.

7 Mariolino Carta, R.M.-E., Croad, M., Rogan, Y. et al. (2013). An Efficient Polymer Molecular Sieve for Membrane Gas Separations. *Science* 339: 303–307.

8 Carta, M., Malpass-Evans, R., Croad, M. et al. (2014). The synthesis of microporous polymers using Tröger's base formation. *Polymer Chemistry* 5: 5267–5272.

9 Olvera, L.I., Zolotukhin, M.G., Hernández-Cruz, O. et al. (2015). Linear, Single-Strand Heteroaromatic Polymers from Superacid-Catalyzed Step-Growth Polymerization of Ketones with Bisphenols. *ACS Macro Letters* 4: 492–494.

10 Lopez, G., Cruz, O.H., Garza, L.I. et al. (2014). Mechanistic aspects of superacid mediated condensation of polyphenols with ketones. Implications for polymer synthesis. *Journal of Molecular Modeling* 20: 2474.

11 Olvera, L.I., Rodríguez-Molina, M., Ruiz-Treviño, F.A. et al. (2017). A Highly Soluble, Fully Aromatic Fluorinated 3D Nanostructured Ladder Polymer. *Macromolecules* 50: 8480–8486.

12 Ghanem, B.S., McKeown, N.B., Budd, P.M. et al. (2009). Synthesis, Characterization, and Gas Permeation Properties of a Novel Group of Polymers with Intrinsic Microporosity: PIM-Polyimides. *Macromolecules* 42: 7881–7888.

13 Ma, X., Swaidan, R., Belmabkhout, Y. et al. (2012). Synthesis and Gas Transport Properties of Hydroxyl-Functionalized Polyimides with Intrinsic Microporosity. *Macromolecules* 45: 3841–3849.

14 Ma, X., Salinas, O., Litwiller, E., and Pinnau, I. (2014). Pristine and thermally-rearranged gas separation membranes from novel o-hydroxyl-functionalized spirobifluorene-based polyimides. *Polymer Chemistry* 5: 6914–6922.

15 Ma, X., Abdulhamid, M., Miao, X., and Pinnau, I. (2017). Facile Synthesis of a Hydroxyl-Functionalized Tröger's Base Diamine: A New Building Block for High-Performance Polyimide Gas Separation Membranes. *Macromolecules* 50: 9569–9576.

16 Liu, S., Jin, Z., Teo, Y.C., and Xia, Y. (2014). Efficient synthesis of rigid ladder polymers via palladium catalyzed annulation. *Journal of the American Chemical Society* 136: 17434–17437.

17 Lai, H.W.H., Benedetti, F.M., Ahn, J.M. et al. (2022). Hydrocarbon ladder polymers with ultrahigh permselectivity for membrane gas separations. *Science* 375: 1390–1392.

18 Lai, H.W.H., Teo, Y.C., and Xia, Y. (2017). Functionalized Rigid Ladder Polymers from Catalytic Arene-Norbornene Annulation Polymerization. *ACS Macro Letters* 6: 1357–1361.

19 Lai, H.W.H., Benedetti, F.M., Jin, Z. et al. (2019). Tuning the Molecular Weights, Chain Packing, and Gas-Transport Properties of CANAL Ladder Polymers by Short Alkyl Substitutions. *Macromolecules* 52: 6294–6302.

20 Cheng, G., Hasell, T., Trewin, A. et al. (2012). Soluble Conjugated Microporous Polymers. *Angewandte Chemie, International Edition* 124: 12899–12903.

21 Cheng, G., Bonillo, B., Sprick, R.S. et al. (2014). Conjugated Polymers of Intrinsic Microporosity (C-PIMs). *Advanced Functional Materials* 24: 5219–5224.

22 Ren, F., Wang, F., Pan, Y. et al. (2018). Flexible and UV Resistant Films Based on Thiophene-Substituted Conjugated Microporous Polymers Bearing Alkyl Chains: Tuning of Rigidity into Soft. *Macromolecular Materials and Engineering* 303: 1700619.

23 Zhao, Y., Zhao, Y., He, Y., and Swager, T.M. (2018). Porous Organic Polymers via Ring Opening Metathesis Polymerization. *ACS Macro Letters* 7: 300–304.

24 He, Y., Benedetti, F.M., Lin, S. et al. (2019). Polymers with Side Chain Porosity for Ultrapermeable and Plasticization Resistant Materials for Gas Separations. *Advanced Materials* 31: 1807871.

25 Benedetti, F.M., Wu, Y.-C.M., Lin, S. et al. (2022). Side-Chain Length and Dispersity in ROMP Polymers with Pore-Generating Side Chains for Gas Separations. *JACS Au* 2.

26 Zhu, X., Tian, C., Mahurin, S.M. et al. (2012). A superacid-catalyzed synthesis of porous membranes based on triazine frameworks for CO_2 separation. *Journal of the American Chemical Society* 134: 10478–10484.

27 Yang, Z., Guo, W., Mahurin, S.M. et al. (2020). Surpassing robeson upper limit for CO_2/N_2 separation with fluorinated carbon molecular sieve membranes. *Chem* 6: 631–645.

28 Wang, Y., Ma, X., Ghanem, B.S. et al. (2018). Polymers of intrinsic microporosity for energy-intensive membrane-based gas separations. *Materials Today Nano* 3: 69–95.

29 Thompson, K.A., Mathias, R., Kim, D. et al. (2020). N-Aryl–linked spirocyclic polymers for membrane separations of complex hydrocarbon mixtures. *Science* 369: 310–315.

30 Zhou, J., Jiao, Z., Zhu, Q. et al. (2021). Biselective microporous Tröger's base membrane for effective ion separation. *Journal of Membrane Science* 627: 119246.

31 Madrid, E., Rong, Y., Carta, M. et al. (2014). Metastable ionic diodes derived from an amine-based polymer of intrinsic microporosity. *Angewandte Chemie. International Edition* 53: 10751–10754.

32 Zhou, J., Liu, Y., Zuo, P. et al. (2021). Highly conductive and vanadium sieving Microporous Tröger's Base Membranes for vanadium redox flow battery. *Journal of Membrane Science* 620: 118832.

33 Tang, H., Geng, K., Lei, W. et al. (2022). Fuel cells with an operational range of –20 °C to 200 °C enabled by phosphoric acid-doped intrinsically ultramicroporous membranes. *Nature Energy* 7: 153–162.

34 Madrid, E., Rong, Y., Carta, M. et al. (2014). Metastable ionic diodes derived from an amine-based polymer of intrinsic microporosity. *Angewandte Chemie, International Edition* 53: 10751–10754.

35 Yang, Z., Guo, R., Malpass-Evans, R. et al. (2016). Highly Conductive Anion-Exchange Membranes from Microporous Troger's Base Polymers. *Angewandte Chemie, International Edition* 55: 11499–11502.

36 Hu, C., Zhang, Q., Lin, C. et al. (2018). Multi-cation crosslinked anion exchange membranes from microporous Tröger's base copolymers. *Journal of Materials Chemistry A* 6: 13302–13311.

37 Yang, Q., Cai, Y.Y., Zhu, Z.Y. et al. (2020). Multiple Enhancement Effects of Crown Ether in Troger's Base Polymers on the Performance of Anion Exchange Membranes. *ACS Applied Materials & Interfaces* 12: 24806–24816.

38 Ishiwari, F., Takeuchi, N., Sato, T. et al. (2017). Rigid-to-Flexible Conformational Transformation: An Efficient Route to Ring-Opening of a Tröger's Base-Containing Ladder Polymer. *ACS Macro Letters* 6: 775–780.

39 Inoue, K., Ishiwari, F., and Fukushima, T. (2020). Selective synthesis of diazacyclooctane -containing flexible ladder polymers with symmetrically or unsymmetrically substituted side chains. *Polymer Chemistry* 11: 3690–3694.

40 Inoue, K., Selyanchyn, R., Fujikawa, S. et al. (2021). Thermal and Gas Adsorption Properties of Tröger's Base/Diaza-cyclooctane Hybrid Ladder Polymers. *ChemNanoMat* 7: 824–830.

41 Ishiwari, F., Miyake, S., Inoue, K. et al. (2021). Two-step Conformational Control of a Dibenzo Diazacyclooctane Derivative by Stepwise Protonation. *Asian Journal of Organic Chemistry* 10: 1377–1381.

42 Ishiwari, F., Ofuchi, M., Inoue, K. et al. (2020). Switching of the conformational flexibility of a diazacyclooctane-containing ladder polymer by coordination and elimination of a Lewis acid. *Polymer Chemistry* 11: 236–240.

43 Wu, L., Chen, X., Zhang, Z. et al. (2021). Enhanced molecular selectivity and plasticization resistance in ring-opened Tröger's base polymer membranes. *Journal of Membrane Science* 634.

44 Ishiwari, F., Sato, T., Yamazaki, H. et al. (2016). An anion-conductive microporous membrane composed of a rigid ladder polymer with a spirobiindane backbone. *Journal of Materials Chemistry A* 4: 17655–17659.

45 Tong Huang, J.Z., Pei, Y., Liu, X. et al. (2021). Mechanically robust microporous anion exchange membranes with efficient anion conduction for fuel cells. *Chemical Engineering Journal* 418: 129311.

46 Li, Z., Guo, J., Zheng, J. et al. (2020). A microporous polymer with suspended cations for anion exchange membrane fuel cells. *Macromolecules* 53: 10998–11008.

47 Zhou, Z., Shinde, D.B., Guo, D. et al. (2022). Flexible Ionic Conjugated Microporous Polymer Membranes for Fast and Selective Ion Transport. *Advanced Functional Materials* 32: 2108672.

48 Du, N., Robertson, G.P., Song, J. et al. (2009). High-Performance Carboxylated Polymers of Intrinsic Microporosity (PIMs) with Tunable Gas Transport Properties. *Macromolecules* 42: 6038–6043.

49 Satilmis, B. and Budd, P.M. (2014). Base-catalysed hydrolysis of PIM-1: amide versus carboxylate formation. *RSC Advances* 4: 52189–52198.

50 Santoso, B., Yanaranop, P., Kang, H. et al. (2017). A Critical Update on the Synthesis of Carboxylated Polymers of Intrinsic Microporosity (C-PIMs). *Macromolecules* 50: 3043–3050.

51 Jeon, J.W., Kim, D.-G., Sohn, E.-h. et al. (2017). Highly Carboxylate-Functionalized Polymers of Intrinsic Microporosity for CO_2-Selective Polymer Membranes. *Macromolecules* 50: 8019–8027.

52 Patel, H.A. and Yavuz, C.T. (2012). Noninvasive functionalization of polymers of intrinsic microporosity for enhanced CO_2 capture. *Chemical Communications* 48: 9989–9991.

53 Mehio, N., Lashely, M.A., Nugent, J.W. et al. (2015). Acidity of the amidoxime functional group in aqueous solution: a combined experimental and computational study. *The Journal of Physical Chemistry. B* 119: 3567–3576.

54 Baran, M.J., Braten, M.N., Sahu, S. et al. (2019). Design Rules for Membranes from Polymers of Intrinsic Microporosity for Crossover-free Aqueous Electrochemical Devices. *Joule* 3: 2968–2985.

55 Tan, R., Wang, A., Malpass-Evans, R. et al. (2020). Hydrophilic microporous membranes for selective ion separation and flow-battery energy storage. *Nature Materials* 19: 195–202.

56 Ye, C., Tan, R., Wang, A. et al. (2022). Long-Life Aqueous Organic Redox Flow Batteries enabled by Amidoxime-Functionalized Ion-Selective Polymer Membranes. *Angewandte Chemie, International Edition* 61.

57 Wang, A., Tan, R., Liu, D. et al. (2023). Ion-Selective Microporous Polymer Membranes with Hydrogen Bond and Salt-Bridge Networks for Aqueous Organic Redox Flow Batteries. *Advanced Materials* 35: e2210098.

58 Kim, B.G., Henkensmeier, D., Kim, H.-J. et al. (2014). Sulfonation of PIM-1 — towards highly oxygen permeable binders for fuel cell application. *Macromolecular Research* 22: 92–98.

59 Zuo, P., Li, Y., Wang, A. et al. (2020). Sulfonated Microporous Polymer Membranes with Fast and Selective Ion Transport for Electrochemical Energy Conversion and Storage. *Angewandte Chemie, International Edition* 59: 9564–9573.

60 Ye, C., Wang, A., Breakwell, C. et al. (2022). Development of efficient aqueous organic redox flow batteries using ion-sieving sulfonated polymer membranes. *Nature Communications* 13: 3184.

61 Dong, H., Zhu, Z., Li, K. et al. (2021). Significantly improved gas separation properties of sulfonated PIM-1 by direct sulfonation using SO_3 solution. *Journal of Membrane Science* 635: 119440.

62 Zhou, S., Lin, X., Hu, Y. et al. (2022). Tailoring Sulfonated Poly(phenyl-alkane)s of Intrinsic Microporosity Membrane for Advanced Osmotic Energy Conversion. *ACS Materials Letters* 4: 1422–1429.

63 Park, H.B., Kamcev, J., Robeson, L.M. et al. (2017). Maximizing the right stuff: The trade-off between membrane permeability and selectivity. *Science* 356: eaab0530.

64 Low, Z.X., Budd, P.M., McKeown, N.B., and Patterson, D.A. (2018). Gas permeation properties, physical aging, and its mitigation in high free volume glassy polymers. *Chemical Reviews* 118: 5871–5911.

65 Zuo, P., Ye, C., Jiao, Z. et al. (2023). Near-frictionless ion transport from rigid nanoconfinement by triazine framework membranes. *Nature.*

66 Service RF (2018). Advances in flow batteries promise cheap backup power. *Science* 362: 508–509.

67 Bo Hu, T.L.L. (2021). Tanking up energy through atypical charging. *Science* 372: 788–789.

7

In Situ Polymerization for IEM Preparation

7.1 Conventional Methods for IEM Preparation

Solution casting is the most widely used method for fabricating IEMs. The ionomer solution (generally in polar solvents such as DMSO and NMP) is uniformly spread onto a heated substrate, and an IEM is formed after the solvent is completely evaporated. Although the solution casting method is popular among these methods, there are some disadvantages. First, the preparation of ionomer solutions requires massive harmful organic solvents, thus resulting in serious environmental pollution and hazards for human beings. Moreover, the evaporation process also wastes considerable energy. In addition, these membranes displayed inferior mechanical properties obtained by this method, and the membranes easily break when they are bent [1].

Apart from the solution casting method, radiation-induced grafting is also commonly used to synthesize membrane materials [2–5]. For instance, Varcoe reported the synthesis of a series of anion exchange membranes (AEMs) containing different quaternary ammonium (QA) groups via radiation-induced grafting onto poly(ethylene-co-tetrafluoroethylene) (ETFE) [2] (Figure 7.1). Among these AEMs, the benzyl-N-methylpyrrolidinium AEM displayed superior performance, including conductivity, alkali stability, and H_2/O_2 fuel cell performance, when compared to the benzyltrimethylammonium AEMs. For instance, the PVB-MPY AEM exhibited an OH^- conductivity of 159 mS/cm with an IEC of 1.66 mmol/g at 80 °C with 95% relative humidity. However, the method can acquire superior AEMs. However, high-energy radiation is desired but not easy to obtain. Moreover, the process also consumes an amount of organic solvent.

Electrospinning is attracting close interest as a membrane fabrication method for producing electrospun nanofibers with nanoscale diameters [6]. Electrospun nanofibers possess absorbing characteristics such as high surface area and aspect ratios, low density, and fully interconnected pores [7]. Xia and Li reported that electrospun nanofibers displayed a tensile modulus superior to that of the bulk [8]. Thus far, there have been comparative studies on the use of electrospinning to prepare CEMs and AEMs with significantly increased performance in several applications. Research has shown that electrospun fibrous membranes have been successfully used in DMFCs with Nafion/poly(vinyl alcohol) (PVA) [9] and

Ion Exchange Membranes: Design, Preparation, and Applications,
First Edition. Tongwen Xu and Yaoming Wang.

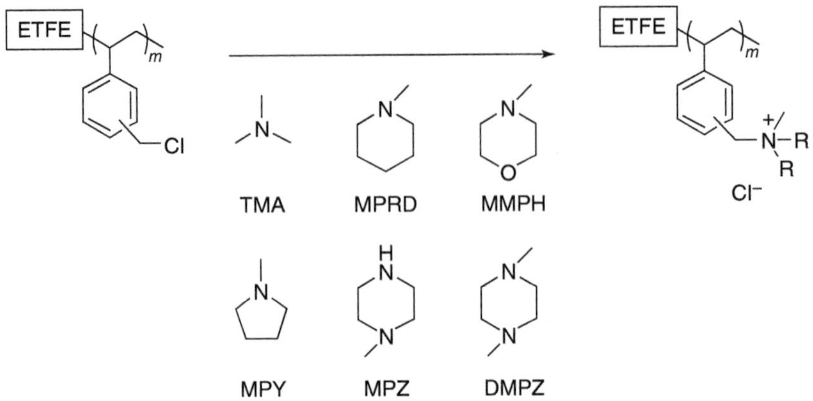

Figure 7.1 The synthesis of AEMs by radiation grafting. Source: Ponce-González et al. [2]/ Royal Society of Chemistry.

Nafion/palladium–silica [10] due to the unique characteristics of fibrous membranes with high specific area, good interconnectivity of pores, and excellent mechanical properties. In addition, the Xu group synthesized nanofiber AEMs composed of nanofibrous mats with an average diameter of 100 nm by electrospinning quaternized poly(2,6-dimethyl-1,4-phenylene oxide)/silicon dioxide (QPPO–SiO_2) mixed solutions [11]. The resulting membranes after hot pressing showed better separation efficiency toward acid and metal ions than other bulk AEMs under similar IEC conditions. Despite its attractive characteristics, the electrospinning strategy has still only been applied on the laboratory scale. It is vital for us to explore in depth to tailor IEMs from the perspective of polymer structures and molecular design.

7.2 Semi-interpenetrating Polymer Network

Semi-interpenetrating polymer networks (semi-IPNs) refer to blends of crosslinked polymers and linear polymers. In 1996, a semi-interpenetrating polymer was defined as "a polymer comprising one or more networks and one or more linear or branched polymer(s) characterized by the penetration on a molecular scale of at least one of the networks by at least some of the linear or branched macromolecules" by the International Union of Pure and Applied Chemistry (IUPAC) [12, 13]. Constructing semi-IPNs is a well-established strategy to synthesize new materials with enhanced properties. Hybrid polymers with semi-interpenetrating networks can combine the superior qualities of each component and complement some of their shortcomings. Such synergistic features of the semi-IPNs precipitated strong research interest in many fields, such as fuel cells and hydrogels. As shown in Figure 7.2, there are two main methods for the preparation of semi-crosslinked interpenetrating networks: in situ synthesis and sequential synthesis [13]. For in situ synthesis, monomer A, crosslinker, and polymer network B were first mixed homogeneously, and then the linear polymer (A) was obtained by polymerization.

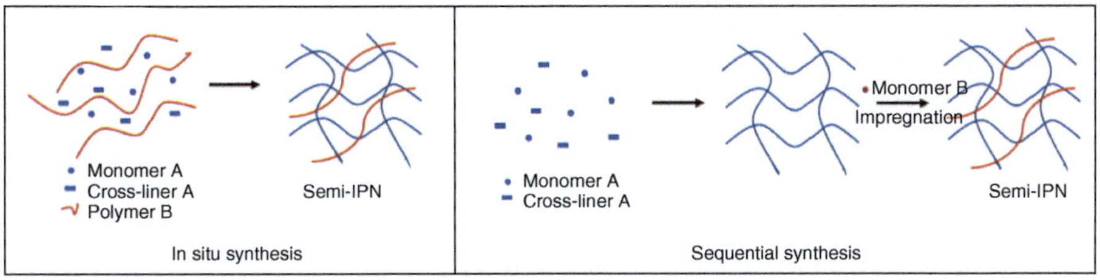

Figure 7.2 Synthesis method of semi-interpenetrating polymer network.

For the sequential pathway, the cross-linked polymer network was first prepared and followed by the mixture of monomers B used for polymerization.

In general, high hydroxide conductivity requires IEMs with a high ion exchange capacity (IEC), which incurs excessive water swelling and destruction of mechanical properties and chemical stability [14–17]. In recent few years, semi-IPNs have been introduced for designing of IEMs for electrochemistry devices [14, 18, 19]. To solve this problem, researchers proposed the construction of a semi-crosslinked interpenetrating network AEM to break through the trade-off between conductivity and mechanical properties. The semi-IPN membrane dissolves in an organic solvent. The cross-linking polymer network partner of the semi-IPN provided improved chemical and mechanical support by entangling the polymer chain in an economical manner, while the conductive polymers served as ion transport [20]. Moreover, the open structure of the semi-IPN polymer network without tortuosity can facilitate ion transport, which can be regulated by microphase separation in the IEM to further promote ion conduction [21].

In view of this, the choice of appropriate crosslinked polymer and linear polymer is one of the most critical issues. As shown in Table 7.1, various polymers have been used to prepare IEMs with semi-interpenetrating networks. Nafion-containing semi-IPNs have been widely reported as hybrid proton exchange membranes (PEMs) for fuel cells, where Nafion, as a linear ion-conducting polymer, shows good

Table 7.1 Various polymers based on semi-IPN architecture.

Name	Cross-linking network	Liner polymer	Reference
SIPN-cQSEBS/QPPO-10	QPPO-10	cQSEBS	[19]
PAA-PVA-Nafion	PAA-PVA	Nafion	[22]
SIPN-AEM	PPO-N_3	CM-PS	[14]
SIPN-xPEG-PAGE/QAPPO	xPEG-PAGE	QAPPO	[23]
OFHDA/S-PEEK	Fluorinated OFHDA	S-PEEK	[24]
FPI/Nafion®212	Fluorinated polyimide	Nafion	[25]
S@N/fEO	sPAEK	Nafion	[26]
Nafion/MPTI	sPES	Nafion	[27]
DVB-Nafion	cross-linked DVB	Nafion	[28]

compatibility with cross-linked polymers such as sPAEK [26], sPES [27], and functionalized polyimide (PI) [25]. Recently, Kim et al. reported a semi-interpenetrating network PEM (NPP) composed of Nafion and a cross-linked network using poly(acrylic acid) (PAA) and PVA [22]. The NPP-95 (Nafion:PAA:PVA = 95:2.5:2.5) membrane displayed enhanced conductivity and fell cell performance (peak power density = 1561 mA/cm^2) compared to the unmodified Nafion (peak power density = 1179 mW/cm^2) with a potential of 0.6 V.

Li et al. designed a series of semi-IPN AEMs composed of a crosslinked polyphenylene oxide (PPO) network and a quaternized poly(styrene) (PS) copolymer (Figure 7.3) [14]. In this system, quaternized PS provides the anion transport moiety, while the crosslinked PPO network of azide–alkyne cycloaddition maintains dimensional stability. When the ion exchange capacity (IEC) was 1.44 mmol/g, the SIPRN-62/30 AEM showed low water absorption (WU = 12 wt%) and a hydroxide conductivity of 37.2 mS/cm at room temperature.

Hicker et al. synthesized a series of semi-IPN AEMs with excellent chemical stability by blending a poly(ethylene glycol)-*co*-poly(allyl glycidyl ether) (xPEG-PAGE) network and quaternized poly(2,6-dimethyl phenylene oxide) (QAPPO) [23] (Figure 7.4a). QAPPO provides the OH^- transport sites and the cross-linked

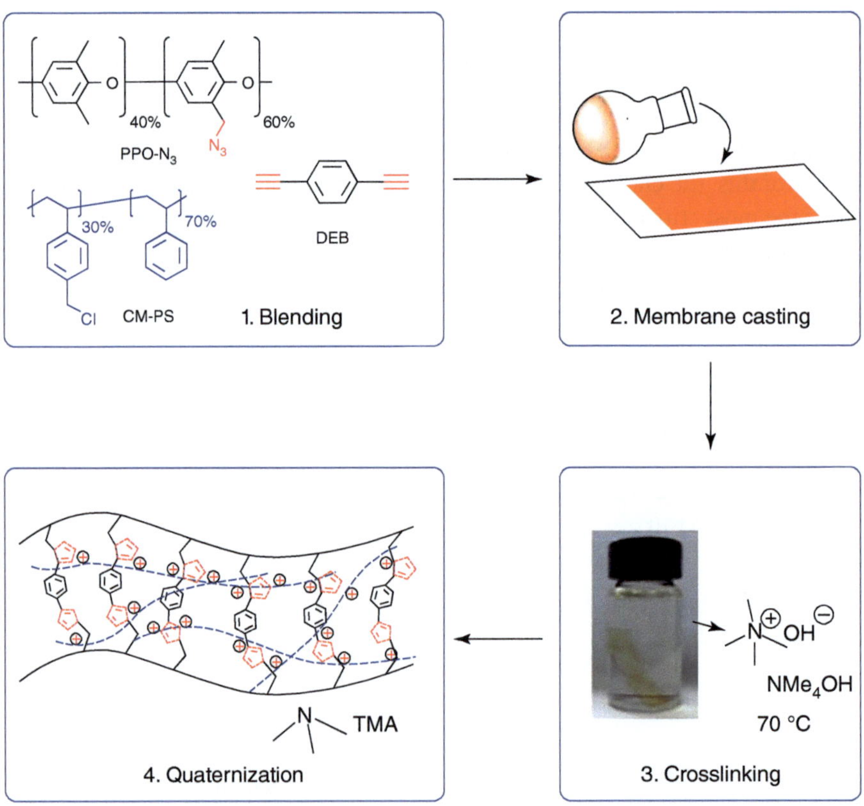

Figure 7.3 Synthesis of semi-IPN AEMs by azide–alkyne cycloaddition. Source: Xue et al. [14]/With permission of Royal Society of Chemistry.

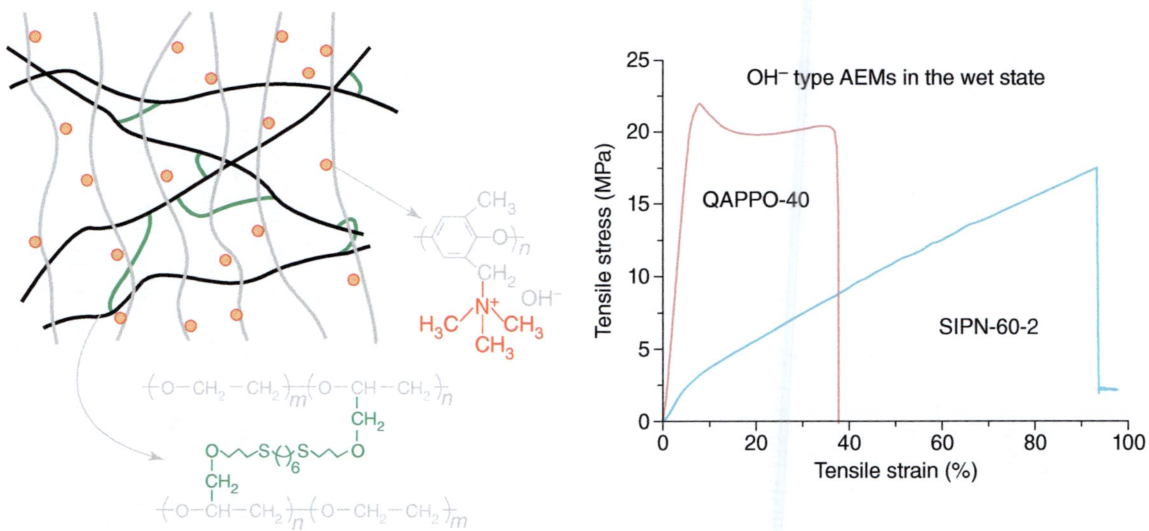

(a) (b)

Figure 7.4 (a) Chemical structure and (b) mechanical properties of the semi-interoperating polymer network (SIPN-60-2) AEM and QAPPO-40 AEM. Source: Pan et al. [23]/American Chemical Society.

xPEG-PAGE network maintains the dimensional stability of the semi-IPN AEM. SIPN-60-2 AEM with an IEC of 1.43 mmol/g displayed a trade-off between OH^- conductivity and swelling at 80 °C. Moreover, SIPN-60-2 AEM showed significantly improved mechanical properties compared to QAPPO AEMs (Figure 7.3b).

7.3 Pore Filling

Pore filling is a promising methodology for developing IEMs with excellent selectivity and swelling suppression [29]. For the pore-filling method, the key is to find suitable porous substrates. Chemical inertia and mechanical stability are indispensable for a porous substrate, which ensures that the inflation of soft electrolyte polymers in pores can be confined by the hard matrix. Various commercially available substrates include polyethylene (PE) [30], polyvinylidene fluoride [31], polypropylene (PP) [32], polyethersulfone (PES) [33], PI [34–36], and porous polyacrylonitrile (PAN) [37]. The pores are generally regulated by track etched and phase inversion methods [35, 38]. In addition to polymeric substrates, some inorganic materials, including porous alumina, can also be utilized for preparing pore-filling IEMs [37]. For inorganic substrates, the pore size distribution is consistent. To obtain better selectivity pore-filling IEMs, the pore is smaller compared to polymer substrates. Filling electrolytes such as sulfonated copolyimide [39] and Nafion [40] have been used to fabricate porous membranes.

As shown in Figure 7.5, pore-filling IEMs are usually obtained by implanting polymer electrolytes into porous substrates [41]. The common method is pouring preformed ionomers on the surface of membranes. The electrolytes will flee into the inert pores, and IEM materials can be formed when the organic solvents are evaporated. To ensure the successful preparation of "defect-free" IEMs, the above process should be repeated several times. In addition, the pore-soaking technique by steeping porous substrates into ionized polymers is also widely used to obtain pore-filling membranes [39]. The foundation of the pore-soaking technique is similar to that of the pore-filling technique. In this method, the Yamaguchi group reported several pore-filling ion exchange membranes, including PAA and poly(acrylic acid-*co*-vinylsulfonic acid) (PAAVS), and porous silica or porous poly(tetrafluoroethylene) (PTFE) was used as the substrate. In subsequent studies, Nakao selected poly(acrylamide-*tert*-butylsulfonic acid) (PATBS) as the ionic electrolyte to obtain improved conductivity. Furthermore, they used a mechanically strong porous cross-linked high-density polyethylene (CLPE) substrate or a PI substrate to restrain the water swelling of the porous substrate [33]. These pore-filling membranes displayed excellent mechanical properties and fuel gas barrier properties. It can be widely utilized for fuel cells.

For IEMs, side-chain type copolymers usually show the anisotropic distribution of the ion transport channels, where the through-plane rather than the in-plane direction is always beneficial for high membrane performance [42]. Thus, the fabrication of isotropic ion nanochannels is highly desired for efficient ion transport [43]. In this situation, many researchers proposed that the pore-filling concept of porous scaffolds with highly ordered, monodispersed ionic channels could be used to facilitate ion conduction. The method of pore filling with ion transport pathways

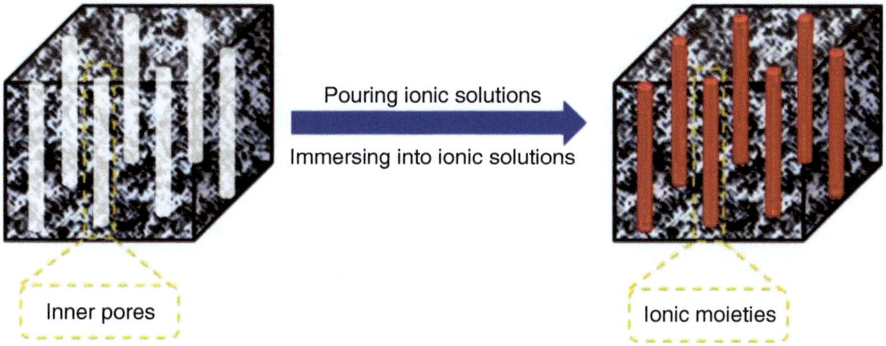

Figure 7.5 The preparation of the pore-filling membrane.

generates aligned nanoscopic channels as ion transport pathways within a robust and durable scaffold. Notably, Yamaguchi et al. proposed a pore-filling membrane with a low equivalent weight (EW) perfluorosulfonic acid (PFSA) polymer and 6 μm ultrahigh molecular weight polyethylene (UHMWPE) [44]. The pore-filling membrane showed high mechanical strength and proton conductivity compared to commercial Nafion 211. A highly packed structure of sulfonic acid groups is present inside the pores of the substrate; thus, it is possible for PEMs to display high proton conductivity and high cell performance under low humidity conditions. Moreover, due to effective swelling suppression, the pore-filling membrane exhibited low hydrogen crossover despite the thickness of this membrane being only approximately one-fourth of that of Nafion 211 (7 μm).

Some researchers have also reported external crosslinking of the polymer inside the preformed pores to restrain extreme membrane swelling and polymer leaching from the pores. Specifically, Kim et al. first prepared a mechanically stable pore-filling membrane system by introducing proton conductive substrates consisting of cross-linked benzoxazine–benzimidazole copolymers developed by the reaction of polybenzimidazole (PBI) with sulfonated benzoxazine [45]. The pore-filling membrane shows more reformative dimensional stability and mechanical strength than the linear sulfonated poly (arylene ether sulfone) membrane. Moreover, the introduction of proton-conductive sulfonic acid groups in the porous substrate can increase hydrophilicity and proton conductivity so that the proton conductivity and cell performance of the pore-filling membrane from the proton-conductive porous substrates are superior to the membrane from nonproton conductive porous substrates.

Hara et al. selected PI aromatic materials as the substrate and sulfonated polyether sulfone (SPES) as the electrolyte (Figure 7.6) [34]. Specifically, the porous PI substrate is mechanically excellent at restraining the swelling of the polymer electrolyte. Moreover, the PI substrate can remain stable when the fabrication process is in polar solvents at approximately 200 °C. Meanwhile, the SPES ionomer can acquire microphase aggregation due to the hydrophobic part in the structure of the block. Thus, the membrane displayed superior performance with low methanol crossover, high durability, and excellent proton conductivity.

To better understand the orientation of ion-exchange groups inside the ionic conductive channel, the Xu group [46] proposed a "covalent bonding-triggered (CBT)

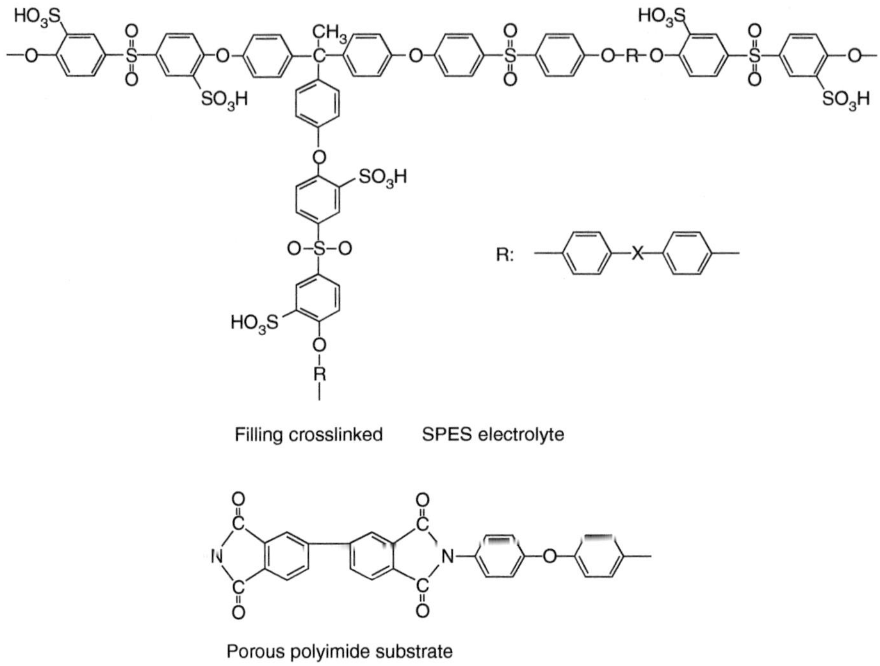

Figure 7.6 The chemical structure of the filling electrolyte and porous substrate. Source: Yamaguchi et al. [34].

pore-filling" strategy to build up high-performance AEMs, in which tertiary amine groups of the filling copolymer anchor with bromomethyl groups in the porous membrane substrate (Figure 7.7). Symmetrically aligned ion transport channels for fast anion conduction were built by QA groups at the inner surface of pores. The CBT pore-filling membrane exhibited a high hydroxide conductivity of 39.8 mS/cm at 30 °C at only 0.97 mmol/g ion-exchange capacity (IEC). Moreover, the AEM displayed excellent chemical stability, with only <6% of OH^- conductivity degraded after 168 h of 2 M NaOH at 60 °C and a significantly high peak power density of 315 mW/cm^2.

In addition to polymeric IEMs, the pore-filling strategy can be used to fabricate organic–inorganic composite membranes. For instance, Wang et al. [37] reported a "pore filling" composite membrane with improved chemical and mechanical stability was obtained by dynamic pressure-driven assembly of a PVA-GO nanohybrid layer poured onto an asymmetric polyacrylonitrile ultrafiltration membrane. Organic–inorganic composite membranes can be used to separate toluene/n-heptane mixtures. The separation performance of the pore-filling membrane can be adjusted by regulating the pressure, filtration time, and polymer and GO concentrations.

Over the past few years, pore-filling and pore-soaking technologies for obtaining IEMs have developed at full speed. Although the approach provided many possibilities for synthesizing IEMs, there were some questions in which few researchers concentrated on how to decrease the divergence of ionic materials in the application of these IEMs.

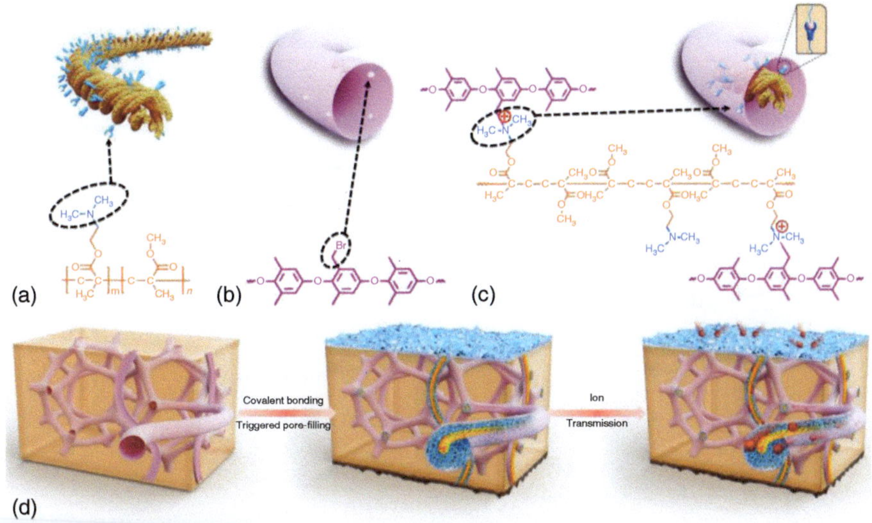

Figure 7.7 Schematic illustrations of the preparation of the CBT pore-filling membrane [46]. (a) Schematic illustration of polymer indicating hydrophobic main chain as yellow cord and hydrophilic tertiary amine groups as blue grappling hooks. (b) Schematic design of the membrane channel containing bromomethyl groups as pink balls. (c) Menshutkin reaction process and schematic diagram representing the covalent bonding between tertiary amine groups of the filling polymer and bromomethyl groups of the pre-formed membrane channel. (d) Schematic diagram of the covalent bonding-triggered pore-filling process and ionic conduction process. The blue region represents the hydrophilic domains formed by the aggregation of the hydrated filling polymer on the membrane surface.

7.4 Solvent-free Strategy

Traditional IEMs were fabricated by modification of polymers and the polymerization of functional monomers. Among these methods, the reaction and membrane fabrication need to consume massive amounts of organic solvent, and the solvent has to be removed during the formation of the IEMs so that it will pollute the environment. Moreover, solvent evaporation requires heat, which acts against the 12 principles of Green Chemistry. Given these problems, it is vital to develop simple, efficient, and eco-friendly approaches to achieve industrial-scale fabrication of IEMs. Huang and coworkers reported manufacturing techniques (continuous foam extrusion technology) used to produce solvent-free polymer IEMs and discussed the applicability and challenges of continuous foam extrusion technology [47]. Given the idea of green and sustainable chemistry [48], relevant researchers have developed a solvent-free strategy to fabricate IEMs [49, 50].

The Xu group reported a one-pot, solvent-free strategy for the preparation of alkaline AEMs by in situ polymerization [51]. As shown in Figure 7.8, the solvent-free strategy refers to dissolving brominated poly(2,6-dimethyl-1,4-phenylene oxide) (BPPO) in a monomer mixture of vinylbenzyl chloride (VBC) and divinylbenzene (DVB) and then introducing BPPO into the network of poly(VBC–DVB) by

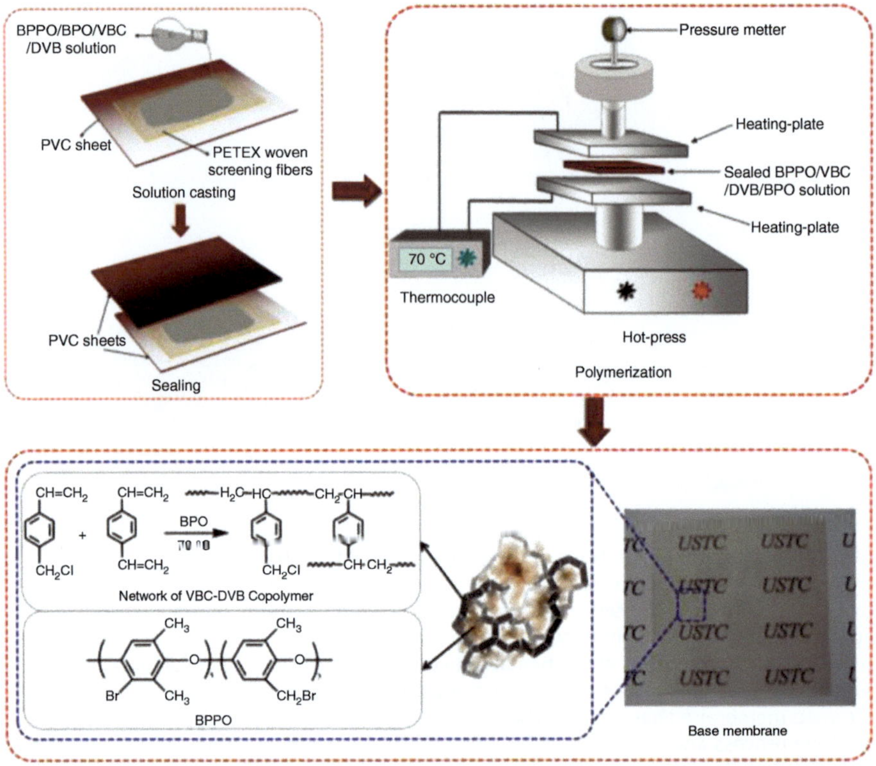

Figure 7.8 Schematic of the strategy to obtain an uncharged membrane. Source: Wu et al. [51]/With permission of Elsevier.

in situ polymerization. A flexible, robust, and transparent AEM was prepared after quaternization. In this system, the AEM displayed excellent water uptake (IEC = 1.55 mmol/g, WU = 20.4%) at room temperature. Moreover, the strategy is not only suited to the BPPO/VBC/DVB system. Research has shown that some compatible systems of (aromatic) polymers and (aromatic) monomers can be applied when homogeneous cast solutions of monomers and polymers are formed. Xu et al. also designed and synthesized cross-linked AEMs with a solvent-free strategy [52, 53]. The cross-linked AEMs were formed by dissolving cardo polyether ketone (PEK-C) into a mixture of VBC and DVB monomers, followed by tetraethylenepentamine (TEPA) as a crosslinker. Furthermore, the crosslinked membrane showed excellent performance.

Paul et al. proposed sulfonated poly(arylene ether sulfone) desalination membranes via solvent-free melt extrusion [54]. Ladewig and coworkers reported high-performance cation exchange membranes (CEMs) prepared by in situ emulsion polymerization without using organic solvents [1]. Polymerization often occurs at approximately 75–90 °C, which consumes energy. It is bothersome for zero-emission manufacturing. Given this problem, more researchers have focused on how to make polymerization occur in relatively suitable situations.

Photopolymerization refers to utilizing ultraviolet (UV)-initiated polymerization, which can enable flexibility in spatial and temporal control. Moreover, the rate of initiation and polymerization can be easily regulated by tuning the light intensity or the concentration of photosensitive compounds [55–57]. For example, the Liu group reported a solvent-free UV-initiated free-radical photopolymerization method to prepare cation-exchange membranes (CEMs) [58]. The solvent-free system refers to dissolving SSTBA (tertrabutylammonium styrene sulfonate) and an initiator in a mixture of styrene and trimethylolpropane triacrylate monomers. The CEM was fabricated, and the process of crosslinking and polymerization progressed at the same time. The CEMs displayed better performance than commercial CEMs, with a desalination ratio of up to 99.3% and a resistance of 0.78 $\Omega\,cm^2$ with an IEC of 2.02 mmol/g.

7.5 In Situ Polymerization

These IEMs in Section 7.4 have been synthesized according to this solution-free strategy by dissolving a polymer in liquid monomers, followed by in situ polymerization to prepare corresponding uncharged membranes [1, 51, 58]. However, these IEMs can be obtained after post-modification [29] (Figure 7.9a). Post-modification, including the quaternization reaction and sulfonation reaction, was performed

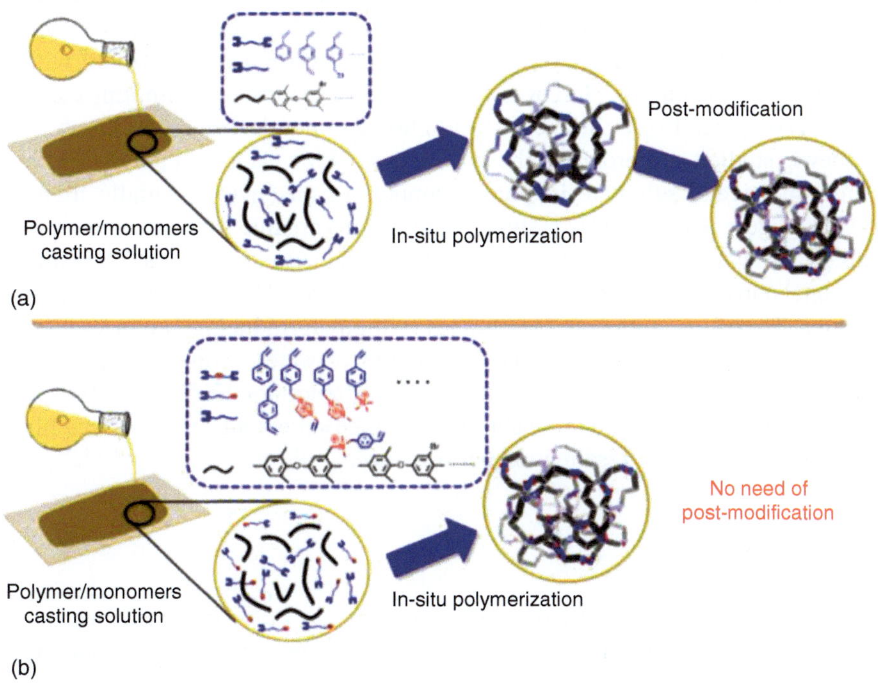

Figure 7.9 Schematic illustrations of (a) conventional and (b) upgraded in situ polymerization strategies for the fabrication of IEMs [29].

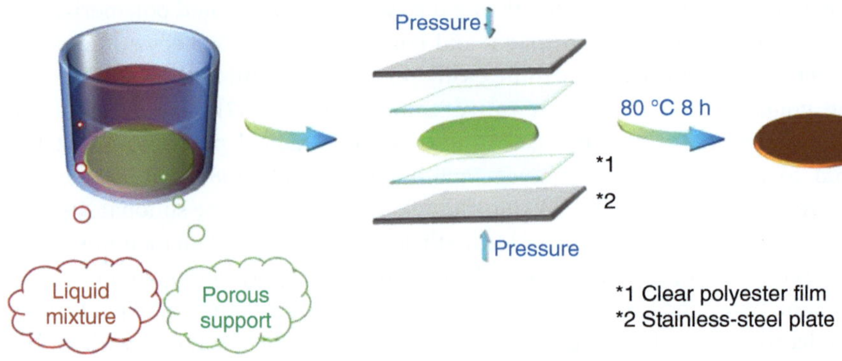

Figure 7.10 Procedures for preparing CEMs by in situ polymerization [1].

to introduce QA and sulfonate anions. Xu and coworkers put forward in situ polymerization to synthesize AEMs without post-functionalization for the first time [59]. As shown in Figure 7.9b, the fabrication process was composed of in situ polymerization and quaternarization. First, brominated poly(2,6-dimethyl-1,4-phenylene oxide) (BPPO) is dissolved in a monomer mixture of VBC and styrene without any organic solvent. Then, a certain amount of *N*-vinylimidazole and *N*-methylimidazole are added to introduce imidazolium cations to the polymer. Finally, strong-mechanical, transparent AEMs could be prepared by thermal crosslinking of the unsaturated moieties during the formation process of the resultant membrane and the imidazolium cations located at the cross-links. These cross-linked membranes displayed a target-low area resistance of 1.4 Ω/cm^2 with a high IEC over 2.0 mmol/g. Moreover, the films displayed improved limiting current densities compared to the commercial AEM tested under the same conditions.

Ladewig studied and proposed the synthesis of CEMs via in situ polymerization of lithium styrene sulfonate in porous support in an environmentally friendly way [1] (Figure 7.10). In this system, there are no organic solvents or corrosive acids. Water was chosen as the solvent in the film fabrication process. The ratio of functional polymers to the membrane support is controlled by adjusting the amount of water. Therefore, high-IEC membranes without introducing cation exchange resins can be obtained by reducing the amount of water. Moreover, the porous support can be used to guarantee the mechanical strength of the films by absorbing the liquid mixture. Based on these above-mentioned statements, the prepared films exhibited excellent high desalination performance.

References

1 Jiang, S. and Ladewig, B.P. (2019). High performance cation exchange membranes synthesized via in situ emulsion polymerization without organic solvents and corrosive acids. *Journal of Materials Chemistry A* 7: 17400–17411.

2 Ponce-González, J., Whelligan, D.K., Wang, L. et al. (2016). High performance aliphatic-heterocyclic benzyl-quaternary ammonium radiation-grafted anion-exchange membranes. *Energy & Environmental Science* 9: 3724–3735.

3 Willdorf-Cohen, S., Zhegur-Khais, A., Ponce-González, J. et al. (2023). Alkaline Stability of Anion-Exchange Membranes. *ACS Applied Energy Materials* 6: 1085–1092.

4 Golubenko, D.V., Pourcelly, G., and Yaroslavtsev, A.B. (2018). Permselectivity and ion-conductivity of grafted cation-exchange membranes based on UV-oxidized polymethylpenten and sulfonated polystyrene. *Separation and Purification Technology* 207: 329–335.

5 Stenina, I., Golubenko, D., Nikonenko, V., and Yaroslavtsev, A. (2020). Selectivity of Transport Processes in Ion-Exchange Membranes: Relationship with the Structure and Methods for Its Improvement. *International Journal of Molecular Sciences* 21.

6 Cavaliere, S., Subianto, S., Savych, I. et al. (2011). Electrospinning: designed architectures for energy conversion and storage devices. *Energy & Environmental Science* 4.

7 Thavasi, V., Singh, G., and Ramakrishna, S. (2008). Electrospun nanofibers in energy and environmental applications. *Energy & Environmental Science* 1.

8 Li, D. and Xia, Y. (2004). Electrospinning of Nanofibers: Reinventing the Wheel? *Advanced Materials* 16: 1151–1170.

9 Mollá, S. and Compañ, V. (2011). Polyvinyl alcohol nanofiber reinforced Nafion membranes for fuel cell applications. *Journal of Membrane Science* 372: 191–200.

10 Thiam, H.S., Daud, W.R.W., Kamarudin, S.K. et al. (2013). Nafion/Pd–SiO_2 nanofiber composite membranes for direct methanol fuel cell applications. *International Journal of Hydrogen Energy* 38: 9474–9483.

11 Pan, J., He, Y., Wu, L. et al. (2015). Anion exchange membranes from hot-pressed electrospun QPPO–SiO_2 hybrid nanofibers for acid recovery. *Journal of Membrane Science* 480: 115–121.

12 Liu, X., Zhang, Y., Deng, S. et al. (2019). Semi-interpenetrating polymer networks toward sulfonated poly(ether ether ketone) membranes for high concentration direct methanol fuel cell. *Chinese Chemical Letters* 30: 299–304.

13 Chikh, L., Delhorbe, V., and Fichet, O. (2011). (Semi-)Interpenetrating polymer networks as fuel cell membranes. *Journal of Membrane Science* 368: 1–17.

14 Xue, J., Liu, L., Liao, J. et al. (2018). Semi-interpenetrating polymer networks by azide–alkyne cycloaddition as novel anion exchange membranes. *Journal of Materials Chemistry A* 6: 11317–11326.

15 Buggy, N.C., Du, Y., Kuo, M.-C. et al. (2020). A Polyethylene-Based Triblock Copolymer Anion Exchange Membrane with High Conductivity and Practical Mechanical Properties. *ACS Applied Polymer Materials* 2: 1294–1303.

16 Chen, C., Arntsen, C., and Tse, Y.S. (2020). Simulation study of the effects of phase separation on hydroxide solvation and transport in anion exchange membranes. *The Journal of Chemical Physics* 152: 094903.

17 Chen, J., Guan, M., Li, K., and Tang, S. (2020). Novel Quaternary Ammonium-Functionalized Covalent Organic Frameworks/Poly(2,6-dimethyl-1,4-phenylene oxide) Hybrid Anion Exchange Membranes with Enhanced Ion Conductivity and Stability. *ACS Applied Materials & Interfaces* 12: 15138–15144.

18 Molavian, M.R., Abdolmaleki, A., Gharibi, H. et al. (2018). Two highly strong semi-IPNs for proton exchange membrane fuel cell (PEMFC) application. *Materials Today Communications* 15: 94–99.

19 Han, J., Liu, C., Deng, C. et al. (2022). Mechanically robust and highly conductive semi-interpenetrating network anion exchange membranes for fuel cell applications. *Journal of Power Sources* 548.

20 Lee, M.-J., Choi, Y.S., Kang, Y.S. et al. (2012). All-solid-state proton conductive membranes prepared by a semi-interpenetrating polymer network (semi-IPN). *Journal of Materials Chemistry* 22.

21 Fang, C., Toh, X.N., Yao, Q. et al. (2013). Semi-interpenetrating polymer network proton exchange membranes with narrow and well-connected hydrophilic channels. *Journal of Power Sources* 226: 289–298.

22 Al Munsur, A.Z., Goo, B.H., Kim, Y. et al. (2021). Nafion-Based Proton-Exchange Membranes Built on Cross-Linked Semi-Interpenetrating Polymer Networks between Poly(acrylic acid) and Poly(vinyl alcohol). *ACS Applied Materials & Interfaces* 13: 28188–28200.

23 Pan, J., Zhu, L., Han, J., and Hickner, M.A. (2015). Mechanically Tough and Chemically Stable Anion Exchange Membranes from Rigid-Flexible Semi-Interpenetrating Networks. *Chemistry of Materials* 27: 6689–6698.

24 Delhorbe, V., Cailleteau, C., Chikh, L. et al. (2013). Influence of the membrane treatment on structure and properties of sulfonated poly(etheretherketone) semi-interpenetrating polymer network. *Journal of Membrane Science* 427: 283–292.

25 Pan, H., Pu, H., Wan, D. et al. (2010). Proton exchange membranes based on semi-interpenetrating polymer networks of fluorine-containing polyimide and Nafion®. *Journal of Power Sources* 195: 3077–3083.

26 Ru, C., Gu, Y., Duan, Y. et al. (2019). Nafion based semi-interpenetrating polymer network membranes from a cross-linkable SPAEK and a fluorinated epoxy resin for DMFCs. *Electrochimica Acta* 324.

27 Sun, S., Ling, L., Xiong, Y. et al. (2020). Trifluoromethanesulfonimide-based hygroscopic semi-interpenetrating polymer network for enhanced proton conductivity of nafion-based proton exchange membranes at low humidity. *Journal of Membrane Science* 612.

28 Matsuguchi, M. and Takahashi, H. (2006). Methanol permeability and proton conductivity of a semi-interpenetrating polymer networks (IPNs) membrane composed of Nafion® and cross-linked DVB. *Journal of Membrane Science* 281: 707–715.

29 Ran, J., Wu, L., He, Y. et al. (2017). Ion exchange membranes: New developments and applications. *Journal of Membrane Science* 522: 267–291.

30 Yamaguchi, T., Kuroki, H., and Miyata, F. (2005). DMFC performances using a pore-filling polymer electrolyte membrane for portable usages. *Electrochemistry Communications* 7: 730–734.

31 S. Alwin, S. D. Bhat, A. K. Sahu, A. Jalajakshi, P. Sridhar, S. Pitchumani, A. K. Shukla. Modified-pore-filled-PVDF-membrane electrolytes for direct methanol fuel cells. *Journal of The Electrochemical Society*, 158 (2), 158 B91.

32 Mohammadi, M. and Mehdipour-Ataei, S. (2020). Robust three-dimensionally printed polypropylene/highly sulfonated polysulfone composites for potential applications in fuel cells. *International Journal of Energy Research* 45: 4224–4238.

33 Yamaguchi, T., Miyata, F., and Nakao, S. (2003). Polymer electrolyte membranes with a pore-filling structure for a direct methanol fuel cell. *Advanced Materials* 15: 1198–1201.

34 Yamaguchi, T., Zhou, H., Nakazawa, S., and Hara, N. (2007). An extremely low methanol crossover and highly durable aromatic pore-filling electrolyte membrane for direct methanol fuel cells. *Advanced Materials* 19: 592–596.

35 Nguyen, T.H., Wang, C., and Wang, X. (2009). Pore-filling membrane for direct methanol fuel cells based on sulfonated poly(styrene-ran-ethylene) and porous polyimide matrix. *Journal of Membrane Science* 342: 208–214.

36 Nguyen, T. and Wang, X. (2010). Multifunctional composite membrane based on a highly porous polyimide matrix for direct methanol fuel cells. *Journal of Power Sources* 195: 1024–1030.

37 Wang, N., Ji, S., Li, J. et al. (2014). Poly(vinyl alcohol)–graphene oxide nanohybrid "pore-filling" membrane for pervaporation of toluene/n-heptane mixtures. *Journal of Membrane Science* 455: 113–120.

38 Gohil, J.M. and Karamanev, D.G. (2013). Novel pore-filled polyelectrolyte composite membranes for cathodic microbial fuel cell application. *Journal of Power Sources* 243: 603–610.

39 Wang, T., Sun, F., Wang, H. et al. (2012). Preparation and properties of pore-filling membranes based on sulfonated copolyimides and porous polyimide matrix. *Polymer* 53: 3154–3162.

40 Gloukhovski, R., Freger, V., and Tsur, Y. (2016). A novel composite nafion/anodized aluminium oxide proton exchange membrane. *Fuel Cells* 16: 434–443.

41 Kim, D.-H., Park, J.-H., Seo, S.-J. et al. (2013). Development of thin anion-exchange pore-filled membranes for high diffusion dialysis performance. *Journal of Membrane Science* 447: 80–86.

42 Gloukhovski, R., Freger, V., and Tsur, Y. (2018). Understanding methods of preparation and characterization of pore-filling polymer composites for proton exchange membranes: a beginner's guide. *Reviews in Chemical Engineering* 34: 455–479.

43 He, Y., Zhang, H., Li, Y. et al. (2015). Synergistic proton transfer through nanofibrous composite membranes by suitably combining proton carriers from the nanofiber mat and pore-filling matrix. *Journal of Materials Chemistry A* 3: 21832–21841.

44 Oshiba, Y., Tomatsu, J., and Yamaguchi, T. (2018). Thin pore-filling membrane with highly packed-acid structure for high temperature and low humidity operating polymer electrolyte fuel cells. *Journal of Power Sources* 394: 67–73.

45 Kihyun, K., Choi, S.-W., Park, J.O. et al. (2017). Proton conductive cross-linked benzoxazine-benzimidazole copolymers as novel porous substrates for reinforced pore-filling membranes in fuel cells operating at high temperatures. *Journal of Membrane Science* 536: 76–85.

46 Xiao, X., Shehzad, M.A., Yasmin, A. et al. (2020). Covalent bonding-triggered pore-filled membranes for alkaline fuel cells. *Journal of Membrane Science* 597.

47 Marino, M.G., Melchior, J.P., Wohlfarth, A., and Kreuer, K.D. (2014). Hydroxide, halide and water transport in a model anion exchange membrane. *Journal of Membrane Science* 464: 61–71.

48 Asveld, L. (2019). Towards including social sustainability in green and sustainable chemistry. *Current Opinion in Green and Sustainable Chemistry* 19: 61–65.

49 Rasool, M.A., Pescarmona, P.P., and Vankelecom, I.F.J. (2019). Applicability of organic carbonates as green solvents for membrane preparation. *ACS Sustainable Chemistry & Engineering* 7: 13774–13785.

50 Ding, X., Hua, M., Zhao, H. et al. (2017). Poly(ethylene oxide) composite membrane synthesized by UV-initiated free radical photopolymerization for CO_2 separation. *Journal of Membrane Science* 531: 129–137.

51 Wu, L., Zhou, G., Liu, X. et al. (2011). Environmentally friendly synthesis of alkaline anion exchange membrane for fuel cells via a solvent-free strategy. *Journal of Membrane Science* 371: 155–162.

52 Lin, X., Liu, Y., Poynton, S.D. et al. (2013). Cross-linked anion exchange membranes for alkaline fuel cells synthesized using a solvent free strategy. *Journal of Power Sources* 233: 259–268.

53 Hugar, K.M., Kostalik, H.A.T., and Coates, G.W. (2015). Imidazolium cations with exceptional alkaline stability: a systematic study of structure-stability relationships. *Journal of the American Chemical Society* 137: 8730–8737.

54 Oh, H.J., McGrath, J.E., and Paul, D.R. (2018). Water and salt transport properties of disulfonated poly(arylene ether sulfone) desalination membranes formed by solvent-free melt extrusion. *Journal of Membrane Science* 546: 234–245.

55 Ho, T.D., Yu, H., Cole, W.T., and Anderson, J.L. (2012). Ultraviolet photoinitiated on-fiber copolymerization of ionic liquid sorbent coatings for headspace and direct immersion solid-phase microextraction. *Analytical Chemistry* 84: 9520–9528.

56 Tibbits, A.C., Mumper, L.E., Kloxin, C.J., and Yan, Y.S. (2015). A single-step monomeric photo-polymerization and crosslinking via thiol-ene reaction for hydroxide exchange membrane fabrication. *Journal of the Electrochemical Society* 162: F1206–F1211.

57 Ahmadian-Alam, L. and Mahdavi, H. (2018). A novel polysulfone-based ternary nanocomposite membrane consisting of metal-organic framework and silica nanoparticles: as proton exchange membrane for polymer electrolyte fuel cells. *Renewable Energy* 126: 630–639.

58 Pan, J., Liu, L., Tao, Y. et al. (2021). Green fabrication of tertrabutylammonium styrene sulfonate cation-exchange membranes via a solvent-free photopolymerization strategy. *Industrial and Engineering Chemistry Research* 60: 17055–17064.

59 Pan, Q., Hossain, M.M., Yang, Z. et al. (2016). One-pot solvent-free synthesis of cross-linked anion exchange membranes for electrodialysis. *Journal of Membrane Science* 515: 115–124.

8

Special IEMs Preparation

8.1 Metal–Organic Framework Membranes

8.1.1 Introduction

A metal–organic framework (MOF) is a type of porous organic–inorganic hybrid crystalline solid that consists of polydentate organic ligands connecting inorganic nodes [1] (Figure 8.1). Specifically, inorganic nodes are composed of metal ions (such as Cr(III), Fe(III), Al(III), Mn(II), Co(II), Cu(II), Cu(I), Zn(II), and Zr(IV)) or metal clusters, known as, inorganic secondary building units (SBUs). These nodes are connected with organic ligands by coordination bonds (usually containing carboxylate, phosphonate, pyridine, imidazole, or other azo functional groups) to form MOFs.

MOFs have been widely used in various fields [2], such as gas storage [3], adsorption and separation [4, 5], catalysis [6], sensing [7, 8], and sustained drug release [9], due to their diverse structure, high porosity, regular pore structure, high specific surface area, controllable pore properties, and rich physical and chemical properties, as shown in Figure 8.2.

8.1.2 Structural Properties of MOFs

8.1.2.1 Structural Diversity

There are many kinds of MOFs, mainly due to the variety of metal nodes (or SBU) and organic ligands used to construct MOFs. In addition, the diversity of self-assembly methods also provides a basis for the formation of multistructured and multifunctional MOF materials [10, 11]. At present, tens of thousands of MOF materials have been synthesized.

Several MOFs have been studied, such as the Iso-Reticular Metal–Organic Framework (IRMOF) series [12], Zeolitic Imidazolate Framework (ZIF) series [13], Materials of the Institut Lavoisier (MIL) series [14, 15], University of Oslo (UiO) series [16], and Porous Coordination Network (PCN) series [17]. IRMOF is a 3D porous mesh material formed by a series of $[Zn_4O]^{6+}$ and organic ligands, for example, MOF-5 [18]; ZIF is synthesized by coordination of transition metals (such as Co, Zn, and Ni) with imidazole and its derivatives, for example, ZIF-8 [19]; MIL is a MOF material formed by trivalent transition metals (such as Cr^{3+}, Fe^{3+}, and Al^{3+})

Ion Exchange Membranes: Design, Preparation, and Applications,
First Edition. Tongwen Xu and Yaoming Wang.

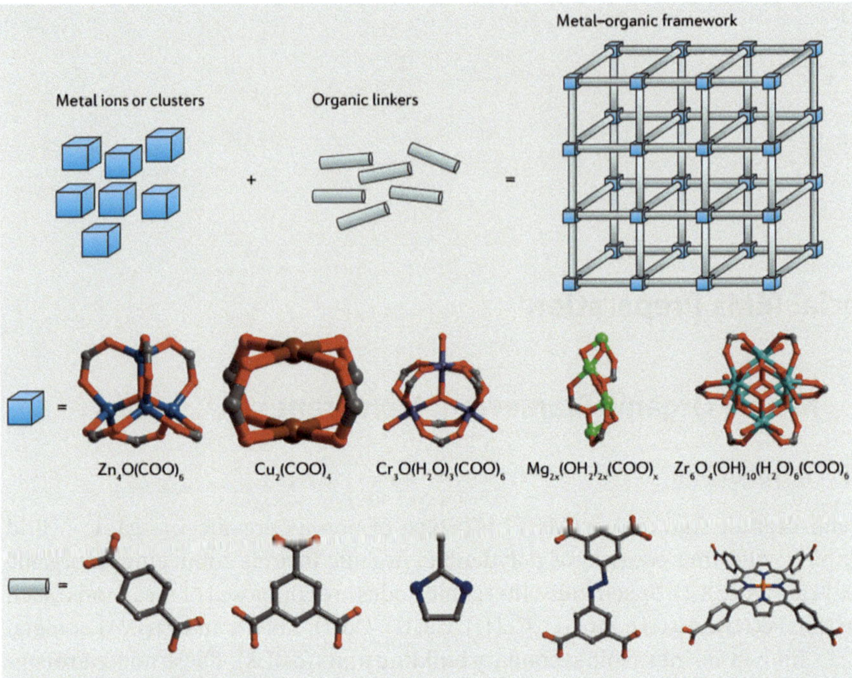

Figure 8.1 Schematic diagram of the synthesis of metal-organic frameworks. Source: Howarth et al. [1]/Springer Nature.

or lanthanide metals (In, Ga) and dicarboxylic acid ligands such as p-benzoic acid and homobenzoic acid, for example, MIL-101 [20]; the UiO series are a super stable class of MOF composed of organic ligands containing carboxyl groups (commonly used ligands include terephthalic acid, biphthalic acid, etc.) and Zr^{4+}, for example, UIO-66 [21]; PCN series are usually synthesized by transition metals (such as Fe, Co, and Ni) and azide organic matter, for example, PCN-222 [17].

One of the most important characteristics of MOFs is their high porosity (fraction of void volume relative to total volume) and high surface area, which has spawned many applications related to gas storage, separation, and catalysis. An important standard of the gas adsorption capacity of MOF is the number of adsorption sites within a given material, so the simplest way to increase the adsorption capacity of MOFs is to use elongated organic ligands that expose the surfaces and edges of the constituent units.

8.1.2.2 Structural Tunability

MOFs have good tunability and designability in structure, which can be attributed to several factors: ① The pore size and structure of the material can be designed and regulated by different metal nodes and organic ligands. ② Organic ligands can be easily designed and modified. For example, specific functional groups such as $-NH_2$, -COOH, and $-SO_3H$ can be introduced into MOF materials to functionalize MOF materials. ③ By regulating the reaction conditions such as solvent, acid, base,

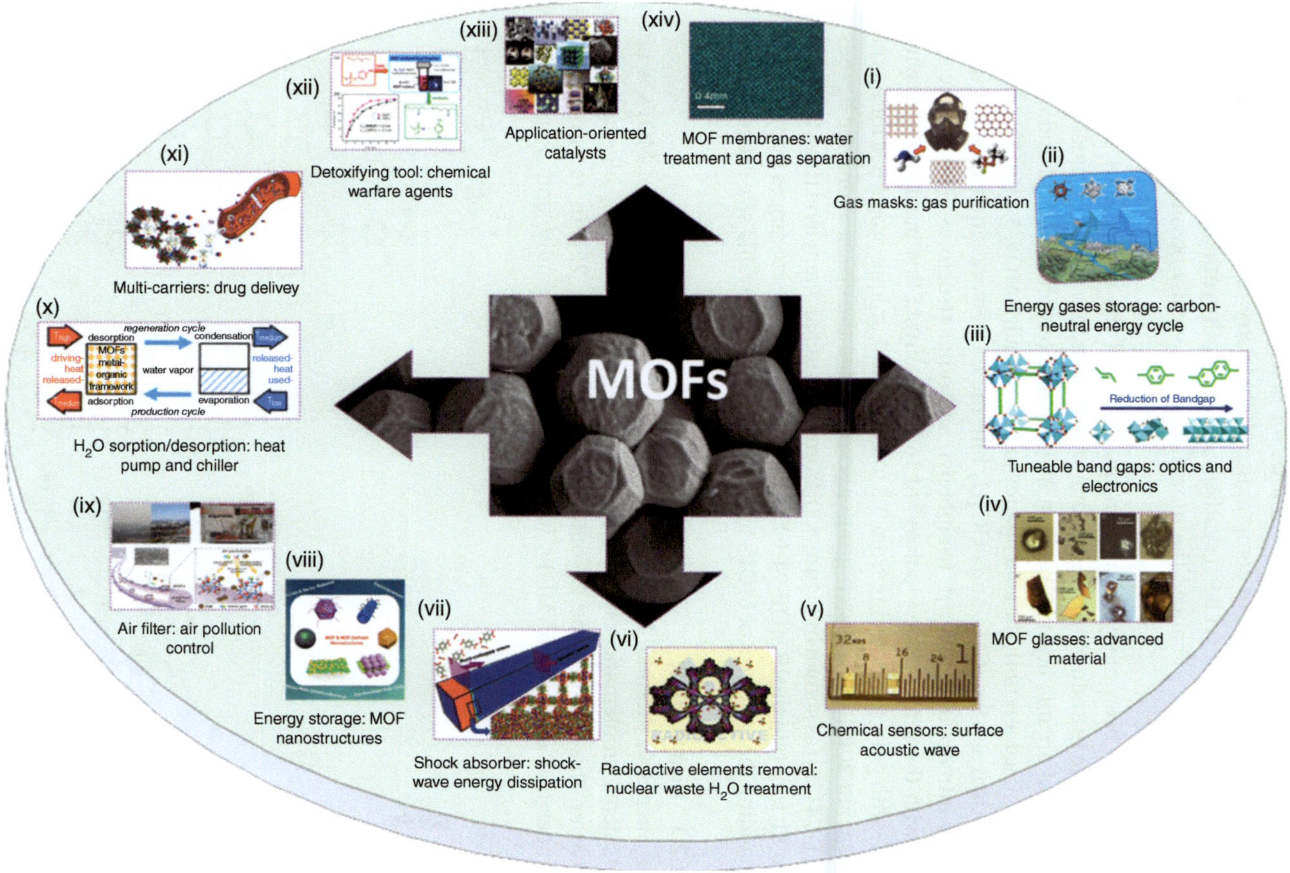

Figure 8.2 Applications of MOF materials in various fields. Source: Ren et al. [2]/with permission of Elsevier.

and temperature, MOF materials with different structures can be formed. ④ MOFs can be modified on metal nodes or organic linker bases by post-synthetic modification, which can generally maintain the original crystal structure. Alternatively, the highly crystalline nature of MOFs allows for precise structural characterization of atoms and allows the use of computational chemistry to predict the properties of skeletons that have not yet been synthesized [22]. Both synthetic and computational developments have enhanced the tunability and diversity possible for the rational design of MOFs.

There have been many reports on the design of the pore size of porous structures and the introduction of functionalized groups into MOFs by prechanging the structure of organic ligands. Yaghi et al. [12] used terephthalic acid and a series of functionalized terephthalic acid ligands to functionalize MOF-5, introduced organic groups such as $-Br$, $-NH_2$, and $-OC_3H_7$ into the framework, and expanded its pore size by introducing biphenyl, tribiphenyl, and other structures. The isopore series (one with the same frame topology) of 16 highly crystalline materials was synthesized, with pore sizes varying from 3.8 to 28.8 Å and unique properties for methane storage. In 2012, the research group also expanded the classical organic ligands of MOF-74 [23] (as shown in Figure 8.3), from the original phenylene ring (I) to ligands II to XI, to obtain the IRMOF-74-I-XI series MOF. Its pore size ranges from 14 to 98 Å, expanding the pore size of MOFs to a previously unachievable size range (>32 Å).

8.1.2.3 High Stability

The chemical, thermal, hydrothermal, and mechanical stability of MOFs is often considered problematic, especially when compared to industrially relevant zeolites; however, this has become an outdated notion [21, 24, 25]. In recent years, the number and diversity of MOFs have increased significantly, and there are now

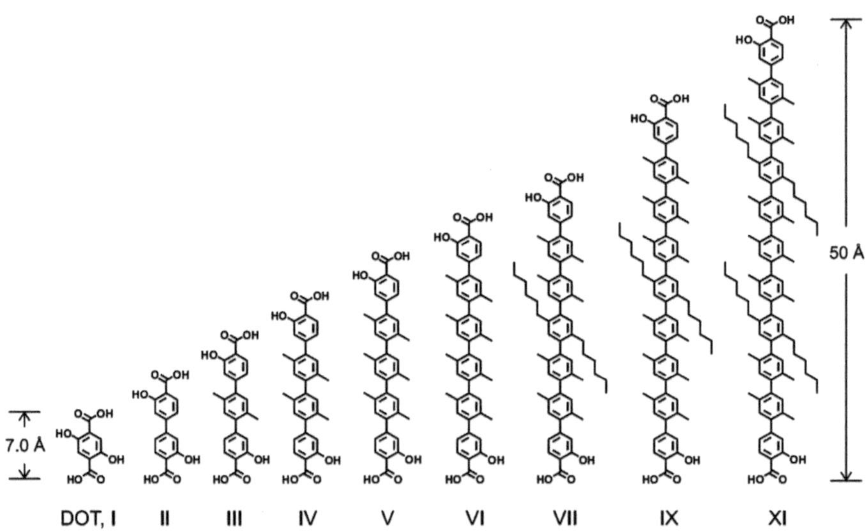

Figure 8.3 IRMOF-74-I-XI series organic ligands. Source: Deng et al. [23]/American Association for the Advancement of Science.

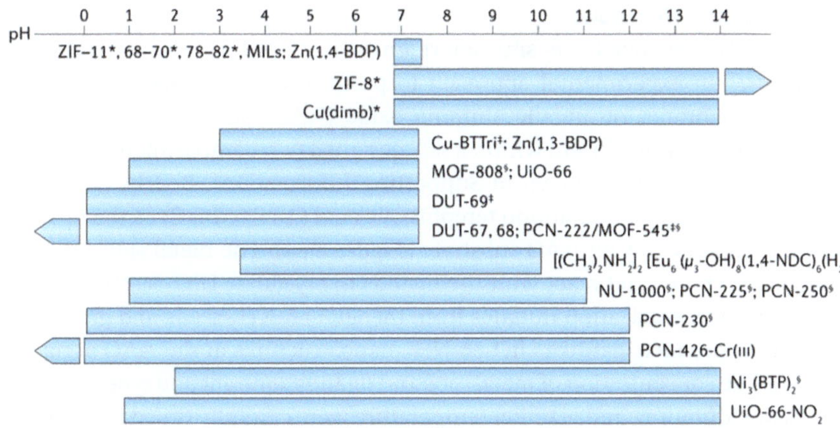

Figure 8.4 Chemical (acid–base) stability of some representative MOFs in the literature. Bar length indicates the pH range tolerated by the MOF. Source: Adapted from Howarth et al. [1].

many water-stable and heat-stable MOFs, which have greatly expanded their application fields.

Depending on the intended application, different functional stabilities are important. For example, chemical (more specifically water stability) and thermal stability are important for industrial applications, including gas separation, water desalination, ion exchange, and mesophilic heat storage, because they are associated with high temperatures, pH changes, or water vapor. Mechanical stability is critical if a compacted form of MOF, such as pellets, is required in an industrial process. The pH ranges for the hydrochemical stability of some representative MOFs are summarized in Figure 8.4. The main MOFS with good stability in water include the UIO-66 series, PSN-222, MIL-101, and ZIF-8.

8.1.3 Preparation of MOF Membranes

There are two main types of MOF membranes. One class is pure MOF membranes, which consist of porous or free-supporting substrates supporting continuous MOF layers. Pure MOF membranes can have high flux and excellent selectivity because of their structural characteristics such as uniform and adjustable pore size. The preparation methods of pure MOF separation membranes mainly include the in situ growth method [26], direct growth on modified supports [27–29], secondary growth [30–32], interfacial growth [33, 34], contradiffusion method [35, 36], and electrochemical deposition method [37, 38]. The other class is MOF hybrid membranes, which mainly mix MOFs and polymers to form mixed matrix membranes (MMMs). In MMMs, MOF particles are used as fillers to improve the separation performance of the polymer matrix. The preparation methods of MOF hybrid membranes mainly include the blending method [39], interfacial polymerization (IP) method [40, 41], polymerization method [42], and electrospinning method [43, 44].

Our group constructed a series of MOF membranes.

8.1.3.1 UiO-66-NH_2 Membrane

We report highly controlled "in situ smart growth of UiO-66-NH_2 as leaf-like nanostructures," an experimental approach to produce ultrathin and defect-free UiO-66-NH_2 membranes on anodic alumina oxide (AAO) substrates. The MOF nanostructures were grown in situ within a self-designed two-compartment reaction cell (experimental details are given in Supporting Information). Briefly, both the $ZrCl_4$ metal salt solution and 2-amino terephthalic acid (2-NH_2-BDC) as the ligand solution were separated by an AAO substrate (Figure 8.5a). The metal salts diffuse through the AAO pores and react with the ligands to crystalize UiO-66-NH_2 seeds on the AAO surface. Nucleation initiates from the UiO-66-NH_2 seeds and grows in such a way as to produce leaf-like UiO-66-NH_2 nanostructures at the AAO surface (Figure 8.5b). The concentration of metal ions and ligands, amount and type of modulators, reaction duration, and reaction temperature were carefully tuned to acquire optimized reaction parameters for reproducible growth of the leaf-like UiO-66-NH_2 nanostructures outwards from the AAO surface. Focusing on the optimized reaction parameters, we fabricated several UiO-66-NH_2 leaf-like membranes on different AAO substrates (henceforth, denoted as LLM-1, LLM-2, LLM-3, and LLM-4 membranes, Figure 3d is a photograph of LLM-4). Here, LLM is abbreviated as leaf-like UiO-66-NH_2 membrane, and 1 to 4 indicate the number of fabricated membranes based on different reaction parameters. For all the LLMs, the thickness of the highly decorated UiO-66-NH_2 surface layer was controlled below 500 nm to boost cation permselectivity. The decoration level (density of the grown UiO-66-NH_2 leaves) was varied to investigate its impact on cation permeation and permselectivity.

The cation-selective separation performance for all the LLMs was investigated using a lab-scale electrodriven cation permselective single cell. All the LLMs exhibit cation permeation in the order of $Na^+ > Li^+ > Mg^{2+}$ (Figure 8.5c,d), which is based on their hydrated diameters. The lowest permeation of the divalent cation is due to its larger hydrated ionic diameter than the inherent angstrom-sized pores in individual UiO-66-NH_2 crystals. The comparatively fastest permeation of Na^+ is due to its smallest hydrated size, which is comparable with the inherent angstrom pores. Therefore, all the LLMs exhibited excellent permselective separation performance for Na^+/Mg^{2+} and Li^+/Mg^{2+} cation systems. Moreover, comparatively higher permselectivity for Na^+/Mg^{2+} (>200) than for Li^+/Mg^{2+} (>60) was achieved owing to relatively faster Na^+ permeation. Notably, among all the fabricated leaf-like UiO-66-NH_2 membranes, LLM-4 exhibited the highest separation performance owing to the relatively denser leaves, which produce smaller interstitial pores and were thus selected for long-life stability evaluation. No obvious damage to the LLM-4 membrane and almost negligible changes in cation permeation and selectivity were observed for five consecutive electrochemical performance cycles (Figure 8.5e). This systematic UiO-66-NH_2 growth produces nanosized interstitial pores among the UiO-66-NH_2 leaves, which enables fast permeation of cations. Moreover, the individual UiO-66-NH_2 crystals possess angstrom-sized inherent pores that can efficiently regulate the cation permselectivity. As a result, controlled and systematically self-assembled UiO-66-NH_2 growth produces dual-sized pores to provide efficient selective separation of mono- over divalent cations (Figure 8.5) [45].

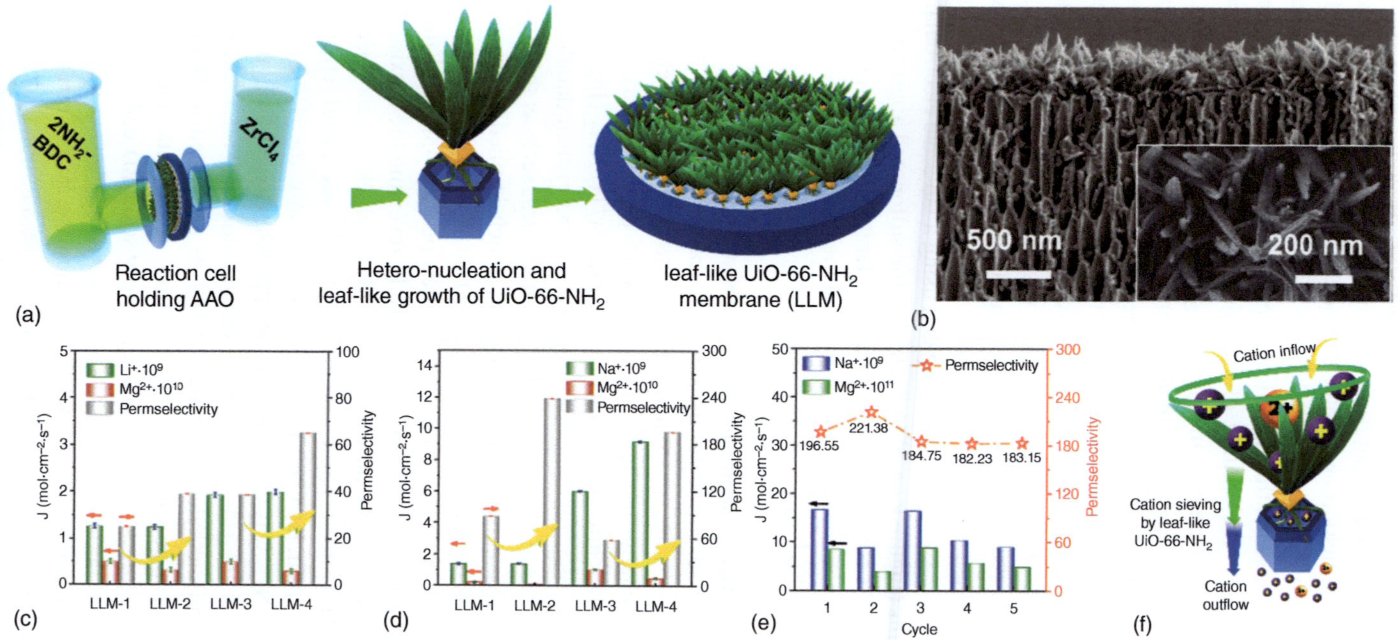

Figure 8.5 (a) Schematic diagram of the preparation process of the UiO-66-NH_2 membrane, (b) SEM images of UiO-66-NH_2 membranes [104]. Tingting Xu et al. 2019/With permission of John Wiley & Sons, (c) Li^+/Mg^{2+} separation of UiO-66-NH_2 membranes, (d) Na^+/Mg^{2+} separation of UiO-66-NH_2 membranes, (e) stability of the separation performance of UiO-66-NH_2 membranes (taking Na^+/Mg^{2+} separation as an example), and (f) ion separation mechanism of the UiO-66-NH_2 membrane.

8.1.3.2 UiO-66-SO_3H Membrane

The mixed ligand method was used to introduce the auxiliary mass transfer group (–SO_3H) into the MOF separation membrane to construct a series of subnanometer-scale ion transport channels (UIO-66-SO_3H membrane) with different sulfonic acid group contents (0%, namely, UIO-66, 10%, 25%, and 33%, denoted as U-SM[X], where X represents the sulfonic acid group content). The relationship between its properties and the content of sulfonic acid groups was investigated (Figure 8.6a). Specifically, we propose in situ smart growth of leaf-like UiO-66-SO_3H membranes within a self-designed two-compartment reaction cell (Figure 8.6b). The solutions containing metal ions and ligands are separated by an AAO substrate. Both solutions diffuse toward opposite directions by passing through the pores of the AAO substrate and meet at the AAO surface to crystallize UiO-66-SO_3H. Nucleation initiates from the seeds and in situ produces ultrathin leaf-like UiO-66-SO_3H membranes (<600 nm). The sulfur content in the UiO-66-SO_3H layer is easily controllable by tuning the feed ratio of the ligands. The cation-selective separation performance for all the U-SM(X)s was investigated using a lab-scale electrodialysis (ED) stack. The results (Figure 8.6c–e) indicate that the sulfonated angstrom-sized ion transport channels in the membranes could accelerate the cation permeation (~3 × faster than nonfunctionalized UiO-66 membrane) and achieve a high ion selectivity (Na^+/Mg^{2+}>140) [46].

8.1.3.3 UiO-66(Zr/Ti)-NH_2/Polyamide Mixed Matrix Membrane

The ion exchange method was used to replace a fraction of the Zr^{4+} ions in UiO-66(Zr)-NH_2 with Ti^{3+} ions, which neutralized some of the positive charge and introduced a negative charge in the porous framework of UiO-66(Zr)-NH_2. The obtained product will hereafter be denoted as UiO-66(Zr/Ti)-NH_2 (Figure 8.7a). The analysis of the SRPES profiles of the samples provided critical details on the post-synthetic process of the UiO-66(Zr)-NH_2 nanoparticles, such as the suppression of the Zr peak and emergence of the Ti peak, which implied the successful replacement of Zr^{4+} ions with Ti^{3+} ions in UiO-66(Zr/Ti)-NH_2. The SRPES profiles of UiO-66(Zr/Ti)-NH_2 in the Ti 2p region show two new peaks at 457.8 and 463.6 eV, corresponding to Ti^{3+} $2p_{3/2}$ and Ti^{3+} $2p_{1/2}$, respectively, revealing the coexistence of Ti^{3+} and Ti^{4+} ions in the structure of UiO-66(Zr/Ti)-NH_2, which was ascribed to the oxidation of $TiCl_3$ to $TiCl_4$ in air. These results indicated the successful replacement of Zr^{4+} ions with Ti^{3+} ions in UiO-66(Zr/Ti)-NH_2. Furthermore, the Ti^{3+}/Ti^{4+} molar ratio of UiO-66(Zr/Ti)-NH_2 was determined to be approximately 6 : 5 (Figure 8.7b).

Following IP, the acyl chloride groups of trimesoyl chloride (TMC) reacted with UiO-66(Zr/Ti)-NH_2 and diethylenetriamine (DETA), as illustrated in Figure 8.7a, and produced a uniform polyamide layer (~170 nm) that contained embedded MOF nanoparticles (Figure 8.7c). A single laboratory-scale ED cell was used to evaluate the monovalent permselectivity of the prepared membranes in Na^+-Mg^{2+} and Li^+-Mg^{2+} mixed solutions (Figure 8.7d). The increase in monovalent cation flux through TFN membranes could be attributed to the additional ion transport channels of the MOF particles embedded in the MOF-containing surface layers. The higher cation flux of the TFN-(Zr/Ti) membranes was attributed to the

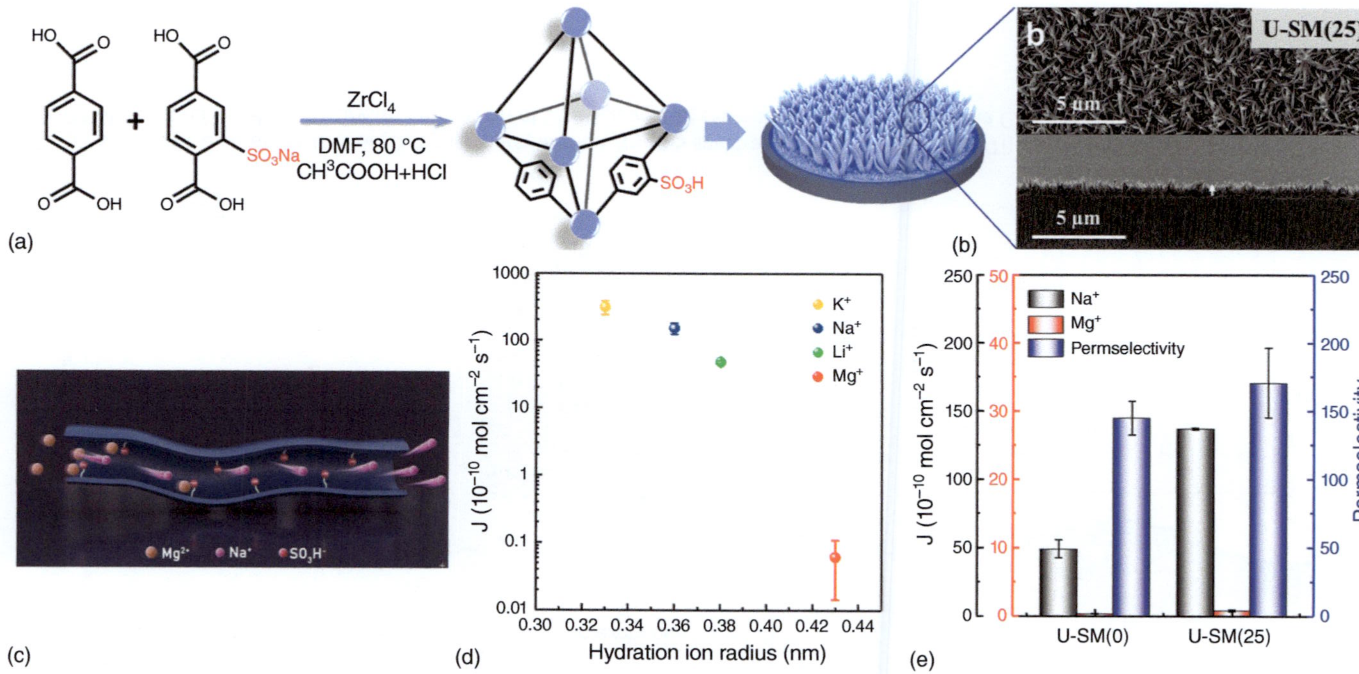

Figure 8.6 (a) Schematic diagram of the synthesis of UiO-66-SO_3H by the mixed ligand method, (b) SEM images of UiO-66-SO_3H membranes, (c) schematic diagram of ion separation of the UiO-66-SO_3H membrane [25]. Haiyan Pan et al. 2010/With permission of Elsevier, (d) ion flux of U-SM(25) in single-component separation performance, and (e) ion separation performance of Na^+/Mg^{2+} of U-SM(0) and U-SM(25).

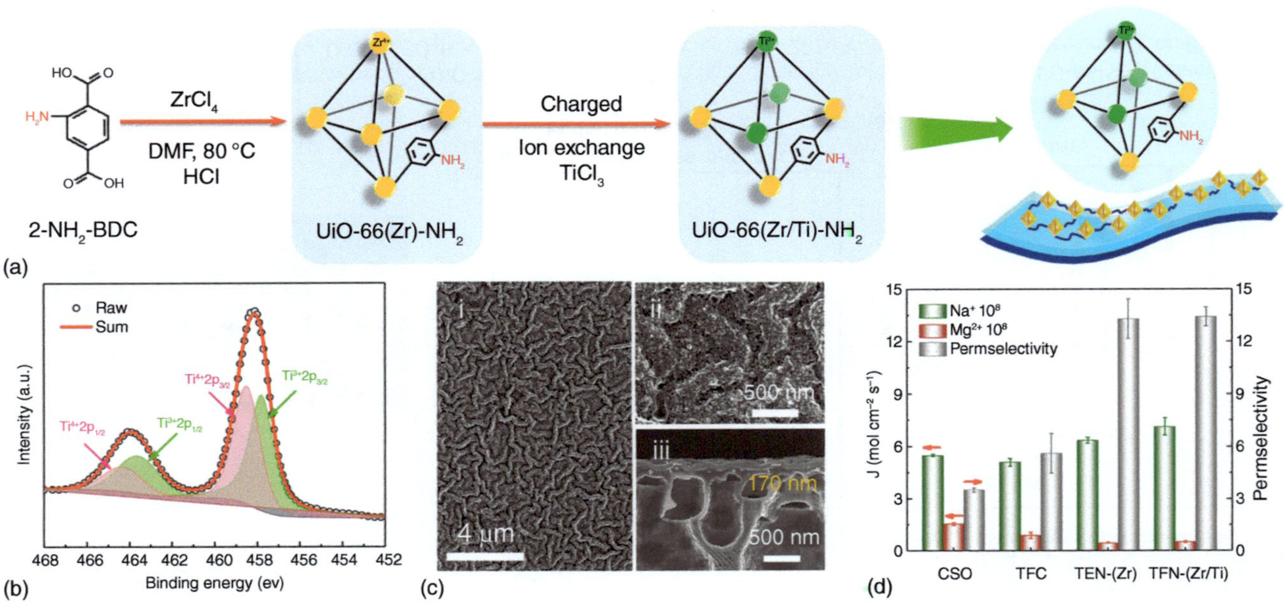

Figure 8.7 (a) Schematic diagram of UiO-66(Zr/Ti)-NH_2 preparation, (b) XPS spectra of Ti (2p) in UiO-66(Zr/Ti)-NH_2, (c) SEM images of TFN-(Zr/Ti) membranes [106]. Tingting Xu et al. 2020/With permission of Elsevier, and (d) separation performance of the CSO, TFC, TFN-(Zr), and TFN-(Zr/Ti) membranes.

electrostatic assistance of the UiO-66(Zr/Ti)-NH_2-containing ions separating the surface layer. Consequently, the combined effect of physicoelectrical properties, such as size-sieving and electrostatic assistance, in UiO-66(Zr/Ti)-NH_2-containing membranes, particularly the TFN-(Zr/Ti)-2 membrane, caused high monovalent ion fluxes ($J_{Na^+} = 7.15 \times 10^{-8}$ mol cm^{-2} s^{-1} and $J_{Li^+} = 5.43 \times 10^{-8}$ mol cm^{-2} s^{-1}). Furthermore, the mono- over divalent cation separation performance of the TFN-(Zr/Ti)-2 membrane ($P_{Na^+/Mg^{2+}} = 13.44$ and $P_{Li^+/Mg^{2+}} = 11.38$) surpassed that of the TFC membrane [47].

8.1.3.4 PolyMOF Membrane

Membrane-based molecular/ionic separations have attracted broad interest in the sustainable development of the modern chemical industry due to the energy-efficient treatment process. Conventional polymers have been extensively used for membrane preparation because they are solution processable. The permeability and selectivity trade-off limitation is ubiquitous in polymeric membranes because of their contorted channels and wide pore size distributions. MOFs have been known as attractive porous crystalline materials since they have ordered and uniform channels with sizes comparable to those of molecules or ions of interest, showing potential applications for gas separation and ion conduction. The fabrication of defect-free MOF membranes remains a great challenge. As an alternative, multiple strategies have been employed to integrate MOFs into a flexible matrix, such as MOF-based MMMs, where MOF nanoparticles are amalgamated into a polymeric matrix to enhance the processing capability. However, the nonselective voids at the MOF/polymer interface hinder the performance of MMMs because of the poor compatibility between MOFs and the matrix phase. Therefore, the lack of solution processability of MOFs prevents them from processing for diverse membrane-based separations to a large extent. The polymerization of ligands or crosslinking of MOFs via organic spacers contributes to boosting MOF processing ability. Recently, a bottom-up approach has been developed to construct porous and crystalline hybrid polymeric MOF (polyMOF) materials using polymer ligands with coordination sites. The designed ligands fold into well-crystalline polyMOFs due to their interaction with metal ions, while the alkyl spacers retain the crystalline morphology, as shown in Figure 8.8.

The discovery of polyMOFs initiated excellent research to convert polymers and MOFs into highly porous crystalline hybrid materials. However, limited studies have been conducted to use polyMOFs as building blocks for membrane construction and related separations. The incorporation of amorphous polymers in MOF frameworks may endow MOFs with distinct features, such as solubility in solvents, which can be employed for the facile fabrication of polycrystalline membranes. In this study, we employed polyligands to construct solution-processable polyMOFs. The as-prepared polyMOFs are soluble in an organic solvent and maintain crystallinity after multiple dissolution and precipitation cycles, as shown in Figure 8.9. Then, polyMOF-based crystalline, durable, and adjustable membranes were fabricated by solution casting on AAO substrates, as shown in Figure 8.10.

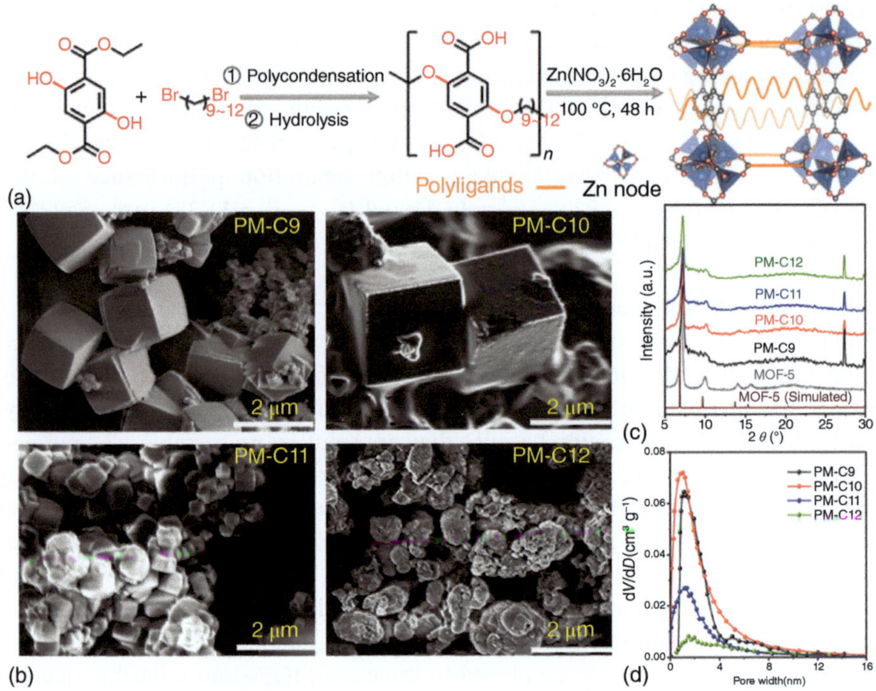

Figure 8.8 (a) The strategy describes the synthesis of polyligands and fabrication of crystalline and porous polyMOFs (PM-C*x* $x = 9, 10, 11, 12$), (b) SEM images of PM-Cx crystals, (c) PXRD patterns of MOF-5 and PM-Cx crystals, and (d) pore size distributions of PM-Cx crystals.

The resulting membranes can achieve high permselectivity for the separation of monovalent and divalent cations through an ED process, in which the permselectivity was 50 for Na^+/Mg^{2+}, 30 for K^+/Mg^{2+}, and 7 for Li^+/Mg^{2+}, as shown in Figure 8.11.

8.2 Porous Organic Cage Membranes

8.2.1 Introduction

A porous cage is a molecule with permanent pores within a rigid structure that allows outside material to enter these pores by their window [48]. Although macrocyclic compounds and cage compounds are already known, porous organic cages (POCs) are a relatively new type of porous material. For example, POC is composed of discrete molecular cages by weak van der Waals forces and hydrogen bonds and connections. In addition, they are soluble in ordinary organic solvents.

POCs were first proposed and developed by Andrew I. Cooper, Michael Mastalerz, and others. POCs are self-assembled by covalent bond stacking, forming a three-dimensional cage-like structure that is stable and durable [49, 50].

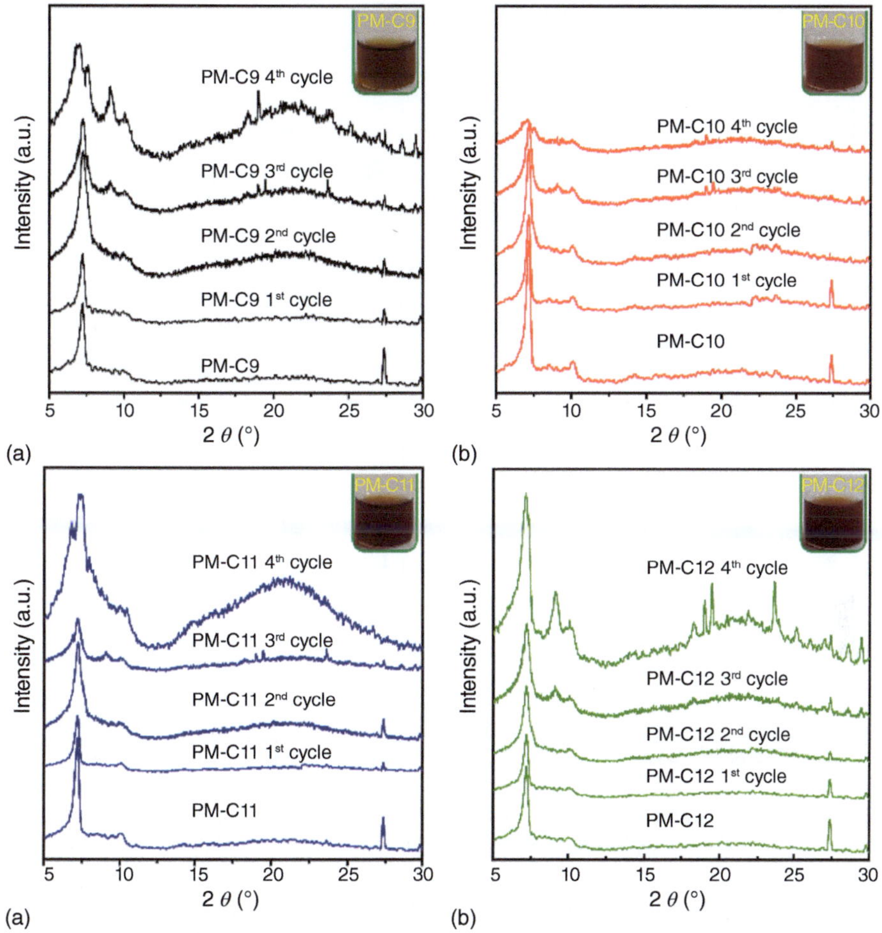

Figure 8.9 PXRD patterns of PM-C*x* crystals after four consecutive cycles of the dissolution-recrystallization process: (a) PM-C9, (b) PM-C10, (c) PM-C11, and (d) PM-C12. The inserted digital photos exhibit homogeneous and transparent solutions of the corresponding PM-C*x* crystals dissolved in NMP.

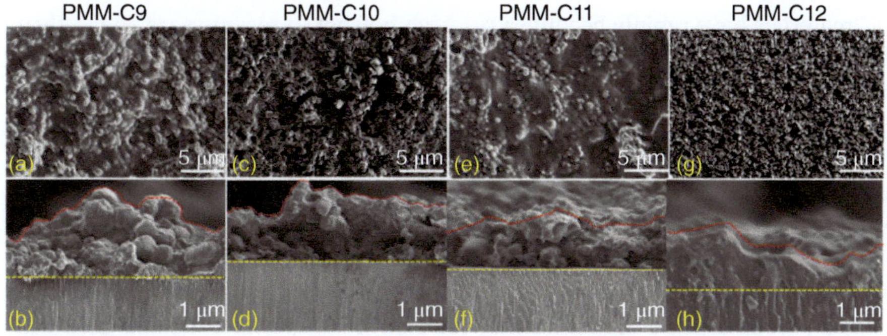

Figure 8.10 SEM surface and cross-sectional images of PMM-C*x* membranes. (a–b) PMM-C9, (c–d) PMM-C10, (e–f) PMM-C11, and (g–h) PMM-12. PMM denotes polyMOFs membranes.

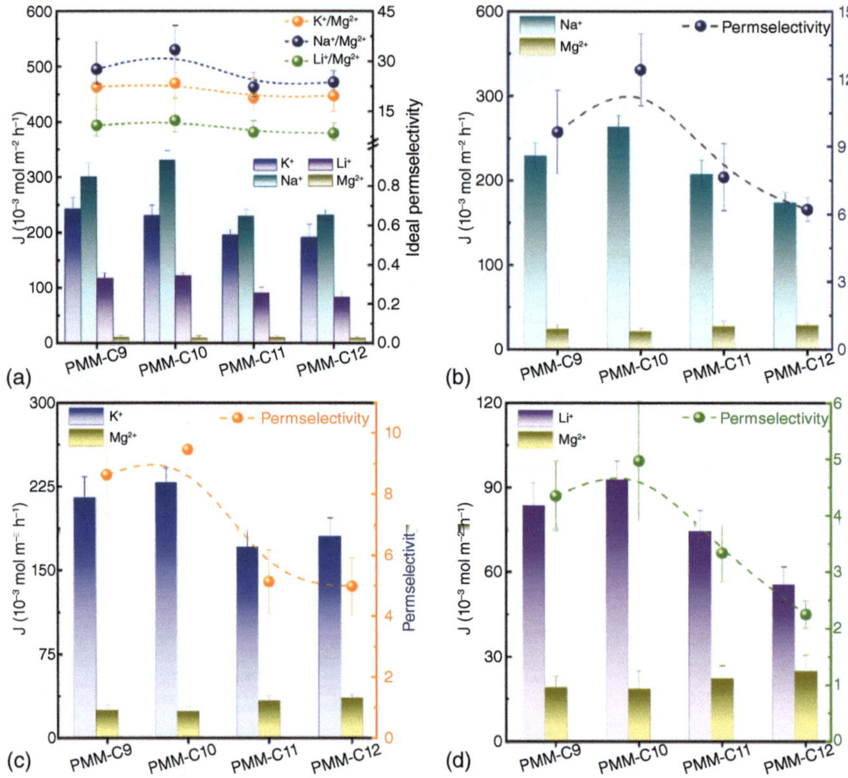

Figure 8.11 Ion flux and permselectivity of PMM-C*x* membranes. (a) Single ion system. Mixed ion solution of the (b) K^+/Mg^{2+} system, (c) Na^+/Mg^{2+} system, and (d) Li^+/Mg^{2+} system.

Imine POCs (e.g. [4 + 6] tetrahedral cages and [2 + 3] triangular prismatic cages) are synthesized by imine condensation/various amine and aldehyde reactions. Figure 8.12 shows the timeline of POC development.

8.2.2 Structural Properties of POCs

POC materials have rapidly become a research hotspot in recent years due to their advantages, such as high specific surface area, good stability, and the ability to dissolve in a variety of organic solvents, and they have potential application value in many fields such as separation analysis and molecular containers [52, 53]. Unlike extended porous crystals such as zeolites and MOFs, POCs are discrete molecules with intrinsic cage structures, and their intrinsic cage structure can be retained in crystals [54], amorphous solids [55], and even in porous liquids formed with specific solvents [56]. However, organic cages still face some challenges in expanding their potential structure and function.

Stable porous organic molecules may have advantages over extended networks for some applications, especially their solubility. The advantages of this approach have recently been illustrated by an inherently microporous polymer, which can be

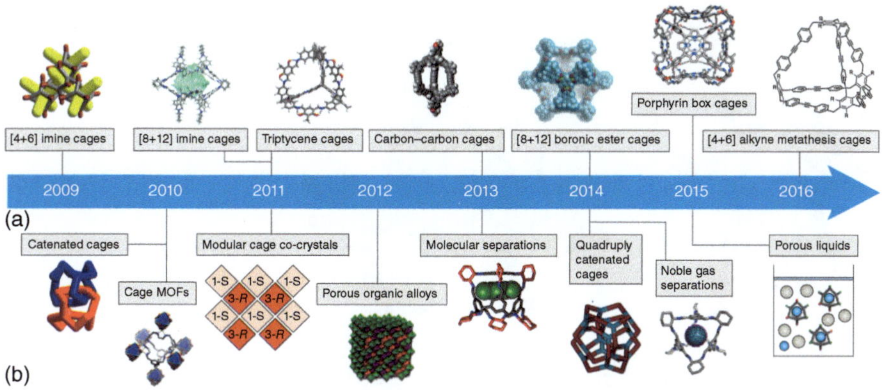

Figure 8.12 Timeline of the development of the porous organic cage. (a) Illustration of a porous organic cage, (b) discoveries related to porous organic cages and their applications. Source: Hasell et al. [51]/Springer Nature.

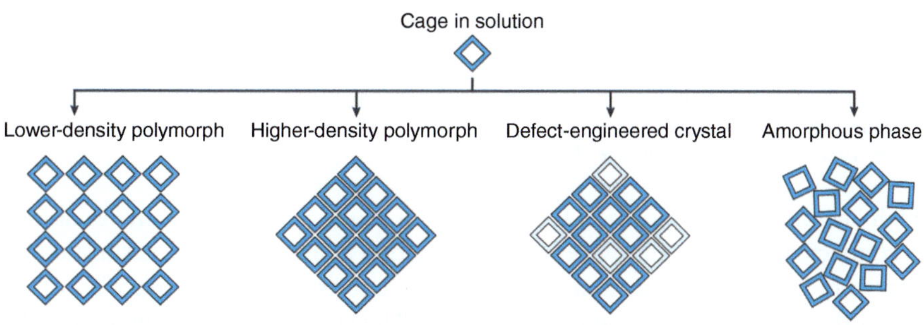

Figure 8.13 Different phases of a single molecular cage. Source: Tozawa et al. [58]/Springer Nature.

cast from solution into a porous membrane [57]. POCs are more processable and operable than insoluble porous solids in some applications due to their solubility. POCs can be crystallized into different polycrystalline forms and switched between solid polycrystalline forms (Figure 8.13) because discrete molecules are combined by relatively weak intermolecular forces [59]. In 2009, Cooper et al. [58] proved the permanent porosity of POCs through gas adsorption experiments. Such a system contains three trihedral organic cages formed by the cyclization of a trialdehyde with a diamine [4 + 6] (four aldehyde units and six diamine units), exhibiting a classical type I gas adsorption isotherm.

The vertices of CC3 consist of relatively large cyclohexyl groups (red) packed in a window-to-window alignment (Figure 8.14a). The interlocking of three cyclohexyl groups on adjacent cages results in the alignment of the three windows. Combined with the cage-like symmetry of the tetrahedron, this results in an interconnected 3D rhomboid channel structure (Figure 8.14b) that passes through the cages rather than between them.

In 2009, Cooper et al. first reported the synthesis of CC1 to CC3 [58]. In principle, water presents two problems for POCs. First, CC1 to CC3 are based on imines,

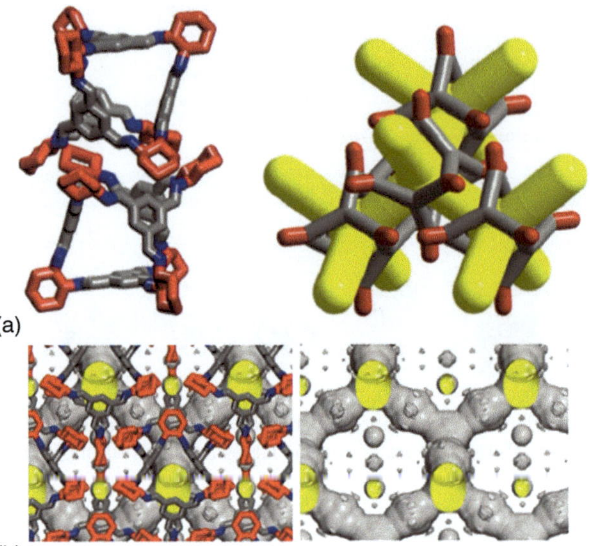

(a)

(b)

Figure 8.14 Cage 3 is directed by the cyclohexyl groups (red) to pack window-to-window, thus generating an interconnected diamondoid-pore network, as illustrated here in yellow. Source: Tozawa et al. [58]/Springer Nature.

which chemically breakdown in the presence of water. Second, these materials are not connected by extended covalent or coordination bonds; therefore, adsorption by water may lead to the transition of POC to a nonporous polycrystalline form [55] or turn the material into an amorphous state. The study of Tom Hasell et al. [60] showed that the solid hydrolytic stability of the CC3 cage was proven by both the chemical stability of imine cage molecules and the structural stability of 3D cubic crystal stacking [61]. In addition, the authors showed that the structure can be carried out reversibly. The material can even be boiled in water for long periods of time without significant decomposition.

8.2.3 Preparation of POC Membranes

POC materials have become potential new membrane materials because of their suitable pore size, easy dissolution in a variety of organic solvents, high chemical stability, and good structural tunability. POC membranes are mainly divided into two categories: pure POC membranes and POC hybrid membranes [62]. The commonly used methods for preparing pure POC membranes include the spin coating method [63], the layer-by-layer assembly method [64], and contradiffusion method. The blending method [65], in situ crystallization method [62], and interface polymerization method [66–68] are commonly used to prepare hybrid POC membranes.

8.2.3.1 POC Membranes of Versatile Channels

We fabricated POC membranes via the contradiffusion growth of CC1, CC3, and CC19 cages on an AAO substrate. The three tetrahedral-shaped POCs contain four

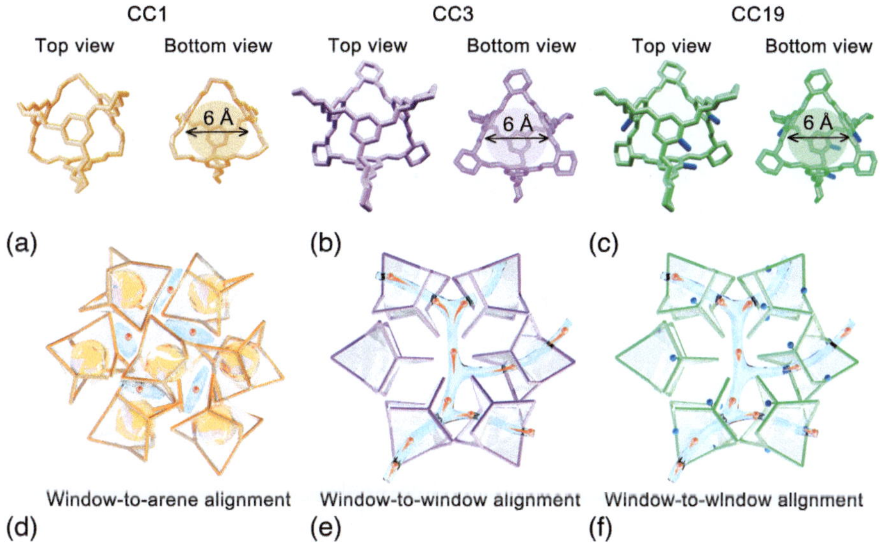

Figure 8.15 Top and bottom views of the (a) CC1 cage, (b) CC3 cage, (c) CC19 cage (hydrogen atoms are set as invisible for clarity), (d) window-to-arene alignment of CC1 cages, (e) window-to-window alignment of CC3 cages, and (f) window-to-window alignment of CC19 cages with hydroxyl groups.

triangular windows of the same size with a central cavity [58, 69]. Different channel features, such as cage arrangement, channel wettability, and surface charge, lead to distinct ion transport behaviors under concentration gradient driving. The window-to-arene alignment of CC1 together with the solvent-occupied window cavity blocks the ion transport channels, resulting in a low ion permeation rate and selectivity for the CC1 membrane (Figure 8.15a,d). The ion pathway formed by the window-to-window alignment of CC3 with a subnanometer-sized window and internal cavity, in addition to the channels formed by the external cavities among CC3 cages, guarantees high ion permeability and selectivity for the CC3 membrane. Moreover, the hydrophobic channel wall of CC3 contributes to fast ion transport (Figure 8.15b,e). The CC19 membrane has the same channel configuration as CC3, except for the hydrophilic channel wall, exhibiting a low ion permeation rate (Figure 8.15c,f).

CC1, CC3, and CC19 membranes were fabricated by the contradiffusion method [70, 71] on an AAO substrate. AAO disks were used as the substrate to separate 1,3,5-triformylbenzene (TFB) and 1,2-ethylenediamine monomers for CC1 membranes, TFB, and (1R,2R)-1,2-cyclohexanediamine (CHDA) for CC3 membranes, and 2-hydroxy-1,3,5-benzenetricarbaldehyde and CHDA for CC19 membranes. During the contradiffusion growth process, crystallization occurred slowly to form a continuous POC layer. After the contradiffusion growth, CC1, CC3, and CC19 crystals were successfully assembled on the surface of the AAO substrate. Scanning electron microscopy (SEM) images of the surface and cross-section of AAO-supported CC1, CC3, and CC19 membranes show their morphologies, confirming the formation of

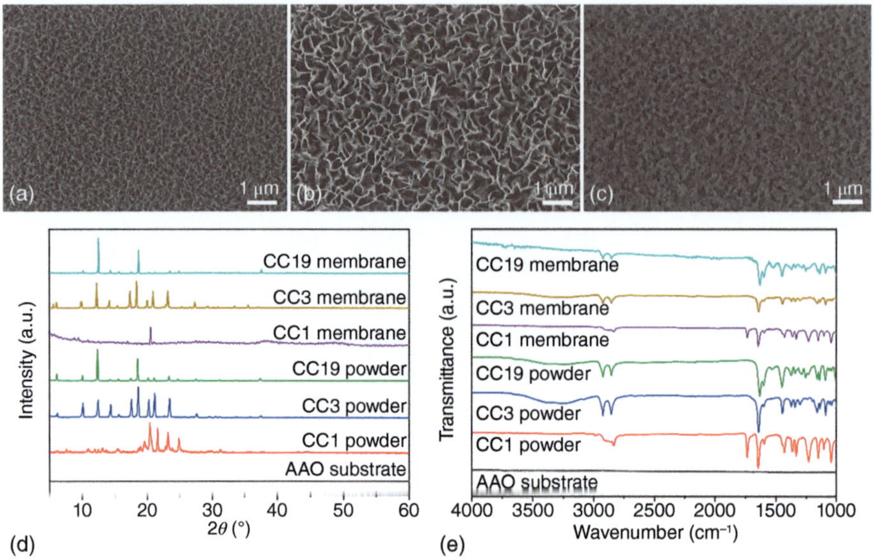

Figure 8.16 SEM images of the surface of the (a) CC1 membrane, (b) CC3 membrane, (c) CC19 membrane, (d) XRD patterns, and (e) FTIR spectra of the AAO substrate, CC1 powder, CC3 powder, CC19 powder, CC1 membrane, CC3 membrane, and CC19 membrane.

continuous CC1, CC3, and CC19 layers (Figure 8.16a–c). X-ray diffraction (XRD) patterns of the CC1, CC3, and CC19 membranes (Figure 8.16d) show high crystallinity [58]. From the FTIR spectra (Figure 8.16e), CC1 exhibits a strong IR vibration at 1740 cm^{-1} corresponding to the carbonyl of the guest solvent [58]. The characteristic peak of C=N at 1654 cm^{-1}, the peak of CH_2 at 1448 cm^{-1}, and the peak of C—N at 1160 cm^{-1} for CC1, CC3, and CC19 demonstrate the successful fabrication of the three POCs [72, 73].

CC1 cages pack in the window-to-arene mode, and these cages are arranged around forming an external cavity, while the internal cavity of CC1 cages is occupied by the solvent, resulting in discontinuous pathways for ion transport. As a result, CC1 membranes exhibit low permeation rates of ~0.07 mol m^{-2} h^{-1} for K^+, 0.04 mol m^{-2} h^{-1} for Na^+, and 0.02 mol m^{-2} h^{-1} for Li^+. The calculated ideal ion selectivity of the CC1 membrane is K^+/Mg^{2+} of ~5.8, Na^+/Mg^{2+} 3.5, and Li^+/Mg^{2+} of ~2.0 (Figure 8.17a). The discontinuous ion channels of CC1 restrict the transport of ions, and the lack of a window-cavity structure diminishes the ion selectivity. CC3 cages pack in a window-to-window arrangement, and the internal cavity is free of occupation, thus generating a continuous ion pathway. More importantly, angstrom-sized windows can sieve ions for selective separation. The external cavities among CC3 also provide pathways for ion transport. In addition, CC3 membranes possess hydrophobic channels contributing to the fast transport of ions because hydrated ions can easily pass through the channel via surface slip with a low diffusion energy barrier [74, 75] The permeation rates of the resultant CC3 membranes are ~0.15 mol m^{-2} h^{-1} for K^+, 0.09 mol m^{-2} h^{-1} for Na^+, and 0.05 mol m^{-2} h^{-1} for Li^+. The ideal selectivities of CC3 membranes for

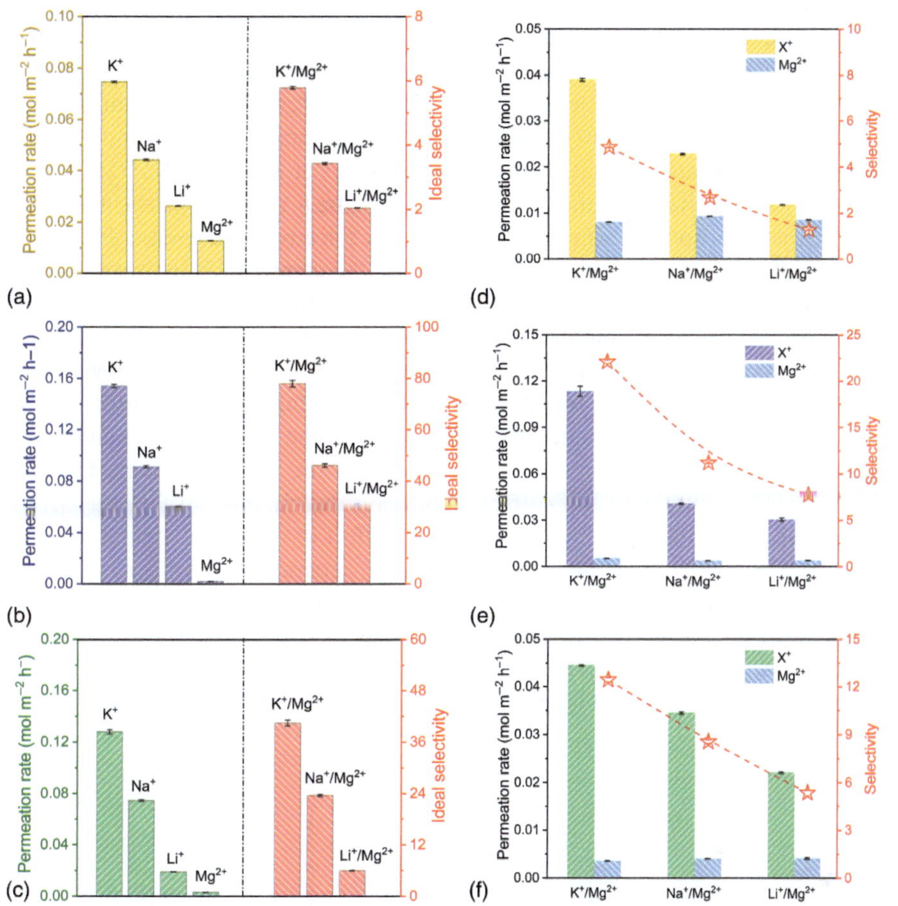

Figure 8.17 Ion separation performance of POC membranes. CC1 membranes (a), CC3 membranes (b), and CC19 membranes (c) in the single-ion system (0.1 M KCl, 0.1 M NaCl, 0.1 M LiCl, or 0.1 M $MgCl_2$). Real ion selectivities of CC1 membranes (d), CC3 membranes (e), and CC19 membranes (f) in the binary-ion system (0.1 M KCl/0.1 M $MgCl_2$, 0.1 M NaCl/0.1 M $MgCl_2$, or 0.1 M LiCl/0.1 M $MgCl_2$ aqueous solutions). Standard deviation values are derived from three tests of one membrane.

K^+/Mg^{2+} are ~78, Na^+/Mg^{2+} at ~46, and Li^+/Mg^{2+} at ~30 (Figure 8.17b). CC19 has a similar structure to CC3 but with additional hydroxyl groups, generating hydrogen-bond interactions with hydrated ions and hampering the fast transport of ions [69]. Therefore, despite having the same channel configuration as that of CC3, CC19 membranes have higher resistance for ion transport and show lower ion permeation and selectivity than CC3 membranes. The permeation rates of the CC19 membranes are ~0.12 mol m^{-2} h^{-1} for K^+, 0.07 mol m^{-2} h^{-1} for Na^+, and 0.02 mol m^{-2} h^{-1} for Li^+. The ideal selectivity of CC19 membranes for K^+/Mg^{2+}, Na^+/Mg^{2+}, and Li^+/Mg^{2+} is ~40, 23, and 6, respectively (Figure 8.17c).

Then, the binary-ion separation performance of POC membranes was investigated. The ion permeation rates obviously decrease for the CC1 and CC19 membranes (Figure 8.17d,f), while there is a slight drop for the CC3 membranes (Figure 8.7e). For CC1 membranes, the real mono–/di-valent ion selectivities remain within the same range as the ideal ones (Figure 8.17d). For CC3 membranes, the real ion selectivities of K^+/Mg^{2+}, Na^+/Mg^{2+}, and Li^+/Mg^{2+} drop to 22, 11, and 7, respectively (Figure 8.17e). The CC19 membranes show a similar trend to the CC3 membranes, with the real ion selectivities of K^+/Mg^{2+}, Na^+/Mg^{2+}, and Li^+/Mg^{2+} dropping to 13, 8, and 5, respectively (Figure 8.17f). The permeation rates of monovalent cations and mono–/di-valent ion selectivities decrease due to the existence of coions in the binary-ion system and the competition between cations. In addition, ion–ion, ion–water, and ion–wall interactions can possibly affect the ion transport property during the separation process [76–79].

The zeta potential results of CC3 membranes at various pH values reveal that the isoelectric point of CC3 membranes is approximately 4.6. The surface of the CC3 membrane is positively charged when the pH value is below 5. When the pH value is above 5, the hydrophobic channel wall of the CC3 membrane tends to adsorb hydroxyl ions onto the surface, rendering the channel surface negatively charged (Figure 8.18a) [80]. To investigate the effect of surface charge on the ion transport properties of CC3 membranes, K^+/Mg^{2+} separation tests were conducted by modulating the pH value of the feed solution from 3 to 8. For the single-ion system, with increasing pH, the permeation rates of both K^+ and Mg^{2+} increase, while Mg^{2+} shows a more obvious increase, resulting in a decrease in the K^+/Mg^{2+} selectivity from 350 to 50 (Figure 8.18b). The positively charged channel surface

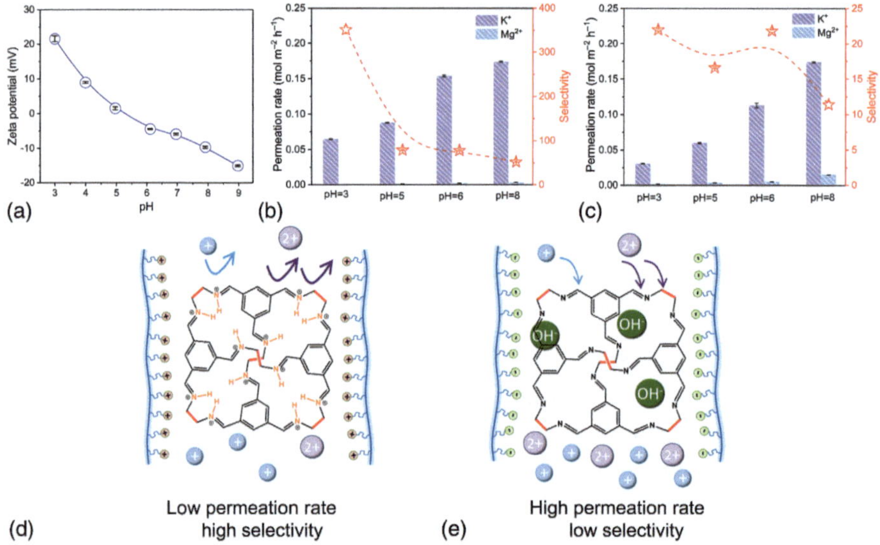

Figure 8.18 (a) Zeta potential of CC3 membranes, (b) ion permeation rates and selectivities of the CC3 membrane under different pH values from 3 to 8 (single-ion system: 0.1 M KCl, or 0.1 M $MgCl_2$), (c) real ion selectivities of the CC3 membrane under different pH values from 3 to 8 (binary-ion system: 0.1 M KCl/$MgCl_2$). Standard deviation values are derived from three tests of one membrane, (d) illustration of positively charged channel surface inhibiting cation transport, and (e) illustration of a negatively charged surface promoting cation transport.

inhibits cation transport because of electrostatic repulsion interactions, and the negatively charged surface promotes cation transport due to electrostatic attraction (Figure 8.18d,e). For the binary-ion system, the ion permeation rates of the CC3 membrane increase as the pH value increases. The ion selectivity fluctuates within a certain range from 10 to 25, considering the complex coion environment in the separation system (Figure 8.18c).

8.2.3.2 High Ion-Permselective CC3 Membrane

We fabricated CC3 membranes on AAO substrates via the contra-diffusion growth of CC3 molecular cages, which was carried out in a self-designed two-compartment reaction cell. AAO disks with a diameter of 25 mm and pore size of 90 ± 10 nm were used as the substrate to separate the TFB monomer and CHDA monomer. During the contradiffusion growth process, TFB diffused through the AAO channels, reacting with CHDA at the substrate surface facing the CHDA chamber, and then CC3 seeds were generated. Subsequently, crystallization occurred slowly to form a continuous CC3 layer (Figure 8.19a). The subnanometer-sized windows of CC3 can sieve ions for selective separation, and the internal cavities within CC3 together with the external cavities among CC3 can provide pathways for fast ion transport (Figure 8.19b). Figure 8.19c is the simplified model of CC3 channels composed of discrete frameworks.

After the contra-diffusion growth, CC3 crystals were successfully assembled on the surface of the AAO substrate, which can be confirmed by scanning electron microscope (SEM) images of the surface and cross-section of AAO-supported CC3 membranes (Figure 8.20a). The SEM images of the CC3 membranes show

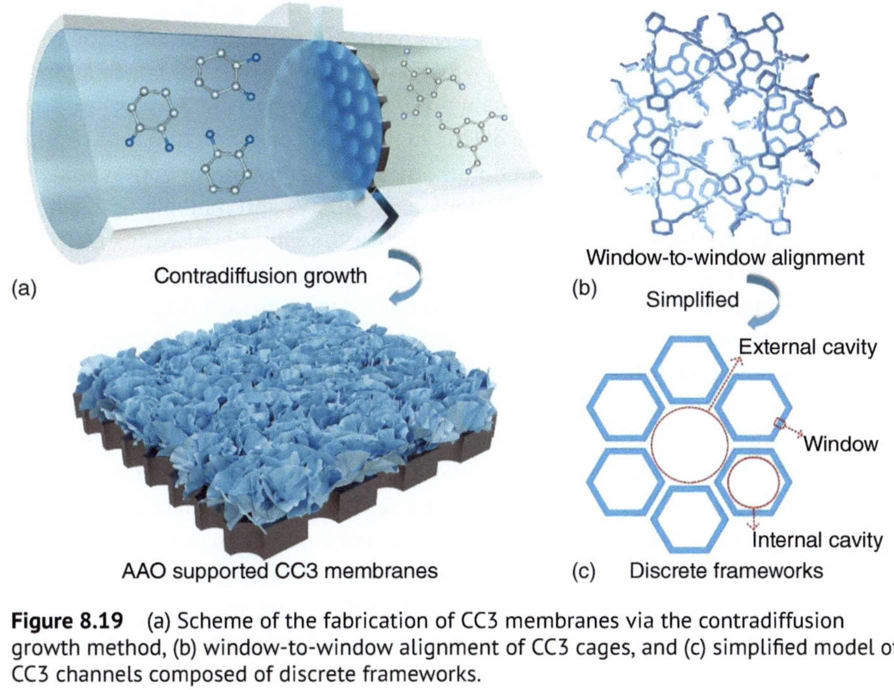

Figure 8.19 (a) Scheme of the fabrication of CC3 membranes via the contradiffusion growth method, (b) window-to-window alignment of CC3 cages, and (c) simplified model of CC3 channels composed of discrete frameworks.

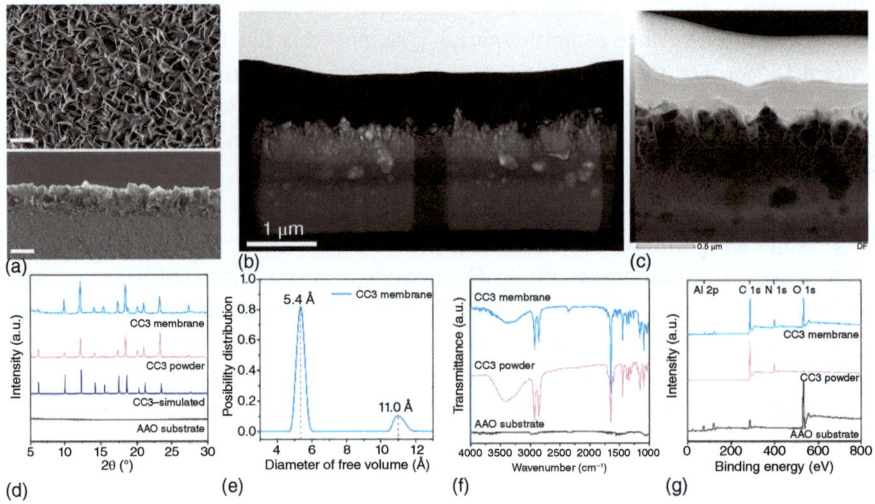

Figure 8.20 (a) SEM images (scale bar 1 μm) of the surface and cross-section of CC3 membranes, (b) TEM image of the CC3 membrane at low magnification, (c) TEM image of the CC3 membrane at high magnification, (d) XRD patterns of the AAO substrate, simulated CC3, CC3 powder, and CC3 membranes, (e) The pore size distribution of CC3 membranes based on PALS measurement, indicating a subnanometer window and a nanometer cavity, (f) FTIR spectra of the AAO substrate, CC3 powder, and CC3 membranes, and (g) XPS spectra of the AAO substrate, CC3 powder, and CC3 membranes.

well-intergrown nanoflakes, confirming the formation of a continuous CC3 layer. In addition, the transmission electron microscopy (TEM) and atomic force microscopy (AFM) images display a uniform layer and ordered microstructure of CC3 on the AAO substrate (Figure 8.20b,c). XRD patterns of the CC3 membranes indicate that CC3 membranes have high crystallinity, which is the same as the simulated patterns (Figure 8.20d). Positron annihilation lifetime spectroscopy (PALS) was used to characterize the pore size distribution of CC3 membranes (Figure 8.20e). The results show that CC3 membranes have two dominant size distributions at approximately 5 and 11 Å, which correspond to subnanometer windows and nanometer cavities, respectively. From the FTIR spectra (Figure 8.20f), the vibrational bands at 1654, 1 448, and 1 160 cm^{-1} are assigned to C=N, CH_2, and C—N, respectively [72, 73]. For the XPS spectra, the featured element (N) of CC3 is confirmed (Figure 8.20g).

The cation transport properties in CC3 membranes were measured using a single laboratory-scale ED cell. The fluxes of monovalent cations of CC3 membranes in the single-ion system are on the same order of magnitude as those of AAO substrates. Simultaneously, CC3 membranes exhibit extremely high ideal ion selectivities for K^+/Mg^{2+} of 1 031, Na^+/Mg^{2+} of 660, and Li^+/Mg^{2+} of 284, which are much higher than those of AAO substrates (Figure 8.21a,b). Then, binary cation flux measurements were carried out. Due to the existence of coions in the binary-ion system, the competition between cations increases, resulting in a decrease in monovalent cation flux. The monovalent ion flux decreases more obviously than that of divalent ions from single-ion to binary-ion systems; thus, the selectivity

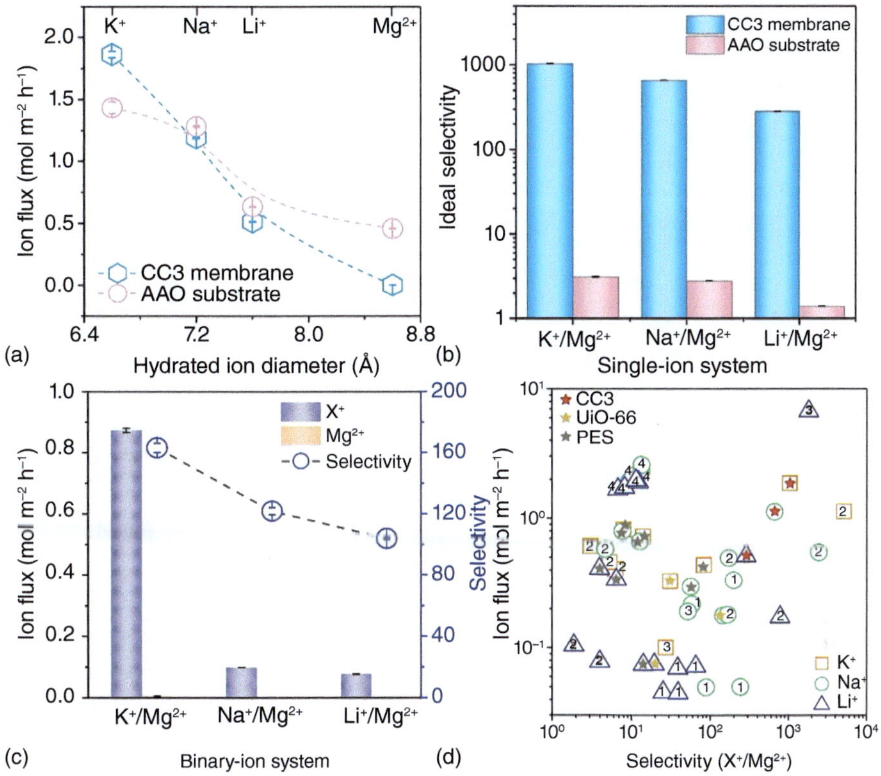

Figure 8.21 (a) Ion fluxes of K^+, Na^+, Li^+, and Mg^{2+} of CC3 membranes and AAO substrates (single-ion system: 0.1 M KCl, 0.1 M NaCl, 0.1 M LiCl, and 0.1 M $MgCl_2$), (b) ideal ion selectivities of CC3 membranes and AAO substrates calculated from single-ion fluxes, (c) real ion selectivities of CC3 membranes under a binary-ion system. (Standard deviation values are derived from three tests of one membrane.), (d) comparison of ion flux and selectivity among membranes of various channel configurations at the subnanometer scale, such as discrete window-cavity structure (CC3 membranes), continuous channels with window-cavity structure (UiO-66 membranes), and etched polymer membranes with aligned channels (PES membranes). K^+ (square), Na^+ (circle), and Li^+ (triangle). CC3 membranes (red star), UiO-66 membranes (yellow star), [81] PES membranes (gray star). [82] UiO-66-NH_2 (1), [45] UiO-66-SO_3H (2), [81] PSS@HKUST-1 (3), [83] Ti-exchanged UiO-66-NH_2 (4) [47].

also decreases. Consequently, the real mono-/di-valent ion selectivities decrease, such as K^+/Mg^{2+} of 163, Na^+/Mg^{2+} of 122, and Li^+/Mg^{2+} of 104 (Figure 8.21 c). Compared to other reported membranes, including MOF channels with continuous window-cavity structures (e.g. UiO-66 membranes) and etched polymer membranes with aligned channels (e.g. PES membranes) under the drive of electrical potential (Figure 8.21d), POC channels with discrete window-cavity structures (e.g. CC3 membranes) show both high ion flux and selectivity.

Molecular dynamics (MD) simulations are employed to explore ion transport mechanisms within CC3 channels. To describe the ion migration process from the external cavity to the internal cavity, a simplified CC3 channel model is adopted (Figure 8.22a). Figure 8.22b exhibits the ion transport routine from one external

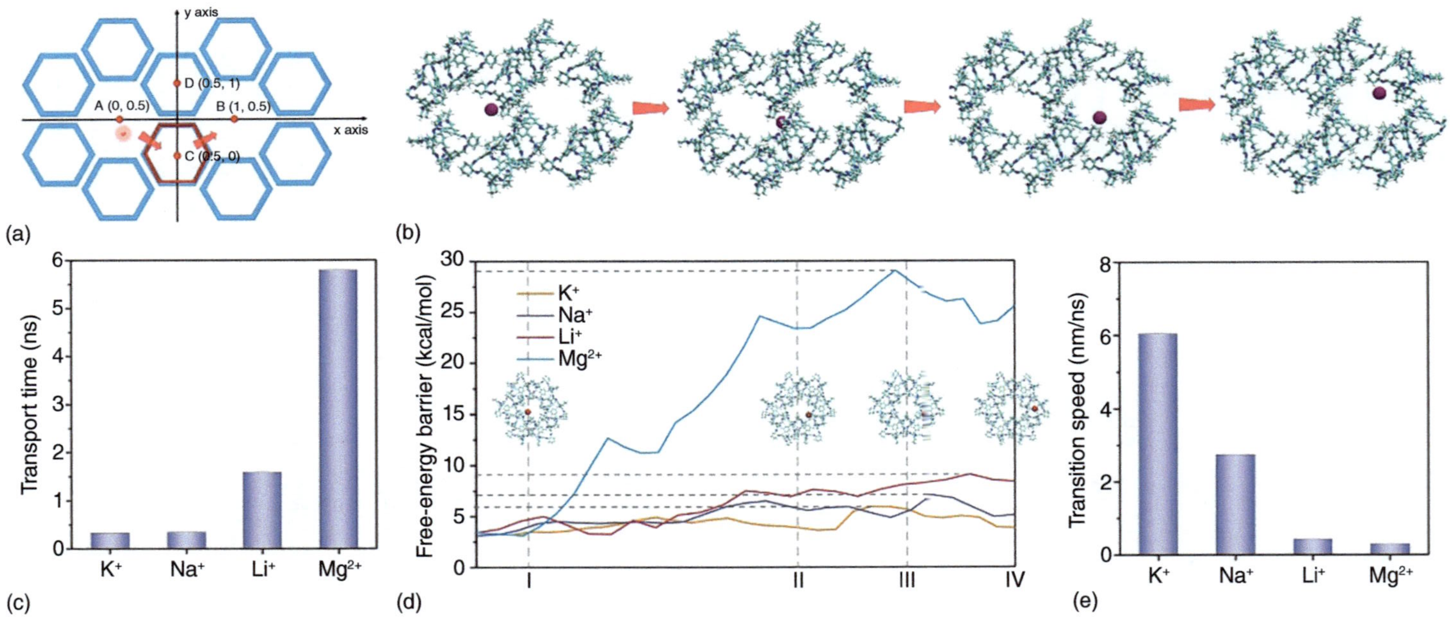

Figure 8.22 (a) Collective variable (CV) coordinates for describing ion migration in CC3 channels. Points A and B are the center of the CC3 external cavity, while C and D are the center of the CC3 internal cavity. The X and Y axes are defined by AB and CD vectors, respectively. The length on the X (Y) axis is normalized by the AB (CD) distance for a stable situation, (b) Schemes of ion transport from one external cavity to another external cavity of CC3 channels, including ion migration from A to C and C to B processes, (c) estimated ion transport time from one external cavity to another external cavity, as shown in b, (d) Ree energy barriers of ion (K^+, Na^+, Li^+, and Mg^{2+}) transport in CC3 channels at four representative positions, and (e) Transition speeds of ions (K^+, Na^+, Li^+, and Mg^{2+}) in CC3 channels.

cavity to another external cavity of CC3 channels, including ion migration from A to C and C to B processes. The corresponding transport times of the four ions are estimated from MD, with a sequence of $K^+ < Na^+ < Li^+ < Mg^{2+}$ (Figure 8.22c). The process of ions passing through the window of the CC3 crystal is further refined. The variation in the free energy barrier when ions pass through the channel is calculated via MD. Four representative positions (I, II, III, IV) during the migration process are selected for the comparison of the energy barriers (Figure 8.22d). From position I to II, ions transport in the external cavity, the energy barriers of the four ions are not significant, and then the energy barriers of monovalent ions (K^+, Na^+, Li^+) increase gradually, while the energy barriers of Mg^{2+} increase sharply. The divalent ion Mg^{2+} has a larger hydrated ion diameter than monovalent ions, leading to a much higher energy barrier for Mg^{2+}. Upon entering (Position II) and leaving (Position IV) the internal cavity, as well as in the internal cavity (Position III), the free energy barrier of ions follows the order of $K^+ < Na^+ < Li^+ < Mg^{2+}$, where K^+ shows the lowest energy barrier (5 kcal/mol) and Mg^{2+} has the highest energy barrier (29 kcal/mol). The calculated ion transition speed in CC3 channels follows the sequence of $K^+ > Na^+ > Li^+ > Mg^{2+}$ (Figure 8.22e, where K^+ moves fastest arising from its lowest free energy barrier during the migration process. MD simulations demonstrate that the excellent ion permselectivity arises from the unique structure of CC3 channels. The subnano-sized windows allow the traverse of monovalent ions with different hydration shells while blocking divalent ions. The discrete CC3 frameworks form ordered and extended external cavities, as ion transport pathways for high ion fluxes.

8.3 Covalent Organic Framework Membranes

8.3.1 Introduction

Covalent organic frameworks (COFs) represent a novel kind of crystalline porous polymer comprising periodically extended and covalently bonded network structures and were first synthesized by Yaghi's group in 2005 [84]. Compared with conventional polymers, COFs have the advantages of predictable chemical compositions, regular pore structure, tunable pore size, and wall nature. COFs have been at the forefront of porous material research in recent years due to their versatile covalent combination of building units, orderly arranged pores, uniform pore size, and high pore density, attracting increasing interest in the fields of membrane-based separation [85–91], energy-storage materials [92–96], optoelectronics [97–100], catalysis [101–103], and many others [104–106] (Figure 8.23).

COFs are synthesized from organic linkers via reversible covalent bond formation. The regular and periodical assembly of these organic linkers by covalent bonds endows COFs with orderly arranged pores, a uniform pore size, and a high pore density. These characteristics also make COFs promising candidates for constructing advanced separation membranes. Moreover, the functionalities of the COF membrane can be specifically tuned through the introduction of various functional sites on the organic linkers to allow a high degree of control over host–guest interactions

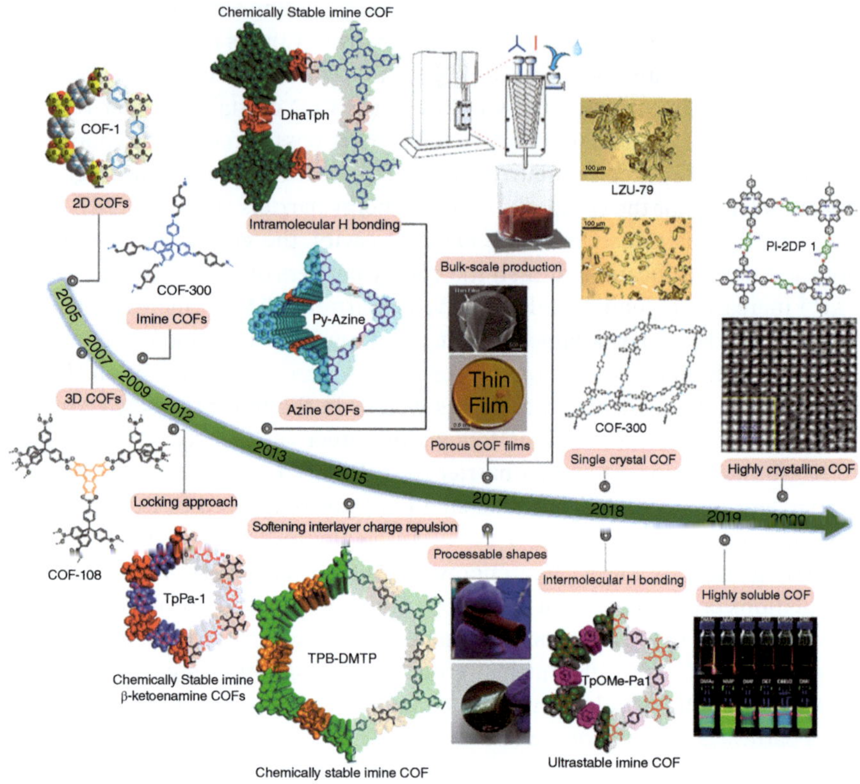

Figure 8.23 Evolution of COF chemistry. Important advances from unstable powder COFs to ultrastable and processable forms. Source: Reprinted with permission from Wang et al. [98]/Reproduced With permission from American Chemical Society. Reprinted with permission from Kandambeth et al. [107]; © 2019, American Chemical Society. Reprinted with permission from Sahabudeen et al. [108]; © 2020, Wiley-VCH.

in an attractive or repulsive way [109]. These superior intrinsic properties of COFs have attracted tremendous attention from the field of membrane technology, making membrane separation one of the key applications for COFs. Recently, many studies have shown that COF membranes also have great potential in highly efficient ion transport and separation. Nonequilibrium molecular dynamics (NEMD) indicate that the in-pore transport of hydrated ions is significantly affected by the polarity of COFs [110]. The rejection of ions can be enhanced without sacrificing the pore size by increasing the in-pore transport effect. Moreover, the results of experiments proved that COFs have high-efficiency separation properties between monovalent and multivalent cations by introducing noncovalent interactions between hydrated ions and the pore wall of COFs [111, 112].

8.3.2 Design Strategies of the COF Structure

COFs are a kind of crystalline porous organic material connected by covalent bonds. In principle, it is a thermodynamically advantageous process for the connection of organic building units to obtain amorphous solids. For example, most disordered conventional polymers can be easily synthesized by the connection of strong covalent bonds between their building units. However, some pore structures can easily form using rigid organic building units or by introducing interactions between polymer chains. However, these pores lack the short- or long-range ordered structure that is necessary for crystalline materials, and the formed pores are also disordered and irregular. To obtain crystalline organic porous materials with regular pore structures, it is necessary to ensure that the covalent bonds formed by the organic building units are reversible and that the reaction rate can be controlled within a certain range so that the building units undergo a self-healing defect process to form skeleton structures by polymerization. Therefore, dynamic covalent chemistry (DCC) is widely used to study the synthesis of crystalline covalent organic frame materials. Since the first covalent organic frame material was synthesized by Yaghi's group in 2005 [84], a number of organic chemical reactions have been developed for the synthesis of covalent organic frame materials. According to the difference in linker groups, covalent organic frame materials with different pore structure types can be formed, including boroxine, boronate ester, imine, imide, hydrazone, azine, triazine, phenazine, and so on (Figure 8.24) [107]. The pore structure of COFs can be well designed, and precise regulation benefits from abundant building units, diverse topological connection types, and tunable channel wall properties.

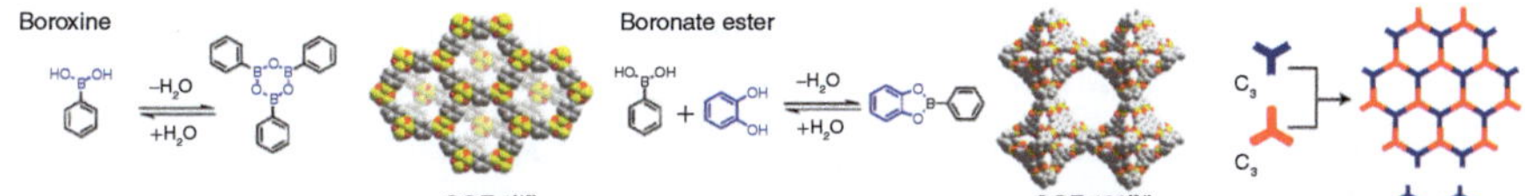

Figure 8.24 Different types of reversible organic reactions used for COF construction. Source: Reproduced with permission from Kandambeth et al. [107]; © 2019, American Chemical Society.

8.3.2.1 Pore Structure Design

The regular pore structure is one of the most fascinating features of COFs, especially for membrane separation. The porous structures of 2D and 3D COFs are clearly different. 2D COFs have open 1D channels that originate from the stacking of 2D layers [113]. These channels are independent of each other and can generate isolated nanospaces with well-defined pore structures. It is a unique characteristic and different from traditional porous materials such as zeolite, active carbons, and amorphous porous polymers. The channel size of COFs can be effectively regulated by the reasonable design of the building units. Most COFs have apertures in the range of 0.5–3 nm, which is suitable for desalination, pervaporation, water treatment, and nanofiltration [109]. Dey et al. [86] constructed a series of β-ketoenamine-linked COFs with pore sizes of 1.4–2.6 nm by changing the structure and length of linear building units, as shown in Figure 8.25a. Mu et al. [116] synthesized COFs with pore sizes up to 10 nm mesopores by further extending the length of linearly building units and minimizing the interlayer nonbonded energy. In addition, the pore size of COFs can also be effectively regulated by introducing side chains or functional groups by the post-modification method. For example, Nagai et al. [114] regulate pore size using different contents of easily functionalized building units to synthesize COF-5 with different azides through a combination of condensation reactions and click chemistry (Figure 8.25a). Moreover, the pore size of COFs can also be decreased by modulating the stacking mode of 2D COF polygonal layers. Li et al. [115] synthesized two free-standing COF membranes with different stacking modes using the IP method by changing the number of phenolic hydroxyl groups on the aldehyde monomer. The pore size of the prepared COF membrane with the AA stacking mode was 1.1 nm, while the pore size of the AB stacking mode decreased to 0.6 nm (Figure 8.25c).

Apart from the size and shape of pores in COFs being well tailored by the above-mentioned method, the introduction of different kinds of pores into a COF is also accessible. For example, Dalapati [117] et al. synthesized TPE-Ph COF with a dual-pore Kagome lattice by the solvothermal condensation reaction of TPE-cored boronic acids (TPEBA) and 1,2,4,5-tetrahydroxybenzene (THB) (Figure 8.26a). Zhou [118] et al. reported the one-step construction of a dual-pore COF bearing two different kinds of pores, in which hexagonal pores (26.9 Å) and triangle pores (7.1 Å) are alternately and periodically arranged (Figure 8.26b). Pang et al. [119] modulated the topological connection mode through reasonable structural design, which changed the substituent groups. When the substituent is a hydroxyl group, it will form a dual-pore COF. However, if the substituent is an oxyethyl group (monomer DETA) or an oxypropyl group (monomer DBTA), the synthesized COFs have rhomboid single pores (Figure 8.26c).

8.3.2.2 Pore Surface Engineering

Due to the variety and designability of the building units, the channel wall properties of COFs can be precisely regulated according to the actual demands, including charge properties and hydrophilicity. For instance, Ma et al. [120] synthesized cationic COFs with high thermal and chemical stability by introducing

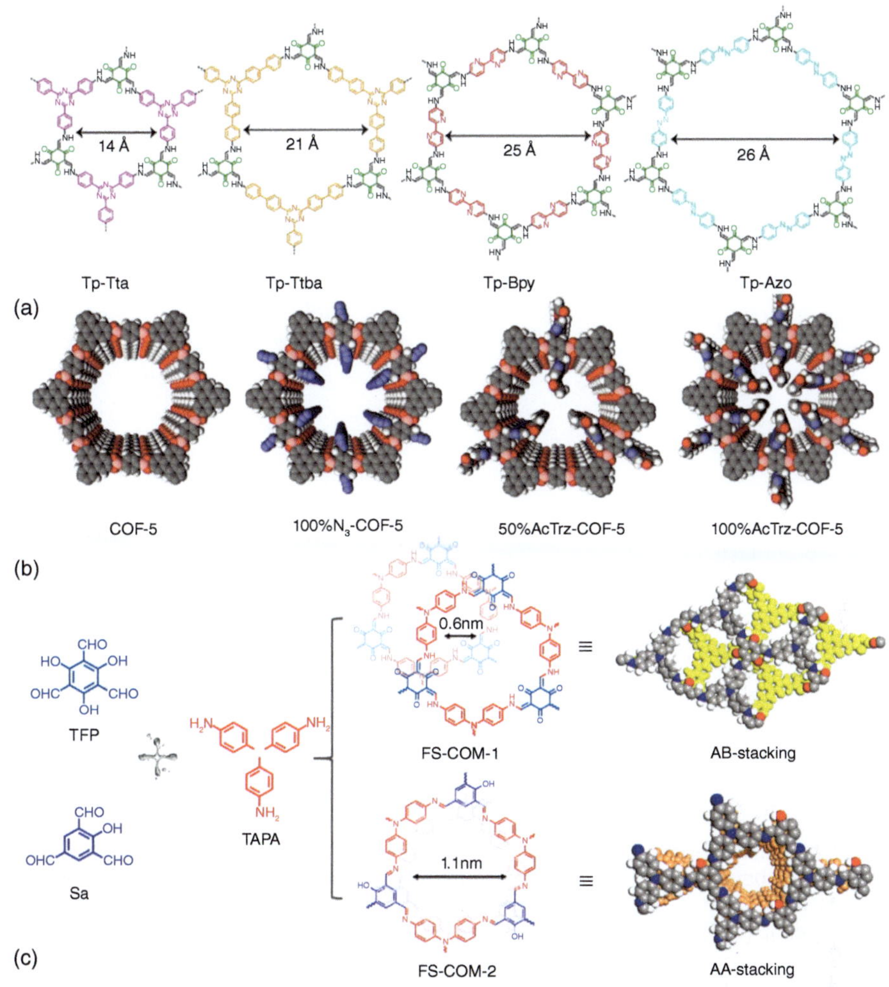

Figure 8.25 (a) Regulating the pore size of COFs by changing the structure and length of linearly building units, (b) regulating the pore size of COFs by introducing side chains or functional groups, (c) regulating the pore size of COFs by modulating the stacking mode of 2D COF layers. Source: Reprinted with permission from Dey et al. [86]; © 2017, American Chemical Society. Reprinted with permission from Refs. [114, 115]; © 2011, 2020, Macmillan Publishers Limited.

a cationic ethidium bromide (EB) monomer (Figure 8.27a). The prepared porous cationic framework shows high proton conductivity. More importantly, the proton conductivity can be enhanced and exceed that of previously reported porous organic materials by introducing $PW_{12}O_{40}{}^{3-}$ into this porous cationic framework. Mitra et al. [123] constructed guanidinium-based cationic covalent organic nanosheets with self-exfoliating properties via Schiff base condensation between triaminoguanidinium chloride and 1,3,5-triformylphloroglucinol monomers. Peng et at. [121] synthesized two sulfonated COFs possessing one-dimensional nanoporous channels using diamine monomers with different sulfonic acid groups.

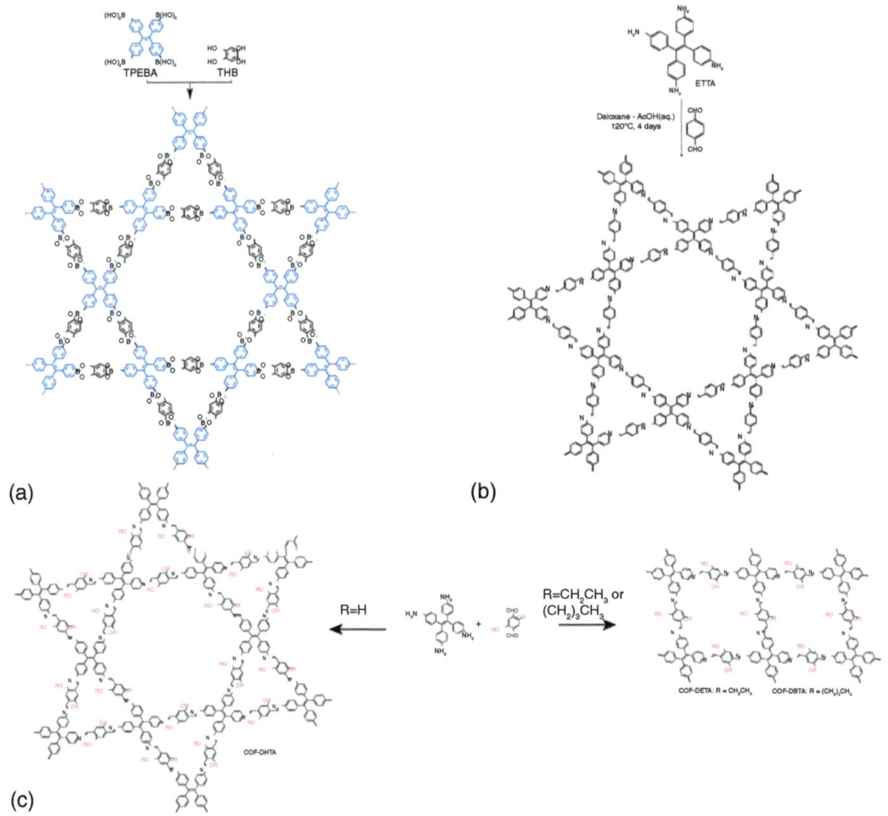

Figure 8.26 (a) Dual-pore COF of boronate ester type, (b) dual-pore COF of imine type, and (c) modulation between dual-pore and single-pore COFs. Source: Reprinted with permission from Refs. [117, 118]; © 2016, 2014, American Chemical Society. Reprinted with permission from Pang et al. [119]; © 2017, The Royal Society of Chemistry.

These sulfonated COFs exhibit high intrinsic proton conductivity, reaching as high as 3.96×10^{-2} S/cm with long-term stability (Figure 8.27b). In addition, Lu et al. [106] constructed a 3D carboxy-functionalized COF through postsynthetic modification of a hydroxy-functionalized COF and explored its application in the selective extraction of lanthanide ions.

The hydrophilicity of the pore is another important parameter for the properties of COFs, especially for water-related applications such as adsorption and membrane separation. For example, Zhang et al. [124] studied COFs with hydrophilic functional groups exhibiting higher water flux than their hydrophobic counterparts of similar aperture sizes using MD simulations and proved the preferential interaction between water and hydrophilic groups resulting in a larger water density at the aperture of the hydrophilic membranes. Zhao et al. [122] reported a strategy to fabricate membranes composed of vertically aligned channels with a hydrophilicity gradient by engineering defects in COF films by the removal of imine bonds. The prepared membrane shows nearly triple the water flux of the state-of-the-art distillation membrane for desalination (Figure 8.27c).

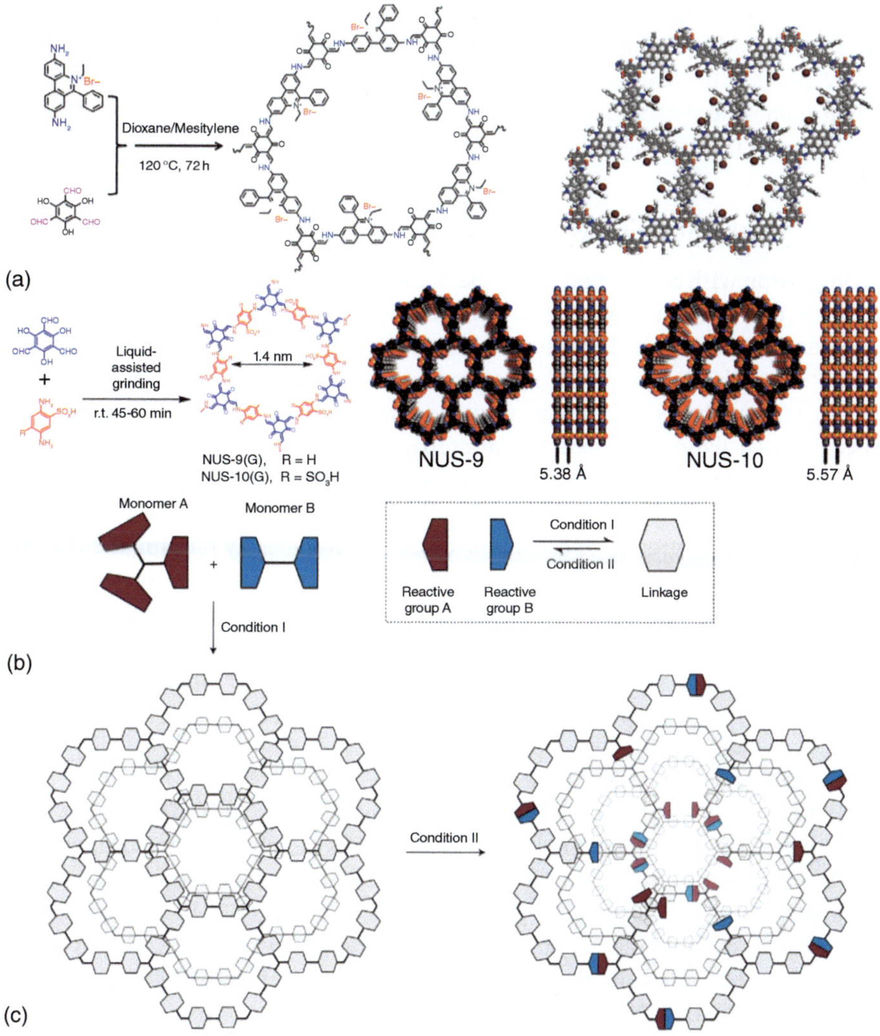

Figure 8.27 (a) Cationic COFs with positively charged nanochannels, (b) COFs with negatively charged nanochannels, and (c) COFs with hydrophilic gradient nanochannels. Source: Reprinted with permission from Refs. [120, 121]; © 2016, 2016, American Chemical Society. Reprinted with permission from Zhao et al. [122]; © 2021, Macmillan Publishers Limited.

In addition, Banerjee's group introduced strong interlayer C—H···N hydrogen bonding among the individual layers, which provides significant steric hindrance and a hydrophobic environment around the imine (—C=N—) bonds, thus preventing their hydrolysis in harsh environments, including H_2SO_4 (18 mol/l), HCl (12 mol/l), and NaOH (9 mol/l) [125]. Bing et al. [112] synthesized COF nanochannels with orderly aligned lithiophilic oligoethers to create ideal membranes for Li^+ extraction. The COF active layer with oligoethers in the pore channels affords a high Li^+/Mg^{2+} separation factor of 64, making it top ranked in terms of extraction efficiency. In our group, we constructed an ultrathin COF membrane with hydrogen bonding sites, which exhibited high monovalent cation permeation rates of 0.1–0.2 mol m^{-2} h^{-1} and extremely low multivalent cation permeabilities, leading to high monovalent over divalent ion selectivities for K^+/Mg^{2+} of ~765, Na^+/Mg^{2+} of ~680, and Li^+/Mg^{2+} of ~217 [111].

8.3.3 Preparation of COF Membranes

In comparison with conventional polymers, COFs have poor solution processability. There are few solvents to dissolve them, so it is a great challenge to prepare COF membranes. Until now, the preparation methods of COF membranes mainly include the following: (i) COF membranes can be fabricated via layer-by-layer stacking of nanosheets or monolayers by the exfoliation of bulk materials [126, 127]. (ii) COF-based hybrid membranes are prepared by integrating COF particles into common polymers [85, 95]. (iii) IP methodology was adapted to prepare thin continuous COF membranes [89, 128, 129]. (iv) COF membranes can also be prepared via in situ growth of continuous COF membranes by refining the current COF synthesis methods [88]. In addition, there are also some other methods used to prepare COF membranes, including the Langmuir–Blodgett (LB) method [129] and chemical vapor deposition (CVD) [130, 131].

Herein, we mainly describe the IP method to prepare three kinds of continuous COF membranes for ion and ionic compound separation [111, 132].

8.3.3.1 COF Membrane with Sub-2-nm Channels

An IP method was employed to fabricate the nanoporous $TpBDMe_2$ membranes. The 1,3,5-triformylphloroglucinol (Tp) monomer was dissolved in dichloromethane (DCM) solution. The 2,6-dimethylbenzidine ($BDMe_2$) monomer was dissolved in water with the assistance of p-toluene sulfonic acid (PTSA) (Figure 8.28a). The $TpBDMe_2$ membrane was formed at the interface between DCM and water after undergoing a reversible Schiff-base reaction and irreversible enol-to-keto tautomerism (Figure 8.28b,c).

As shown in Figure 8.29a, a defect-free $TpBDMe_2$ layer with a thickness of ~20 nm formed on the surface of the AAO substrate. The high-resolution transmission electron microscopy (HRTEM) imaging shows a highly ordered arrangement with an independent diffraction direction, confirming the high crystallinity of the as-prepared COF. The distance between adjacent lattice fringes is 3.4 Å, which is consistent with the theoretical interlamellar spacing of $TpBDMe_2$. The pore size distributions reveal that $TpBDMe_2$ has a uniform pore size of 1.4 nm (Figure 8.29b), which is in accordance with the value reported in the literature [127]. To demonstrate the hydrogen bonding interaction between the $TpBDMe_2$ polar channel wall and water molecules, FTIR spectra of $TpBDMe_2$ at different temperatures were measured under air and N_2 atmospheres (Figure 8.29c). An obvious shift in the C–N bands under air conditions coupled with the absence of a shift in the N_2 atmosphere reveals the formation of hydrogen bonds between the –NH moiety of the $TpBDMe_2$ channel and water molecules. For the stretching vibration peak of

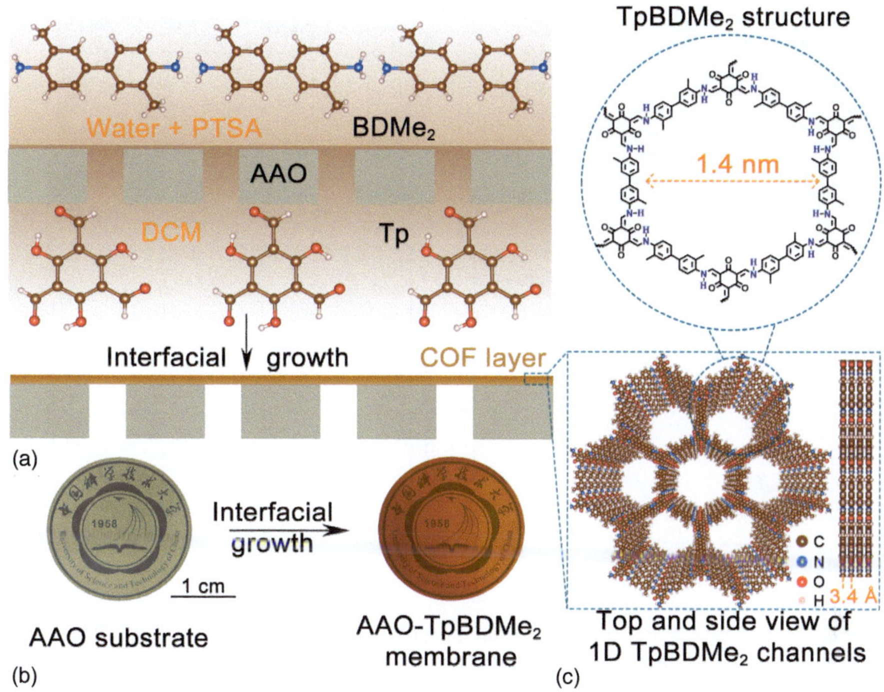

Figure 8.28 (a) Schematic of the fabrication of TpBDMe$_2$ membranes, (b) photographs of AAO substrate and AAO-supported TpBDMe$_2$ membranes with a diameter of ~2 cm, and (c) the structure of TpBDMe$_2$ membranes with 1D nanochannels of 1.4 nm and hydrogen bonding sites (–NH) on the channel wall. Source: Reproduced with permission from Sheng et al. [111]; © 2021, Wiley-VCH.

C–N at 1250–1260 cm^{-1}, the electron density of C–N decreases with the formation of hydrogen bonds between nitrogen (hydrogen bonding acceptor) of C–N and hydrogen (hydrogen bonding donor) of water molecules, and the peak shifts to a high wavenumber. As the temperature increases, the hydrogen bonding interactions weaken gradually, and the peak shifts to a low wavenumber.

The ion transport properties in TpBDMe$_2$ membrane nanochannels were investigated via concentration-driven diffusion. Aqueous solutions including 0.1 M LiCl, NaCl, KCl, $MgCl_2$, $CaCl_2$, $NiCl_2$, $CuCl_2$, $LaCl_3$, and $CeCl_3$ are used as the feed solution, while the permeate side is deionized water. The transport of ions was monitored by recording the ion conductivity of the permeate side. The concentration of various ions in the permeate solution was detected by inductively coupled plasma–optical emission spectrometry (ICP–OES). For TpBDMe$_2$ membranes, the ion conductivities of the permeate side display a linear increase. The permeation rates follow the sequence of $K^+ > Na^+ > Li^+ >> Ca^{2+}$, Cu^{2+}, Mg^{2+}, Ni^{2+}, La^{3+}, and Ce^{3+}, indicating a sharp cut-off channel size of ~8 Å (Figure 8.30a), which is much smaller than the measured TpBDMe$_2$ membrane channel size of ~1.4 nm. The calculated diffusion coefficients of metal ions in the COF channels are lower than those in the AAO channel, which are quite close to those of the bulk solution (Figure 8.30b).

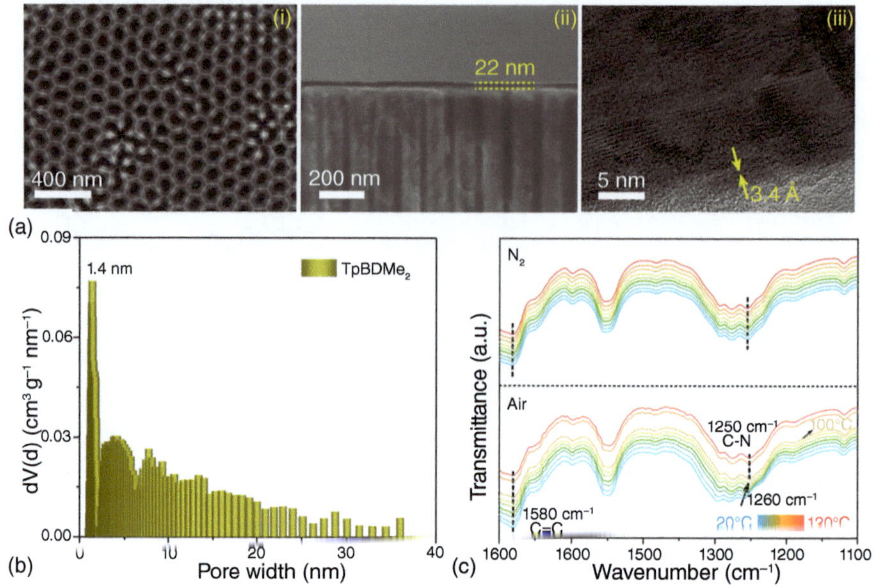

Figure 8.29 (a) Surface and cross-sectional morphologies of AAO-supported $TpBDMe_2$ membranes with a COF layer of ~20 nm; HRTEM of $TpBDMe_2$ membranes showing a distance of 3.4 Å between the $\pi-\pi$ stacking layers, (b) pore size distribution of $TpBDMe_2$ membranes, and (c) FTIR spectra of $TpBDMe_2$ membranes under N_2 and air atmospheres at different temperatures revealing the formation of hydrogen bonds between the –NH moiety of the $TpBDMe_2$ channel and water molecules. Source: Sheng et al. [111]/Reproduced with permission from John Wiley & Sons.

This finding indicates that there should be certain interactions, such as electrostatic interactions and hydrogen bonding, between the COF channel wall and metal ions. The $TpBDMe_2$ membrane displays a distinctive selectivity for monovalent cations over multivalent cations. Specifically, $TpBDMe_2$ membranes exhibit selectivities for K^+/Mg^{2+} of ~765, Na^+/Mg^{2+} of ~680, and Li^+/Mg^{2+} of ~217 and maintain a high monovalent cation permeation rate of 0.1–0.2 mol m^{-2} h^{-1}, much higher than those of previously reported membranes (Figure 8.30c). Furthermore, for the binary-ion system, the permeation rates of monovalent cations decrease slightly, while the selectivities for monovalent cations over divalent cations drop to ~213, ~96, and 36 for K^+/Mg^{2+}, Na^+/Mg^{2+}, and Li^+/Mg^{2+}, respectively (Figure 8.30d).

To gain further insight into the mechanism of ion transport through the TpBDMe channels, MD simulations and density functional theory (DFT) calculations were used to analyze the interaction between hydrated metal ions and hydrogen bonding sites in the confined nanochannels. Based on the characterization of FTIR, we confirmed the existence of hydrogen bonding sites in TpBDMe2 channels, which can form hydrogen bonding interactions with water molecules. Ions can attract surrounding water molecules, forming hydration shells due to the electrostatic interaction between charged ions and water dipoles in bulk solution [133]. Therefore, ions with diverse hydration shells can form hydrogen bonds with the

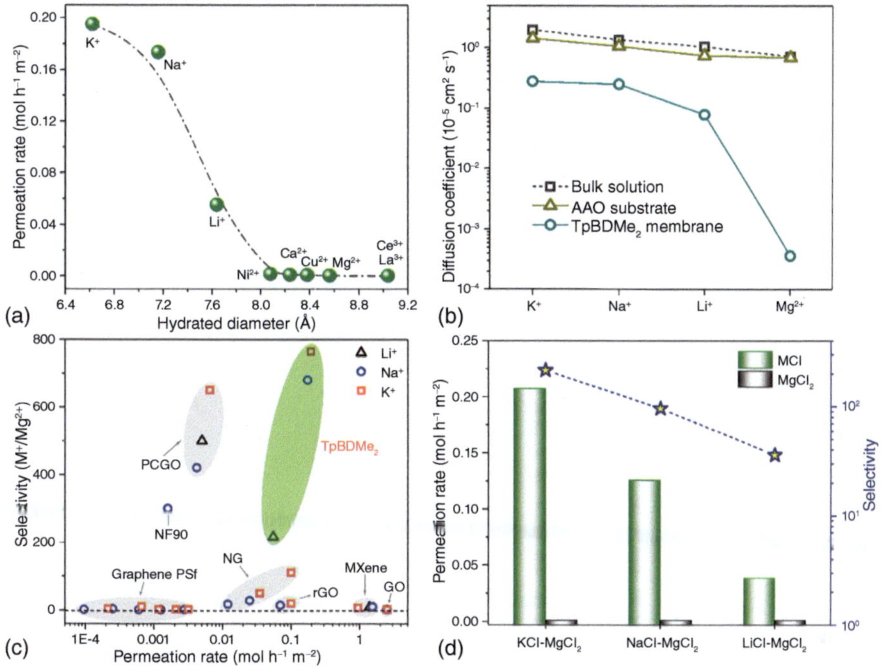

Figure 8.30 (a) Permeation rates of metal ions as a function of hydrated ion diameters, (b) the calculated diffusion coefficients of metal ions in the COF channels, AAO channels, and bulk solution, (c) Ion permeation rate versus selectivity (K^+/Mg^{2+}, Na^+/Mg^{2+}, and Li^+/Mg^{2+}) of reported nanoporous membranes under concentration gradient driving, and (d) The mono/divalent cation permeation rate and selectivity in binary ion solution systems. Source: Reproduced with permission from Sheng et al. [111]; © 2021, Wiley-VCH.

$TpBDMe_2$ channel wall. Water molecules of the ion hydration shell prefer to form hydrogen bonds with the hydrogen bonding sites (−NH) distributed along the nanochannels of $TpBDMe_2$ membranes. Thus, four hydrated metal ions (K^+, Na^+, Li^+, and Mg^{2+}) were employed to investigate the effects of hydrogen bonds on ion transport properties in COF nanochannels. As shown in Figure 8.31a, the hydrogen bonding energies of the four metal ions with the channel wall follow the order of $K^+ < Na^+ < Li^+ < Mg^{2+}$ according to the MD simulations. The bare ion diameters of the studied ions follow the sequence of $Li^+ < Mg^{2+} < Na^+ < K^+$. Considering the charge of each ion, the calculated charge density follows the order of $K^+ < Na^+ < Li^+ < Mg^{2+}$. Consequently, Mg^{2+} can bind the oxygen of H_2O in the hydration shell more tightly due to the stronger electrostatic attraction, making the hydrogen of H_2O more positively charged, which leads to higher bonding energy with −NH sites of the $TpBDMe_2$ nanochannels. The energy barrier value of K^+ is the lowest, whereas the value of Mg^{2+} is the highest, which displays the same trend as the hydrogen bonding energy (Figure 8.31b).

Based on the difference in hydrogen bonding interactions between the ions and COF channels, the diffusion coefficient of K^+ is 9.76×10^{-6} m^2 s^{-1}, nearly an order

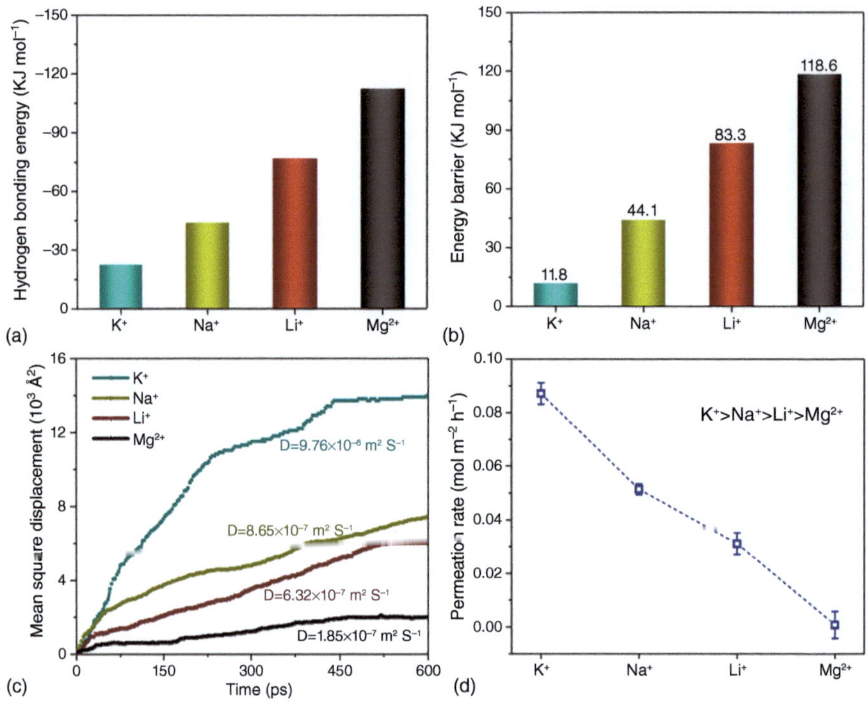

Figure 8.31 (a) The hydrogen bonding energy between hydrated metal ions and the channel wall, (b) the energy barrier for metal ion transport through $TpBDMe_2$ nanochannels, And (c, d) the calculated diffusion coefficients and permeation rates of metal ions according to MD simulation. Source: Reproduced with permission from Sheng et al. [111]; © 2021, Wiley-VCH.

of magnitude higher than those of Na^+ and Li^+ and two orders of magnitude higher than that of Mg^{2+} (1.85×10^{-7} m^2/s) over a period (Figure 8.31c). As a result, Mg^{2+} is transported much slowly in the COF channel than K^+ since the hydrogen bonding interaction of divalent cations is stronger than that of monovalent cations. The simulated ion permeation rates are shown in Figure 8.31d, with the order of $K^+ > Na^+ > Li^+ > Mg^{2+}$, leading to a selectivity for K^+/Mg^{2+} of 115, Na^+/Mg^{2+} of 68, and Li^+/Mg^{2+} of 41, which show a similar trend to the experimental results.

8.3.3.2 Cationic COF Membrane

Wastewater containing organic dyes and a high salt content has posed a serious challenge to global water security. Membrane separation technologies have been widely applied in dye wastewater treatment due to their outstanding features, including high efficiency, energy savings, no phase transitions, and environmentally friendly operations. However, traditional NF membranes show high rejections of both dyes and salts because of their dense structure, limiting their applications in dye/salt separation [134–136]. Subsequently, loose NF membranes have been developed to achieve a low rejection of salts [137–140]. Although loose NF membranes show noteworthy separation performance for mixtures of dyes and salts, the trade-off

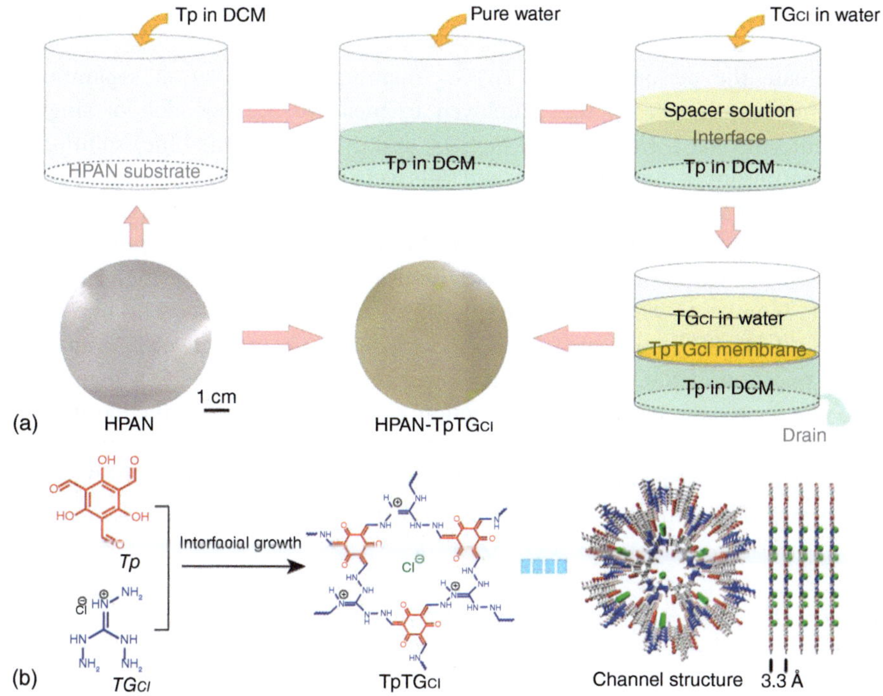

Figure 8.32 (a) Schematic diagram of the fabrication procedure of COF membranes via an interfacial growth and transferring strategy, (b) the synthetic diagram and channel structure of $TpTG_{Cl}$ membranes with eclipsed stacking mode and $\pi-\pi$ stacking between consecutive 2D layers. Source: Sheng et al. [132]/Reproduced with permission from Elsevier.

effect between dye/salt selectivity and water permeability restricts the widespread application of NF membranes. More importantly, the irregular membrane structure makes it difficult to precisely control the channel size and surface chemistry for the mechanistic study of dye/salt separation.

A cationic COF membrane was prepared using an IP method, followed by transfer onto the surface of a hydrolyzed polyacrylonitrile (HPAN) substrate. As shown in Figure 8.32a, the Tp monomer was dissolved in DCM solution (0.75 mmol/l) and used as the organic phase, while the aqueous phase dissolved the TG_{Cl} monomer (3.75 mmol/l). First, the HPAN substrate was fixed in the bottle of the device. Twenty milliliters of the Tp solution were poured into the device, and then 10 ml of pure water was added to the surface of the Tp solution and acted as a spacer solution. After 30 minutes, 10 ml of TG_{Cl} aqueous solution was dropped slowly on top of the spacer solution. The whole system was kept for 72 hours at room temperature under static conditions. The $TpTG_{Cl}$ nanoporous membrane was formed at the interface between the organic and aqueous phases. The aqueous phase solution was removed from the top of the device, and the organic phase solution was drained from the bottom of the device. The formed $TpTG_{Cl}$ membrane was transferred onto the surface of the HPAN porous substrate. The synthetic diagram and channel structure of the prepared $TpTG_{Cl}$ nanoporous membrane are shown in Figure 8.32b. $TpTG_{Cl}$

has an eclipsed stacking mode and π–π stacking distance between consecutive two-dimensional (2D) layers.

To evaluate the performance of $TpTG_{Cl}$ membranes for dye/salt separation, a dead-end filtration setup was employed to measure the rejection of single-component salt (NaCl, Na_2SO_4, and $MgSO_4$) and dye (methylene blue) solutions. The concentrations of methylene blue and salts were determined by UV–vis absorption spectra and a conductivity meter, respectively. The UV–vis absorption spectra of methylene blue solution before and after filtration are shown in Figure 8.33a.

As shown in Figure 8.33b, $TpTG_{Cl}$ membranes exhibit a high rejection rate for methylene blue up to 99.5%, and extremely low rejection for salts. The rejection rates of NaCl, Na_2SO_4, and $MgSO_4$ are 3.2%, 2.6%, and 9.4%, respectively. Compared with the rejection of NaCl, the relatively lower rejection rate of Na_2SO_4 is mainly due to the increased electrostatic attraction between SO_4^{2-} and the positively charged nanochannels of $TpTG_{Cl}$ membranes. The higher rejection rate of $MgSO_4$ is due to the stronger electrostatic repulsion between Mg^{2+} and the positively charged nanochannel. To confirm the repeatability of the prepared $TpTG_{Cl}$ membrane,

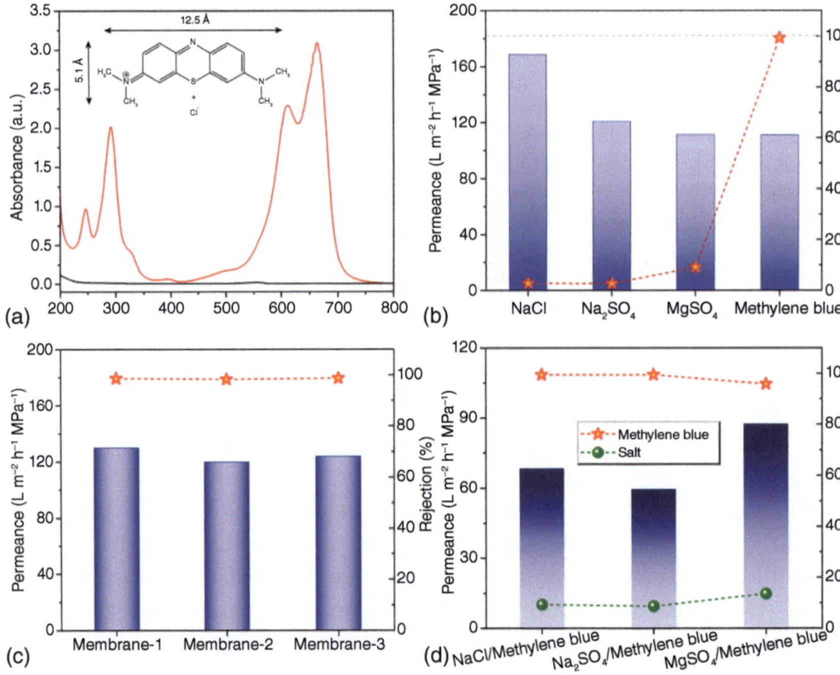

Figure 8.33 (a) UV–vis absorption spectra of methylene blue solution before (red) and after (black) filtration, (b) the rejection rates of single component methylene blue and different salts (NaCl, Na_2SO_4, and $MgSO_4$), (c) reproduced results of three individual $TpTG_{Cl}$ membranes for the rejection of methylene blue, and (d) the rejection rates of methylene blue and salt mixed solutions (NaCl/methylene blue, Na_2SO_4/methylene blue, and $MgSO_4$/methylene blue). Source: Reproduced with permission from Sheng et al. [132]; © 2022, Elsevier B.V.

three individual $TpTG_{Cl}$ membranes prepared under the same conditions were used to conduct methylene blue filtration experiments. As shown in Figure 8.33c, the water flux and methylene blue rejection rate changed slightly, which indicates the good repeatability of the $TpTG_{Cl}$ membrane. To further evaluate the dye/salt separation properties of the $TpTG_{Cl}$ membrane, filtration experiments were conducted using methylene blue/salt (NaCl/methylene blue, Na_2SO_4/methylene blue, and $MgSO_4$/methylene blue) mixed solutions. The results showed that the methylene blue rejection rates for the three mixed solutions were higher than 96%, while the rejection rates of NaCl, Na_2SO_4, and $MgSO_4$ were 9.3%, 8.7%, and 13.6%, respectively (Figure 8.33d). The high dye/salt separation performance for mixed solutions further proves that the $TpTG_{Cl}$ membrane has good potential in dye/salt separation applications.

The zeta potential of the $TpTG_{Cl}$ membrane shows the positively charged property of the nanochannel (Figure 8.34a). To further explore the mechanism of dye rejection, four kinds of dyes of different charges and molecular sizes were used to conduct filtration experiments (Table 8.1). As shown in Figure 8.34b, $TpTG_{Cl}$ membranes show high rejection rates of over 92% for cationic dyes, including methylene blue

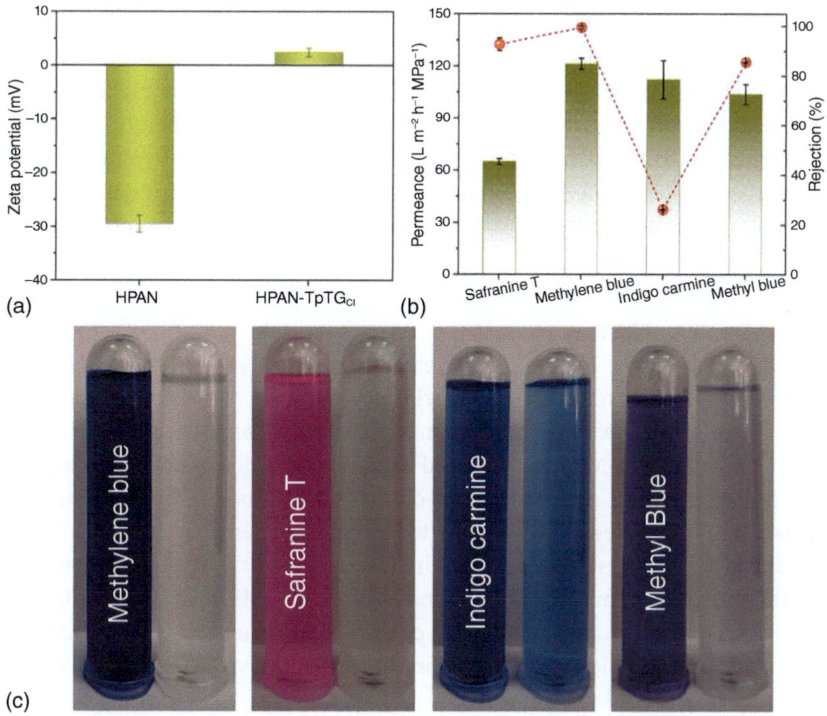

Figure 8.34 (a) Zeta potential of HPAN substrate and $TpTG_{Cl}$ membranes supported by HPAN, (b) the rejection rates of dyes with different molecular sizes and charges, and (c) digital figures of methylene blue, safranine T, indigo carmine, and methyl blue before (left) and after (right) the filtration of $TpTG_{Cl}$ membranes. Source: Reproduced with permission from Sheng et al. [132]; © 2022, Elsevier B.V.

Table 8.1 Properties of the studied dye molecules.

Dye molecule	Dye type	λ_{max} (nm)	MW	Molecular size
Methylene blue	Cationic	664	319.1	1.25 nm × 0.51 nm
Safranine T	Cationic	553	350.9	1.13 nm × 0.84 nm
Indigo carmine	Anionic	609	466.4	1.71 nm × 0.57 nm
Methyl blue	Anionic	580	799.8	2.36 nm × 1.74 nm

Note: The molecular sizes of methylene blue and methyl blue are quoted from reference [88]. The molecular sizes of other dyes were measured by Materials Studio 2017. Structure optimization is performed using the Gaussian 16 program at the B3LYP/6-311G(d) level and B3LYP/6-311+G(d) level for Safranine T and indigo carmine, respectively.
Source: Reprinted with permission from Sheng et al. [132]; © 2022, Elsevier B.V.

and safranine T. Compared with cationic dyes, the rejection rate of indigo carmine with negative charges but a larger molecular size dramatically decreases to 26%, suggesting that the mechanism of methylene blue rejection is governed by electrostatic repulsion. As a comparison, the rejection rate of negatively charged methyl blue increases to 86%. The markedly increased rejection rate is because the molecular size of methyl blue is larger than the channel size of $TpTG_{Cl}$. These results illustrate that the mechanism of dye rejection is dominated by size sieving when the molecular size of the dye is larger than the channel size. Figure 8.34c shows the digital figures of methylene blue, safranine T, indigo carmine, and methyl blue before and after filtration, revealing that $TpTG_{Cl}$ membranes have high rejection for dyes of positive charges or larger molecular size. The concentrations of safranine T, indigo carmine, and methyl blue were determined by the UV–vis absorption spectra of standard dye solutions.

8.3.3.3 Self-Standing COF Membrane

Acid recovery has drawn intensive interest due to the tremendous consumption of acidic chemicals in various industrial applications, which generate large amounts of acidic wastewater and inevitably result in serious environmental pollution [125, 141–144]. Membrane-based separation is considered an alternative for pollutant removal and resource recovery due to its energy-efficient and process-sustainable features [145–147]. Anion exchange membrane (AEM)-based diffusion dialysis has attracted attention in the recovery of acidic compounds. However, the transport of protons (H^+) in conventional AEMs via diffusion dialysis is a low-level efficiency process due to the electrostatic repulsion of AEM toward H^+ [148]. In addition, the low selectivity for H^+ over monovalent cations restricts the precise recovery of acid.

Hence, a self-standing COF (TpTAPA) membrane with high acid stability was prepared by the IP of aldehyde and amine monomers at the interface (Figure 8.35a). The TpTAPA membrane exhibits good crystallinity with an ordered lattice fringe and independent diffraction direction according to the HRTEM image. Subsequently, the nanochannels of the TpTAPA membrane were decorated with charges by the

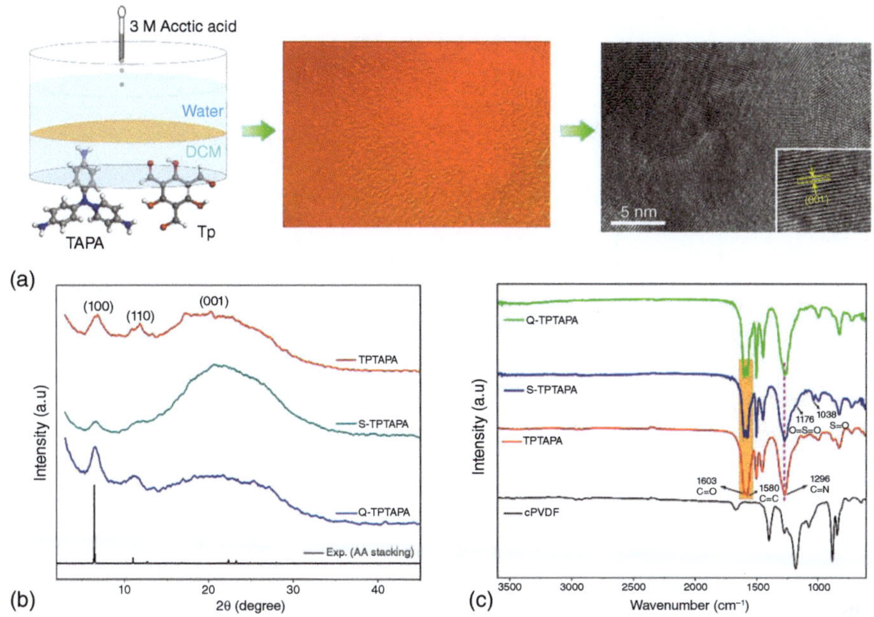

Figure 8.35 (a) Synthetic scheme of the self-standing TpTAPA membrane, (b) XRD patterns of TpTAPA, S-TpTAPA, and Q-TpTAPA membranes, and (C) FTIR spectra of TpTAPA, S-TpTAPA, and Q-TpTAPA membranes.

post-functionalization strategy. The XRD patterns of the TpTAPA membrane show an intense peak at ~6.5°, which corresponds to the (100) plane with a theoretical pore size of ~1.36 nm (Figure 8.35b). The minor peak at ~11.3° is attributed to the (110) plane, and the broad peak at ~22.4° is assigned to the (001) plane representing π–π stacking between consecutive 2D layers. The experimental XRD pattern is in good agreement with the simulated pattern of the AA stacking mode. In addition, the XRD patterns of the S-TpTAPA and Q-TpTAPA membranes with charge modification show no obvious changes, which means that the functionalization process has no effect on the crystallinity of the TpTAPA membrane. FTIR spectroscopy was employed to identify the chemical structure of TpTAPA derivative membranes (Figure 8.35c). The bands at 1603, 1 580, and 1 296 cm^{-1} represent the stretching vibrations of C=O, C=C, and C–N groups, respectively, confirming the formation of β-ketoenamine linkages in the TpTAPA membrane [149]. After modification with negative charges, an obvious change in the spectrum can be observed for S-TpTAPA. The band at 1038 cm^{-1} can be assigned to the stretching vibration of the S–O group, and the band appearing at 1176 cm^{-1} represents the symmetric stretching modes of the O=S=O group [150].

XPS spectra were obtained to identify the surface chemical composition of TpTAPA derivative membranes. As depicted in Figure 8.36a, there are three major peaks for TpTAPA derivative membranes observed at 285.2, 399.8, and 532.0 eV corresponding to C 1 s, N 1 s, and O 1 s, respectively. Compared with the TpTAPA membrane, the S-TpTAPA membrane with negatively charged nanochannels shows

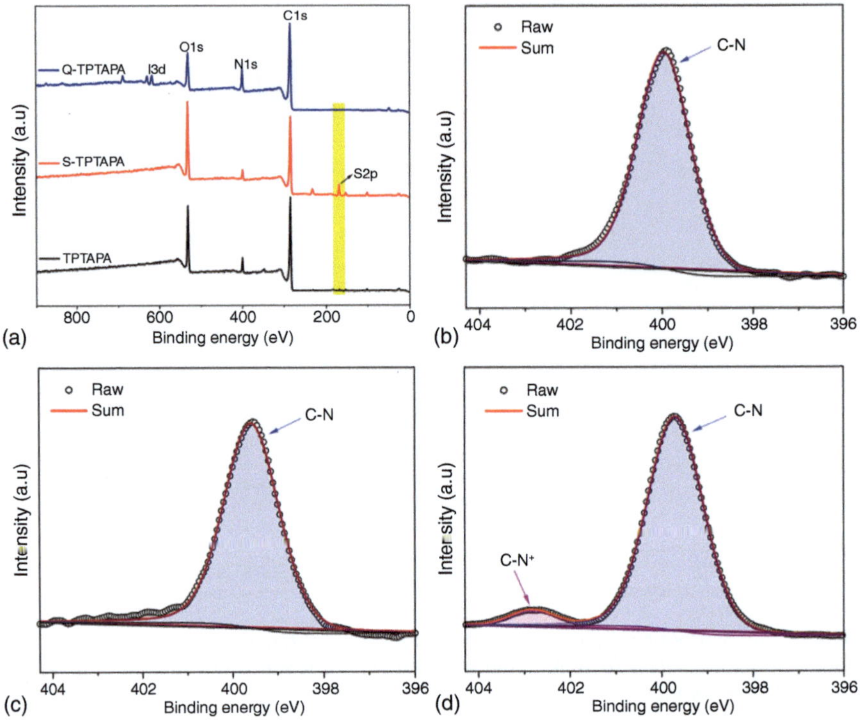

Figure 8.36 (a) XPS spectra of TpTAPA, S-TpTAPA, and Q-TpTAPA membranes, (b–d) N1s high-resolution spectra of the Q-TpTAPA membrane.

a new peak at 169.1 eV representing S2p, while the Q-TpTAPA membrane with positively charged modification has two new peaks at 618.5 and 630.6 eV representing I3d, which indicates the successful decoration of the TpTAPA membrane with charges. The high-resolution spectra of N1s further prove that the Q-TpTAPA membrane was modified with positive charges (Figure 8.36b–d). Moreover, the modification degree of S-TpTAPA and Q-TpTAPA membranes can be estimated as 35% and 7.0%, respectively, according to the ratio of S/N and $C–N^+/(C–N + C–N^+)$ (Table 8.2).

To evaluate the surface charge of TpTAPA nanochannels on the permeability of H^+ and the selectivity for H^+ over other cations, the permeation rates of H^+, K^+,

Table 8.2 The element contents of TpTAPA derivative membranes.

Membranes	Atomic conc. (mol%)					Modification degree (%)
	C1s	N1s	O1s	S2p	I3d	
TpTAPA	74.10	5.90	20.00	—	—	—
S-TpTAPA	73.11	2.59	23.39	0.91	—	35.0
Q-TpTAPA	74.56	10.90	14.11	—	0.43	7.0

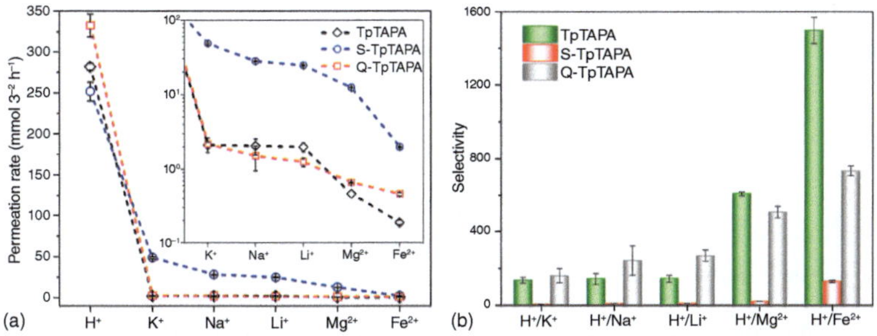

Figure 8.37 (a) Permeation rates of various electrolytes (0.1 M) across TpTAPA, S-TpTAPA, and Q-TpTAPA membranes (the inset of (A) is the enlarged figure of ion permeation rates), and (b) the corresponding ideal H^+/M^{n+} selectivity.

Na^+, Li^+, Mg^{2+}, and Fe^{2+} were measured using 0.1 mol/l chlorine salt solution under concentration gradient driving. As shown in Figure 8.37a, the TpTAPA membrane exhibits highly selective transport for H^+ over other cations. The selectivities of H^+/K^+, H^+/Na^+, H^+/Li^+, H^+/Mg^{2+}, and H^+/Fe^{2+} are 136, 143, 145, 608, and 1499, respectively (Figure 8.37b). As a comparison, the S-TpTAPA membrane with negatively charged nanochannels shows a lower permeation rate of H^+, which is attributed to the electrostatic repulsion between the S-TpTAPA membrane and Cl^-. Moreover, the electrostatic attraction between the S-TpTAPA membrane and other cations (K^+, Na^+, Li^+, Mg^{2+}, and Fe^{2+}) results in a large amount of leakage of these chlorine salts. As a result, the S-TpTAPA membrane has a lower permeation rate of H^+ and selectivity for H^+ over other cations. In contrast, the H^+ permeability of the Q-TpTAPA membrane with positively charged nanochannels increases due to the electrostatic attraction between the Q-TpTAPA membrane and Cl^-. For other monovalent cations, the leakage of K^+ has no obvious changes due to the similar properties between hydrated K^+ and Cl^-, while the leakage of Na^+ and Li^+ decreases slightly due to their larger hydrated diameters compared with Cl^-. However, the leakage of multivalent cations is enhanced compared with the non-charged nanochannels of the TpTAPA membrane because of the increased electrostatic attraction (higher Cl^- concentration) for Cl^-. The selectivities of Q-TpTAPA membranes for H^+/K^+, H^+/Na^+, H^+/Li^+, H^+/Mg^{2+}, and H^+/Fe^{2+} are 161, 243, 270, 507, and 733, respectively.

The dialysis coefficient and H^+/Fe^{2+} selectivity of TpTAPA derivative membranes were measured to assess their acid recovery performance in a mixed solution of 1 M HCl/0.267 M $FeCl_2$ during the DD process (Figure 8.38a). The TpTAPA membrane shows a H^+ dialysis coefficient ($U_H{}^+$) of $\sim 3.5 \times 10^{-3}$ m h^{-1}, while the selectivity for H^+ over Fe^{2+} is 1 193. After introducing negative charges, the S-TpTAPA membrane shows a lower $U_H{}^+$ of 2.0×10^{-3} m h^{-1}, and the selectivity for H^+/Fe^{2+} dramatically decreases to 146 due to the large amount of leakage of Fe^{2+} caused by the electrostatic attraction between the negative charges of the S-TaTAPA membrane and Fe^{2+}. In addition, the Q-TpTAPA membrane with positively charged nanochannels shows an increased $U_H{}^+$ of 4.9×10^{-3} m h^{-1}, and the selectivity of H^+/Fe^{2+} decreases

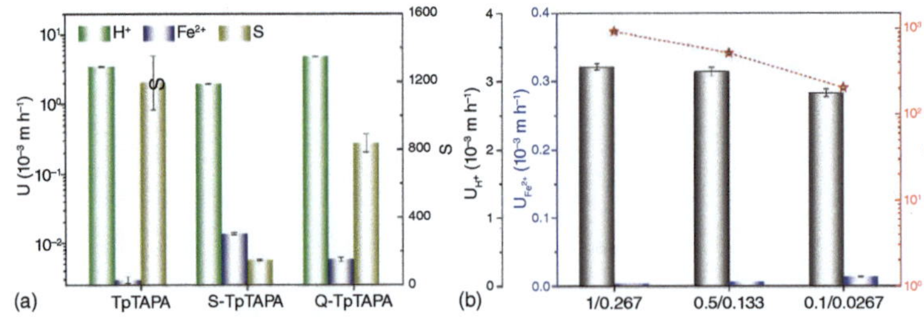

Figure 8.38 (a) Acid recovery performance of TpTAPA, S-TpTAPA, and Q-TpTAPA membranes using a mixed solution of 1 M HCl/0.267 M $FeCl_2$, (b) acid recovery properties of the TpTAPA membrane with feed solutions of different concentrations.

to 836. The increased $U_H{}^+$ is mainly because of the electrostatic attraction between the Q-TpTAPA membrane and Cl^-, which accelerates the transport of H^+. Meanwhile, the electrostatic attraction for Cl^- also leads to an increase in Fe^{2+} leakage. As a result, the positively charged nanochannels are beneficial for the transport of H^+, and the negatively charged nanochannels are adverse for the transport of H^+ when compared to pristine nanochannels. However, both negatively and positively charged nanochannels increase the leakage of Fe^{2+}, which is in agreement with the results of single-salt diffusion experiments.

As shown in Figure 8.38b, three feed solutions with different acid concentrations (i.e. $HCl/FeCl_2$: 1 M/0.267 M, 0.5 M/0.133 M, 0.1 M/0.0267 M) were chosen to measure the acid recovery performance. The results show that the feed concentration has a great effect on the acid recovery properties. The $U_H{}^+$ rises with increasing feed concentration, while the dialysis coefficients of Fe^{2+} ($U_{Fe}{}^{2+}$) decrease, which leads to the separation factor increasing notably. This is mainly due to the competitive effect between H^+ and Fe^{2+} during the transmembrane process, where H^+ is more competent to transport through the membrane than Fe^{2+} with increasing feed concentration.

Practical acidic wastewater, containing various cations, especially monovalent cations, restricts the recovery of exquisite acidic compounds. To demonstrate that the TpTAPA membrane possesses good performance in acid recovery even for complicated feed solutions, a quaternary mixture solution (1 M HCl, 0.267 M KCl, 0.267 M NaCl, 0.267 M $FeCl_2$) was used as the feed to conduct the DD experiment, while the permeate chamber was filled with DI water. As shown in Figure 8.39a, the dialysis coefficients of H^+, K^+, Na^+, and Fe^{2+} are 4.88×10^{-3}, 1.84×10^{-5}, 1.50×10^{-5}, and 0.53×10^{-5} m h^{-1}, respectively, and the corresponding separation factors of H^+/K^+, H^+/Na^+, and H^+/Fe^{2+} are 304, 364, and 916, respectively. These results further confirm that the TpTAPA membrane has great potential in the precise recovery of acidic compounds from complex feed systems. In addition, we further conducted a diffusion dialysis experiment using a more complicated feed solution containing 0.1 M HCl, 0.1 M HNO_3, 0.1 M H_2SO_4, 0.1 M KCl, 0.1 M NaCl, 0.1 M $FeCl_2$, and 0.1 M $AlCl_3$. As shown in Figure 8.39b, the TpTAPA membrane exhibits a high selectivity for H^+, NO^{3-}, and Cl^- over K^+, Na^+, Fe^{2+}, Al^{3+}, and $SO_4{}^{2-}$. The separation factors of H^+/K^+, H^+/Na^+, H^+/Fe^{2+}, H^+/Al^{3+}, $NO^{3-}/SO_4{}^{2-}$, and $Cl^-/SO_4{}^{2-}$

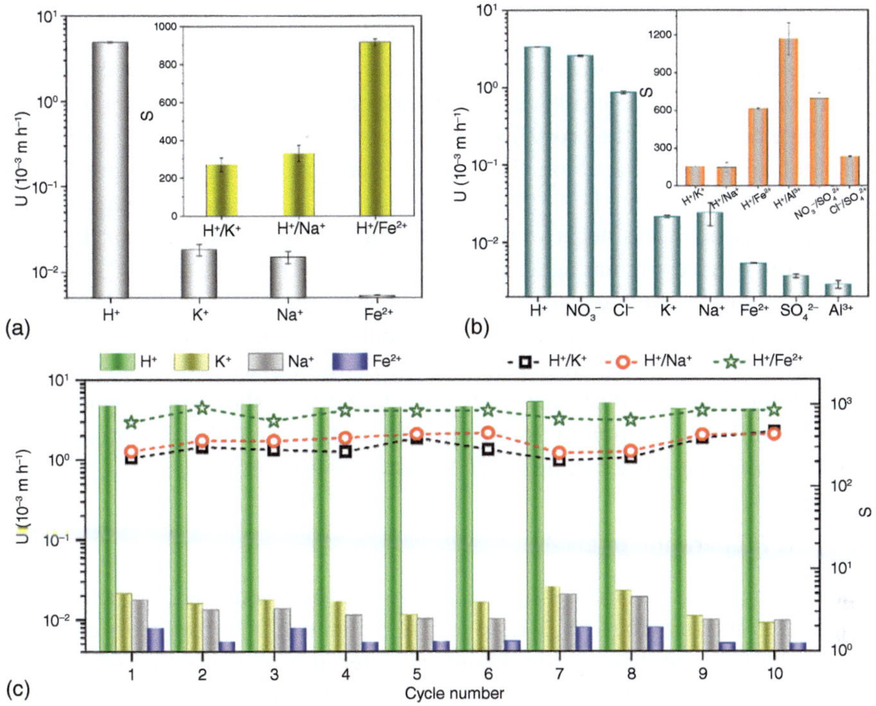

Figure 8.39 (a) Acid recovery performance of the TpTAPA membrane for the quaternary mixture solution (1 M HCl, 0.267 M KCl, 0.267 M NaCl, 0.267 M $FeCl_2$), (b) acid recovery performance of the TpTAPA membrane for a multicomponent mixture solution (0.1 M HCl, 0.1 M HNO_3, 0.1 M H_2SO_4, 0.1 M KCl, 0.1 M NaCl, 0.1 M $FeCl_2$, and 0.1 M $AlCl_3$), and (c) the dialysis coefficients of H^+ and separation factor for H^+ over other cations for the TpTAPA membrane with 10 consecutive cycles.

are 155, 147, 617, 1 170, 696, and 235, respectively. The results demonstrate that the TpTAPA membrane shows preferential transport of monobasic acids and negligible permeability for other ions. From Figure 8.39c, the slight fluctuations in the dialysis coefficients of H^+, K^+, Na^+, Fe^{2+}, and the separation factors of H^+/K^+, H^+/Na^+, and H^+/Fe^{2+} within 10 consecutive cycles confirm the operational stability of the TpTAPA membrane.

8.4 Electro-Nanofiltration Membranes

8.4.1 Introduction

The IP membranes, which are constructed by the polymerization reaction between the amino monomers and aryl chloride monomers, are named the electronanofiltration membranes (ENMs) by our group because the IP membranes have been applied maturely in the nanofiltration field, and we applied them in the ED process for their excellent mono-divalent ion separation performance [151–153]. As mentioned above, the key mechanism of ion separation in the ED process is the size effect and electrostatic repulsion effect between the membrane surface and ions [154, 155]. The ENMs are suitable for the key mechanism of mono-divalent ion separation due to the structure of the polyamide crosslinking network of the functional layer [156, 157]. The average pore size of the crosslinking network of the functional layer is in the range of the different hydrated ion diameters or even smaller so that the size-sieving effect of the ENFs is strong [158, 159]. At the same time, the surface charge of the functional layer is always negative due to the high density of the carboxylic acid groups on the surface originating from the unreacted acyl chloride groups [160]. For anion separation, the high density of the negative charge on the surface leads to a strong repulsion effect on the divalent anion so that the anion selectivity is enhanced [161]. For cation separation, the negative charge on the surface assists cation transport so that the ion flux is increased [162]. Therefore, ENMs show great potential for mono-divalent ion separation in the ED process, and our group's work proves that ENMs have excellent ion separation performance (Figure 8.40).

8.4.2 The Preparation of ENMs

The fabrication process of the ENMs is similar, while the reacted monomers are different. Amino monomers such as 1,3-phenylenediamine (MPD), ethylenediamine (EDA), and DETA are dissolved in deionized water at a concentration of approximately 1–2 wt%. The aryl chloride monomer, 1,3,5-benzenetricarbonyl trichloride (TMC), is dissolved in hexane at a concentration of approximately 0.1–0.2 w/v%. The chosen porous substrate was a commercial or made-in-lab ultrafiltration porous membrane and was washed with deionized water before the fabrication of the ENMs. First, in the fabrication process, the substrate is fixed on the plate glass or in the frames, and all the water droplets are removed. Then, the substrate surface is coated with the amino monomer solution for a certain time for the substrate to adsorb the amino monomers. The remaining amino monomer solution was removed, and dust-free paper and a squeegee roller were used to remove the residual solution droplets. Finally, the substrate was coated with the TMC solution for a certain time, and the rest of the solution was removed. Before storage in deionized water, the obtained ENMs were washed with hexane several times, followed by a drying step (Figure 8.41). [152, 163]

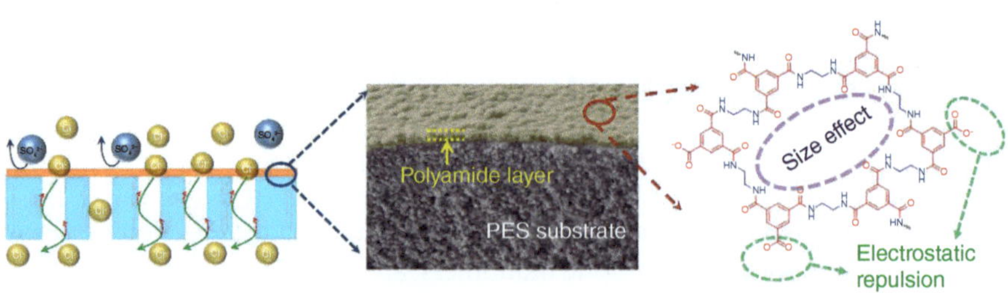

Figure 8.40 Schematic diagram of the ion separation mechanism of ENMs. Source: Zhao et al. [163]/Reproduced with permission from American Chemical Society.

Figure 8.41 Schematic illustration of the preparation of ENMs. Source: Reproduced with permission from Hou et al. [152]; © 2018, Elsevier.

8.4.3 The Performance of ENMs

Only a few works have been published about the fabrication and ion separation performance in the ED process of ENMs. In this section, we introduce these works to show the great potential of ENMs ED applications.

Ge and his coworkers first applied ENMs to mono-divalent anion separation in the ED process [151]. They fabricated ENF membranes with the reaction between MPD and TMC monomers on a PES substrate, and these membranes were applied to H^+/Zn^{2+} and Na^+/Mg^{2+} separation in the ED process. As shown in Figure 8.42a, FTIR spectra of the ENMs at different preparation stages indicated the success of the fabrication of the polyamide layer. Compared to the spectra of the PES substrate, new peaks appear after the reaction between the monomers. The peak at 1540 cm^{-1} represents the stretching vibration of the C—N bonds of amide groups, while the peak

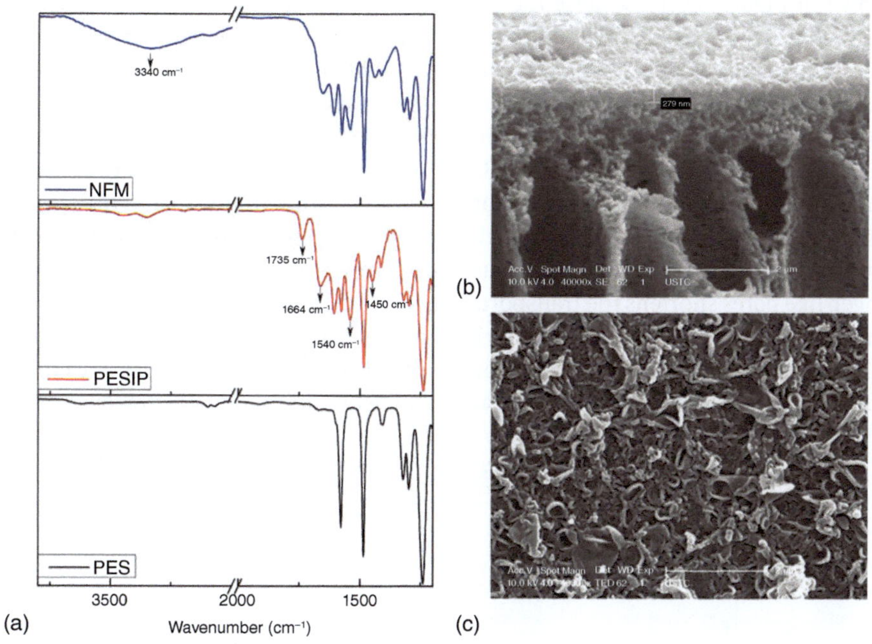

Figure 8.42 (a) FTIR spectra of the ENMs at different preparation stages. Morphology of the cross-section (b) and surface (c) of the ENM. Source: Reproduced with permission from Ge et al. [151]; © 2016, Elsevier.

at 1664 cm^{-1} indicates the stretching vibration of the C=O bonds of amide groups. The fabrication of the amide groups on the substrate shows the success of the IP. Moreover, the peak at 1664 cm^{-1} indicates the stretching vibration of C=O bonds in unreacted aryl chloride groups, and it disappears after the membrane is immersed in water, as the aryl chloride groups become carboxylic acid groups. The cross-sectional morphology of the ENMs is shown in Figure 8.42b, which indicates that NFMs have an asymmetric membrane structure consisting of a thin, dense functional layer and a finger-like porous substrate. The surface morphology of the ENMs is shown in Figure 8.42c, and the special ridge-and-valley structure shows the successful fabrication of the functional layer.

The ion separation performance and I–V curve of the ENMs and commercial membrane CSO are shown in Figure 8.43, and the results indicate the excellent performance of the ENMs. As shown in Figure 8.42a, for the H^+/Zn^{2+} system, the selectivity of the ENMs (named NFM) is approximately 354, which is much higher than that of the CSO membrane, which is approximately 15. The H^+ flux of the NFM is 3.4×10^{-7} mol cm^{-2} s^{-1}, which is two times higher than that of the CSO membrane. The higher flux is attributed to the large pores interconnected with finger-like macrovoids on the support layer, and the higher selectivity was caused by the small pores rejecting the larger hydrated Ca^{2+} ions. For the Na^+/Mg^{2+} system, the selectivity of the NFM is approximately 7, which is higher than that of the CSO membrane, which is approximately 4. The Na^+ flux of the NFM is 2.2×10^{-7} mol cm^{-2} s^{-1}, which is also higher than that of the CSO membrane, as shown in Figure 8.42b. Because the diversity of the hydrated diameter of Na^+ and Mg^{2+} is less than the diversity between H^+ and Zn^{2+}, the selectivity of Na^+/Mg^{2+} is lower. Moreover, the limited current density of the NFM is also much higher than that of the CSO membrane, while the area resistance is lower than that of the commercial CSO membrane, which indicates the better potential of the NFM to be applied in the ED process (Figure 8.43c). Compared with the commercial membrane, the I–V curve of the NFM is more similar to a straight line due to the special structure of the pores and negative charges of the PES substrate combined with ultrathin polyamide. The designed structure featuring an ultrathin surface layer and finger-like pores shortens the transportation pathway, which facilitates ion diffusion during the ED process and hence alleviates concentration polarization [151].

Based on Ge's work, Hou and her coworkers fabricated IP membranes with three different amine monomers (EDA, tetraethylenepentamine, and polyethyleneimine)

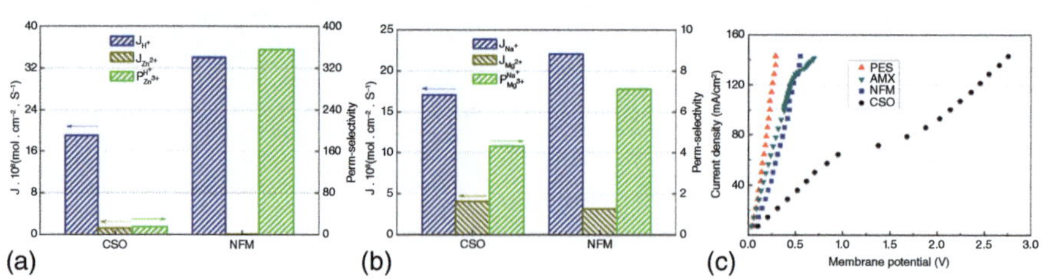

Figure 8.43 (a) The H^+/Zn^{2+} separation performance of the CSO membrane and NFM, (b) Na^+/Mg^{2+} separation performance of the CSO membrane and NFM, and (c) the I–V curves of the membranes. Source: Reproduced with permission from Ge et al. [151]; © 2016, Elsevier.

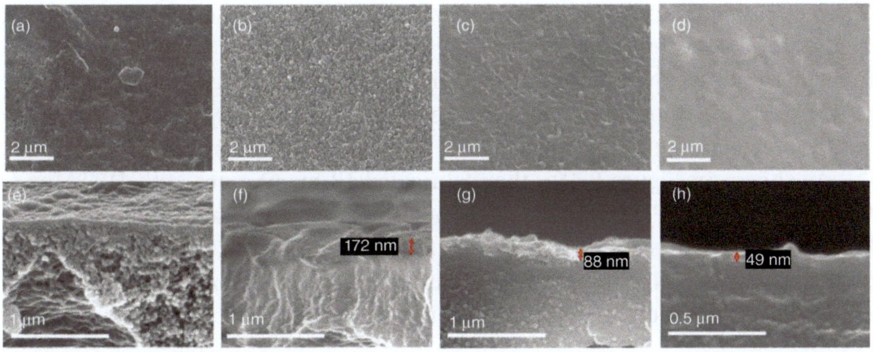

Figure 8.44 Surface SEM images of (a) HPAN porous substrate, (b) E-HPAN, (c) T-HPAN, and (d) P-HPAN; cross-section images of (e) HPAN porous substrate, (f) E-HPAN, (g) T-HPAN, and (h) P-HPAN. Source: Hou et al. [152]/Reproduced with permission from Elsevier.

and systematically investigated the characterization and ED performance of these membranes named E-HPAN, T-HPAN, and P-HPAN [152]. The morphology of the surface and cross section is shown in Figure 8.44 Different amine monomers cause different surface morphologies. As the molecular size increases, the surface morphology becomes smoother, and the height of the functional layer decreases because it is more difficult for larger amine monomers to pass through the nascent polyamide layer so that further reaction is inhibited. As the EDA molecule is the shortest in these three chosen monomers, the surface morphology of E-HPAN is the roughest, while the functional layer is the thickest.

As shown in Figure 8.45a, the Na^+/Mg^{2+} separation performance of these three membranes was also measured and compared with the commercial CSO membrane, and the results show that the IP membranes have higher selectivity, while E-HPAN has the best separation performance due to the smallest pore size and the thickest functional layer. Moreover, the long-term stability of the ED performance of the ENMs is also measured, and the results are shown in Figure 8.45b. As the flux and

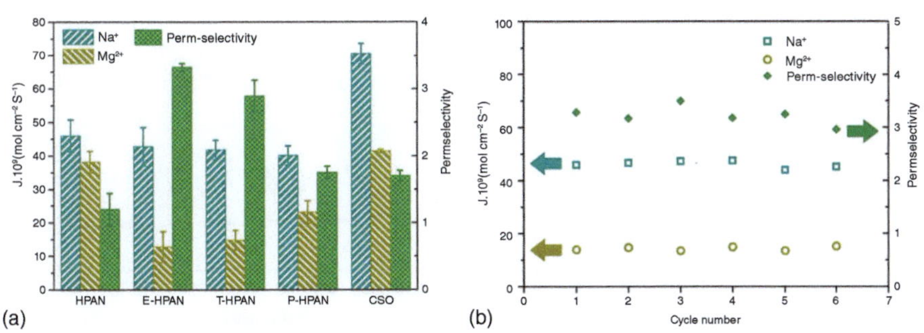

Figure 8.45 (a) Ion flux and perm-selectivity in the ED process of the ENMs and commercial membrane, (b) the variation of flux and perm-selectivity of E-HPAN with cycles. Source: Reproduced with permission from Hou et al. [152]; © 2018, Elsevier.

permselectivity remain stable in the cycle measurement, ENMs imply the potential for long-term ED processes.

To further improve the cation permselectivity of ENMs, Sheng and his coworkers used a modified method to increase the positive charge density on the surface to increase the electronic repulsion between the divalent cations and ENM surface [153]. As shown in Figure 8.46a, they fabricated IP membranes with DETA and TMC named the ENF membrane and modified the membrane with a 10 wt% iodomethane/methanol solution under alkaline conditions at 25 °C for 12 h, 24 h, and 48 h to obtain final membranes with different quaternization degrees to adjust the Li^+/Mg^{2+} separation performance in the ED process (modified membranes

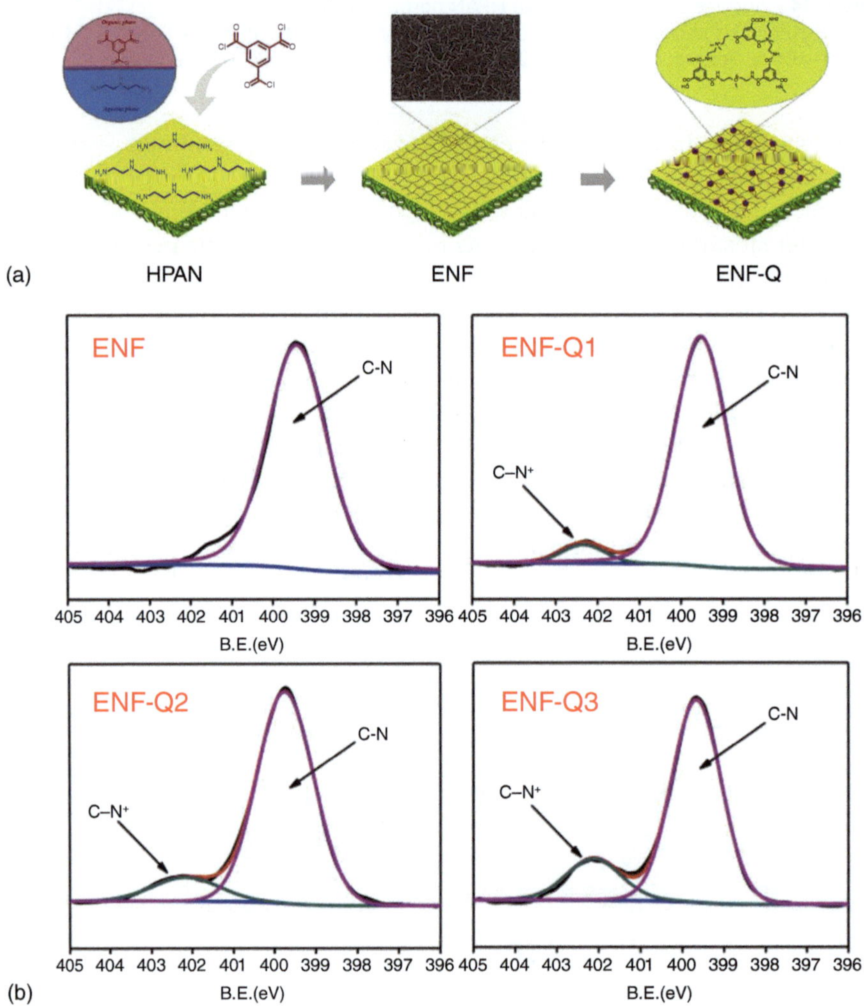

Figure 8.46 (a) Schematic representative of the preparation of modified ENMs, (b) XPS N1s spectra of the unmodified and modified ENMs. Source: Reproduced with permission from Sheng et al. [153]; © 2020, Elsevier.

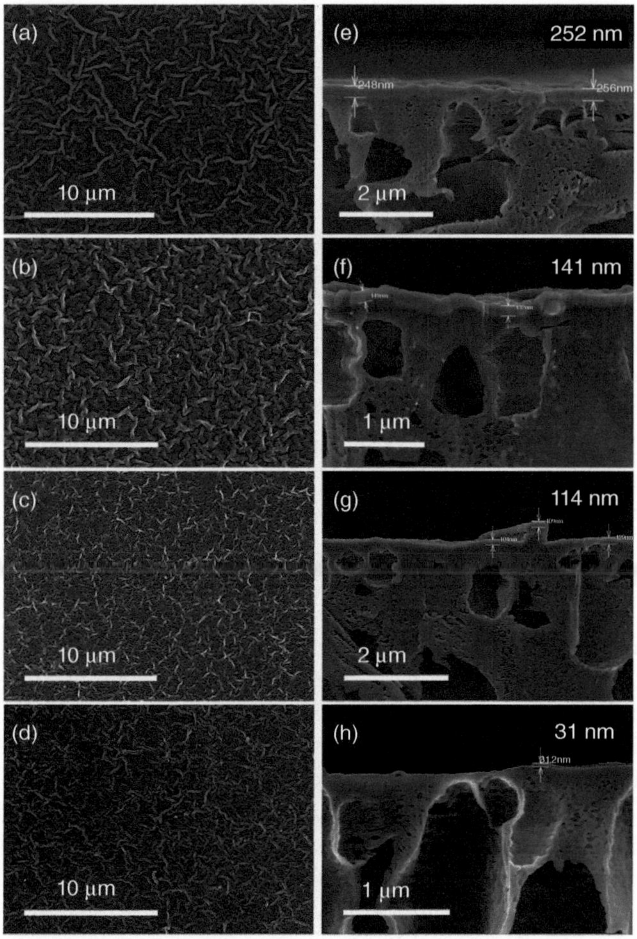

Figure 8.47 Surface (a–d) and cross-sectional (e–h) morphology of the prepared membranes: (a, e) ENF, (b, f) ENF-Q1, (c, g) ENF-Q2, and (f, h) ENF-Q3. Source: Reprinted with permission from Sheng et al. [153]/Reproduced with permission from Elsevier.

were denoted ENF-Q1, ENF-Q2, and ENF-Q3). As the XPS N1s spectra shown in Figure 8.46b, after the modification step, a new peak at 402.2 corresponding to $C–N^+$ species appears, which shows the formation of quaternary ammonium groups in the functional layer so that the positive charge on the surface increases.

The morphology of the surface and cross-section was also characterized to investigate the influence of the modification step. As shown in Figure 8.47a, the ridge-and-valley structure on the surface indicated the fabrication of the IP layer. After the modification step, there are no obvious pores on the surface, and the ridge-and-valley structure still exists, which indicates that the modification step does not destroy all the functional layers and that the functional layer is still a dense layer. However, the height of the functional layer decreases after the modification step, and the functional layer becomes thinner with increasing modification time.

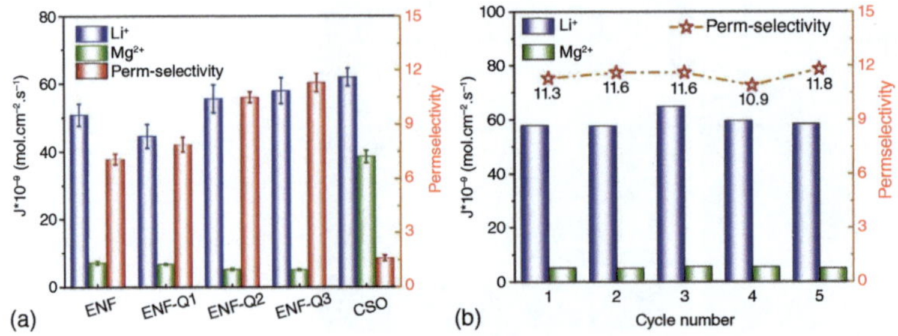

Figure 8.48 (a) Ion flux and permselectivity in the ED process of the ENMs and commercial membrane, (b) the variation in flux and perm-selectivity of ENF-Q3 with 5 cycles. Source: Reproduced with permission from ShengReproduced w et al. [153]; © 2020, Elsevier.

The possible reason is the partial hydrolysis of the amide bond connected with a quaternary ammonium group under alkaline conditions.

The ED performance of the ENMs was also measured. As shown in Figure 8.48a, the ENF-Q3 membrane, with the highest quaternization degree has the highest selectivity of 11.3, which is seven times greater than that of the commercial CSO membrane, while the Li^+ flux is 5.8×10^{-8} mol cm^{-2} s^{-1}, which is at the same level as that of the CSO membrane. The increasing selectivity after the modification step is related to the higher quaternization degrees resulting in the higher positive charge density and electrostatic repulsion between the ions and membrane surface. The greater electropositivity hindered the Mg^{2+} flux significantly compared to Li^+ so that the selectivity increased with higher quaternization degrees. Moreover, the ENMs also have excellent ED performance stability, as the flux and selectivity changed slightly in the five sequential cycles (Figure 8.48b) [153].

As ENMs have shown excellent potential for mono-divalent cation separation in the ED process, Zhao and coworkers thought ENMs had excellent anion separation performance due to the high negative charge density on the surface, and they fabricated ENFs with EDA and TMC monomers for anion separation. The influence of the TMC concentration and reaction time was also investigated systematically in this work [163]. The physicochemical properties of the ENMs under different synthesis conditions were measured, and the results are shown in Figure 8.49. With increasing reaction time and TMC concentration, the area resistance of the ENMs increased due to the thicker functional layer, while it was still lower than that of the commercial membrane (Figure 8.49a,d). The zeta potential also increases because the density of the carboxylic acid groups on the surface increases, and this phenomenon indicates that the surface charge becomes more negative with increasing reaction time and TMC concentration (Figure 8.49b,e). The transport number also increases due to the increasing negative charge density on the surface, as it assists cation transport through the membrane (Figure 8.49c,f).

As the densely cross-linked network structure and the strong electrostatic repulsion to anions of high valence by a high density of carboxylic residues lead to the low permeation of SO_4^{2-} and high permselectivity of Cl^-, the ENMs show excellent anion separation performance. The performance of the optimized ENM shows that

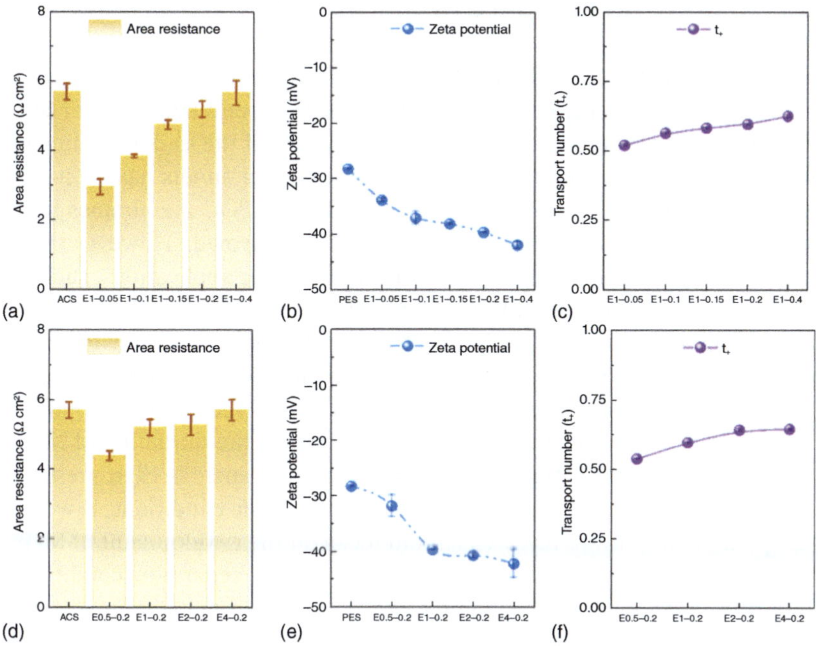

Figure 8.49 The area resistance of the commercial ACS membrane and the ENMs with different TMC concentrations (a) and different reaction times (b). The zeta potential of the PES substrate and the ENMs with different TMC concentrations (c) and different reaction times (d); the transport number of the ENF membranes with different TMC concentrations (e) and different reaction times (f). Source: Reproduced with permission from Zhao et al. [163]; © 2022, Wiley.

the selectivity is 98, which is 19 times higher than the selectivity of the commercial ACS membrane, while the Cl^- flux is 2.31 mol h^{-1} $m^{-2,}$ which is at the same level as the ACS membrane. Moreover, the membrane also shows excellent cycle stability, as the permselectivity remains stable in a 10-consecutive ED process and better separation performance at different current densities compared with the ACS membrane (Figure 8.50a,b) [163].

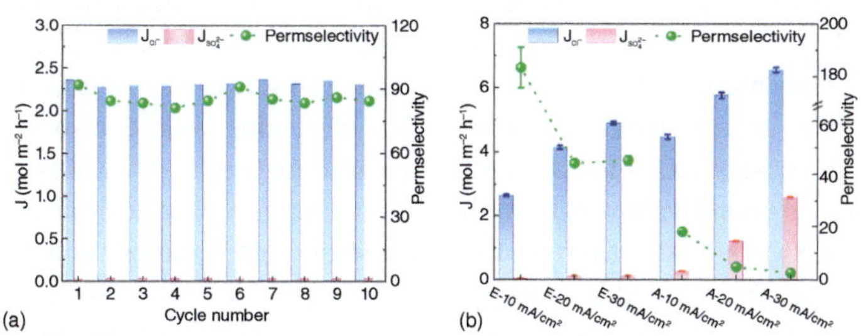

Figure 8.50 (a) The long-term stability of the chosen ENM, (b) the Cl^-/SO_4^{2-} separation performance of the ACS and chosen ENM at different current densities. Source: Reproduced with permission from Zhao et al. [163]; © 2022, Wiley.

8.5 Conclusion and Perspective

The successful development of MOF and POC membranes provides promising ways to broaden the applications of selective ion separations. Many new strategies for constructing MOF and POC membranes have been studied, and the understanding of the ion-selective transport mechanisms in MOF and POC channels has been further improved. Strategies for the construction of MOF and POC membranes have been explored, including in situ growth and contradiffusion growth. However, there are some drawbacks of these strategies that limit their further applications, such as elaborate fabrication procedures and limited membrane area. More efforts should be devoted to exploring facile strategies, such as soluble polyMOFs, for the construction of industry-scale membranes. In addition, economical synthetic strategies for MOF and POC membranes need to be developed that do not rely on complex organic ligands or expensive metal ions. More importantly, the durability of MOF and POC membranes under extreme operation conditions, such as high pressure or electrical potential, high salt concentrations, or acidic/basic conditions, is worth investigating. Therefore, future research should focus on the development of MOF and POC membranes with desirable durability under extreme separation conditions and cost-effective large-scale fabrication processes.

COFs are a novel kind of crystalline porous polymer consisting of orderly arranged pore structures, which endows them with a low density, high porosity, and large surface area. Moreover, versatile building units for the synthesis of COFs enable them to have facilely tailored functionalities. All these characteristics make COFs as promising candidate for constructing advanced membrane separation. In the past decade, many methods have been developed to prepare COF membranes that exhibit excellent properties in separation applications, breaking the trade-off effect of conventional membranes between flux and selectivity. In recent years, COF membranes have also shown prominently in highly efficient ion transport and separation. However, challenges remain for the future development and application of COF membranes. First, more methodologies for fabricating COF membranes need to be further developed, especially for free-standing COF membranes with high strength and crystallinity. Then, the pore size of COFs with subnanometers is highly required to construct ion/molecular separation membranes with smaller sizes, including gas separation, monovalent ion separation, organic solvent nanofiltration, and desalination. Such COFs could be achieved by using relatively small and short organic linkers or changing the stacking mode between 2D COF layers. In addition, long-term stability, low-cost COFs, and cost-effective fabrication strategies should also be developed to control the cost of COF-based membranes in industrial applications.

ENMs have excellent mono-divalent cation and anion separation performance. The selectivity of the ENMs is better than that of commercial membranes such as CSO and ACS. Moreover, there are still some other advantages of ENMs, such as a simple preparation method, high limited current density, low cost, and chemical stability, so ENMs have great potential to be scaled up and applied in industry. However, there are still some limitations preventing ENMs from being applied in the ion separation field. For instance, the low ion transport number is due to the ultrathin functional layer, so the current efficiency is low, indicating a higher energy cost. The other limitation is that the upper limit of the concentration of the ions in the concentrated chamber is low, as the functional layer is too thin to prevent the ions from going back to the dilute chamber when the concentration of the solution in the concentration chamber is much higher than that in the dilute chamber. These limitations exist due to the ENM structure. Although ENMs have excellent separation performance, the technology of the ED process should be optimized to avoid the limitations of ENMs and make full use of the advantages.

References

1 Howarth, A.J., Liu, Y., Li, P. et al. (2016). Chemical, thermal and mechanical stabilities of metal–organic frameworks. *Nature Reviews Materials* 1: 1–15.

2 Ren, J., Dyosiba, X., Musyoka, N.M. et al. (2017). Review on the current practices and efforts towards pilot-scale production of metal-organic frameworks (MOFs). *Coordination Chemistry Reviews* 352: 187–219.

3 Kitagawa, S., Kitaura, R., and Noro, S. (2004). Functional porous coordination polymers. *Angewandte Chemie* 43: 2334–2375.

4 Li, J.-R., Sculley, J., and Zhou, H.-C. (2012). Metal–organic frameworks for separations. *Chemical Reviews* 112: 869–932.

5 Hu, Y., Wei, J., Liang, Y. et al. (2016). Zeolitic Imidazolate framework/graphene oxide hybrid nanosheets as seeds for the growth of ultrathin molecular sieving membranes. *Angewandte Chemie* 55: 2048–2052.

6 Corma, A., García, H., and Llabrés i Xamena F. (2010). Engineering metal organic frameworks for heterogeneous catalysis. *Chemical Reviews* 110: 4606–4655.

7 Chen, B., Wang, L., Xiao, Y. et al. (2009). A luminescent metal–organic framework with Lewis basic pyridyl sites for the sensing of metal ions. *Angewandte Chemie International Edition* 48: 500–503.

8 Kreno, L.E., Leong, K., Farha, O.K. et al. (2011). Metal–organic framework materials as chemical sensors. *Chemical Reviews* 112: 1105–1125.

9 Orellana-Tavra, C., Baxter, E.F., Tian, T. et al. (2015). Amorphous metal–organic frameworks for drug delivery. *Chemical Communications* 51: 13878–13881.

10 Rowsell, J.L. and Yaghi, O.M. (2004). Metal–organic frameworks: a new class of porous materials. *Microporous and Mesoporous Materials* 73: 3–14.

11 Lillerud, K.P., Olsbye, U., and Tilset, M. (2010). Designing heterogeneous catalysts by incorporating enzyme-like functionalities into MOFs. *Topics in Catalysis* 53: 859–868.

12 Eddaoudi, M., Kim, J., Rosi, N. et al. (2002). Systematic design of pore size and functionality in isoreticular MOFs and their application in methane storage. *Science* 295: 469–472.

13 Banerjee, R., Phan, A., Wang, B. et al. (2008). High-throughput synthesis of zeolitic imidazolate frameworks and application to CO_2 capture. *Science* 319: 939–943.

14 Riou-Cavellec, M., Albinet, C., Livage, C. et al. (2002). Ferromagnetism of the hybrid open framework K $[M_3\ (BTC)_3]\cdot 5H_2O$ (M= Fe, co) or MIL-45. *Solid State Sciences* 4: 267–270.

15 Millange, F., Serre, C., and Férey, G. (2002). Synthesis, structure determination and properties of MIL-53as and MIL-53ht: the first Cr iii hybrid inorganic–organic microporous solids: Cr iii $(OH)\cdot\{O_2C–C_6H_4–CO_2\}\cdot\{HO_2\ C–C_6\ H_4–CO_2\ H\}_x$. *Chemical Communications* 822–823.

16 Hu, Z. and Zhao, D. (2015). De facto methodologies toward the synthesis and scale-up production of UiO-66-type metal–organic frameworks and membrane materials. *Dalton Transactions* 44: 19018–19040.

17 Feng, D., Gu, Z.Y., Li, J.R. et al. (2012). Zirconium-metalloporphyrin PCN-222: mesoporous metal–organic frameworks with ultrahigh stability as biomimetic catalysts. *Angewandte Chemie International Edition* 51: 10307–10310.

18 Li, H., Eddaoudi, M., O'Keeffe, M., and Yaghi, O.M. (1999). Design and synthesis of an exceptionally stable and highly porous metal-organic framework. *Nature* 402: 276–279.

19 Pan, Y., Liu, Y., Zeng, G. et al. (2011). Rapid synthesis of zeolitic imidazolate framework-8 (ZIF-8) nanocrystals in an aqueous system. *Chemical Communications* 47: 2071–2073.

20 Ma, X.-H., Yang, Z., Yao, Z.-K. et al. (2017). A facile preparation of novel positively charged MOF/chitosan nanofiltration membranes. *Journal of Membrane Science* 525: 269–276.

21 Cavka, J.H., Jakobsen, S., Olsbye, U. et al. (2008). A new zirconium inorganic building brick forming metal organic frameworks with exceptional stability. *Journal of the American Chemical Society* 130: 13850–13851.

22 Colón, Y.J. and Snurr, R.Q. (2014). High-throughput computational screening of metal–organic frameworks. *Chemical Society Reviews.* 43: 5735–5749.

23 Deng, H., Grunder, S., Cordova, K.E. et al. (2012). Large-pore apertures in a series of metal-organic frameworks. *Science* 336: 1018–1023.

24 Canivet, J., Fateeva, A., Guo, Y. et al. (2014). Water adsorption in MOFs: fundamentals and applications. *Chemical Society Reviews* 43: 5594–5617.

25 Burtch, N.C., Jasuja, H., and Walton, K.S. (2014). Water stability and adsorption in metal–organic frameworks. *Chemical Reviews* 114: 10575–10612.

26 Liu, Y., Ng, Z., Khan, E.A. et al. (2009). Synthesis of continuous MOF-5 membranes on porous α-alumina substrates. *Microporous and Mesoporous Materials* 118: 296–301.

27 Huang, A., Bux, H., Steinbach, F., and Caro, J. (2010). Molecular-sieve membrane with hydrogen permselectivity: ZIF-22 in LTA topology prepared with 3-aminopropyltriethoxysilane as covalent linker. *Angewandte Chemie International Edition* 49: 4958–4961.

28 Huang, A., Dou, W., and Caro, J. (2010). Steam-stable zeolitic imidazolate framework ZIF-90 membrane with hydrogen selectivity through covalent functionalization. *Journal of the American Chemical Society* 132: 15562–15564.

29 Zhu, Y., Gupta, K.M., Liu, Q. et al. (2016). Synthesis and seawater desalination of molecular sieving zeolitic imidazolate framework membranes. *Desalination* 385: 75–82.

30 Zhang, F., Zou, X., Gao, X. et al. (2012). Hydrogen selective NH2-MIL-53 (Al) MOF membranes with high permeability. *Advanced Functional Materials* 22: 3583–3590.

31 Shamsaei, E., Lin, X., Wan, L. et al. (2016). A one-dimensional material as a nano-scaffold and a pseudo-seed for facilitated growth of ultrathin, mechanically reinforced molecular sieving membranes. *Chemical Communications* 52: 13764–13767.

32 Hu, Y., Dong, X., Nan, J. et al. (2011). Metal–organic framework membranes fabricated via reactive seeding. *Chemical Communications* 47: 737–739.

33 Ameloot, R., Vermoortele, F., Vanhove, W. et al. (2011). Interfacial synthesis of hollow metal–organic framework capsules demonstrating selective permeability. *Nature Chemistry* 3: 382–387.

34 Makiura, R. and Konovalov, O. (2013). Interfacial growth of large-area single-layer metal-organic framework nanosheets. *Scientific Reports* 3: 1–8.

35 Yao, J., Dong, D., Li, D. et al. (2011). Contra-diffusion synthesis of ZIF-8 films on a polymer substrate. *Chemical Communications* 47: 2559–2561.

36 Barankova, E., Tan, X., Villalobos, L.F. et al. (2017). A metal chelating porous polymeric support: the missing link for a defect-free metal-organic framework composite membrane. *Angewandte Chemie* 56: 2965–2968.

37 Liu, J. and Wöll, C. (2017). Surface-supported metal–organic framework thin films: fabrication methods, applications, and challenges. *Chemical Society Reviews* 46: 5730–5770.

38 Li, M. and Dincă, M. (2014). Selective formation of biphasic thin films of metal–organic frameworks by potential-controlled cathodic electrodeposition. *Chemical Science* 5: 107–111.

39 Denny, M.S. Jr., and Cohen, S.M. (2015). In situ modification of metal–organic frameworks in mixed-matrix membranes. *Angewandte Chemie International Edition* 54: 9029–9032.

40 Sorribas, S., Gorgojo, P., Téllez, C. et al. (2013). High flux thin film nanocomposite membranes based on metal–organic frameworks for organic solvent nanofiltration. *Journal of the American Chemical Society* 135: 15201–15208.

41 Ma, D., Peh, S.B., Han, G., and Chen, S.B. (2017). Thin-film nanocomposite (TFN) membranes incorporated with super-hydrophilic metal–organic framework (MOF) UiO-66: toward enhancement of water flux and salt rejection. *ACS Applied Materials & Interfaces* 9: 7523–7534.

42 Zhang, Y., Feng, X., Li, H. et al. (2015). Photoinduced postsynthetic polymerization of a metal–organic framework toward a flexible stand-alone membrane. *Angewandte Chemie International Edition* 54: 4259–4263.

43 Ostermann, R., Cravillon, J., Weidmann, C. et al. (2011). Metal-organic framework nanofibers via electrospinning. *Chemical Communications* 47: 442–444.

44 Zhang, Y., Yuan, S., Feng, X. et al. (2016). Preparation of nanofibrous metal–organic framework filters for efficient air pollution control. *Journal of the American Chemical Society* 138: 5785–5788.

45 Xu, T., Shehzad, M.A., Yu, D. et al. (2019). Highly cation permselective metal–organic framework membranes with leaf-like morphology. *ChemSusChem* 12: 2593–2597.

46 Xu, T., Shehzad, M.A., Wang, X. et al. (2020). Engineering leaf-like UiO-66-SO_3H membranes for selective transport of cations. *Nano-Micro Letters* 12: 51.

47 Xu, T., Sheng, F., Wu, B. et al. (2020). Ti-exchanged UiO-66-NH_2–containing polyamide membranes with remarkable cation permselectivity. *Journal of Membrane Science* 615: 118608.

48 Holst, J.R., Trewin, A., and Cooper, A.I. (2010). Porous organic molecules. *Nature Chemistry* 2: 915–920.

49 Zhang, G. and Mastalerz, M. (2014). Organic cage compounds–from shape-persistency to function. *Chemical Society Reviews* 43: 1934–1947.

50 Mastalerz, M. (2010). Shape persistent organic cage compounds by dynamic covalent bond formation. *Angewandte Chemie International Edition* 49: 5042–5053.

51 Hasell, T. and Cooper, A.I. (2016). Porous organic cages: soluble, modular and molecular pores. *Nature Reviews Materials* 1: 16053.

52 Evans, J.D., Sumby, C.J., and Doonan, C.J. (2015). Synthesis and applications of porous organic cages. *Chemistry Letters* 44: 582–588.

53 Xiong, M., Ding, H., Li, B. et al. (2014). Porous organic molecular cages: from preparation to applications. *Current Organic Chemistry* 18: 1965–1972.

54 Jones, J.T.A., Hasell, T., Wu, X. et al. (2011). Modular and predictable assembly of porous organic molecular crystals. *Nature* 474: 367–371.

55 Tian, J., Thallapally, P.K., Dalgarno, S.J. et al. (2009). Amorphous molecular organic solids for gas adsorption. *Angewandte Chemie* 48: 5492–5495.

56 Giri, N., Popolo, M.G.D., Melaugh, G. et al. (2015). Liquids with permanent porosity. *Nature* 527: 216–220.

57 Budd, P.M., Ghanem, B.S., Makhseed, S. et al. (2004). Polymers of intrinsic microporosity (PIMs): robust, solution-processable, organic nanoporous materials. *Chemical Communications* 230–231.

58 Tozawa, T., Jones, J.T., Swamy, S.I. et al. (2009). Porous organic cages. *Nature Materials* 8: 973–978.

59 Jones, J.T.A., Holden, D., Mitra, T. et al. (2011). On–off porosity switching in a molecular organic solid. *Angewandte Chemie International Edition* 50: 749–753.

60 Hasell, T., Schmidtmann, M., Stone, C.A. et al. (2012). Reversible water uptake by a stable imine-based porous organic cage. *Chemical Communications* 48: 4689–4691.

61 Simard, M., Su, D., and Wuest, J.D. (1991). Use of hydrogen bonds to control molecular aggregation. Self-assembly of three-dimensional networks with large chambers. *Journal of the American Chemical Society* 113: 4696–4698.

62 Han, R. and Wu, P. (2018). Composite proton-exchange membrane with highly improved proton conductivity prepared by in situ crystallization of porous organic cage. *ACS Applied Materials & Interfaces* 10: 18351–18358.

63 Song, Q., Jiang, S., Hasell, T. et al. (2016). Porous organic cage thin films and molecular-sieving membranes. *Advanced Materials* 28: 2629–2637.

64 Jiang, S., Song, Q., Massey, A. et al. (2017). Oriented two-dimensional porous organic cage crystals. *Angewandte Chemie* 56: 9391–9395.

65 Bushell, A.F., Budd, P.M., Attfield, M.P. et al. (2013). Nanoporous organic polymer/cage composite membranes. *Angewandte Chemie International Edition* 52: 1253–1256.

66 Zhai, Z., Jiang, C., Zhao, N. et al. (2020). Polyarylate membrane constructed from porous organic cage for high-performance organic solvent nanofiltration. *Journal of Membrane Science* 595: 117505.

67 Zhai, Z., Zhao, N., Liu, J. et al. (2020). Advanced nanofiltration membrane fabricated on the porous organic cage tailored support for water purification application. *Separation and Purification Technology* 230: 115845.

68 Xu, X., Shao, Z., Shi, L. et al. (2020). Enhancing proton conductivity of proton exchange membrane with SPES nanofibers containing porous organic cage. *Polymers for Advanced Technologies* 31: 7.

69 Yuan, Y.D., Dong, J., Liu, J. et al. (2020). Porous organic cages as synthetic water channels. *Nature Communications* 11: 4927.

70 Zhou, Y., Zhang, X.-F., Yao, J., and Wang, H. (2022). Contra-diffusion synthesis of metal-organic framework separation membranes: a review. *Separation and Purification Technology* 300: 121837.

71 Du, Y., Zhang, X., Yang, J. et al. (2020). Ultra-thin graphene oxide films via contra-diffusion method: fast fabrication for ion rejection. *Journal of Membrane Science* 595: 117586.

72 Zhu, G., Hoffman, C.D., Liu, Y. et al. (2016). Engineering porous organic cage crystals with increased acid gas resistance. *Chemistry - A European Journal* 22: 10743–10747.

73 Martínez-Ahumada, E., He, D., Berryman, V. et al. (2021). SO_2 capture using porous organic cages. *Angewandte Chemie, International Edition* 60: 17556–17563.

74 Zhu, Z., Wang, D., Tian, Y., and Jiang, L. (2019). Ion/molecule transportation in nanopores and nanochannels: from critical principles to diverse functions. *Journal of the American Chemical Society* 141: 8658–8669.

75 Wang, H., Wang, M., Liang, X. et al. (2021). Organic molecular sieve membranes for chemical separations. *Chemical Society Reviews* 50: 5468–5516.

76 Wang, P., Wang, M., Liu, F. et al. (2018). Ultrafast ion sieving using Nanoporous polymeric membranes. *Nature Communications* 9: 569.

77 Beckstein, O., Tai, K., and Sansom, M.S.P. (2004). Not ions alone: barriers to ion permeation in nanopores and channels. *Journal of the American Chemical Society* 126: 14694–14695.

78 Zhao, W., Jia, C., Zhang, T. et al. (2021). Effects of nanopore geometry on confined water flow: a view of lattice Boltzmann simulation. *Chemical Engineering Science* 230: 116183.

79 Zhang, Z., Chen, B., Zhang, H. et al. (2022). Numerical simulation of ion transport across monovalent ion perm-selective membranes. *Chemical Engineering Science* 264: 118193.

80 Tandon, V., Bhagavatula, S.K., Nelson, W.C., and Kirby, B.J. (2008). Zeta potential and electroosmotic mobility in microfluidic devices fabricated from hydrophobic polymers: 1. *The Origins of Charge. Electrophoresis* 29: 1092–1101.

81 Xu, T., Shehzad, M.A., Wang, X. et al. (2020). Engineering leaf-like UiO-66-SO_3 H membranes for selective transport of cations. *Nano-micro Letter* 12: 1–11.

82 Wu, S., Cheng, Y., Ma, J. et al. (2021). Preparation and ion separation properties of sub-nanoporous PES membrane with high chemical resistance. *Journal of Membrane Science* 119467.

83 Guo, Y., Ying, Y., Mao, Y. et al. (2016). Polystyrene sulfonate threaded through a metal–organic framework membrane for fast and selective lithium-ion separation. *Angewandte Chemie, International Edition* 128. 13344–13348.

84 Côté, A.P., Benin, A.I., Ockwig, N.W. et al. (2005). Porous, crystalline, covalent organic frameworks. *Science* 310: 1166–1170.

85 Biswal, B.P., Chaudhari, H.D., Banerjee, R., and Kharul, U.K. (2016). Chemically stable covalent organic framework (COF)-Polybenzimidazole hybrid membranes: enhanced gas separation through pore modulation. *Chemistry* 22: 4695–4699.

86 Dey, K., Pal, M., Rout, K.C. et al. (2017). Selective molecular separation by interfacially crystallized covalent organic framework thin films. *Journal of the American Chemical Society* 139: 13083–13091.

87 Kandambeth, S., Biswal, B.P., Chaudhari, H.D. et al. (2017). Selective molecular sieving in self-standing porous covalent-organic-framework membranes. *Advanced Materials* 29.

88 Fan, H., Gu, J., Meng, H. et al. (2018). High-flux membranes based on the covalent organic framework COF-LZU1 for selective dye separation by nanofiltration. *Angewandte Chemie (International Ed. in English)* 57: 4083–4087.

89 Khan, N.A., Zhang, R., Wu, H. et al. (2020). Solid-vapor Interface engineered covalent organic framework membranes for molecular separation. *Journal of the American Chemical Society* 142: 13450–13458.

90 Shinde, D.B., Sheng, G., Li, X. et al. (2018). Crystalline 2D covalent organic framework membranes for high-flux organic solvent nanofiltration. *Journal of the American Chemical Society* 140: 14342–14349.

91 Yang, H., Yang, L., Wang, H. et al. (2019). Covalent organic framework membranes through a mixed-dimensional assembly for molecular separations. *Nature Communications* 10: 2101.

92 Chen, H., Tu, H., Hu, C. et al. (2018). Cationic covalent organic framework nanosheets for fast Li-ion conduction. *Journal of the American Chemical Society* 140: 896–899.

93 DeBlase, C.R., Silberstein, K.E., Truong, T.T. et al. (2013). beta-Ketoenamine-linked covalent organic frameworks capable of pseudocapacitive energy storage. *Journal of the American Chemical Society* 135: 16821–16824.

94 Chandra, S., Roy Chowdhury, D., Addicoat, M. et al. (2017). Molecular level control of the capacitance of two-dimensional covalent organic frameworks: role of hydrogen bonding in energy storage materials. *Chemistry of Materials* 29: 2074–2080.

95 Guo, Z., Zhang, Y., Dong, Y. et al. (2019). Fast ion transport pathway provided by polyethylene glycol confined in covalent organic frameworks. *Journal of the American Chemical Society* 141: 1923–1927.

96 Hu, Y., Dunlap, N., Wan, S. et al. (2019). Crystalline lithium imidazolate covalent organic frameworks with high Li-ion conductivity. *Journal of the American Chemical Society* 141: 7518–7525.

97 Shi, G.Q., Zhou, J.F., Li, Z. et al. Graphene-like covalent organic framework with a wide band gap synthesized on surface via stepwise reactions. *Angewandte Chemie-International Edition* 59 (37): 15958–15962.

98 Wang, L., Zeng, C., Xu, H. et al. (2019). A highly soluble, crystalline covalent organic framework compatible with device implementation. *Chemical Science* 10: 1023–1028.

99 Yu, F., Liu, W., Li, B. et al. (2019). Photostimulus-responsive large-area two-dimensional covalent organic framework films. *Angewandte Chemie (International Ed. in English)* 58: 16101–16104.

100 Zhang, M., Lu, M., Lang, Z.L. et al. (2020). Semiconductor/covalent-organic-framework Z-scheme heterojunctions for artificial photosynthesis. *Angewandte Chemie (International Ed. in English)* 59: 6500–6506.

101 Zhi, Y.F., Wang, Z.R., Zhang, H.L., and Zhang, Q.C. (2020). Recent Progress in metal-free covalent organic frameworks as heterogeneous catalysts. *Small* 16 (24): e2001070.

102 Zhu, Y.Y., Wang, W.D., Sun, X. et al. (2020). Palladium nanoclusters confined in MOF@COP as a novel Nanoreactor for catalytic hydrogenation. *ACS Applied Materials & Interfaces* 12: 7285–7294.

103 Zion, N., Cullen, D.A., Zelenay, P., and Elbaz, L. (2020). Heat-treated aerogel as a catalyst for the oxygen reduction reaction. *Angewandte Chemie-International Edition* 59: 2483–2489.

104 Huang, N., Chen, X., Krishna, R., and Jiang, D. (2015). Two-dimensional covalent organic frameworks for carbon dioxide capture through Channel-Wall functionalization. *Angewandte Chemie International Edition* 54: 2986–2990.

105 Kaleeswaran, D., Vishnoi, P., and Murugavel, R. (2015). [3+3] imine and β-ketoenamine tethered fluorescent covalent-organic frameworks for CO_2 uptake and nitroaromatic sensing. *Journal of Materials Chemistry C* 3: 7159–7171.

106 Lu, Q., Ma, Y., Li, H. et al. (2018). Postsynthetic functionalization of three-dimensional covalent organic frameworks for selective extraction of lanthanide ions. *Angewandte Chemie (International Ed. in English)* 57: 6042–6048.

107 Kandambeth, S., Dey, K., and Banerjee, R. (2019). Covalent organic frameworks: chemistry beyond the structure. *Journal of the American Chemical Society* 141: 1807–1822.

108 Sahabudeen, H., Qi, H., Ballabio, M. et al. (2020). Highly crystalline and semiconducting imine-based two-dimensional polymers enabled by interfacial synthesis. *Angewandte International Edition Chemie* 132: 6084–6092.

109 Yuan, S., Li, X., Zhu, J. et al. (2019). Covalent organic frameworks for membrane separation. *Chemical Society Reviews* 48: 2665–2681.

110 Xu, F., Wei, M., Zhang, X., and Wang, Y. (2019). Ion rejection in covalent organic frameworks: revealing the overlooked effect of in-pore transport. *ACS Applied Materials & Interfaces* 11: 45246–45255.

111 Sheng, F., Wu, B., Li, X. et al. (2021). Efficient ion sieving in covalent organic framework membranes with Sub-2-nanometer channels. *Advanced Materials* 33: e2104404.

112 Bing, S., Xian, W., Chen, S. et al. (2021). Bio-inspired construction of ion conductive pathway in covalent organic framework membranes for efficient lithium extraction. *Matter* .

113 Liu, R., Tan, K.T., Gong, Y. et al. (2021). Covalent organic frameworks: an ideal platform for designing ordered materials and advanced applications. *Chemical Society Reviews* 50: 120–242.

114 Nagai, A., Guo, Z., Feng, X. et al. (2011). Pore surface engineering in covalent organic frameworks. *Nature Communications* 2.

115 Li, Y., Wu, Q., Guo, X. et al. (2020). Laminated self-standing covalent organic framework membrane with uniformly distributed subnanopores for ionic and molecular sieving. *Nature Communications* 11: 599.

116 Mu, Z., Zhu, Y., Li, B. et al. (2022). Covalent organic frameworks with record pore apertures. *Journal of the American Chemical Society* 144: 5145–5154.

117 Dalapati, S., Jin, E., Addicoat, M. et al. (2016). Highly emissive covalent organic frameworks. *Journal of the American Chemical Society* 138: 5797–5800.

118 Zhou, T.Y., Xu, S.Q., Wen, Q. et al. (2014). One-step construction of two different kinds of pores in a 2D covalent organic framework. *Journal of the American Chemical Society* 136: 15885–15888.

119 Pang, Z.-F., Zhou, T.-Y., Liang, R.-R. et al. (2017). Regulating the topology of 2D covalent organic frameworks by the rational introduction of substituents. *Chemical Science* 8: 3866–3870.

120 Ma, H., Liu, B., Li, B. et al. (2016). Cationic covalent organic frameworks: a simple platform of anionic exchange for porosity tuning and proton conduction. *Journal of the American Chemical Society* 138: 5897–5903.

121 Peng, Y., Xu, G., Hu, Z. et al. (2016). Mechanoassisted synthesis of sulfonated covalent organic frameworks with high intrinsic proton conductivity. *ACS Applied Materials & Interfaces* 8: 18505–18512.

122 Zhao, S., Jiang, C., Fan, J. et al. (2021). Hydrophilicity gradient in covalent organic frameworks for membrane distillation. *Nature Materials* 20: 1551–1558.

123 Mitra, S., Kandambeth, S., Biswal, B.P. et al. (2016). Self-exfoliated guanidinium-based ionic covalent organic nanosheets (iCONs). *Journal of the American Chemical Society* 138: 2823–2828.

124 Zhang, K., He, Z., Gupta, K.M., and Jiang, J. (2017). Computational design of 2D functional covalent–organic framework membranes for water desalination. *Environmental Science: Water Research & Technology* 3: 735–743.

125 Halder, A., Karak, S., Addicoat, M. et al. (2018). Ultrastable imine-based covalent organic frameworks for sulfuric acid recovery: an effect of interlayer hydrogen bonding. *Angewandte International Edition Chemie* 57: 5797–5802.

126 Berlanga, I., Ruiz-Gonzalez, M.L., Gonzalez-Calbet, J.M. et al. (2011). Delamination of layered covalent organic frameworks. *Small* 7: 1207–1211.

127 Chandra, S., Kandambeth, S., Biswal, B.P. et al. (2013). Chemically stable multilayered covalent organic nanosheets from covalent organic frameworks via mechanical delamination. *Journal of the American Chemical Society* 135: 17853–17861.

128 Kaushik Dey, Manas Pal, Kanhu Charan Rout, Shebeeb Kunjattu H., Anuja Das, Rabibrata Mukherjee, Ulhas K. Kharul, and Rahul Banerjee. Selective molecular separation by Interfacially crystallized covalent organic framework thin films. 2017.

129 Shinde, D.B., Huang, K.-W., and La, Z. (2018). Crystalline 2D covalent organic framework membranes for highFlux organic solvent nanofiltration. *Journal of the American Chemical Society* 140 (43): 14342–14349.

130 Medina, D.D., Rotter, J.M., Hu, Y. et al. (2015). Room temperature synthesis of covalent–organic framework films through vapor-assisted conversion. *Journal of the American Chemical Society* 137: 1016–1019.

131 Smith, B.J., Parent, L.R., Overholts, A.C. et al. (2017). Colloidal covalent organic frameworks. *ACS Central Science* 3: 58–65.

132 Sheng, F., Li, X., Li, Y. et al. (2022). Cationic covalent organic framework membranes for efficient dye/salt separation. *Journal of Membrane Science* 644: 120118.

133 Corry, B. (2017). Mechanisms of selective ion transport and salt rejection in carbon nanostructures. *MRS Bulletin* 42: 306–310.

134 Cao, X.-L., Yan, Y.-N., Zhou, F.-Y., and Sun, S.-P. (2020). Tailoring nanofiltration membranes for effective removing dye intermediates in complex dye-wastewater. *Journal of Membrane Science* 595: 117476.

135 Abdi, G., Alizadeh, A., Zinadini, S., and Moradi, G. (2018). Removal of dye and heavy metal ion using a novel synthetic polyethersulfone nanofiltration membrane modified by magnetic graphene oxide/metformin hybrid. *Journal of Membrane Science* 552: 326–335.

136 Ormanci-Acar, T., Tas, C.E., Keskin, B. et al. (2020). Thin-film composite nanofiltration membranes with high flux and dye rejection fabricated from disulfonated diamine monomer. *Journal of Membrane Science* 608: 118172.

137 Zhu, J., Tian, M., Hou, J. et al. (2016). Surface zwitterionic functionalized graphene oxide for a novel loose nanofiltration membrane. *Journal of Materials Chemistry A* 4: 1980–1990.

138 Liu, F., Ma, B.-r., Zhou, D. et al. (2015). Positively charged loose nanofiltration membrane grafted by diallyl dimethyl ammonium chloride (DADMAC) via UV for salt and dye removal. *Reactive and Functional Polymers* 86: 191–198.

139 Kang, Y., Jang, J., Kim, S. et al. (2020). PIP/TMC interfacial polymerization with electrospray: novel loose nanofiltration membrane for dye wastewater treatment. *ACS Applied Materials & Interfaces* 12: 36148–36158.

140 Xing, L., Guo, N., Zhang, Y. et al. (2015). A negatively charged loose nanofiltration membrane by blending with poly (sodium 4-styrene sulfonate) grafted SiO_2 via SI-ATRP for dye purification. *Separation and Purification Technology* 146: 50–59.

141 Agrawal, A. and Sahu, K.K. (2009). An overview of the recovery of acid from spent acidic solutions from steel and electroplating industries. *Journal of Hazardous Materials* 171: 61–75.

142 Yuan, Z., Zhu, X., Li, M. et al. (2016). A highly ion-selective zeolite flake layer on porous membranes for flow battery applications. *Angewandte International Edition Chemie* 55: 3058–3062.

143 Naidu, G., Ryu, S., Thiruvenkatachari, R. et al. (2019). A critical review on remediation, reuse, and resource recovery from acid mine drainage. *Environmental Pollution* 247: 1110–1124.

144 Dong, Q., Guo, X., Huang, X. et al. (2019). Selective removal of lead ions through capacitive deionization: role of ion-exchange membrane. *Chemical Engineering Journal* 361: 1535–1542.

145 Feng, J., Chen, J., Wei, B. et al. (2019). Series-connected hexacations cross-linked anion exchange membranes for diffusion dialysis in acid recovery. *Journal of Membrane Science* 570-571: 120–129.

146 Ji, W., Wu, B., Zhu, Y. et al. (2020). Self-organized nanostructured anion exchange membranes for acid recovery. *Chemical Engineering Journal* 382: 122838.

147 Kamali, M., Suhas, D.P., Costa, M.E. et al. (2019). Sustainability considerations in membrane-based technologies for industrial effluents treatment. *Chemical Engineering Journal* 368: 474–494.

148 He, Y., Pan, J., Wu, L. et al. (2015). Facile preparation of 1,8-Diazabicyclo [5.4.0]undec-7-ene based high performance anion exchange membranes for diffusion dialysis applications. *Journal of Membrane Science* 491: 45–52.

149 Liu, J., Han, G., Zhao, D. et al. (2020). Self-standing and flexible covalent organic framework (COF) membranes for molecular separation. *Science Advances* 6: eabb1110.

150 Ruan, H., Pan, N., Wang, C. et al. (2021). Functional UiO-66 series membranes with high perm selectivity of monovalent and bivalent anions for electrodialysis applications. *Industrial & Engineering Chemistry Research* 60: 4086–4096.

151 Ge, L., Wu, B., Li, Q. et al. (2016). Electrodialysis with nanofiltration membrane (EDNF) for high-efficiency cations fractionation. *Journal of Membrane Science* 498: 192–200.

152 Hou, L., Wu, B., Yu, D. et al. (2018). Asymmetric porous monovalent cation perm-selective membranes with an ultrathin polyamide selective layer for cations separation. *Journal of Membrane Science* 557: 49–57.

153 Sheng, F., Hou, L., Wang, X. et al. (2020). Electro-nanofiltration membranes with positively charged polyamide layer for cations separation. *Journal of Membrane Science* 594: 117453.

154 Soyekwo, F., Wen, H., Liao, D., and Liu, C. (2022). Nanofiltration membranes modified with a clustered multiquaternary ammonium-based ionic liquid for improved magnesium/Lithium separation. *ACS Applied Materials & Interfaces* 14: 32420–32432.

155 Shi, L., Liu, S., Hung, W.-S. et al. (2023). The tailoring of nanofiltration membrane structure for mono/divalent anions separation via precisely adjusting the reaction site distance. *Journal of Membrane Science* 668: 121252.

156 Liu, Z., Mi, Z., Meng, L. et al. (2023). Quaternary ammonium salts modification preparing charged Janus nanofiltration membrane for the simultaneous separation of divalent anions and cations. *Journal of Membrane Science* 672: 121440.

157 Wang, Q., Wang, Y., Chen, B.-Z. et al. (2019). Designing high-performance Nanofiltration membranes for high-salinity separation of sulfate and chloride in the Chlor-alkali process. *Industrial & Engineering Chemistry Research* 58: 12280–12290.

158 Yuan, B., Zhao, S., Xu, S. et al. (2022). Aliphatic polyamide nanofilm with ordered nanostripe, synergistic pore size and charge density for the enhancement of cation sieving. *Journal of Membrane Science* 660: 120839.

159 Liu, C., Yang, J., Guo, B.-B. et al. (2021). Interfacial polymerization at the alkane/ionic liquid Interface. *Angewandte Chemie, International Edition* 60: 14636–14643.

160 Xu, Y., Peng, H., Luo, H. et al. (2022). High performance Mg^{2+}/Li^{+} separation membranes modified by a bis-quaternary ammonium salt. *Desalination* 526: 115519.

161 Huang, B.-Q., Tang, Y.-J., Zeng, Z.-X. et al. (2020). High-performance zwitterionic nanofiltration membranes fabricated via microwave-assisted grafting of betaine. *ACS Applied Materials & Interfaces* 12: 35523–35531.

162 Zhao, J., You, X., Wang, G. et al. (2022). Mix-charged polyamide membranes via molecular hybridization for selective ionic nanofiltration. *Journal of Membrane Science* 644: 120051.

163 Zhao, Z., Li, X., Zhang, H. et al. (2022). Polyamide-based electronanofiltration membranes for efficient anion separation. *Industrial & Engineering Chemistry Research* 61: 9869–9878.

9

Applications

9.1 Diffusion Dialysis (DD)

Diffusion dialysis (DD) uses an ion exchange membrane as the separation medium, takes the concentration difference of the solution on both sides of the ion exchange membrane as the mass transfer driving force, relies on the difference in the diffusion coefficient of ions with different hydration radii in the ion membrane, and realizes the removal and recovery of the target ions from the mixed solution based on the theory of conservation of charge, which is currently widely used in the resource recovery of waste acid [1–7] and waste alkali [8]. DD is the earliest membrane separation process discovered and studied, and it is a naturally occurring physical phenomenon. With the difference in the diffusion coefficient of ions in the membrane, the membrane process that allows various solutes to separate is called dialysis, also known as DD or natural dialysis, because the driving force of the process is the concentration difference, so it is also called concentrated diffusion. In this process, the entropy of the system increases and the Gibbs free energy decreases, which is favorable in thermodynamics. The main advantages of DD over other membrane separation methods are (i) low energy consumption, usually with no additional energy application (except for equipment, such as the energy consumption of pumps); (ii) low installation and operation costs, the cost of DD membranes is lower than that of electric drive ion exchange membranes, and the composition of DD equipment is relatively simple, and the operating cost is low; (iii) stable, reliable, easy to operate, fixed operating conditions, and stable composition of liquid; and (iv) environmentally friendly and pollution-free.

9.1.1 The Basic Theory of Diffusion Dialysis

At present, there are two theoretical models used to explain the transport mechanism of ions in DD membranes: the dissolution-diffusion model and the three-phase membrane model [9, 10]. The dissolution-diffusion model holds that the solute to be separated will first be adsorbed on the surface of the membrane on the side with a high solute concentration. Under the action of a concentration difference, the molecules or ions of the solute will spontaneously diffuse to the low concentration side, and the diffusion rate of different molecules or ions in the membrane is quite

Ion Exchange Membranes: Design, Preparation, and Applications,
First Edition. Tongwen Xu and Yaoming Wang.

different to realize the separation of ions. This diffusion process follows Fick's law. In the DD process, hydroxide ions (or hydrogen ions) in solution have a smaller hydration radius than acid anion ions (or metal ions), have a faster diffusion rate in solution and the membrane phase, and will preferentially adsorb and diffuse through the ion membrane to the low concentration side. However, this model is not suitable for the separation process of organic acids or bases because organic acids or organic bases mostly exist as weak species, the degree of dissociation is low, and there is interaction with the membrane, resulting in a relatively complex diffusion process in the membrane phase, which is also a limitation of the dissolution–diffusion model [10]. The three-phase membrane theory holds that the homogeneous membrane is composed of a hydrophobic polymer backbone phase, an active zone composed of functional groups and counterions, and an interstitial zone in the intermediate region of the two [11]. In a solution, ions exist and move in the form of hydrated ions, so the diffusion process in the membrane phase should also occur in the active zone and the interstitial zone [12]. In the $FeCl_2$/HCl system as an example, under the action of electrostatic attraction, chloride ions are mainly transferred in the positively charged active zone and follow the hopping mechanism; that is, they migrate between ion exchange groups. To maintain the requirement of electrical neutrality, cations (H^+ ions and Fe^{2+} ions) are transferred through the region with small charge repulsion and follow the dragging mechanism. The small charge and hydration radius of hydrogen ions can also be transferred in the positively charged active phase. During the whole process, the hydrophobic polymer does not have ion transmission but only plays a supporting role.

DD also has certain limitations: the concentration gradient as a driving force leads to low membrane flux and the processing capacity and acid–base recovery rate of membrane stacks are relatively low. Second, DD cannot completely separate the ions in the waste liquid, so secondary treatment is needed. The acid or base recovery rate is proportional to the flow rate ratio (pure water side/feed side), and to obtain a higher recovery rate, a higher flow rate ratio is also needed, resulting in the consumption of a large amount of fresh water. At present, the flow rate ratio in the industry is often between 1.0/1.0–1.2/1.0, and 1.0 tons or even 1.2 tons of pure water are consumed to treat 1.0 tons of waste liquid [13].

9.1.1.1 High-performance Diffusion Dialysis Membranes

The ion exchange membrane is the core component of the DD stack, and the DD membrane with excellent performance can effectively improve the recovery rate of waste acid and alkali and the separation efficiency. An excellent diffusion DD membrane should have the following characteristics [14]:

(a) High H^+/OH^- ion flux and selectivity. DD has advantages that other technologies cannot match, such as low energy consumption, simple equipment, and a high degree of automation. The low membrane flux during operation limits its processing capacity, and it is difficult to meet the waste acid and alkali treatment needs of large-scale industries. Adopting appropriate membrane production strategies to improve membrane flux is of great significance for the realization of DD industrial applications.

(b) Excellent mechanical properties and thermostable stability. The pressure is not high during the DD process. It is affected by factors such as fluid erosion and temperature during long-term operation, which will also lead to a decrease in stability and affect the operation of the DD.
(c) Good chemical stability. The complex composition of the waste liquid, which includes a large number of heavy metal ions and high concentrations of acids and alkalis affect the functional groups and even the main chain of the membrane. To ensure that the membrane can operate for a long time, it should have high acid/alkali resistance, solvent resistance, oxidation resistance, anti-fouling ability, etc.
(d) Simple preparation methods and cost-effective. At present, the DD process is mostly used in metallurgical and other industries in waste acids and alkali recovery. Due to the low added value of the product, investment cost is a key factor restricting its large-scale application. Therefore, it is particularly important to develop a high-performance diffusion dialysis membrane that is simple to prepare and cost-effective.

By adjusting the ion exchange capacity (IEC) of the membrane, the exchange capacity of the membrane can be effectively improved, which is beneficial to the improvement of the membrane flux. However, too high of an IEC will lead to swelling of the membrane, which is not conducive to the retention of metal ions or acid ions. Therefore, it is of great significance to rationally optimize the charged group content and water content of the membrane and simultaneously improve the pore size and crosslinking degree of the main chain of the base membrane material to prepare a DD membrane with high flux, selectivity, and stability. Based on this, the current main strategies for preparing membranes include organic–inorganic hybridization [15–19], surface modification [20–22], phase inversion [2, 23–28], and electrospinning [29–32].

The organic–inorganic hybrid method is based on the polymer organic phase as the main component, taking advantage of the good solvent resistance, acid and alkali resistance, and other advantages of the inorganic phase. High-performance DD membranes were prepared by improving the hydrophilicity, IEC, and density of the main body. Among them, the commonly used organic phases are polyphenylene oxide (PPO) [2, 27, 33], polyethylene glycol (PEG) [2, 27], polyvinylidene fluoride (PVDF) [34–36], and polyvinyl alcohol (PVA) [21, 32, 37–40]. The inorganic phase includes silicon dioxide (SiO_2) [29, 41–43], titanium dioxide (TiO_2) [44, 45], alumina (Al_2O_3), etc. The combination of organic and inorganic materials can also be divided into hybrid membrane preparation methods with weak interactions and covalent bonds or ionic bonds.

The interface modification method is a simple and efficient method that improves the interception capacity of the polymer membrane to acid anion ions or metal ions by modifying the membranes interface. Polystyrene, divinylbenzene, and other monomers are polymerized in the pores of the polyethylene (PE) microporous membrane [9, 20]. By virtue of the quaternary ammonium reaction and surface modification, the anion exchange groups, and a surface pyridine layer are introduced, which effectively improve the selectivity and stability of the PE porous membrane and show excellent performance in the DD test [20, 46]. Commercial anion exchange membranes (AEMs) can also be modified, and some scholars have introduced a dense polypyrrole layer on the surface of the commercial anion Neosepta-AFX and obtained good DD performance. Compared with the base membrane, the optimized polypyrrole-modified membrane can significantly inhibit the transport of Fe^{2+} and improve the separation efficiency of the membrane. However, this method also has certain drawbacks, although the modification of the polypyrrole layer greatly improves the selectivity of the membrane; however, the flux of the membrane also decreases.

In addition, the use of blending and crosslinking between polymers is also a common preparation method for DD membranes, which is simple in steps, easy to operate, and easy to achieve for large-scale preparation. For example, DF series DD membranes (developed by China Shandong Tianwei Membrane Technology Co., Ltd.) is amine brominated with polyphenylene ether, and the prepared membrane has good mechanical strength and chemical stability. The above-mentioned preparation methods mainly inhibit the transport of metal or acid anion group ions by changing the density, hydrophilicity, or charge density of the membrane matrix or surface. In recent years, the membrane preparation strategy to construct ion transport channels has become a research hotspot. This type of method optimizes the design of the polymer structure, introduces corresponding hydrophilic auxiliary groups, and constructs specific ion channels at the same time, which can effectively reduce the defects of low flux and low selectivity in traditional preparation methods [47–52], thereby improving DD performance.

9.1.2 Diffusion Dialysis Components

Static diffusion cells are often used in laboratory research. Usually, made from plexiglass, they consist of two compartments, that is, diffusion compartment and the dialysis compartment, which are separated by DD membranes. The device is mainly used for rapid testing of the dialysis coefficient and separation factor, which is convenient for theoretical research on the DD process.

In the plate-frame DD component, the device includes a certain number of repeating units, the DD membrane, and the grid are alternately arranged to form a diffusion compartment and a dialysis compartment, and the components also include a liquid distribution plate, reinforcement plate, and liquid flow plate frame. The advantages include simple structure, easy disassembly, and assembly [53]. As shown in Figure 9.1, the dialysate and diffusion fluid flow countercurrent inside the membrane stack.

The spiral wound diffusion dialysis (SWDD) component [4] overcomes the shortcomings of plate-frame DD components, such as large volume, unsuitable transportation, and insufficient mass transfer. The components are independently developed by the University of Science and Technology of China, Xu, and others. The main features of this component are its compact structure, high filling density, high mass transfer efficiency, large processing capacity per unit membrane area, and easy integration with other equipment. The assembly adds an ion exchange

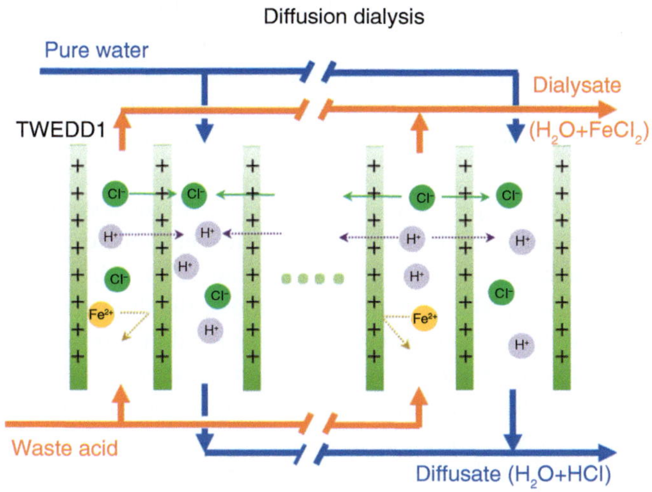

Figure 9.1 Diagram of the mechanism of diffusion dialysis recovery of waste acid. TWEDD1 presents one type of diffusion dialysis anion exchange membrane.

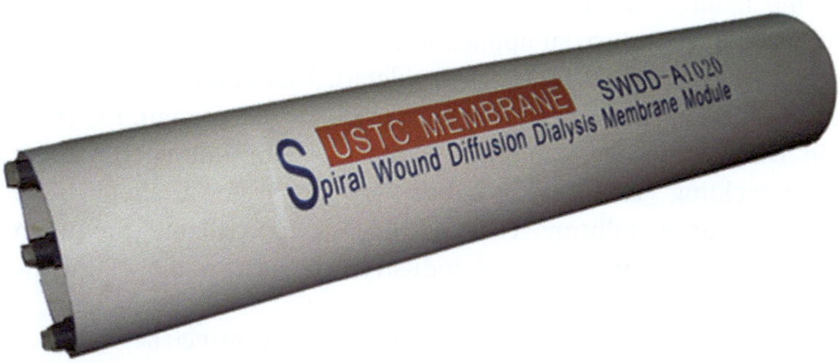

Figure 9.2 Digital image of the spiral wound diffusion dialysis.

membrane between two semi-circular central tubes with small holes on the surface, lining the two sides of the membrane with a channel grid, and then the membrane and the partition mesh are rolled around the central tube to form a cylindrical membrane module body with two flow channels. Finally, the body is sealed in a cylindrical PVC shell. Figure 9.2 is a schematic view of the spiral DD membrane module section and the appearance of the module.

9.1.3 Diffusion Dialysis Application Field

9.1.3.1 Recovery of Waste Acid

Figure 9.1 is a schematic diagram of the DD recovery of steel waste acid ($FeCl_2/HCl$). The waste acid liquid and fresh water are on both sides of the membrane, and under the force of the concentration difference, Fe^{2+}, Cl^-, and H^+ all have a tendency to diffuse to the water side, but the AEM only allows Cl^- ions to be permeable and has a strong charge repulsion for Fe^{2+}, so Fe^{2+} cannot penetrate the AEM. However, because H^+ has a small ion hydration radius, low charge density, and low steric resistance across membranes, it is relatively easy to reach the water side through the membrane. Combine with Cl^- to ensure the electrical neutrality requirements of the concentration chamber. The implementation principle is similar to other waste acid systems.

The performance of DD to recover spent acid is mainly related to the concentration difference between the two sides of the membrane and the performance of the ion exchange membrane. In practice, the waste acid concentration is fixed, and the acid recovery rate is changed by adjusting the flow rate ratio of the diffusion side and the dialysis side. To achieve high acid recovery, increasing the flow rate of fresh water is necessary. However, this also reduces the acid concentration recovered and consumes a large amount of fresh water. At present, the industrial flow rate ratio of the diffusion side to the dialysis side is 1.1/1–1.2/1.

Second, it is equally important to develop DD ion exchange membranes with high H^+ flux and low metal ion leakage. The DF [6, 54, 55] series anion membranes mentioned above are based on polyphenylene ether (PPO) poly (2,6-dimethyl-1, 4-phenylene oxide) further bromination. With the help of PPO's excellent water resistance, good mechanical strength, and good film formation after bromination, it has successfully achieved industrial application and has become the main force of DD commercial films. Table 9.1 provides a detailed summary of commonly used commercial DD ion exchange membranes worldwide. At present, the industries involved in DD acid recovery mainly include the aluminum foil industry ($HCl/AlCl_3$), steel pickling ($HCl/FeCl_2$), battery (H_2SO_4 system), chemical foil ($H_2SO_4/Al_2[SO_4]_3$), hydrometallurgy ($HNO_3/CuNO_3$), polysilicon production industry (HCl + HF system), and graphite industry. The integration of the two

Table 9.1 Properties of some commercially available ion exchange membranes for the diffusion dialysis process.

Membrane	Type	Thickness (mm)	Rupture strength (MPa)	Area resistance[a)] ($\Omega \cdot cm^2$)
Selemion DSVN	AEM	0.095	≥0.15	1.1
Neosepta AFX	AEM	0.17	≥0.25	1.0
Neosepta AFN	AEM	0.16	≥0.25	0.5
SB-6407	AEM	0.152	–	0.3–1.2
DF120–1	AEM	0.30–0.32	–	–
DF120–2	AEM	0.20–0.30	–	–
TWDDC1 (Standard)	ACM	0.10–0.13	–	≤ 4.00
TWDDC1 (Anti-fouling)	ACM	0.10–0.13	–	≤ 4.00

Note. The data were directly collected from the product brochure provided by the manufacturers.
a) Area resistances were measured in 0.5 mol/l NaCl, 25 °C.

processes of electrodialysis (ED) and DD can realize the recovery and concentration of waste acid. The process first introduces the waste acid solution (aluminum foil waste acid) into SWDD and then concentrates the recovered acid by ED [56]. This process proves that DD coupled with ED can not only save water consumption but also increase the concentration of DD recovered acid, which is convenient for direct reuse. In addition to coupling DD with ED [57], coupling DD with bipolar membrane electrodialysis (BMED) can realize waste acid recovery at the same time and alkalinize aluminum ions to produce polyaluminum chloride products to achieve high-value recovery of aluminum.

Selective electrodialysis (SED) uses the electric field force as the driving force and predicts that its flux is higher than that of DD. The acid recovery rate, pure water usage, and treatment cost are used as indicators to compare DD, SED, and DD + SED coupling processes [13]. The H^+ performance of recycling $HCl/FeCl_2$ is shown in Figure 9.1. Under the same feed concentration (1.5 M/1.0 M, $HCl/FeCl_2$), the SED membrane flux is up to approximately 20 times that of DD, and in the SED process, when the current density is greater than 10 mA/cm^2, the acid recovery rate can reach more than 80%, and the Fe^{2+} leakage rate is less than 3.15%. As the current density increases, the separation efficiency first increases and then decreases. At the optimal current density of 25 mA/cm^2, the acid recovery rate of SED was 90.75%, the iron leakage rate was 1.57%, and the separation efficiency was the highest at 95.73%. The coupling of the DD-SED process results in higher acid recovery, lower disposal costs, and less water consumption. With a flow ratio of 0.5 : 1 and a current density of 25 mA/cm^2, the total process cost was 17.60 $/t, and the acid recovery rate was as high as 92.38%. This coupling process costs much less than single DD (20.66 $/t) and single SED (18.24 $/t) processes. SED and NF coupling [13] also shows superiority in molybdenum metallurgy waste acid recycling [58]. Under the optimal conditions, the sulfuric acid recovery rates of the single SED and NF processes are 89.6% and 59.1%, and the retention rates of MoO_2^{2+} are 98.1% and 96.3%, respectively. SED–NF coupling achieves up to 94.2% spent acid recovery. Through the integration of membrane separation processes such as DD, ED, and BMED, its processing capacity and separation efficiency have been greatly improved, and DD is still a separation technology with considerable practical application value.

9.1.3.2 Alkali Recovery

The principle of DD alkali recovery is consistent with acid recovery, but due to the large hydration radius of OH^-, the steric hindrance in the ion exchange membrane is much higher than that of H^+, and alkali resistance has always been one of the challenges in preparing ion exchange membranes. Therefore, there are relatively few cation exchange membranes (CEMs) for DD compared to AEMs. Xu's team developed a series of alkali recovery CEMs based on sulfonated polyphenylene ether [43] and a mixed matrix membrane composed of PVA and various substances [59–61].

At present, DD alkali recovery is mainly used in the metallurgical industry, and the extraction of tungsten, aluminum, and vanadium in tungsten, aluminum, and vanadium ore is usually used by alkali leaching, resulting in the production of a large amount of waste alkali solution. In view of this [8], DD and membrane electrolysis can be used to recover the alkali solution from alkaline sodium metavanadate (Na + $NaVO_3$), and the preliminary separation of spent alkali can be realized by the CEM DD device hybridized by PVA. The recovered alkali solution is then concentrated to 1.445 mol/l again by membrane electrolysis, and the pH of the spent alkali solution can be reduced to approximately 6. After membrane electrolysis, the dialysate of DD can be returned directly to the vanadium precipitation process without the need to introduce acid for neutralization. In addition, the hydrogen and oxygen produced during membrane electrolysis can be reused as clean energy in the calcination process in the V_2O_5 production process, further reducing energy consumption.

However, DD still has limitations, such as low membrane flux leading to low treatment capacity and low acid–base recovery efficiency. A single 512 m^2 DD industrial device can only dispose of 6 tons of waste acid per day. In addition, the acid–base concentration of the recovered solution is also limited by the equilibrium concentration, resulting in a low concentration of recovered acid–base, and there is still a gap in the attention given to membrane separation processes such as ED and reverse osmosis (RO). In recent years, with the continuous advancement of synthesis methods and the preparation of membrane materials, the types of homogeneous ionic membranes have also been expanded. SED has shown advantages in the field of acid recovery and is composed of mono/divalent ion exchange membranes. Driven by the force of the electric field, the membrane flux of SED is nearly 20 times that of DD. By coupling DD and SED, the treatment capacity of waste acid is increased, and the treatment cost is also greatly reduced.

9.2 Reverse Electrodialysis (RED)

Salinity gradient power (SGP), as a clean energy source that can be used by humans in nature, is called osmotic energy or "blue energy," which refers to the chemical potential difference energy between seawater and freshwater or two solutions with different salt concentrations, mainly in the estuaries of rivers. Unlike intermittent energy sources, salt differential energy can be obtained continuously and is less limited by climatic and environmental conditions. Theoretically, the salt differential energy density at the river–sea intersection is approximately 0.8 kWh/m^3, and the total salt differential energy reserves in the global estuarine areas are as high as 30 and 2.6 TW may be utilized. In addition, high-salt wastewater from the process industry and brine from natural systems can be used as important sources of salt differential energy [62]. Since 1939, people have used salt differential energy to convert and use it for power generation, and with the maturity of technology in the past 10 years, research on salt differential energy technology in various countries has gradually increased. The main technologies to achieve salt differential energy conversion are pressure retarded osmosis (PRO) [63–65], reverse electrodialysis (RED) [66–68], capacitive mixing (CapMix) [69, 70], and the vapor pressure method (VPD) [71]. These four types of salt differential energy conversion technologies have their own advantages in the fields they use, but they are subject to their

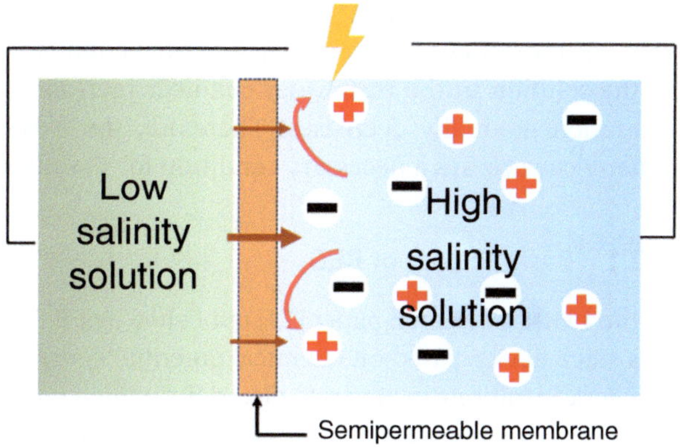

Figure 9.3 A schematic diagram of the principle of the PRO process technology.

respective technical characteristics. For example, the VPD energy plant is too large, and the initial investment is expensive. The CapMix method is periodic, and the voltage drops significantly over time due to the self-discharge of the battery and the nonideal behavior of the film. Both methods remain in the experimental stage. Relatively speaking, RED and pressure-retarded osmosis (PRO) with high power density and energy recovery rate are relatively common and promising conversion technologies.

The PRO operating mechanism diagram is shown in Figure 9.3. The principle is to place the asymmetric porous membrane between different concentrations of salt solutions, and the chemical potential difference between the two solutions is manifested as osmotic pressure. The osmotic pressure difference between fresh water and salt water is used as the driving force so that the water penetrates from the low-concentration side to the high-concentration side, with the increase in the volume flow of the solution on the high-concentration side to drive the turbine generator to generate electricity to achieve the conversion between salt difference energy, potential energy, mechanical energy, and electrical energy.

The RED concentration difference energy power generation process is the reverse process of ED, which uses the directional and selective migration of different ions in the ion exchange membrane to directly convert chemical potential energy into electrical energy. It has the advantages of low investment cost and high energy density. The technical principle is shown in Figure 9.4. CEM arranged in a staggered manner, grids and AEMs are used to form flow channels so that seawater and freshwater with a certain concentration difference flow through. Under the force of the concentration difference, Na^+ and Cl^- in the seawater channel pass through the CEM and AEM, respectively, and migrate into the adjacent fresh water, forming a directional charge flow. The electrode solution carrying ions circulates in the electrode compartment formed by the electrode plate and the ion exchange membrane, and oxidation and reduction reactions occur. The migration of ions in the stack leads to the generation of internal current, and the electromotive force is formed through the oxidation–reduction reaction on the electrodes. The electrons flow through the external circuit to do work, and the concentration difference energy (chemical potential energy) between seawater and freshwater is converted into electrical energy in the RED manner. As the process progresses, the concentration of ions in the seawater chamber decreases, the concentration of the solution in the freshwater chamber increases, and the driving force for ion migration gradually decreases. Therefore, the existence of a salt concentration difference becomes a necessary condition for the sustainable RED process [72].

9.2.1 Basic Theory of RED

In the 1950s, a research paper [73] published by PE Pattle in Nature proposed the use of a stack device based on chemical potential energy between solutions of different salt concentrations to generate electricity, which was a pioneering work in the field of RED based on salt differential energy. The core of RED technology is the use of ion exchange membranes to extract energy from two solutions with salinity gradients. The solution source can be river water and seawater or a combination of river water and industrial wastewater. Although the potential of using RED technology to convert salt differential energy is huge, it was not until the 1970s and early 1980s of the twentieth century that RED technology began to attract attention, especially since 2010, and the number of research papers on RED technology has increased year by year, which is also inseparable from the development of ion membranes in the past few years.

The subject device of RED is that the RED membrane stack (Figure 9.4) is mainly composed of anodes, alternately arranged AEM/CEMs, and cathodes. The adjacent ion exchange membranes are mainly separated by the grid, which results in the formation of alternating high-salinity (HS) stream compartments and low-salinity (LS) stream compartments. Two adjacent ion exchange membranes constitute a single

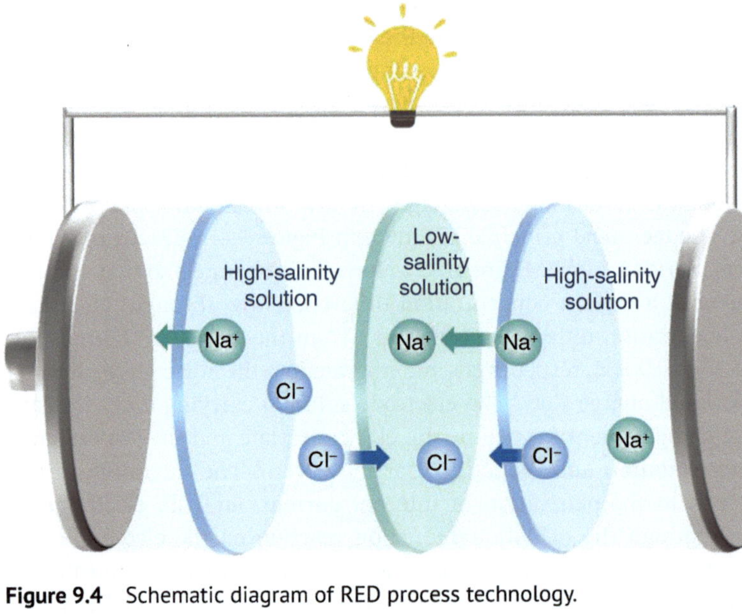

Figure 9.4 Schematic diagram of RED process technology.

membrane pair ("cell" or "cell pair"), and a cell includes an AEM, HS, CEM, and LS [66, 74–76]. A RED membrane stack usually contains a large number of membrane cells. When the anode and cathode at both ends are connected to the load to form a complete circuit, driven by the concentration difference, the anions and cations in the HS chamber pass through the AEM and the CEM, respectively, and migrate directionally to the LS chamber to form an internal current. Then, through the electrochemical reaction (redox reaction) of the electrode, the ion conductor is converted into an electrical conductor, and the internal current of ion migration can be converted into the external circuit current of electron migration to supply power to the load [77–79].

When the salt differential energy is converted by the RED process, the output power (W) is the most important parameter, and the equation is obtained by Kirchhoffs law [67, 80]; see Eq. (9.1):

$$W = I^2 \times R_{\text{load}} = \frac{E^2}{(R_{\text{stack}} + R_{load})^2} \times R_{\text{load}} \tag{9.1}$$

I is the current; R_{load} is the external resistance; E is the potential difference of the membrane stack; and R_{stack} is the resistance of the membrane stack. The output power of the RED process mainly depends on two types of factors: the potential energy difference E and the total resistance R_{stack} of the membrane stack (composed of ohmic resistance and nonohmic resistance). The greater the potential energy difference E, that is, the greater the concentration difference of the salt solution on both sides of the membrane, the smaller the total resistance of the membrane stack, and the greater the output power W, indicating a higher efficiency of the salt difference energy recovery.

The transmembrane potential energy difference (E) is generated by different concentrations of saline solutions on both sides of the membrane. In an ideal case, the total potential E generated by the RED stack can be calculated using the Nernst equation, as shown in Eq. (9.2):

$$E = N(\alpha_{\text{CEM}} + \alpha_{\text{AEM}})\frac{RT}{zF} \ln \frac{\gamma_{\text{HC}} C_{\text{HC}}}{\gamma_{\text{LC}} C_{\text{LC}}} \tag{9.2}$$

N represents the number of membrane cells; α indicates the permeation coefficient; R is the gas constant; T is the absolute temperature; z is the ionization valence; F is the Faraday constant; γ is the ion activity coefficient; and C is the ion concentration in the solution.

E mainly depends on the permeability coefficient of the membrane, the concentration gradient, and the valence state of the ion. The permeation coefficient of the membrane represents the ability of the membrane to counterions and exclude coions. Improving the permeability selectivity coefficient and concentration gradient is conducive to achieving a high potential energy difference.

The R_{stack} consists of three parts: ohmic resistance, nonohmic resistance, and electrode resistance. The membrane stack ohmic resistance R_{ohm} is mainly composed of the resistance of the membrane and the solution; nonohmic resistance is attributed to diffusion boundary layer resistance (R_{BL}) and electric double-layer resistance (R_{DE}) [81, 82]. The values of R_{BL} and R_{DE} can be determined by electrochemical impedance spectroscopy (EIS) to calculate [83]. In general, in highly concentrated brine systems, nonohmic resistors contribute little to the stack resistance, and the influence of nonohmic resistors on the system will only be considered in regard to operations in low-concentration solutions. The electrode resistance is attributed to the Nernst voltage, overpotential, and ohmic portion of the electrode solution of the redox reaction. The diffusion boundary layer formed on the electrode surface can be adjusted by the current density, electrolyte temperature, concentration, and flow rate. In RED large-scale applications, if there are enough membrane cells, the influence of electrode resistance on the resistance of the membrane can be ignored. Therefore, the stack resistance R_{stack} can be calculated by Eq. (9.3):

$$R_{\text{stack}} = R_{\text{ohm}} + R_{\text{BL}} + R_{\text{DE}} \tag{9.3}$$

where the ohmic resistance is composed of membrane resistance and a solution resistance, which can be calculated by Eq. (9.4):

$$R_{\text{ohm}} = N \times (R_{\text{AEM}} + R_{\text{CEM}} + R_{\text{HC}} + R_{\text{LC}}) \tag{9.4}$$

where R_{AEM} and R_{CEM} are the AEM and CEM resistances, respectively, and R_{HC} and R_{LC} are the resistances of the HS and LS compartments, respectively. In the RED stack, according to the ion concentration gradient distribution, the salt concentration of the diffusion boundary layer on the membrane surface of the HC side is lower than that of the bulk solution, while the opposite is true on the other side of the LS solution, and the membrane surface concentration is higher than that of the LC bulk solution. Therefore, the ohmic resistance in the RED stack is mainly determined by the LS concentration, and the thin enrichment layer on the LC side and the thin depletion layer on the HC side hardly affect the total ohmic resistance [84].

The grid influence is not considered, the solution ohmic resistance can be calculated from the specific conductivity, membrane area, and compartment thickness of the saline solution, or the effect of the grid on the ohmic resistance can be expressed f_v using correction factors. The corrected ohmic resistance calculation formula is:

$$R_{\text{HC}} = \frac{\delta_{\text{HC}}}{f_v \sigma_{\text{HC}} A_{\text{cell}}} \tag{9.5}$$

$$R_{\text{LC}} = \frac{\delta_{\text{LC}}}{f_v \sigma_{\text{LC}} A_{\text{cell}}} \tag{9.6}$$

In the formula, δ_{HC} and δ_{LC} indicate the thickness of the compartment [85]; σ_{HC} and σ_{LC} indicate the concentration chamber conductivity; A is the effective area of the membrane; and f_v is the grid correction factor. It can be seen from the above-mentioned equation that increasing the conductivity of the solution and reducing the thickness of the compartment are conducive to reducing the resistance of the membrane stack, thereby improving the power of the RED process.

9.2.2 The Main Components of RED

As shown in the basic schematic diagram of the RED (Figure 9.4), the RED device is similar to the ED device, consisting of three parts: electrode| ion exchange membrane |spacers| electrode. Specifically, the AEMs and CEMs are alternately arranged and spaced by the grid. The electrode compartments at both ends of the stack include the electrode material and electrolyte solution (generally suitable redox pairs, such as Fe^{2+} and Fe^{3+}, or $[Fe(CN)_6]^{4-}/[Fe(CN)_6]^{3-}$) [86, 87]. The anions in the anode chamber undergo an oxidation reaction on the electrode surface and lose electrons, the electrons are transferred from the anode to the cathode through an external circuit, and the cations in the cathode chamber pass through the electrode surface to obtain electrons and undergo a reduction reaction. Inside the membrane stack, the ions in the HS chamber migrate directionally through the ion exchange membrane to achieve a closed circuit.

From the above-mentioned analysis of the electrochemical performance of RED, it can be seen that the performance of RED directly depends on the stack voltage and stack resistance. The membrane stack voltage is mainly determined by the permeation coefficient of the membrane and solution properties, and the other is the membrane stack resistance, which is composed of the surface resistance of the ion exchange membrane and solution resistance. That is, ion exchange membranes, differential concentration solution systems, grid structures, and electrode systems affect the efficiency of energy capacity. At present, the optimization of the capacity performance of RED equipment is mainly focused on these aspects.

9.2.2.1 Ion Exchange Membrane

The ion exchange membrane is the core of the RED system, and its performance is a decisive factor in the energy recovery efficiency of RED technology. By Eqs. (9.1)–(9.4), it can be seen that the permeation and surface resistance of the ion membrane are the key factors in achieving high-performance RED, but there is a trade-off effect between the permeation coefficient and surface resistance. In the process of preparing ion exchange membranes, the membrane resistance is reduced by adjusting the IEC and reducing the thickness of the membrane, but the increase in IEC will lead to an increase in the hydrophilicity of the membrane and obvious swelling. Excessive swelling leads to a decrease in the compactness of the membrane, which in turn reduces the permeability coefficient and mechanical strength of the membrane. In terms of practical large-scale applications, membranes also need to be dimensionally stable (low swelling), mechanically and chemically stable, and cost-effective.

In recent years, with the advancement of material chemistry, there have been an increasing number of types of ion exchange membranes suitable for RED, which can be roughly divided into three types: new ion exchange membrane materials (polymer skeleton, fixed group, etc.), new preparation methods, and innovative membrane structures.

To reduce the high cost of current commercial ion exchange membranes, low-priced hydrocarbon polymers are considered to be suitable materials for the preparation of RED membranes. Take halogenated polyether materials, such as polyepichlorohydrin (PECH), as an example. A tertiary diamine (1,4-diazabicyclo [2.2.2]octane, DABCO) was used to introduce the ion-exchange groups by amination and for simultaneous cross-linking of the polymer membrane. The area resistance of the membranes ranged from 0.82 to 2.05 Ω/cm^2, and the permselectivity ranged from 87% to 90%. Depending on the excellent performance of the membrane, the high-power density of RED is 1.27 W/m, which exceeds the power output obtained with commercially available AMX membranes [88]. After that, the CEM was prepared by sulfonated polyetherketone (SPEEK) with different sulfonation degrees, and the AEM prepared by the above PECH was composed of a RED membrane stack. The results showed that when the chemical composition of the membrane was the same, the membrane thickness had no obvious effect on IEC and selective permeability; however, reducing the thickness could shorten the ion transport path and accelerate ion transport [89]. Qing Zhu et al. [62] designed and prepared a sulfonated ultramicroporous polymer membrane (SPX) for the extraction of osmotic energy stored in different concentrations of solution. SPX membranes have hydrophilic micropores with sizes of 5–9 Å, exhibiting obvious surface charge-controlled ion transport and excellent cation selectivity. Under the 50-fold concentration gradient of simulated seawater and river water mixing, a power density of 0.62 W/m^2 is achieved, and the energy conversion efficiency is maintained above 38.5%. By utilizing the synergistic effect of the thermal gradient and concentration gradient, its performance is further improved, reaching 1.23 W/m^2 and 48.7%, respectively (very close to the theoretical limit of 50%). This is the highest efficiency reported thus far at a 50-fold NaCl gradient. At the same time, the size sieving effect in subnanometer channels is revealed, which enables SPX membranes to extract Gibbs free energy from mixed solutions at equal molar concentrations, realizing a completely new concept of an energy conversion system. At 0.5 M HCl/0.5 M LiCl, the power output reaches 0.55 W/m^2. It extends the concept of concentration power generation to diffusion power generation, which may be used in the future to extract energy from industrial wastewater. This work also has certain implications for the application of unsourced microporous materials and the design of advanced permeation transduction membranes.

In addition to the above homogeneous membranes, two-dimensional materials with abundant surface charges (boron nitride, molybdenum disulfide, and graphene oxide), and other heterogeneous membranes, which show great potential in terms of output power. The ultrathin and ion-selective Janus membrane fabricated by hybridizing two block copolymer (BCP) membranes exhibits extraordinary performance [90]. Janus membrane is fabricated by hybridizing two BCP membranes with different chemical compositions. The thickness of the tailor-made hybrid membrane is on the submicron scale (~500 nm), allowing low fluidic resistance and rapid mass transport. The system is capable of high-performance osmotic energy conversion with a power density of approximately 2.04 W/m^2 by mixing natural seawater and river water. A power output of more than 10 W/m^2 is envisaged by increasing the utilization rate of the pores in the functional layer. The large-scale preparation of nanoporous membranes is still a great challenge in design and manufacturing. By simplifying the synthesis steps, another three/two-dimensional polymer/graphene oxide heterogeneous nanoporous membrane was prepared by modifying existing porous materials by introducing a positively charged pyridine moiety into the engineered polymer polyphenylsulfone. The multiphysical–chemical properties of the membrane enable it to work as a nanogate switch with synergy between wettability and surface charge change in response to pH. By adjusting the charge density of the copolymer via chemical copolymerization through a controlled design route, the rectifying ratio of this asymmetric membrane could be amplified 4 times, and the highest power density of 0.76 W/m^2 was obtained [91].

In addition to modifying the microstructure of the membrane surface [74, 92, 93], innovative preparation methods have been adopted, such as CEMs with directional movement ion channels with pulsed electric fields and CEMs with very low resistance by radiation chemical grafting polymerization, and the preparation of a new mono/divalent ion exchange membrane has been reported. Nevertheless, how to achieve large-scale preparation of low-cost RED films with low resistance and high selective permeability for industrial production is still a great challenge.

9.2.2.2 Spacers

The grid is also an important component of RED. The grid provides flow channels for the fluid in the chamber and supports the membrane. As shown in Eqs. (9.1)–(9.3), (9.1)–(9.5), and (9.1)–(9.6), the spacer is directly related to the diffusion boundary layer resistance, compartment thickness, and effective area of the membrane. Therefore, the geometric structure of the spacers, such as thickness, porosity, and flow channel structure, will affect the solution resistance and the hydrodynamic parameters of the chamber solution; on the other hand, the material type of the spacer will also affect the resistance of the membrane stack and disturb the movement of ions.

The conductivity of the LC chamber is relatively low [66, 94], which has the greatest impact on the solution resistance of the membrane stack. Therefore, reducing the thickness of the spacer is one of the main means to reduce the solution resistance. Studies have shown that for the same type of spacer, by only reducing the thickness from 0.5 to 0.2 mm, the power density correspondingly increases from 0.56 to 0.87 W/m^2 [66], and the thickness of the partition is reduced to 100 μm. The membrane can also reach a total power of 2.2 W/m^2 [94]. However, excessively reducing the thickness of the mesh will not only lose the mechanical strength but also increase the energy consumption of the pump, resulting in a decrease in the net output power density of the overall RED system. Too thin a spacer (thickness $\leq$ 60 μm) will increase the pressure difference between the liquid and the inlet. In addition, thin spacers are more likely to cause membrane fouling, reduce the life of the membrane, and affect the performance of the membrane stack.

In addition to the thickness of the spacer, the presence of nonconductive spacers reduces the active membrane area for ionic conduction (which is known as the spacer shielding effect), and the shielding effect might increase internal resistance by a factor of 1.5 and possibly contribute to 30–40% less energy consumption compared to the theoretically available energy [95]. Increasing the aperture ratio is one of the methods, and designing ion-conductive spaces can also reduce its negative impact. The CEM (Neosepta CMX) and the AEM (Neosepta AMX) were cut using a thermocompressor to produce the cation and anion conductive spaces, respectively [96]. The results clearly show that the use of ion conductive spacers in RED leads to a tremendous increase in power density by a factor of 3–4 and a decrease in membrane stack resistance by a factor of 2 compared to the use of nonconductive spacers with the same open area and shape. Based on theoretical calculations, a power density higher than 4 W/m^2 can be easily obtained with low-resistance membranes and spacers with thicknesses of only 0.15 mm. Increasing the open area of the spacers is also conducive to reducing the shielding effect. An interesting concept of profiled membranes has been proposed in the recent literature. These novel RED membranes, also termed "spacerless membranes," are designed with tailored microstructures (e.g. ridges, waves, pillars, and reliefs) that avoid the use of spacers [74, 93, 97]. The profiled membrane refers to the straight ridge rough structure parallel to the direction of fluid movement that is extruded on one surface of the membrane by hot pressing [74]. The ohmic resistance of the RED stack assembled with the profiled membrane and the inlet pressure of the solution is significantly reduced, but the boundary layer resistance is higher due to the concentration polarization phenomenon, and the power density increases only slightly. The net power density of the columnar structure of the profiled membrane is increased by approximately 20% compared with common spacers; however, the nonohmic resistance is significantly higher, which may result in increased boundary layer thickness due to lower solution turbulence [98].

9.2.2.3 Electrode System

The electrode system consists of two parts: electrode material and electrolyte. The electrode system is used to convert ion current into electron current by generating a redox reaction at the interface between the electrode material and the electrolyte and realize the direct energy conversion of chemical potential energy to electrical energy. Therefore, the electrode system is also one of the constraints of RED technology. Depending on whether an electrode reaction occurs, the most commonly used electrode systems can be divided into two categories [86]: one does not undergo an electrode reaction, and a redox reaction (hydrogen evolution reaction on the surface of the cathode electrode, oxygen evolution reaction on the surface of the anode electrode) usually involves gas formation (H_2, Cl_2, or O_2) [79]. These electrode systems typically have high reaction activation energy losses and require additional equipment to collect the gas. The other type of electrode system reacts at the cathode and anode, and the solution composition does not change after circulation. This electrode system can be roughly divided into two types: one is a system with active electrolytes (such as the Cu–$CuSO_4$ system [73] and Zn–$ZnSO_4$ [99]), and the other is a system with a homogeneous redox pair of inert electrodes (such as $FeCl_3/FeCl_2$, $[Fe(CN)_6]^{3-}/[Fe(CN)_6]^{4-}$, and Fe(III)-EDTA/Fe(II)-EDTA). The inert electrode is generally loaded with Ru—Ir metal oxide titanium mesh, graphite electrode, etc. [77] This is also the most commonly used type of RED in current research and electrode reactions:

$$\text{Cathode}: [Fe(CN)_6]^{3-} + e \rightarrow [Fe(CN)_6]^{4-} \tag{9.7}$$

$$\text{Anode}: [Fe(CN)_6]^{4-} \rightarrow [Fe(CN)_6]^{3-} + e \tag{9.8}$$

The electrode system was evaluated from the perspectives of technical feasibility, economy, safety, health, and environment, and it was found that the redox electropair (Fe^{2+}/Fe^{3+} or $[Fe(CN)_6]^{3-}/[Fe(CN)_6]^{4-}$) using an inert DSA (dimensionally stable anode) electrode is better than traditional electrolytes such as Na_2SO_4 [86]. Scholars have compared the performance of $FeCl_3/FeCl_2$, exacyanoferrate(III)/hexacyanoferrate(II), and Fe(III)–EDTA/Fe(II)-EDTA on graphite and DSAs, and the results show that the exacyanoferrate(III)/hexacyanoferrate(II) system has good stability and excellent performance. It does not pass through the Nafion membrane to other compartments. These studies demonstrate the superiority of iron-based complex-like redox pairs as electrolytes, but their potential to precipitate on the membrane and electrode surfaces leads to fouling and current drop. The first choice of inert electrodes corresponding to iron-based redox couples may be carbon electrodes rather than the most commonly used noble metal oxide electrodes because carbon electrodes have higher gas evolution potential. Second, increasing the specific surface area of the electrode can further improve the power density. Although energy consumption on the electrode is unavoidable, the influence of the electrode can be reduced by using a large number of cells in the reverse dialysis stack.

Although the use of RED as a sustainable salt differential energy conversion process has many advantages, it also exposes some shortcomings in industrial applications. (i) The cost of the ion exchange membrane as the core of RED is too high, which makes the initial investment of the membrane stack too large. (ii) During the operation of RED, the membrane is easily polluted by water or the natural rivers are complex, resulting in a decrease in the output energy density. (iii) The actual output voltage of RED is low. Taking river/sea water as an example, the theoretical voltage of a single membrane pair is approximately 0.17 V, the actual open circuit voltage is much lower than the theoretical value, and the conversion rate is low when the salt difference can be converted to it. Exploring the energy conversion form of salt differential energy through integration with other technologies is an area where salt differential energy needs to be supplemented. Taking hydrogen production as an example, hydrogen production by RED combined with the hydrolysis process is a better energy storage strategy, but the theoretical potential and overpotential of the hydrogen evolution reaction of the cathode and the oxygen evolution reaction of the anode are large, so the hydrogen production efficiency is low, and the stability of the required catalyst in the corresponding system still needs to be further investigated. In addition, we can also explore the new integration of RED, deeply tap the superiority of salt differential energy resources, rationally use salt differential energy, aim at its direct conversion and utilization, improve process economy and practicality, reduce RED costs, and achieve environmental protection.

9.3 Donnan Dialysis

Donnan dialysis is a membrane separation technology that relies on concentration difference (or chemical potential difference) as the driving force without the need for an external driving force. As early as 1924, F.G. Donnan discovered that the electrolyte solution was balanced on both sides of the semipermeable membrane and later named this phenomenon Donnan dialysis [100]. It uses the selective permeability of ion exchange membranes to different ions to achieve separation and concentration [101]. The Donnan dialysis process has low energy consumption and is simple to operate. Therefore, Donnan dialysis is often used for ion enrichment, concentration, and purification of drinking water, especially in hard water softening. For example, in softened hard water, brine flow is introduced on one side of the CEM, and hard water is introduced on the other side. The ions in the solution on both sides have chemical potentials, but the CEM only allows cations to penetrate, so some Ca^{2+}/Mg^{2+} ions diffuse to the brine side through the CEM, and Na^+ ions diffuse to the hard water side. Donnan dialysis tends to be a continuous, countercurrent process that achieves the process of desalination or concentration of a certain number of ions. The principle of Donnan dialysis is very similar to that of DD. Both are accomplished via a dialysis process through ion-exchange membranes with a concentration gradient as the main driving force. A direct difference between the two technologies is that ions in the former case transfer in the opposite direction through the membrane, while ions in the latter case migrate in the same direction. Donnan dialysis is an energy-saving process since no electricity is consumed. This process also has the advantage that it needs regeneration and can be operated continuously.

9.3.1 Basic Theory of Donnan Dialysis

The salt solution is balanced on both sides of the membrane, which is due to the thermodynamic properties of the ions. Under equilibrium conditions, the chemical potential of any dissolved species is the same in each term [102]:

$$\mu_i = \mu_i^\circ + RT \ln \gamma_i \tag{9.9}$$

μ_i° presents the chemical potential of species i in the standard state; R presents the gas law constant; T presents the absolute temperature; γ_i presents the activity of the particular chemical species i being considered. The change in free energy during ionization cannot be ignored, so the chemical potential of ion i in solution is calculated:

$$\varphi_i = \mu_i^\circ + RT \ln \gamma_i + z_i F \varepsilon \tag{9.10}$$

where z_i represents the ionic charge, F represents Faraday's constant, and ε represents the electrical potential. When ion i is in equilibrium in solution 1 and solution 2 on both sides of the membrane, the electrochemical potential of any mobile species i in the two phases can be equated.

$$\varphi_{i1} = \varphi_{i2}, \text{or } RT \ln \gamma_{i1} + z_i F \varepsilon_1 = RT \ln \gamma_{i2} + z_i F \varepsilon_2 \tag{9.11}$$

μ_i° terms cancel because the same standard state exists in both liquid solutions. Whenever salt concentrations on opposite sides of the membrane differ, there will be a potential difference across the membrane caused by the concentration difference. This potential difference called the "Donnan potential," E_{Don} is described by rearrangement of Eq. (9.12):

$$E_{\text{Don}} = \varepsilon_2 - \varepsilon_1 = {}^{RT}/_{z_i F} (\ln \gamma_{i1} - \ln \gamma_{i2}) = {}^{RT}/_{F} \ln \left(\frac{\gamma_{i1}}{ln\gamma_{i2}} \right)^{\frac{1}{z_i}} \tag{9.12}$$

Since the Donnan potential acts on all mobile ionic species, the value of $(\gamma_{i1}/ln\gamma_{i2})^{1/z_i}$ is the same for all of the counterions in the system. In other words, the concentration difference of the coions causes an electrical potential that acts on the counterions.

As Donnan pointed out in his seminal description of the theory, a precise treatment of the equilibria would require the use of activities rather than concentrations of ions in the solutions. However, the use of molar concentrations greatly simplifies the presentation of the theory, so that is the approach taken here. For the experiment described by Donnan where $z_i = +1$ for both Na^+ and K^+ ions, the equilibrium written with concentrations would be:

$$\frac{[Na^+]_1}{[Na^+]_2} = \frac{[K^+]_1}{[K^+]_2} \text{ or } \frac{[Na^+]_1}{[K^+]_1} = \frac{[Na^+]_2}{[K^+]_2} \tag{9.13}$$

As shown in Figure 9.5, a CEM separates two dilute salt solutions, and the CEM is permeable to cations (Na^+ and K^+) and blocks anions (Cl^-) ions. The ions in the solution have a potential difference on both sides of the membrane, but due to the strong charge repulsion of the cation to the anion, only Na^+ and K^+ ions pass transport the CEM, and finally, the solutions on both sides reach an equilibrium state [103, 104].

The equilibrium concentrations can be expressed in terms of the initial molar concentrations c_1 of NaCl on the left and c_2 of KCl on the right, x moles transported through the membrane (still the same for both cations), and the volumes V_1 and V_2 of the solutions [105, 106]:

$$(c_1 - x/V_1)/(x/V_1) = (x/V_2)/(c_2 - x/V_2) \tag{9.14}$$

$$x = c_1c_2/(c_1/V_2 + c_2/V_1) \tag{9.15}$$

Donnan dialysis is particularly effective for the recovery or removal of multivalent ions. The Donnan equilibrium for a divalent Ca^{2+} ion and a univalent K^+ is described by the equation:

$$\left([Ca^{2+}]_1/[Ca^{2+}]_2\right)^{1/2} = [K^+]_1/[K^+]_2 \tag{9.16}$$

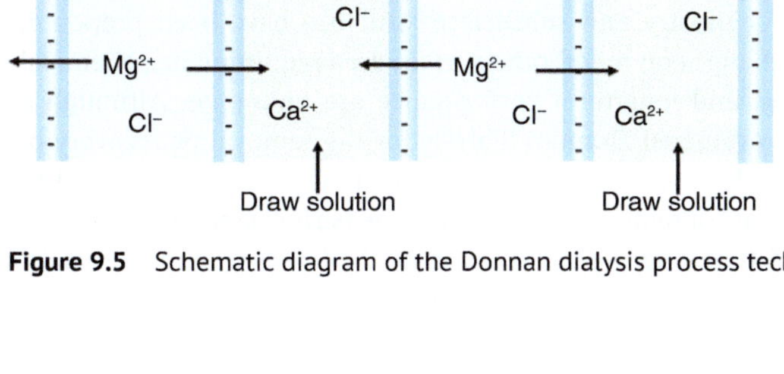

Figure 9.5 Schematic diagram of the Donnan dialysis process technology.

To maintain electroneutrality in the system, the transport of x moles of Ca^{2+} ions through the membrane requires the transport of $2x$ moles of K^+ ions in the opposite direction. Thus, the equilibrium is described by:

$$(c_1 - x/V_1)/(x/V_1)^{1/2} = (2x/V_2)/(c_2 - 2x/V_2) \quad (9.17)$$

For $V_1 = 10$, $V_2 = 1$ and initial concentrations of c_1=0.01 and $c_2 = 1$, the value of $x = 0.095$ is calculated by Eq. (9.17), which means that more than 95% of the calcium would be driven through the membrane. The effect of valence is even more dramatic when the c_1/c_2 ratio is much lower than the value used in this example.

9.3.1.1 The Parameters Affecting Donnan Dialysis

The target ion separation efficiency is only dependent upon the solution pH [107–110], the draw ion type [111–113], and the properties of the ion exchange membrane [114–116]. For the draw ion type, three parameters are important: (i) The valence of the draw ions; (ii) The hydrated radius and dehydration energy of the draw ion; and (iii). The concentration of the draw solution. The first two parameters affect the diffusion coefficient of ions in the membrane and affect the processing efficiency of Donnan dialysis. The concentration is the driving force of the diffusion of ions into the chamber, that is, the potential difference of ions on both sides of the membrane.

While Donnan equilibrium is not a function of membrane properties, the rate of target ion removal is highly dependent on diffusion coefficients and the thickness of the membrane [115, 117]. The target ion concentration gradient in the membrane is governed by (i) the selectivity coefficients for the target ion in the draw and feed solutions, (ii) the concentrations of competing species, and (iii) membrane thickness [117]. Almost all commercial AEMs contain quaternary ammonium sites (e.g. $-R_3N^+$), whereas CEMs generally have sulfonate groups (e.g. $-RSO_3^-$). Although the functional groups are well conserved, different membranes exhibit variable crosslinking, porosity, water uptake (hydration), ion exchange capacity, and permselectivity [114, 116]. Ion exchange membranes with lower thickness, greater hydration, less crosslinking, higher selectivity coefficients for the target ion in the feed solution, and higher ion exchange capacity provide faster rates of target ion removal. Few membranes exhibit all of the ideal characteristics; therefore, future efforts should consider new approaches to membrane design and production for Donnan dialysis applications. Importantly, the draw solution composition can affect membrane swelling and, therefore, change the diffusion coefficients for target ions in a particular membrane. For this reason, we recommend a systematic evaluation of multiple membranes with different draw solutions. Interesting modifications to the membrane surface chemistry and subsurface structure have been proposed, but the effects of these changes on membrane surface area requirements, chemical and mechanical stability, and long-term performance are unknown. Although a number of studies have evaluated Donnan dialysis for the removal or recovery of specific ions, few reports have calculated the fundamental parameters needed to appropriately model the rate of removal and compare the performance against other studies. In this regard, future studies should calculate the (i) separation factors for the target ion over the draw ion in both the feed and draw solutions and (ii) diffusion coefficients of the target and draw ions in the ion-exchange membrane.

Donnan dialysis is an energy-saving process since no electricity is consumed. This process also has the advantage that it needs regeneration and can be operated continuously. At present, Donnan dialysis is widely used in the fields of recovery/removal of metal ions, recovery of nutrients, and other inorganic or organic ions in water. Although Donnan dialysis is similar to DD, both use one type of ion exchange membrane to separate and recover ions. In principle, Donnan dialysis is more similar to ion exchange resin processes. Both are ion exchanges in two phases: Donnan dialysis is the exchange of ions on both sides of the membrane, while ion exchange resin is the exchange of ions in the resin beads with ions in solution. Donnan dialysis can be regarded as a continuous ion exchange process that only requires the addition of draw salt, eliminating the process of resin regeneration. In the water preparation section, with the help of the low treatment cost of Donnan dialysis, hard water is pretreated, which can effectively extend the service life of the RO membrane or nanofiltration membrane in the next process. However, this process system is not economical because the driving force caused by the concentration gradient and resultant potential differences is weak. Therefore, until now, Donnan dialysis has not been used on a large industrial scale. The major applications of Donnan dialysis were found in water softening and the removal of certain heavy metal ions from wastewater [118]. It seems that the Donnan dialysis process has no significant technical or economical advantage over the conventional ion exchange process.

9.4 Electrodialysis (ED)

ED, one of the earliest and most fundamental applications of ion exchange membranes, has been used for desalination for more than 50 years and has become the most commercially available and relevant ion exchange membrane process by far. As Figure 9.6 shows, the membrane stack is the important core of ED, which consists of

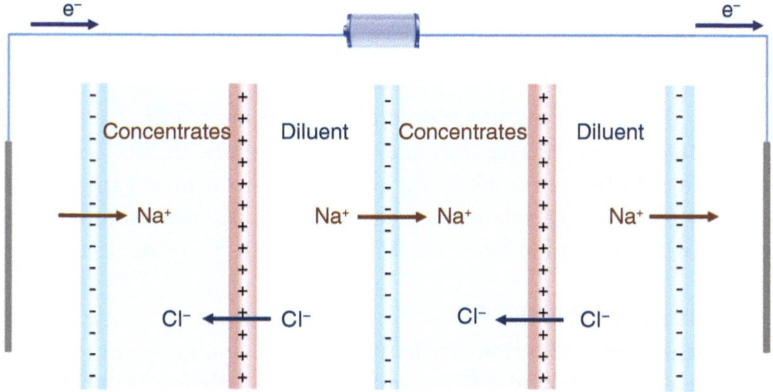

Figure 9.6 The working principle of the electrodialysis process.

CEMs and AEMs alternatively arranged in one pair of electrodes. In the ED cell, the anions pass through the AEM, and the cations migrate through the CEM, moving toward the anode and cathode under the influence of the electric field. The charged ions are not allowed to pass through the membrane when similarly charged ions (coions) are fixed due to the Donnan exclusion effect. ED is an efficient technology widely used in many fields, such as food, the chemical industry, the environment, and biology, including wastewater desalination, chemical concentration, wastewater resource conversion, and carbon dioxide capture [119–121].

The rapidly increasing demand for freshwater has become a great challenge for the development of technologies to produce clean water under huge consumption in agriculture, industry, and households. Water desalination at low energy consumption and capital expense is a feasible solution to the problem of freshwater shortages. ED accounts for only 4% of the world's capacity for desalination. There are also many other pressure-driven technologies applied in water desalination, including RO, disc-tube reverse osmosis (DTRO), high-pressure reverse osmosis (HPRO), and forward osmosis (FO). The RO process is known as the most competitive technology in the application of water desalination compared with ED. The comparisons of both processes are shown below.

The advantages of electrodialysis are as follows:

(1) The desalinated water of ED contains all other substances, except the extracted salts, which causes the constituent to be close to the original water. However, RO filters out all chemicals in the desalinated water, leaving only pure water.
(2) ED produces concentrated solution as well as desalinated water. This concentrated solution can be used for salt production, whereas RO concentrates the brine to approximately 8–10 wt% NaCl.
(3) ED demands less than 1.5 m^3 H_2O/t, whereas RO consumes 2–3 m^3 H_2O/t in desalination.
(4) ED is more competitive than RO for water desalination when the salt concentration is less than 10 g/L, such as brackish water or water extracted from briny sources.
(5) ED allows the selective removal of undesired components such as nitrate in water.

The disadvantages of electrodialysis are as follows:

(1) Before applying the ED process, pretreatment processes are required to remove multivalent metallic ions such as iron and manganese ions in the feed solution. Maintaining multivalent irons at low concentrations is one of the limitations of the ED process. Post-treatment procedures are also needed to remove some neutral toxic compounds, such as viruses and bacteria, from desalinated water.
(2) The chlorine gas produced in the oxidation reaction occurs at the anode, which will lead to corrosion and pollution problems in the desalination plant.
(3) Considering the energy demand, ED technology is especially less suitable than RO in desalinating salt water with high concentrations.

In summary, ED is a promising process to minimize the energy demand for water desalination, while RO is favorable for producing pure water from brine with high concentrations. There is much research investigating the application of water desalination by comparing the processes of ED and RO. The standard ED, a simple hybrid ED–RO, and a recirculated hybrid ED–RO focused on energy consumption and water recovery are systematically compared by McGovern [122]. A simple hybrid refers to the concentrate from the RO process being transferred into the ED stack as the feed solution and blending the products from both units. A recirculated hybrid ED–RO refers to the ED product recirculated back to the RO unit. The results show that ED possesses the advantages of high water recovery and high brine concentrations above 10%, which exceed the osmotic pressures that can be achieved by current RO systems [123]. The authors conclude that hybrid ED–RO is beneficial unless the water consumed from a stand-alone ED system is more than 40% of that from the RO unit. Comparing the expense of ED and RO–ED integration technology, the authors find that the cost of treating water by independent ED and RO–ED integrated technology is almost the same if the concentration cost by RO is 60–70% of that by the ED device. However, RO can be used to concentrate low-salinity wastewater to a higher salt content when the cost of water treatment by RO is low. If the cost of water treatment is too high for RO units, it is more economical to use separate ED technologies. For example, industrial refractory high salt wastewater can be discharged near zero by the combination of ED and RO processes. To deal with ammonium chloride-containing wastewater, the traditional process adopts the direct evaporation and crystallization method to treat wastewater in the industry. On account of the high investment and high energy consumption in the traditional process, fresh water with NH_3–N < 1 mg/l is finally returned to the workshop for reuse by using the ED system coupled with the RO system. The concentrated water can be enriched to more than 15% by ED for the mechanical vapor recompression (MVR) evaporation system. After evaporation and crystallization by MVR, the semifinished product is finally centrifuged to obtain ammonium chloride crystals with moisture content <5% for sale. It is regarded as an efficient method to realize the double recovery of water resources and ammonium salt, by which both the water resource recovery rate and ammonium chloride recovery are up to 98%. The method not only realizes zero discharge of ammonium salt wastewater but also achieves the full utilization of resources and clean production.

9.5 Application of ED

9.5.1 Desalination

In addition to producing pure water, it is a significant application for ED to deal with the increasing amount of other wastewater around the world. It is estimated that approximately 1 000 km^3 of wastewater is discharged worldwide each year, including more than 300 km^3 of municipal wastewater and more than 600 km^3 of industrial wastewater [124, 125]. Wastewater treatment plants consume nearly ∼3%

of global electricity consumption and contribute ~1.6% of greenhouse gas emissions [126, 127], which not only increases the environmental burden but also aggravates the greenhouse effect. With stricter regulations on the disposal of wastewater, zero liquid discharge (ZLD) and water loop closure (WLC) are widely recommended because they not only valorize inorganic salts but also eliminate the disposal cost of waste salts. There are some processes for industrials to achieve the goal of ZLD, such as filtration, precipitation, oxidation, and ion exchange membrane separation. ED is one of the separation processes based on an ion exchange membrane equipped with many advantages and disadvantages for desalinating wastewater displayed below:

The advantage of electrodialysis is as follows:

(1) ED is environmentally friendly, and there is no secondary pollution in the application of wastewater treatment.
(2) Only a partial addition of chemicals is needed, and the operation is convenient during the ED process.
(3) The treatment can be operated at elevated temperatures and various pH ranges by using more thermally and chemically stable membranes.
(4) Some useful products can be recycled through the ED process, which is different from the removal and discharge of all contaminants in conventional processes.
(5) ED has high concentration efficiency, and there are no complicated pretreatment steps.

The disadvantage of electrodialysis is as follows:

(1) Ionic components can only be removed, and the additives are usually neutral in a galvanic bath and cannot be removed.
(2) ED has been extensively used in the electroplating industry to remove and recycle metal ions such as nickel, zinc, cobalt, chromium, cadmium, and cyanide. It is noted that the current density should be properly controlled to prevent the precipitation of metal hydroxide on the surface of membranes caused by water dissociation.
(3) In addition, the membranes will possibly become polluted if the organic substances in the solution are added to the electroplating bath. A pretreatment process is required to remove these organic compounds. Interestingly, applying a suitable complexing agent is a feasible method to separate two mixed metal ions and avoid membrane fouling.

There are several state-of-the-art applications of ED to deal with various kinds of wastewater. In the wet metallurgy industry, large amounts of acidic wastewater are produced in the traditional process, including evaporation concentration, resin adsorption, and DD. The traditional method can realize the recycling of acid to a certain extent as well as some obvious shortcomings. First, the water cannot be recycled due to high concentrations of chloride and fluoride. Large energy consumption exists in the evaporation concentration step. Second, the resin adsorption process is complicated and produces secondary pollution. Third, the DD step has a small processing capacity per unit of time and is only suitable for high-concentration acid recovery. ED with a homogeneous membrane process is employed in industry to separate sulfuric acid and fluorine chlorine solutions from wastewater first and concentrate the acid concentration from 2.8% to 10%. After that, the acid concentration is enriched to 70% through concentration, catalytic oxidation, and an air-blowing process. High-purity concentrated sulfuric acid is obtained after removing the fluorine and chlorine directly. The integrated ED process is simple and efficient to achieve the complete treatment of waste acid without introducing other salts. Additionally, recyclable acid, valuable metals, and water resources can be recovered efficiently through integrated ED with a blowing process, which greatly reduces the production of solid waste and overcomes the drawbacks of conventional treatment methods.

9.5.2 Concentration

As another aspect of wastewater desalination, concentration is conducive to reducing seawater quantification, decreasing treatment costs, and recovering valuable compounds. Currently, brine concentration technologies consist of thermal and membrane methods. Thermal methods, including mechanical vapor recompression (MVR), multi-effect distillation (MED), and multistage flash (MSF), have been extensively used in water evaporation and seawater desalination [128]. Membrane processes include RO and ED desalination processes. RO concentrates are often further concentrated with ED to increase the salt content and reduce the volume of wastewater, thereby reducing the energy consumption of the system and increasing the water recovery rate. For the same concentration of high-salt wastewater, the energy consumption of thermal concentration is approximately 20–25 kWh/m^3, while the ED technology only needs 7–15 kWh/m^3. Those studies have indicated that ED is more competitive in concentrating RO brines to improve the overall water recovery rate.

The ED process has low requirements on the water quality of the feed solution, fewer pretreatment procedures, and less scaling degree than thermal methods. With the advantages of high concentration, low energy consumption, and mild operating conditions, ED has been applied in the field of concentration and reduction of high-concentration brine. By increasing the salt content of wastewater and controlling the migration of organic substances, the whole wastewater treatment system and stability can be improved, and the cost of investment and operation can be reduced. High-concentration products of approximately 20% can be obtained by ED concentration, which is capable of directly producing kosher salt and is employed as the raw material for the chlor-alkali industry or BMED. It will be helpful to extend the application of ED if the treatment of a higher concentration solution with ED is achieved.

9.5.3 The Influencing Parameters on ED Concentration

It is conducive to increasing the salt limiting concentration (the maximum concentration of salt) by optimizing parameters such as the applied voltage, current density, flow rate of the solution, and volume ratio between the dilute and concentrate chambers. The salt concentration limitation is restricted by two main factors.

One is water migration from the dilute to the concentrate chamber, and the other is ion back diffusion from the concentrate to the dilute chamber. Both electro-osmosis and concentration-gradient osmosis are significant elements for water migration to decrease the salt concentration and improve the concentration limitation in ED.

(1) **Water migration.** Water migration is a common phenomenon in ED to limit the salt concentration caused by electro-osmosis and concentration-gradient osmosis. Electro-osmosis means that water molecules combine with salt ions in the form of hydrated ions and are transported through the ion exchange membrane under the application of an electric field. The increasing salt concentration becomes restricted as the solution volume rises. Structures with a small free volume of the ion exchange membrane contribute to alleviating water migration in electro-osmosis.
The concentration-gradient osmosis is related to the concentration gradient in different concentration streams. This means that the water molecules migrate from the dilute chamber to the concentration chamber to balance the osmotic pressure in different concentration solutions.
(2) **Coion diffusion.** In addition, concentration-gradient osmosis can also be achieved by coion diffusion. Coion diffusion is a vital phenomenon in ED that influences the salt concentration, desalination rate, and salt purity due to the properties of the ion exchange membrane. Generally, the ion exchange membrane is able to selectively allow the migration of counterions (oppositely charged ions with fixed charges in the membrane) but reject coions (similarly charged ions with fixed charges in the membrane) according to Donnan exclusion. However, some coions pass through the ion exchange membrane and enter the dilute or concentration chamber to decrease the salt concentration or the desalination rate because the ion perm-selectivity of the ion exchange membrane cannot reach 100%. High-concentration gradients can result in substantial back diffusion of ions from the concentrate to the dilution chamber. As an ineluctable phenomenon, improving the fixed active group concentration of the membrane and decreasing the solution concentration on both sides of the membrane are regarded as efficient methods to increase the permselectivity of the ion exchange membrane and decrease the back diffusion of ions. A higher perm-selectivity of the ion exchange membrane is capable of preventing a larger concentration gradient and lowering the concentration limitation.
(3) **Limit current density.** It is an essential factor to estimate the feasibility of ED in terms of investment cost. Controlling the employment and consumption of electricity allows for achieving the goal of energy conservation and emission reduction.

However, it is noted that there still is a large potential for further development for ED, such as avoiding the polarization phenomena, reducing the membrane resistance, and producing ion-conduction spacers.

9.5.3.1 Approaches to Improve the Concentration on ED

To increase the enrichment of the ED process, there are available choices to concentrate the salt to extremely high content by high-volume ratio concentration and multistage batches of ED. Improving the volume ratio between the dilute and concentrate chambers is capable of increasing the initial salt content several times due to the concentration difference. Zhou et al. [129] investigated the influence of different volume ratios on the mass fraction between the concentrate and dilute chamber. The result demonstrates that the highest Li_2SO_4 mass fraction of approximately 15.8 wt% is obtained when the volume of the dilute chamber is 10 times that of the concentration chamber ($V_c : V_d = 1:10$). The mass fraction of Li_2SO_4 is only 9.2 wt% with $V_c : V_d = 1:1$. More Li_2SO_4 in the dilute chamber is transported into the concentrate chamber to obtain a higher concentration with a high-volume ratio. The solvent permeation is overwhelmed by salt migration even though a small portion of water will migrate to the concentrate chamber. As a consequence, it is feasible to concentrate the ionic electrolyte by improving the volume ratio of diluent and concentrates. However, water migration will be faster than salt migration as the concentration in the concentration chamber increases, indicating that water transport due to concentration-gradient osmosis becomes a dominant factor when the limiting concentration is coming.

The multistage batches enrich the salt content stage by stage to maintain a low concentration gradient between the dilute and concentration compartments. The concentrate from the former stage is flowed into the next stage as the diluent to obtain a higher concentration in this method. There are two advantages to the application of multistage batches. The back diffusion of salt from the concentration chamber to the dilution chamber is mitigated because of the relatively low concentration ratio between the concentrate and the dilution chamber in this mode of operation. In addition, a particular advantage of multistage ED over conventional ED is that it is possible to decrease the number of membrane units and the capital cost of multistage ED because the high salt concentration in multistage batches is conducive for ED to operate at a high current density. However, the disadvantage of multistage concentration is the high energy consumption. Yan et al. [130] proposed a multi-stage batch ED system for salt concentration, as shown in Figure 9.7. The results indicate that the concentration gradient between the dilute and the concentrated solutions can be maintained at a low value in each stage-batch of ED, and that the flux ratio of water to salt decreases as the stage-batch advances. The salt content of the concentrated solution can be increased from 3.5% to 17.9% and 20.6% by two- and three-stage-batch ED, respectively. The total energy consumption based on the material balance for two- and three-stage-batch ED is 0.31 and 0.45 kWh/kg, respectively. However, the flux ratio of water and salt decreases with increasing multistage batches.

9.5.4 Resource Conversion

The lithium resources of saline lakes account for approximately 60% of the world's lithium resources, and lithium extraction from lake brine is an important route to isolate lithium in China. However, lake brine in China has the characteristics of a large Mg^{2+}/Li^{+} ratio, high magnesium content, and low lithium content. Due to the similar physical and chemical properties of lithium and magnesium, it is difficult to extract lithium from lake brines. Different lithium extraction technologies,

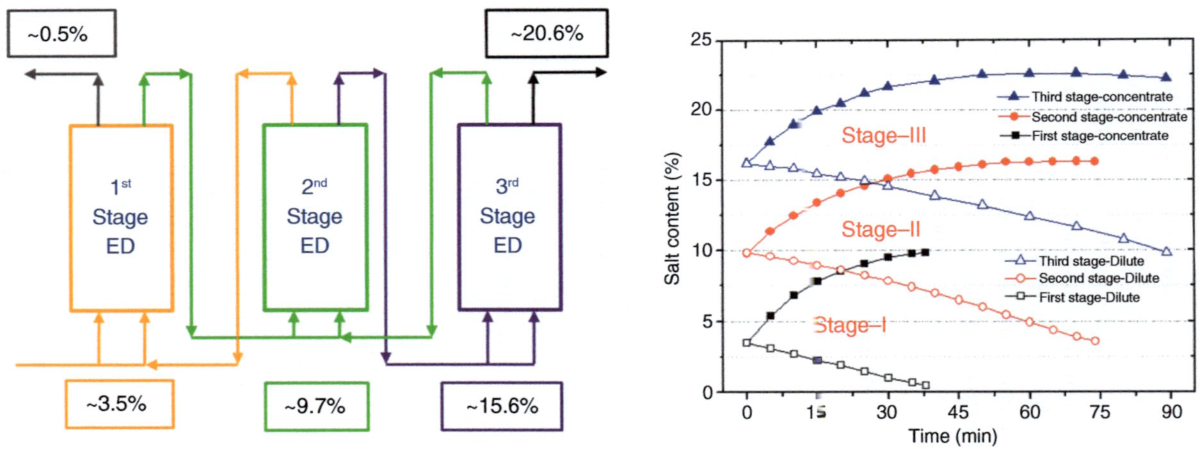

Figure 9.7 Three-stage batch ED system for salt concentration. Source: Adapted from Yan et al. [130].

including nanofiltration, adsorption, crystallization, and ED processes, can be selected based on the different qualities of lake brine resources and geographical environments. One traditional ED-like selectivity electrodialysis technology (SED) is suitable for brine with high lithium content. Unfortunately, the selectivity of lithium isolation by SED is only 20.3–33, which is limited by the selectivity and permeability of the ion exchange membrane. To improve lithium selectivity, state-of-the-art membrane materials such as MOF or COF have been developed to obtain an excellent lithium selectivity of approximately 1815 by using ED. It is difficult to further increase the lithium selectivity by traditional ED. Inspired by distillation, Jiang et al. [131] reported a novel multistage ion-"distillation" technique to effectively isolate lithium from natural lake brine. It is an ingenious process to arrange the membranes according to the principle of "identical membranes at the same side," which is capable of isolating lithium from lake brine with a selectivity of Li^+/Mg^{2+} up to 26 177. The Li selectivity is amplified by orders of magnitude due to the electro-striping mechanism for the transportation of lithium ions. In addition, this technique has great potential to make a huge breakthrough to improve selectivity by employing new membrane materials, and it is also suitable to separate species in other applications of isotopic and metallurgical compounds.

9.5.5 CO_2 Capture

According to statistics, the global CO_2 concentration has reached 36.3 billion tons, and the CO_2 concentration in the atmosphere will increase to an unprecedented peak of 419 ppm in 2021, which is capable of increasing the global temperature by at least 2.7 °C at the end of this century [127, 132, 133]. Capturing CO_2 is an effective method to decrease the CO_2 concentration. However, it consumes much energy to directly capture CO_2 from the atmosphere because of the low CO_2 concentration. Seawater is the largest carbon source, containing a large number of carbon elements. As the largest carbon sink, seawater has absorbed 40% of CO_2 since the industrial period. The effective concentration of CO_2 reaches 2.1 mmol/kg, which is 120 times that in the atmosphere [134]. To achieve the goal of carbon neutrality, it is feasible to directly extract CO_2 from seawater to decrease the CO_2 concentration and alleviate the greenhouse effect by using ED technology. The principle of the ED process is extracting CO_2 from seawater by moving the bicarbonate equilibrium reaction toward the direction of dissolving and releasing CO_2 after acidifying seawater through the ED process. As shown in Figure 9.8, there are two CEMs assembled between a pair of electrodes to form a feed compartment, an anode compartment, and a cathode compartment. Natural seawater containing a large amount of HCO_3^- and CO_3^{2-} flows in the feed compartment, and the Na^+ in the feed compartment combines with OH^- of the cathode compartment to form NaOH in the cathode compartment. The H_2 from the hydrogen evolution reaction (HER) is simultaneously produced in the cathode compartment. The H^+ generated by the oxygen evolution reaction (OER) passes through the CEM to supplement the migration of Na^+, which can also acidify the seawater in the feed chamber. The acidified seawater can be

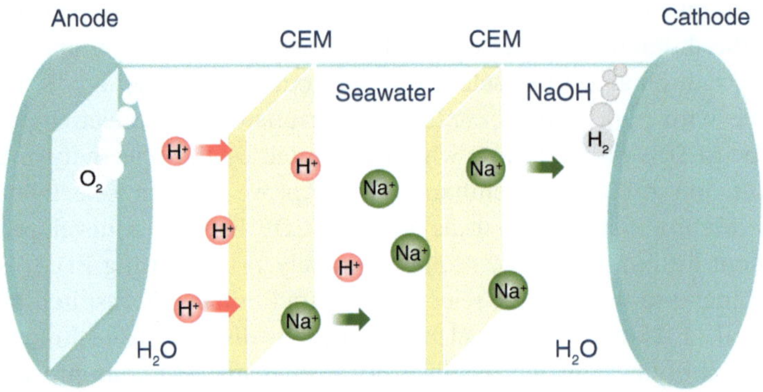

Figure 9.8 Schematic diagram of the electrodialysis (ED) process for the extraction of carbon dioxide from seawater.

recombined with NaOH in the cathode chamber to maintain the outflow of seawater at the initial pH value [135].

As a consequence, capturing CO_2 can be regarded as one of the raw materials for the production of hydrocarbons. The electricity consumed to capture CO_2 from seawater can be used as power for the production of H_2. In this way, it is an available technology to capture CO_2 and produce H_2 at the same time. As a novel and challenging application for ED, it is necessary to improve the efficiency and reduce the energy consumption of capturing CO_2 by further optimizing the process of ED.

9.6 Electrodialysis with Bipolar Membranes (BMED)

BMED is a new type of membrane technology combining the properties of ED and the water dissociation of the bipolar membrane. For a typical BMED stack, the bipolar membranes are inserted into the membrane stack based on ED, which is alternately arranged with anion and CEMs together with gaskets between a pair of electrodes to convert the salt solution into the corresponding acid and base. As shown in Figure 9.9, there are three kinds of configurations in the BMED, including BP-A-C-BP, BP-A-BP, and BP-A-BP. The difference between those configurations is not only the number of membrane pieces but also the employment of various applications. It is a transformation of the feed solution MX into corresponding HX and MOH under the action of the external electric field. The BP-A-C-BP configuration is a composite of three compartments: the feed compartments, the acid compartments, and the base compartments. In the BP-A-C-BP configuration, the cations (M^+) in the feed compartment migrate toward the cathode and pass through the cation-exchange membranes into the base compartment, and form MOH with OH^- produced by the bipolar membrane. Similarly, the anions (X^-) in the feed solution pass through the anion-exchange membranes into the acid compartment and form HX with the H^+ produced by the bipolar membrane. In this way, the ionic composition of the salt is depleted, while an acid and a base are concentrated in the alternate compartments after a period of operation. Replacing

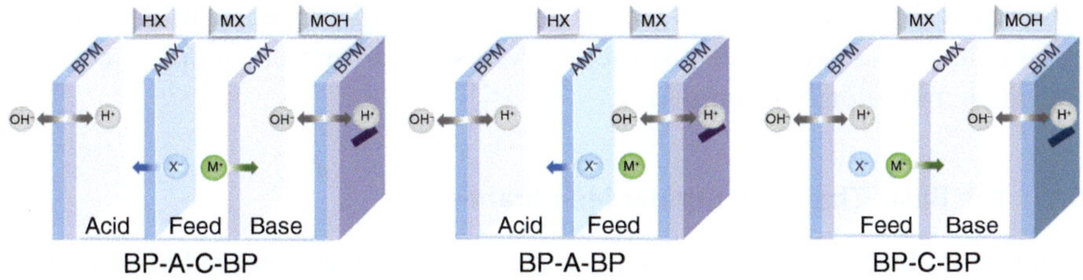

Figure 9.9 Schematic diagram of three kinds of membrane configurations in the bipolar membrane electrodialysis process (BMED).

anion- and cation-exchange membranes in membrane stacks with monovalent selective AEM–CEMs is capable of producing acid and base when using saline solution mixed with divalent ions. The divalent ions are trapped and separated in the feed compartment due to the selectivity of the ion-exchange membrane.

At present, the membrane configuration of BP-A-C-BP has many applications in the preparation of inorganic acids and bases, the preparation of organic acids and bases, and the zero discharge of industrial wastewater. The biggest feature is that it can convert salts into corresponding acids and bases at the same time. It is suitable for treating salt solutions that are easily dissociated and produce acids and bases with higher purity. However, the structure of the three compartments has a relatively large resistance due to a large number of membranes.

The two-compartment configuration has greater advantages in reducing energy consumption and capital expenses when the feed solution only needs to be acidified or alkalized, which are available in the BP-A-BP and BP-C-BP configurations. In addition, the benefits of high efficiency and low energy consumption in the two-compartment configurations are more obvious due to the weak dissociation characteristics of weak salts in many fermentation broths with low conductivity and high resistance.

The membrane stacks are divided into two compartments in the BP-A-BP and BP-C-BP membrane configurations. The mixture of salt and base chambers and the individual acid chamber belong to the BP-A-BP membrane configuration, which is usually used to prepare organic bases and inorganic acids with strong acid-weak base salts. The BP-C-BP membrane configuration possesses a blend of salt and acid chambers and an individual base chamber, which is suitable for the production of organic acids and inorganic bases with strong base-weak acid salts. In the BP-A-BP membrane configuration, the X^- ions in the feed chamber migrate to the acid chamber through the anion-exchange membrane and combine with H^+ dissociated from water in the bipolar membrane to form inorganic acids (HX), while the M^+ ions combine with OH^- dissociated from water in the bipolar membrane to form organic bases (MOH). The concentration and purity of organic bases in the feed chamber gradually increase as the electric field is applied. However, a small amount of salt preserved in the later stage of the experiment leads to a reduction in the current efficiency, which is mainly attributed to the competitive migration of OH^- and X^-.

In the BP-C-BP membrane configuration, the migration behaviors of ions are similar to the BP-A-C-BP and BP-A-BP membrane configurations that produce the inorganic base MOH and organic acid HX. The phenomenon of decreasing current efficiency is also caused by the reserved salts in the feed chamber and the competitive migration of H^+ and M^+.

9.6.1 The Basic Theory of Bipolar Membranes

As a core component of BMED, the bipolar membrane is a new type of ion exchange composite membrane with ion exchange characteristics, which is composed of an anion-exchange layer (P-type membrane) and a cation-exchange membrane (N-type membrane). Similar to the characteristics of N–P junctions in semiconductors, bipolar membranes with different properties can be made using different materials under different conditions. When forward voltage is applied to the bipolar membrane, the anode is on the cation exchange layer side of the bipolar membrane, and the cathode is on the anion exchange layer side of the bipolar membrane. The cations and anions in the salt solution will pass through the cation-exchange layer and anion-exchange layer of the bipolar membrane, respectively, and reach the bipolar membrane interface due to the perm selectivity of the anion- and cation-exchange layers. The electrolyte concentration in the bipolar membrane intermediate interface will increase, leading to a nonsignificant change in the resistance of the bipolar membrane. In the process of reverse voltage on the bipolar membrane, the cations pre-adsorbed in the bipolar membrane intermediate interface will pass through the cation-exchange layer and transport toward the cathode, and the anions will pass through the anion-exchange layer and migrate to the anode, resulting in a decrease in the electrolyte concentration in the bipolar membrane intermediate interface. The membrane resistance is constant, and the current–voltage curve is a straight line at this time. As the applied voltage increases to a special value, the number of ions entering the bipolar membrane intermediate interface due to the diffusion of coions is less than the ions migrating out of the bipolar membrane intermediate interface. When the salt ions are depleted, the water molecules in the intermediate interface of the bipolar membrane are dissociated into H^+ and OH^-. At this time, the current density is named the limit current density.

The phenomenon of water dissociation occurs in the intermediate interface of BPM, which converts water into hydrogen ions and hydroxide ions under the influence of the electric field force. H^+ and OH^- enter the adjacent sides of the bipolar membrane through the cation exchange layer and the anion exchange layer, respectively [136, 137]. The required theoretical potential is 0.828 V, and the energy consumption is 79.5 kJ/mol in the bipolar membrane water dissociation process, which is much lower than the theoretical potential of 2.057 V and energy consumption of 198.5 kJ/mol for the electrolysis of water. This is attributed to the generation of hydrogen and oxygen in water electrolysis. Additionally, the water dissociation rate of the BPM can reach 5×10^7 times that of the conventional solution [138]. According to theoretical calculations, the normal water dissociation rate can only bear a maximum current density of 3.5×10^{-5} A/m^2, while the bipolar membrane can even be used normally at a current density of 2000 A/m^2 in experiments [139]. Therefore, there must be some effects that accelerate the water dissociation process in the BPM.

At present, the mechanisms for the water dissociation of BPM mainly consist of the second Wien effect and the chemical reaction model. The second Wien effect assumes that the water dissociation of BPM occurs in the intermediate interface of BPM, which is equivalent to the dissociation of a weak electrolyte driven by a high electric field. In addition, the water dissociation rate of BPM increases with increasing electric field since the water dissociation process is promoted by the electric field. The effect of the electric field on the water dissociation rate can be expressed by the following formula:

$$K_E = \left(1 + b + \frac{b^2}{3} + \frac{b^3}{18} + \frac{b^4}{180} + \frac{b^5}{2700} + \frac{b^6}{56700} + \dots\right) K \tag{9.18}$$

where

$$b = 0.09636 \frac{E}{\varepsilon_r T^2} \tag{9.19}$$

In the above formula, K_E is the constant of the water dissociation rate under the influence of the electric field, E is the electric field intensity, ε_r is the relative permittivity, and T is the temperature. According to the calculation, the calculated result of K_E/K will be lower than 10^7 when the relative permittivity is greater than 10 [139]. Therefore, the second Wien effect is only applicable to the low electric field lower than 10^7 V/m and the electrolyte solution with a relative permittivity less than 10. In addition, the water dissociation phenomenon in the anion-exchange membrane is faster and more pronounced than that in the cation-exchange membrane, which is different from the assumption of the second Wien effect. By methylating the tertiary amines into quaternary amines, it is beneficial to eliminate the water dissociation in the AEM. Finally, weak acids such as pyridine, acrylic acid, phosphoric acid, and acetic acid are capable of contributing to the acceleration of water dissociation as well as some heavy metal ions such as Mg^{2+}, Sn^{2+}, Fe^{2+}, Fe^{3+}, Sn^{4+}, Ti^{4+}, Zr^{4+}, Ru3+ and Pd^{2+}.

The limitations are not explained in the second Wien effect, and the chemical reaction model is developed. The chemical reaction model mainly believes that the water dissociation in BPM is caused by the reversible protonation and deprotonation of weak acid and weak base groups. The reaction for water dissociation is proposed by Simons as follows:

$$2H_2O \leftrightarrow H_3O^+ + OH^- \tag{9.20}$$

$$B + H_2O \leftrightarrow BH^+ + OH^- \tag{9.21}$$

$$BH^+ + H_2O \leftrightarrow B + H_3O^+ \tag{9.22}$$

$$A^- + H_2O \leftrightarrow AH + OH^- \tag{9.23}$$

$$AH + H_2O \leftrightarrow A^- + H_3O^+ \tag{9.24}$$

where B and AH are bases and acids, respectively. Many experimental phenomena of water dissociation of ion exchange membranes can be explained by chemical reaction models when the membrane contains weakly acidic groups or heavy metal ions and weak acids exist in the neighboring solutions of the membrane surface. Compared with the second Wien effect, this model considers the influence of the BPM structure on water dissociation, which explains why water molecules are prone to polarization on the surface of anion-exchange membranes. However, this does not explain why the water dissociation rate of BPM increases with increasing electric field strength.

BMED is considered the most promising technology to produce acid and base with many advantages. There are no undesirable products or oxidizing or reducing species generated in the operation of BMED compared with other technologies. Additionally, the water splitting of bipolar membranes can be achieved by only one pair of electrodes in the BMED process, which not only simplifies operation and installation but also saves space and investment. Based on these advantages, BMED technology has been widely used in environmental protection, waste salt resource utilization, and carbon capture and utilization.

9.6.2 Application of the BMED Process

9.6.2.1 The Production of Alkali

One of the most vital applications of the BMED technique is for the production of alkali. Alkali, such as sodium hydroxide, sodium carbonate, tetramethyl ammonium hydroxide (TMAH), etc., are widely used in the light, food, textile, metallurgy, and petrochemical industries. The chlor-alkali industry is one of the most basic chemical industries for producing the essential product of caustic soda. There are several main alkali production processes, including diaphragm electrolysis, ion-exchange membrane electrolysis, and mercury. The mercury method is almost abandoned because of the high cost of caustic material, low yield, and pollution of the environment. Electrolysis is the main NaOH production technology due to its advantages of simplicity and practicability. The refined saturated salt solution in the electrolysis cell flows into the anode chamber, and pure water (adding a certain amount of NaOH solution) is added to the cathode chamber. Applying the current field, hydrogen is generated in the cathode, and chlorine is generated in the anode, while a caustic soda solution is obtained by electrolyzing a saturated NaCl solution.

In diaphragm electrolysis, asbestos is introduced in the cell as the diaphragm to separate the cathode product NaOH and the anode gas Cl_2. However, a small amount of gas is mixed because the separation efficiency of asbestos is not high enough. The employment of asbestos also greatly pollutes the environment. Compared with the diaphragm, the dense ion-exchange membrane allows for preventing the mixture and explosion of H_2 and Cl_2 produced by the electrode reaction, which also avoids the reaction of Cl_2 and NaOH solution to generate sodium hypochlorite ($NaClO_3$), which affects the purity of caustic soda. Additionally, the high perm-selectivity of the cation-exchange membrane only allows the passage of Na^+ ions but rejects Cl^- and OH^- ions, which is beneficial for obtaining high-purity NaOH. In addition, the NaOH concentration can reach 30–32%, resulting in the products being sold directly. Electrolysis has less investment per unit capacity, less environmental pollution, low energy consumption, and high purity of caustic soda. In recent years, the proportion of the capacity of ion-exchange membrane electrolysis has reached more than 99.7% with the increasing capacity of the ion-exchange membrane method and the decreasing capacity of the diaphragm method. In summary, ion-exchange membrane electrolysis technology possesses the advantages of simple equipment, large production capacity, less pollution, and so on. Further reducing electrolytic energy consumption has become an important direction of the development of the chlor-alkali industry.

However, there are still many shortcomings in NaOH production by this technology. First, high energy is required in the chlor-alkali industry. The total energy consumption of caustic soda by electrolysis accounts for 0.5% of China's energy consumption, and the power cost accounts for 60% of the total cost of caustic soda. Second, this electrolysis in the chlor-alkali industry produces serious pollution when using the calcium carbide method to produce downstream PVC. It is difficult to address the dust problems from the calcium carbide reaction and the mercury pollution caused by the use of mercury-containing catalysts. The continuous pressure of environmental protection brings a higher threshold to the new production capacity of the industry in the future. Third, the balance between caustic soda and chlorine gas has always been a constant contradiction in the development of the chlor-alkali industry. The output of caustic soda is largely restricted by the downstream consumption of chlorine gas. The construction of caustic soda enterprises should take into account the balance of chloralkali. To recycle the waste salt and decrease the base energy consumption economically and environmentally, the BMED process is introduced to NaOH production. In particular, the product of HCl and NaOH as raw materials for downstream production is beneficial to maintain the balance of supply and demand in this process.

In this sense, BMED is a significant technology for electro-basification. NaOH is a common base produced in BMED from organic and inorganic wastewater containing sodium salts such as sodium chloride, sodium sulfate, sodium gluconate, and sodium aniline as the feed solution. Many studies have presented the feasibility and sustainability of base production with satisfactory energy consumption and current efficiency. However, it is rarely reported that the maximum concentrated base by BMED technology can only be reached at 4.5 mol/l, which is still lower than the ion-exchange membrane electrolysis in the chlor-alkali industry. The commercial application to produce a base with high purity is still a challenge for BMED technology due to the limited perm-selectivity of ion exchange membranes.

Organic base production can also be achieved by BMED technology, such as choline hydroxide, quaternary ammonium hydroxide [140], and di-quaternary ammonium hydroxide [141].

Choline plays an important role in the metabolism of organisms as one of the essential nutrients. Choline hydroxide has extensive applications in the formation of acetylcholine in the body, the synthesis of organic intermediates, and growth promoters in animal feed. It is mainly obtained by the direct reaction of triethylamine with water and ethylene oxide through the phase separation process. However, there are many byproducts produced by this method, such as mono ethoxy choline, diethoxycholine, and glycol, which decrease the recovery rate and pollute the environment [142]. Another technology named electro-electrodialysis (EED) also shows the potential to produce choline hydroxide. The process is the integration of an ion-exchange membrane and the electrolysis of water. The choline ions combine with hydroxide from the HER to produce choline hydroxide, and the chloride ions combine with protons from the OER to obtain hydrochloric acid. Both of the products are separated under the influence of cation exchange membranes. Then, the high-purity choline hydroxide can be purified by a strong basic anion exchange resin. Unfortunately, more waste acid and alkali are produced in this process. Both methods also have the characteristics of low yield, low purity, and serious environmental pollution. To overcome these shortcomings, choline chloride can be used as the feed solution by using BMED technology to achieve high efficiency and low consumption production of choline hydroxide and hydrochloric acid. The study shows that 321.75 ppm choline hydroxide could be obtained by treating a 0.4 M choline chloride water solution with an applied 12.5 V voltage drop. Only 0.74 kWh/kg energy is consumed, and the current efficiency can reach 84.6% in this process. In addition, the continuous operation of BMED is beneficial to maintain the concentration gradient and inhibit the reverse diffusion between the feed and base compartments, which can effectively achieve a high conversion rate of up to 90.4% and obviously bring economic and environmental benefits [143].

9.6.2.2 The Production of Acid

It is noted that the production of acid by BMED is more sufficient compared with the base production. There are numerous studies in BMED to produce various acids, especially organic acids, such as gluconic acid [144], citric acid [145], acetic acid [146], malic acid [147], lactic acid [148], and amino acids [149]. Organic acids are an important type of chemical that are widely used in food, chemical, pharmaceutical, and material fields [150]. The main production methods of organic acids include chemical synthesis and microbial fermentation. Chemical synthesis will be replaced by other technologies because it requires non-renewable petrochemical resources. Fermentation is the traditional method to produce the majority of organic acids. However, calcium salts are always used as acid–base regulators to maintain a pH-neutral environment in fermentation. In this way, a large number of organic salts with alkaline regulators are produced along with the production of organic acids in the fermentation process, causing many drawbacks in this process. Therefore, to recycle the organic acid from these salts, many chemicals need to be consumed to convert those organic salts into organic acids through acidification steps, which increases the consumption of large energy. Additionally, calcium salt will bring a large amount of solid waste calcium sulfate by reacting with sulfuric acid, which places a greater burden on the environment.

To decrease the requirements of complex postprocessing steps in the fermentation process, it is feasible to recycle the organic salt with the combination of BMED and the fermentation process. As an environmentally friendly technology, the BMED process enables the one-step production of organic acids by using organic salt as a feed solution. In addition to producing acid, the byproduct NaOH is a great substitution as an acid–base regulator to generate salt without any pollution compared with calcium salt. The salt produced in the adjusting pH is capable of being regarded as the feed solution for BMED, which forms a closed cycle with the fermentation method to produce acid. The whole BMED process is clean and efficient, which plays an important role in the fermentation industry. For instance, Wang et al. [151] investigated the BMED technique for gluconic acid production and proved that it is feasible to one-step conversion of gluconate salt into the gluconic acid product with the highest current efficiency of 88% and energy consumption of 0.25–0.40 kWh/kg-gluconic acid. Besides, it is also possible to obtain a high NaOH concentration of 4.58 mol/l with purity of 96.6%. This high-concentration, high-purity base product can be returned to the upstreaming procedures such as fermentation or enzymatic catalyst, leading to completely closed-loop production of this organic acid production.

However, BMED and fermentation methods also have some challenges in the actual coupling of organic acid production. Primarily, there are compatibility problems between the membrane stacks of BMED and the fermentation vessel due to the complex and diverse composition of the fermentation broth in the actual process. Second, some organic acids, such as salicylic acid, have low solubility in 100 °C water, and it is difficult for BMED to sustain such high temperatures, low concentrations, and high energy consumption [152]. Finally, although the highest organic acid concentration of BMED can reach 5 mol/l compared with only 1–2 mol/l organic acid concentration produced by the fermentation method, the problem of coion leakage in this process leads to the low purity of organic acid. Therefore, actual research in this field needs to be explored in depth.

Many inorganic acids, such as hydrochloric acid and sulfuric acid, are obtained by BMED to treat high-concentration inorganic salt wastewater, seawater desalination, or recovery of heavy metals [153–155]. Compared with the traditional method, BMED technology is a desirable method to produce high purity or concentrations of acids and bases, which is conducive to improving the quality and value of products. However, the economic advantages are not obvious due to the high investment cost of inorganic acid production in the BMED process compared with the traditional large-scale acid production process. In addition, proton leakage has become very common in the application of BMED because the mobility of water and protons is higher than that of other ions and because of the quality gap of the commercial membrane between reality and theory [156].

According to the investigation, water consumption will reach 3 800 km^3/year by 2025. The huge consumption of water resources leads to a shortage of water for daily life, which is an increasingly serious problem facing human society all over the world. Developing new desalination technology is a sensible approach to solving the problem of the water crisis [157, 158]. Seawater reverse osmosis (SWRO) is an effective method of solving the freshwater crisis. The main methods used for seawater desalination include distillation, RO, multistage flash evaporation (MSF), multi-effect distillation (MED), and ED. As an emerging technology developed in recent decades, it has largely eased the crisis of water scarcity. In addition to alleviating human water demand, SWRO has brought new problems. (i) Acidification pretreatment problem. Due to the complex composition of seawater, strict pretreatment is necessary before SWRO. To prevent the scaling problem of the system during the desalination process, it is often necessary to add a large amount of strong acid to acidify the seawater, which causes acidified water. (ii) Concentrated seawater treatment problems. A large amount of concentrated seawater will be generated after desalination. Taking the recovery rate of 40% in RO as an example, 10 000 tons of seawater will produce 6 000 tons of concentrated seawater after desalination [159]. The discharge of concentrated seawater directly into the sea is not only a serious waste of resources but also serious damage to the environment. (iii) Posttreatment of seawater desalination. Unlike surface water, water treated by RO has low TDS and unstable water quality due to its high desalination rate. The municipal pipe network and ancillary facilities can be corroded, and hidden dangers will be brought to human health if the water from SWRO is supplied directly without treatment [160].

Converting NaCl into HCl and NaOH using seawater as a feed solution by BMED technology is considered to be an economical and environmentally friendly choice. It has the following advantages: (i) The seawater can be efficiently converted into corresponding acids and alkalis under optimal conditions, which preliminarily proves the feasibility of using concentrated seawater for acid–base preparation. (ii) BMED systems can produce higher concentrations (approximately 2 M) of NaOH and HCl at much lower energy consumption. (iii) The magnesium resources in seawater can be recovered and purified by the acid and base prepared by BMED, in which the recovery rate is more than 80% and the hydroxide purity can reach more than 98%. However, SWRO brine contains other divalent ions, such as Ca^{2+} and Mg^{2+}, which can reduce the performance of BMED membranes due to membrane fouling. Therefore, the integration of SED with BMED, where the BMED system contains a monovalent selective ion exchange membrane that facilitates the separation of primary and divalent ions in seawater by integrating SED with BMED. Moreover, the selectivity of membranes for brine pretreatment needs to be improved for the high purity of the product.

9.6.2.3 Production of CO_2 Conversion

Another feasible application of BMED technology is found in CO_2 capture and utilization to alleviate the greenhouse effect. Similar to the CO_2 capture by the ED process above, it is effective for BMED to capture CO_2 by acidifying seawater. However, the H^+ used to acidify seawater is from the characteristics of BPM rather than the electrolysis of the water process. As the BMED process shows in Figure 9.9, the BPM produces H^+ after water molecules dissociate inside the BPM when an electric field is applied. The proton migrates into the seawater compartment to release CO_2 and obtain acidified seawater by reacting with HCO_3^- and CO_3^{2-} ions. The obtained acidified seawater containing CO_2 is divided into two parts. One part enters the base compartment in the BMED to form HCO_3^- and alkalized seawater after reacting with OH^- from BPM [134]. The other part is treated by gas cooling and a vacuum pump to obtain pure CO_2, which can be directly transported into the electrocatalytic reaction to produce CO. The successful application of extracting CO_2 from seawater to the electrocatalytic production of organic carbon fuels confirms the feasibility of CO_2 recovery by BMED [161]. On the other hand, BMED technology can realize the recovery of CO_2 from organic wastewater as well as inorganic wastewater. For example, CO_2 is absorbed by the mixed solution of the intermediate sodium methionine and sodium carbonate to produce a feed solution for BMED. and the feed solution is flowed into the BMED to generate amino acids, reusable bases, and pure CO_2. Obtaining byproduct amino acids and organic matter, such as methionine [162] and aniline [163], is beneficial to alleviate the environmental burden caused by chemical synthesis. In addition, capturing CO_2 from seawater is conducive to achieving "negative emissions," but energy consumption has always been an urgent challenge in carbon capture and utilization technology. The theoretical energy requirement for CO_2 recovery from an alkaline carbonate solution is approximately 0.8 MJ/kg CO_2. However, the actual energy consumption is 4.4 MJ/kg, which is an order of magnitude lower than the energy requirement of CO_2 recovery by ED reported in the literature. To reduce the energy consumption of CO_2 recovery in the BMED ED process, the parameters of BMED should be optimized. The results show that only 2.1 MJ/kg CO_2 is required to recover CO_2 when a current density of 2.4 mA/cm^2 is applied in a 1 M $NaHCO_3$ feed solution by BMED. The process can further reduce energy consumption by reducing the coion leakage in the BMED [164]. The energy consumption of this process is estimated using a mathematical model, which found that the theoretical energy consumption of CO_2 recovery is 2 MJ/kg CO_2 using $NaHCO_3$ by BMED [165]. The difference is because parts of the CO_2 stored in the membrane stack are not released in time, which causes the actual energy consumption to be slightly higher than the calculated theoretical energy consumption. It is worth noting that CO_2 is easily trapped in the membrane stack due to excessive temperature to form "local hot spots" in the process of BMED, which not only damages the membrane surface but also reduces the membrane life. Therefore, placing the membrane stack at a pressure of 20 atm is conducive to eliminating the "hot spot" problem, protecting the ion exchange membrane, and reducing energy consumption during the experiment. [166].

9.6.3 The Limitations of BMED

In the BMED process, purity, energy consumption, and current efficiency are important parameters to measure whether BMED is suitable for this application. However, they are restricted by the following limitations in the BMED process, although the BMED process has the advantages of being environmentally friendly and cost-effective. Primarily, it is necessary to ensure the purity of effluent by removing the polyelectrolytes, some colloidal matter, large organic anions or di- or multivalent ions, and some components through pretreatment of effluent. As the base concentration increases, the metal ions and polyelectrolytes will precipitate or attach to the membrane surface, causing severe fouling of the membrane and a decrease in membrane performance. Generally, the concentration of multivalent ions should be less than 1 ppm before flow into the BMED membrane stack. Similar to the ED process, the ion-exchange membrane with a limited perm-selectivity of less than 100% will contribute to the reduction in product purity, which decreases with increasing product concentration. The phenomenon of coions, water electro-osmosis, and water osmosis by the concentration gradient also exists in the BMED process. This is not only the reason for the decrease in purity and concentration of the product but also the cause of the rise in energy consumption and reduction in current efficiency. Moreover, a low electrolyte concentration contributes to the high resistance of the whole BMED stack. However, a high electrolyte concentration will lead to the low perm-selectivity of the ion-exchange membrane [167]. Therefore, a minimum conductivity of 20 mS/cm is required in the salt compartment [168], and the destination of salt conductivity in laboratory research is always set at 8 mS/cm according to the industry. The HER and OER that occurred in the BMED also increased the energy. Additionally, the anion-exchange layer of bipolar membranes is sensitive to high-concentration bases and easily swells in the presence of aromatic compounds, which shows poor chemical stability and decreases the base and acid purity [169]. Ultimately, the HER and OER produced in the electrode chamber will lead to a loss of voltage, increasing the energy consumption of the BMED process.

Focused on the limitation in the BMED process, some phenomena can be retarded, such as membrane fouling and electrode reactions, by some treatment measures except the pre-treatment of salt solution. In the case of membrane contamination, chemicals such as $CaCO_3$ and $Mg(OH)_2$ can be eliminated by the pickling membrane stack method based on the dissolution of precipitates by some organic or inorganic acid solutions. Various precipitates, such as $CaSO_4$, can also be eliminated by the reverse electrode method. The impregnation effect of some surfactant and foaming agents on the membrane surface sediment can also be employed to peel the sediment on the surface of the membrane. With regard to the electrode reactions, the HER and OER that occurred on the electrodes during the BMED process are capable of being replaced by reversible oxidative coupling reactions with minimal thermodynamic and kinetic voltage loss. The current in this process can be loaded by water dissociation in the intermediate interface of the bipolar membrane [134]. Moreover, a small portion of the product can be added in the product chamber, such as an acid or base chamber, at the beginning, which is beneficial to lower the resistance of the solution and decrease the energy consumption of the whole process. Finally, the development of new materials to prepare ion exchange membranes and bipolar membranes with high performance is an important way to alleviate the limited selectivity of ion penetration.

BMED technology possesses environmental benefits and inherently economic value that deserves wide attention and sustainable development. Based on the water-splitting characteristic of the bipolar membrane, numerous applications of BMED have been explored in chemical synthesis, bio-food processing, and environmental protection. Unfortunately, the BMED is not well implemented at the large-scale, industrial level due to the high price of the bipolar membrane and some intrinsic limitations of this technology such as low acid and base product concentrations and purities. For the limited concentration in the BMED process, continuous saturation feeding at a high current density ($>100\ mA/cm^2$) is regarded as a credible choice to improve the base and acid concentrations. In addition, the symmetric plate and frame modules of the ion exchange membrane can be changed to an asymmetric design to reduce energy consumption and alleviate concentration polarization, which can more effectively match the migration of ions in the ion exchange membrane. The capital and operating costs of ion exchange-related processes are further reduced by means of process integration.

9.7 Electrodialysis Metathesis (EDM)

Electrodialysis metathesis (EDM) has attracted much attention as a novel technology combining the function of ED and the double decomposition reaction. In contrast to conventional ED, the cation- and anion-exchange membranes are arranged alternately in the membrane stacks to form four compartments in the EDM, which include two feed chambers and two product chambers. As Figure 9.10 shows, the cations and anions of the first salt in one feed chamber move to the different product chambers and combine with the anions and cations of the second salt in the other feed chamber to form two new salts under the influence of the electric field.

EDM is often used in the production of ionic liquids, the transformation of a small amount of soluble salt into a soluble solution, the production of inorganic salt fertilizer, and other aspects. It is equipped with the properties of both concentration and desalination and double decomposition [170, 171]. EDM has a lower operating cost, high current efficiency, and higher concentration ratio than the RO process [172, 173]. However, coion leakage is also a disadvantage of EDM. In contrast to the traditional complex reaction, EDM is not an equilibrium process and contributes to the separation of products from the reaction mixture. In theory, any metathesis reaction between two electrolytes can be accomplished in an EDM stack, even those that cannot occur spontaneously according to the laws of thermodynamics.

9.7.1 Application of the EDM Process

9.7.1.1 The Production of Ionic Liquid

Ionic liquids are important organic salts composed of cationic and anionic solvents. It is a kind of ionic compound with a melting point lower than 100 °C. There are enormous applications in chemical synthesis catalysts, liquid–liquid extraction, lubricants, and electrochemical applications because of their versatility [174]. Ionic liquids are usually prepared by a two-step process: cation synthesis and a metathesis step. Cations are synthesized by quaternary amination of the organic bases with haloalkanes, and the anions are obtained by adding Bronsted acid (HY) or group 1 metal/ammonium/silver salt (MY) to the alkylated Lewis base or its halide salts or by anion exchange materials [175]. To obtain a product in higher purity, an expensive and unsafe metathesis process is required to separate the target ions. Unfortunately, the product is mixed with high halide and silver impurities, which is a potential threat to human health and pollution to the environment. In some cases, ion-exchange methods also require high-purity products, but the frequent regeneration of ion-exchange resins requires intermittent operations, resulting in additional steps that are both costly and complex to perform. Therefore, EDM has superior technical and economic advantages in the production of ionic

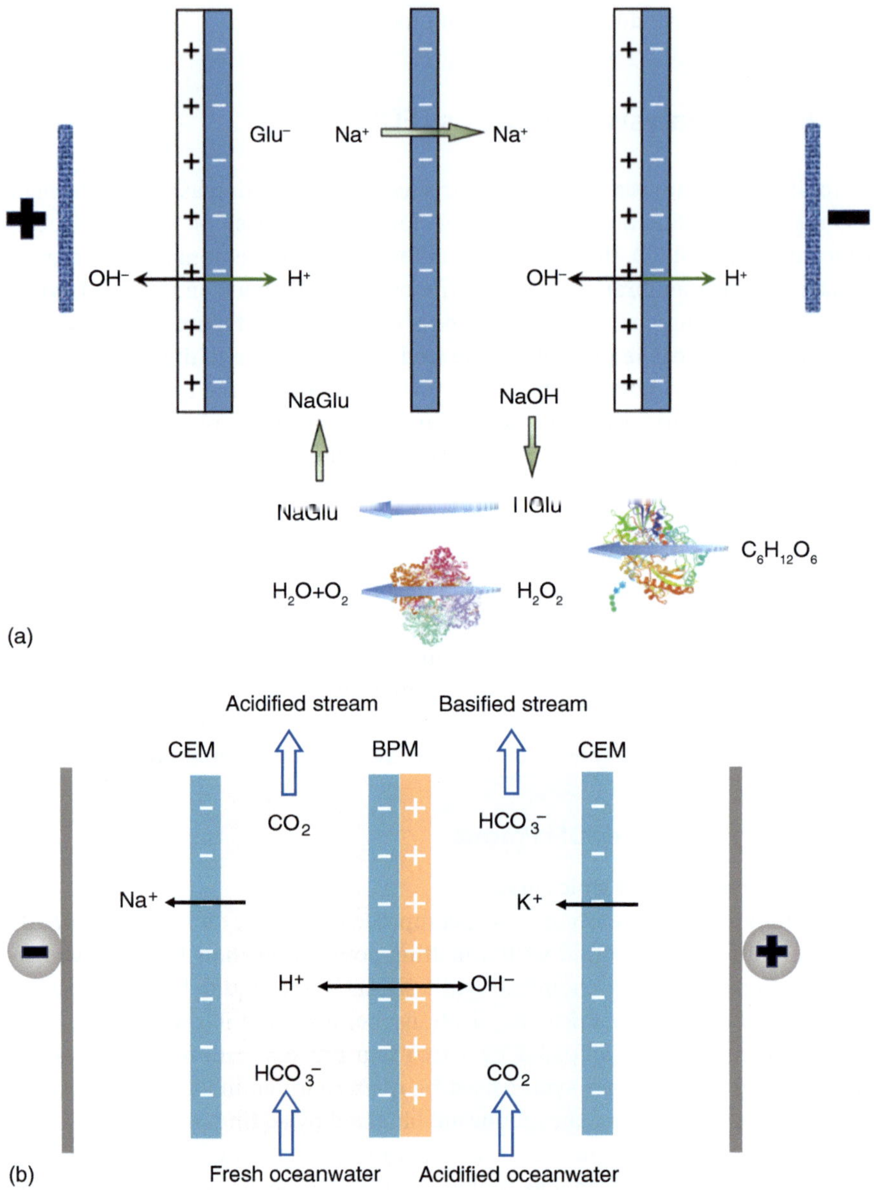

Figure 9.10 (a) Schematic diagram of electrodialysis (BMED) and enzyme catalysis: cation-exchange membrane (CEM), bipolar membranes (BP-1), catalase (CAT, PDB: 1DGB), and glucose oxidase (GOD, PDB: 1GAL). (b) Schematic diagram of the bipolar membrane electrodialysis membrane stack configuration for seawater desalination. (c) Schematic principles of the metathesis electrodialysis (EDM) process to prepare potassium chloride. Source: Wang et al. [151]/American Chemical Society.

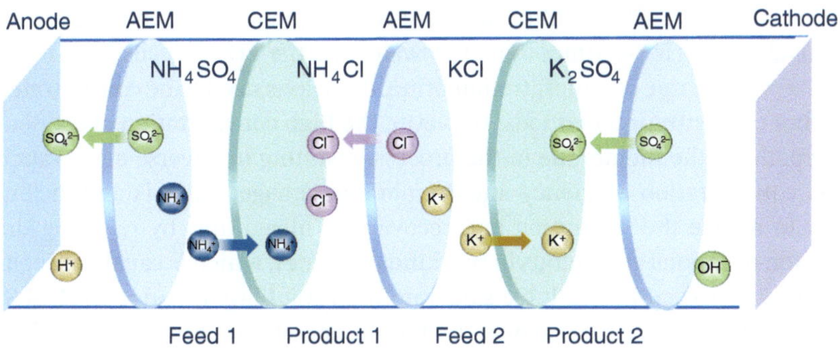

(c) $2KCl + (NH_4)_2SO_4 = K_2SO_4 + 2NH_4Cl$

Figure 9.10 (*Continued*)

liquids in high purity. For example, choline dihydrogen phosphate is regarded as one of the most promising choices for employment as a medium to stabilize protein [176] and DNA triplet formation [177]. Choline dihydrogen phosphate is produced by the reaction of choline chloride and sodium dihydrogen phosphate by ED metathesis with the membrane configuration of CACA [178]. The choline ions pass through the cation-exchange membrane and enter the concentrating chamber, forming choline dihydrogen phosphate with the dihydrogen phosphate ions from the diluting chamber of sodium dihydrogen phosphate. Sodium chloride is produced in the other concentrating chamber. After analyzing the effect of operation parameters, it was found that EDM could produce the highest product purity at a high current density and low concentration of choline dihydrogen phosphate. The current efficiencies of choline chloride and dihydrogen phosphate reach 94% and 58.6%, respectively, under the optimal conditions, which also leads to the total cost of chloride dihydrogen phosphate by EDM being lower than 0.06€ per mole. In addition, purity is also a vital parameter in the process of EDM, which is mainly influenced by coion leakage. Due to the permeability selectivity of the ion exchange membrane used in EDM, the purity of the product will be reduced when the coions pass through the ion exchange membrane. The influence of operation parameters on the flux of the coions during EDM operation is studied. The results show that the flux of the coion is mainly determined by the difference in ion concentration between the two sides of the membrane, which increases with increasing ion concentration on both sides but decreases with increasing counterion concentration. In conventional ED, cationic impurities are mainly caused by concentration diffusion, while anions are mainly caused by coion leakage. Generally, impurity ions produced by concentration diffusion are higher than coion leakage. Moreover, current density has little effect on coion flux, so high current density is beneficial to reducing coion flux and improving product purity in EDM [179].

9.7.1.2 High-Salinity Wastewater Conversion

The treatment of high-salinity wastewater from nanofiltration is considered another vital application of EDM. The obtained nanofiltration brine contains a relatively high concentration of Ca^{2+} and Mg^{2+} with greater hardness, in which the concentration of bivalent ions is 3–4 times that of seawater and the TDS exceeds 7 000 mg/l. To achieve zero discharge of nanofiltration brine, these ions need to be concentrated through other concentration methods. However, the high concentration of divalent ions easily pollutes the membrane in the process, resulting in adverse effects such as reduced concentration efficiency and membrane damage. EDM is an efficient technology to realize the separation and recovery of mixed ions by realizing the recombination of cations and anions in two kinds of brines, which is capable of converting slightly dissolved or insoluble salts into high solubility liquid salts (HSLS) by softening brine. This process avoids the fouling membrane caused by bivalent ion scaling while improving water recovery and effluent concentration in seawater desalination. The highly soluble Na^+ salts and Cl^- salts are used to react with the anions and cations in the SWRO waste brine by the employment of EDM. The insoluble calcium salts and magnesium salts transform into highly soluble chlorine salts, reducing the phenomenon of membrane pollution. The TDS value of the concentrated solution gradually increased from 0 to approximately 200 g/l after 72 h, and no scaling occurred in the multibatch preconcentration process [180]. EDM technology is not only a new method to enhance SWRO water recovery from SWRO brine but also provides a potential possibility for the effective enrichment of high-solubility liquid salts. Without the risk of membrane pollution, the concentrations of Na^+ salts and Cl^- salts reached 200 and 170 g/l, respectively, after 7 batches of treatment. In addition, the two concentrated streams can be combined to precipitate $CaSO_4$ and remove approximately 60% of Ca^{2+}, which provides a feasible way for zero discharge and recycling HS wastewater widely existing in desalination engineering and industrial wastewater treatment. In addition, the experiment shows that a homogeneous membrane can promote the migration of counterions, suitable for zero discharge of high hardness brine because of high fixed ion density and permeability selectivity as well as low resistance and water absorption [181].

9.7.1.3 The Production of Potassium Fertilizers

Potassium is an essential nutrient in plant life that promotes photosynthesis, resists harsh weather, and benefits the growth of crops. However, common chlorine-containing potassium fertilizers are poisonous to chlorine-allergic crops and easily increase salinity and acidification in soil. In addition, chloride ions are capable of combining with nitrogen fertilizers such as NH_4NO_3 to form environmentally harmful compounds. Therefore, the production of chlorine-free potassium fertilizers is a promising choice for chlorine-containing potassium fertilizers. The traditional methods for producing chlorine-free potassium fertilizers such as K_2SO_4 and KNO_3 are composed of double-decomposition reactions, the Mannheim process [182], the conversion method of gypsum [183], the conversion method of mirabilite [184], and the picromerite method [185]. These methods require a complex separation and purification process, in which a large amount of energy is consumed. The practicability of those methods is limited by the low conversion rate, high equipment requirements, and low product purity. Therefore, EDM can be a sensible technology for producing chlorine-containing potassium fertilizers

Table 9.2 Data reported [186–192] on the production of inorganic potassium salts using EDM.

Product	Concentration (mol/l)	Purity (%)	Energy consumption (kWh/kg)	Capital cost ($/t)
K_2SO_4	0.57	–	–	71.4
K_2SO_4	0.71	–	0.43	200
KNO_3	1.27	–	0.29	119.1
KNO_3	0.97	99	0.82	–
KNO_3	2.16	96.43	0.36	–
K_2SO_4	0.46	96	1.28	–
KNO_3	1.4–1.47	92.35–95.91	0.41–0.64	160

due to its low cost and high purity. Inorganic salts such as Na_2SO_4, $(NH_4)_2SO_4$, and $NaNO_3$ are great candidates for a double-decomposition reaction with KCl to produce commonly used chlorine-free potassium fertilizers such as K_2SO_4 and KNO_3 [186, 187], which are listed in Table 9.2. The current density has a great influence on the experimental cost. The production cost of K_2SO_4 is 71.4 $/t when the current is 150 mA/cm^2 [188]. However, the calculated data show that the production expense of K_2SO_4 is 200 $/t at a current density of 15–20 mA/cm^2 when the byproduct of the petrochemical industry is selected as the feed solution to prepare K_2SO_4 [189]. In addition, the type of membrane also has a certain influence on the EDM production of potassium fertilizers. The heterogeneous membrane is more effective than the homogeneous membrane in blocking water migration during the osmosis process. The transmission of hydration ions during the process reduces the product concentration. Membranes fixed with high ion groups and low water content are beneficial to reduce the problems of reverse diffusion, osmosis, and electromigration of water in EDM [190]. Therefore, the selection of membranes with high selectivity, high conductivity, and low hydropower transport capacity is conducive to the achievement of low energy consumption of EDM. The kinds of potassium salts have different demands on energy consumption. Han et al. [191] prepared chlorine-free potassium fertilizer with KCl with nitrate, carbonate, and sulfate by EDM, and the purity of KNO_3 increased to 95.91% after increasing the current density to 25 mA/cm^2. The average energy consumption among these different anions displays that $-E_{K_2CO_3} > E_{K_2SO_4} \approx {}^{-}E_{KNO_3}$, which significantly indicates that the production of KNO_3 is relatively pure and economic compared with that of K_2SO_4 and K_2CO_3. In summary, EDM has great application potential in the preparation of chlorine-free potash fertilizers, especially for KNO_3 and K_2CO_3.

Generally, EDM is a high-productivity, low-cost, easy-to-operate, environmentally friendly technology that is used in the application of converting waste resources into value-added products. However, there are several problems among these three processes, such as ion leakage, low product purity, and membrane pollution. This is because the main properties of ion-exchange membranes are closely related to the surrounding environment. For example, membrane selectivity decreases sharply with increasing electrolyte concentration. Therefore, it is sensible to improve the performance of ion exchange membranes with low resistance and high selectivity. There still has a long road ahead to promote the EDM to industrial-scale application.

9.8 Ion-Distillation Technology

Inspired by the working mechanism in the phase-balance-based distillation process and rate-transfer-based chromatography process, Jiang et al. proposed an ion-'distillation' technique for precisely isolating monovalent ions from its mixed solution [131], i.e. Li^+ ions from lake brine. Distillation process internal-integrated single evaporation process. In distillation, the separation result is from the difference in vapor- and liquid-phase compositions arising from the partial vaporization of a liquid mixture or the partial condensation of a vapor mixture. The vapor phase becomes enriched in the more volatile components, while the liquid phase is depleted of those same components. In practice, distillation would be carried out by either of two principal methods, i.e. (i) separating the vapor by boiling up the liquid mixture (feed) and condensing the vapors without allowing any liquid (feed) to return to the still. There is then no reflux; and (ii) the return of part of the condensate to the still and brought into intimate contact with the vapors on their way to the condenser. By externally integrating the simple distillation operation, multistage distillation would then be formed. Otherwise, when separating mixtures with two components that boil at a wide range of temperatures, flash distillation (multistage distillation) is used, which is common in industrial large-scale production. According to the phase balance on the serial plates, continuous flash distillation would produce nearly pure products at both the bottom and top of the tower. Flash distillation is recognized as internally integrated multistage distillation without reflux. In comparison to external integration, flash distillation operation efficiently separates the different components, decreases the capital cost and energy consumption, and incorporates single operations into one continuous process. Generally, external integrated multistage distillation increases the yield of products and elevates the isolation efficiency in separating components of comparable volatility using reflux [193]. However, distillation relies on heat as the energy for resource separation and has been recognized as an energy-intensity separation process that depends on the combustion of large amounts of conventional fossil fuels and the release of enormous amounts of CO_2 gas into the atmosphere.

Benefiting from the development of irreversible thermodynamics in the separation process, the physics and chemistry of membrane processes have been clarified. Membrane-based separation methods, or other nonthermal methods, can be an order of magnitude more energy efficient than heat-driven separations that use distillation. They follow the transition rate differences between different resources, according to their chemical properties or size differences, under the driving force of pressure (ΔP), electric (ΔE), and concentration difference (ΔC) [194–197]. However, the nonideal isolation efficiency is one critical issue that limits its large-scale application. It was proposed that "...*Hybrid membranes can be used for bulk separation and cryogenic distillation for 'polishing' the product. Such approaches would reduce the energy intensity of alkene production by a factor of 2 or 3, until membranes become good enough to replace distillation entirely*...," and was recognized as one strategy of "*Seven chemical separations to change the world.*" [198]

However, the design of the internal integrated membrane separation process is limited by the separating driven force ΔP. If the semipermeable membrane was assembled in a multistage module, the driving force ΔP supplied on the 1st membrane would not effectively conduct to the following stages (see Figure 9.11) because the pressure is applied based on direct contact. Multifiltration (MF) beds, which deliver pressure differences from the top to bottom of columns, are a practical alternative when treating complex solutions. For instance, for the separation of complex resources and system intensification, the water recovery and management (WRM) system implemented at the International Space Station (ISS) adopts multi-filtration (MF) beds to recycle freshwater from crew urine, humidity condensate, and Sabatier product water for the implications of crew drinking and hygiene activities, oxygen generation, urinal flush water, and various payloads. [199] Therefore, external multistage integration or the encoded procedure "*multistage membrane cascade*" is an effective approach to integrate pressure-driven membrane processes [197, 200] (Figure 9.11).

The electric driving force (ΔE) would trans-across the solutions with ion conductivity and drive the migration of ions across the ion exchange membranes under a reversible thermodynamic state. According to the ions' physicochemical properties, i.e. the charge (cation/anion), the valence (mono–/multivalent cation/anion), and the radius (ionic crystal radii/effective hydrated radii), different ions in the solutions would be isolated correspondingly by using the definite ion exchange membrane. For instance, the cation exchange membrane allows the transition of cations and block anions, the monovalent cation-selective membrane (MVCs) allows the transition of monovalent cations and blocks multivalent cations, and the anion-, ion-imprint- or functional material-optimized membranes allow the perm-selective transition of definite ions while blocking other ions. Incorporating the two classical characteristics and under the role of the driving force from the electric field and the selectivity of the ion exchange membranes, electro-membrane processes were proposed, i.e., ED (for water desalination), selectrodialysis (for monovalent ion isolation), and BMED (for transforming salt into the corresponding acid and base) [203]. However, all these proposed electro-membrane processes were designed according to the original ED process that was first reported in 1958 and were trapped into the "cell-repeating" rule. For instance, in the selectrodialysis membrane stack [204], one monovalent cation-selective membrane (MVC) and one monovalent anion selective membrane (MVA) would configure into one cell (see Figure 9.12) and then simultaneously separate the monovalent cation and monovalent anion from the complexing ion-composited solutions, i.e. isolating lithium from the brine, extracting sodium chloride from seawater, and separating the rare earth metal ions from the ore leachate. By incorporating the cells into hundreds/thousands of repeating units, the processing capacity would thus be increased. However, the

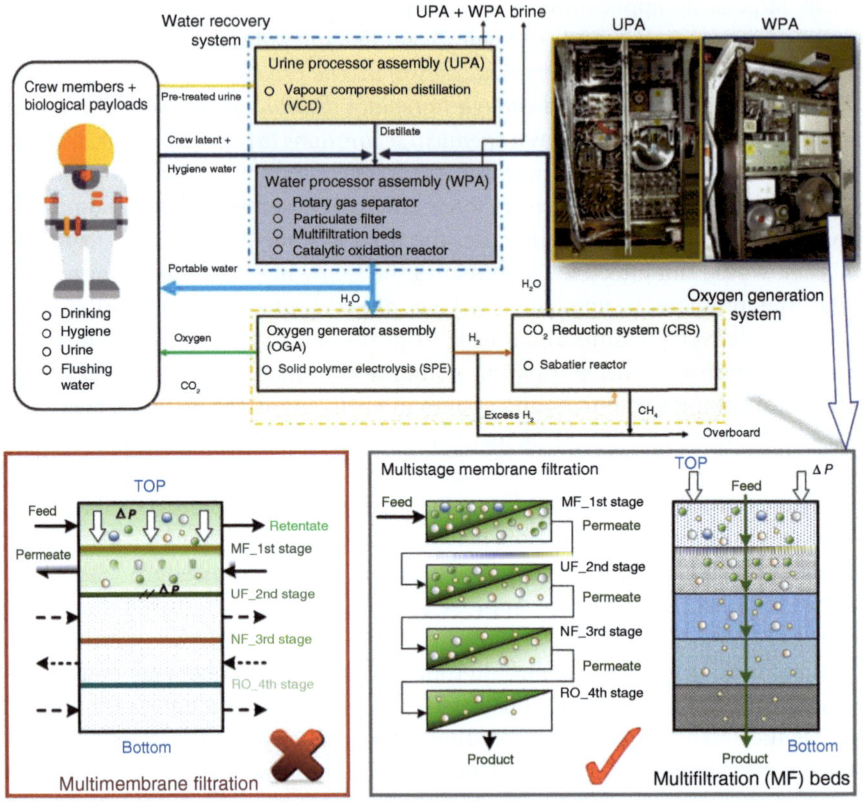

Figure 9.11 Schematic of the urine processor assembly (UPA) and the water processor assembly (WPA) installed at the International Space Station (ISS). Source: Adapted and modified from Refs. [199, 201, 202] (Copyright © 2020 MDPI).

monovalent ion's selectivity (S) as the critical feature was limited by the nonideal selectivity of the MVC/MVA. The external integration of the selectrodialysis stack, which was called multistage operation as that generally adopted in the distillation and membrane filtration process, was the unique option. Staged electro-membrane processes are also called multistage ED and are used in SWRO brine management and bio-fermentation broth [123]. Inspired by flash distillation and taking advantage of the working mechanism in chromatography, for the internal integration of the electro-membrane stack and efficient separation of the definite products from the complexing ion composite solution, we proposed a novel ion-"distillation" technique that is different from conventional electro-membrane processes, such as the selectrodialysis process [204]. As shown in Figure 9.12 and Figure 9.13, we reconfigured the ion exchange membranes in the selectrodialysis cell and assembled the same type of ion exchange in sequence between the cathode and anode according to the "Identical Membranes at the Same Side (IMSS)" principle. The working mechanism and electro-striping principle of the ion-"distillation' process were described in detail in the main work. The proposed ion-"distillation" operation would efficiently

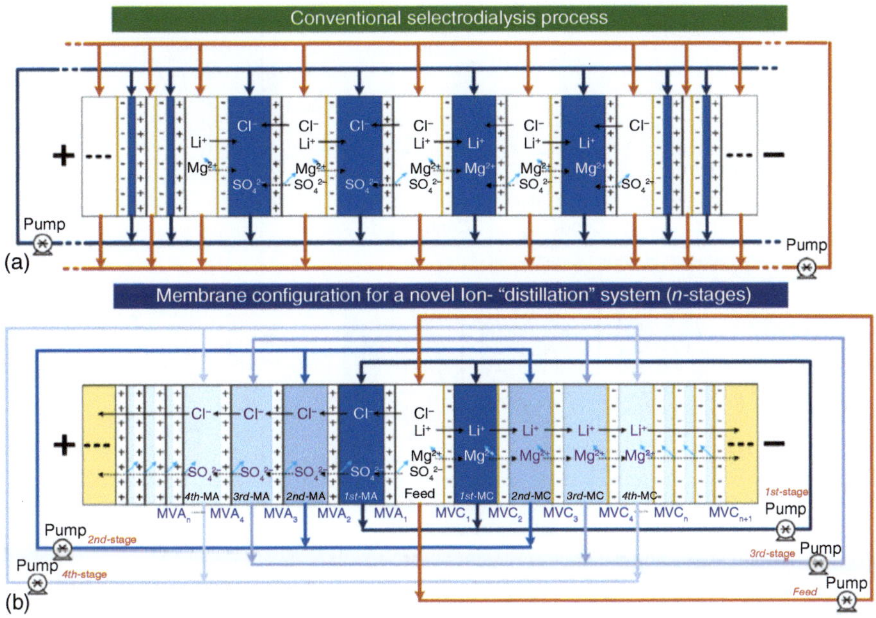

Figure 9.12 a Membrane configuration for selectrodialysis, b membrane configuration for ion-distillation membrane stack. The membrane configuration is Anode (+) MVA_{n+1}| *n*th-MA stage | MVA_n| MVA_4| 3rd MA stage | MVA_3| 2nd MA stage | MVA_2| 1st MA stage | MVA_1| Feed | MVC_1| 1st-MC stage | MVC_2| 2nd-MC stage | MVC_3| 3rd-MC stage | MVC_4| MVC_n| n^{th}-MC stage | MVC_{n+1}| (–) Cathode.

separate different compositions with the desired selectivity (*S*) while overcoming the constraint from the hysteresis development of commercially available ion-selective membranes. Such a configuration of in-series stacking increases the selectivity of membranes without sacrificing ion flux. The cations migrate across the MVCs labeled MVC_1, MVC_2, MVC_3, MVC_4, ·····MVC_n, and MVC_{n+1} (*n* is the membrane number in the ion-"distillation" module) toward the cathode. Each two consecutive membranes are separated by an ion-stripping chamber (*n*th-MC, where n is the chamber number in the ion-"distillation" module); for example, MVC_1 and MVC_2 are separated by the 1st-MC (1st-Monovalent Cation, i.e. 1st-Li^+) stage. The monovalent anions are separated via the same mechanism as the cations, and the MVAs between the feed chamber and the anode are assembled in the same sequence.

The ion-distillation process was first used to isolate lithium from lake brine with a complex ion composition, especially magnesium and lithium ions coexisting in brine with a high Mg^{2+}/Li^+ mass ratio. Simulated brine with a Mg^{2+}/Li^+ ratio of 35 was treated to evaluate the feasibility of the ion-distillation concept. The results demonstrate that the selectivity of lithium and magnesium increases from 30 (one stage) to 1104 (two stages), 3297 (three stages), and 26177 (four stages) for the different ion-distillation processes. The purity of lithium products achieved via the second-stage and four-stage ion-distillation process exceeds the industrial-grade (99.69%) and battery-grade (99.98%), respectively. In the fourth

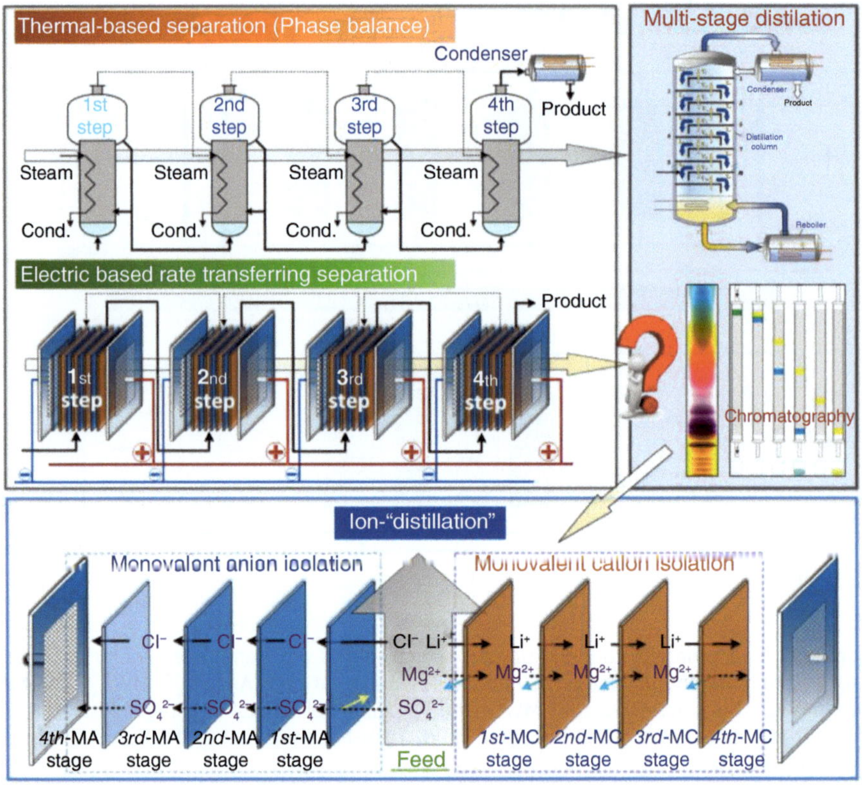

Figure 9.13 The design principle for ion "distillation."

step, the selectivity of chloride ions and sulfate ions reaches 27 000 at the same time. Figure 9.14 demonstrates the concentration changes for the Li^+ and Mg^{2+} ions in the *four*-stage ion-"distillation" process when treating practical lake brine. A current density of 2.5 mA/cm^2 was selected in the process to prevent ion precipitation, which may be induced by the overlimiting current density. In addition, pretreatment of lake brine is generally required to remove coexisting ions, i.e. Na^+, K^+, and B. Then, the obtained brine is processed in the next critical section: Li^+ and Mg^{2+} separation to produce lithium products. For the lake brine, the Li^+ ion in the 4th stage increases from the initial 0.1 to 0.5 M, while that of Mg^{2+} ions increase from the initial 8E-5 to 3E-4 mol/l with a Li^+ selectivity $\left(P^{4\text{th}}_{Li^+/Mg^{2+}}\right)$ of 13 323. The Li^+ ion flux is 3.45E-7 mol/cm^2 A with a current efficiency of 82.5%. The products of Li_2CO_3 and Li_3PO_4 were finally obtained through a precipitation procedure.

Electroneutrality was realized by connecting the respective cation-distillation and anion-distillation chambers. In an n-stage system, the feed solution containing Li^+, Mg^{2+}, Cl^-, and $SO_4{}^{2-}$ (0.1 M LiCl and 0.1 M $MgSO_4$) circulated in the feed compartment between the stacked monovalent ion exchange membranes, i.e. MVA_1 and MVC_1. The MA and MC stages were fed with supporting electrolytes (0.1 M LiCl) using the pumps. For instance, when the *1* MA and *1st*-MC chambers were

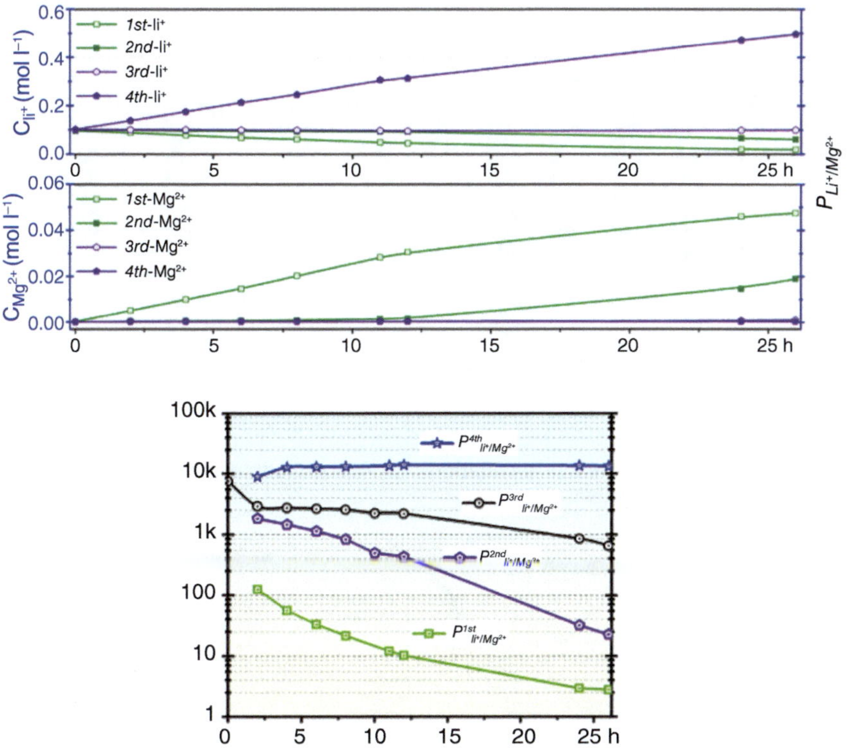

Figure 9.14 The concentration changes for the Li^+ and Mg^{2+} ions in the procedure of ion "distillation" for treating lake brine from East Taiji Nai'er Lake, China.

connected, the Cl^- ions migrated across MVA_1 into the *1* MA stage, while the Li^+ ions migrated across MVC_1 into the *1st*-MC stage. Correspondingly, the LiCl product was produced under the principle of electroneutrality. *nth*-MA and the *nth*-MC stages were connected following the same mechanism as *1st*-stages. Figure 9.15 illustrates the 3D scheme of the ion-"distillation" membrane module and the solution circulating manner in the system.

Selectrodialysis, nanofiltration, and membrane capacitive deionization are generally adopted membrane processes for lithium mining from lake brine (see Figure 9.16). The proposed selectivity (S) values for Li^+ isolation from lake brine are 20.2–33, 3.5, and 2.95 for the proposed three techniques. To improve Li^+ selectivity, researchers have produced cutting-edge membrane materials such as MOF-based membranes (S = 1815), bioinspired COF-based membranes (S = 64), MXene composite membranes (S = 28), and polymer inclusion membranes (S = 893.8). Despite the fact that these membranes significantly improve lithium selectivity, their applications are restricted due to the trade-off between selectivity and permeability. The multistage ion-distillation approach is capable of separating lithium from lake brine with S exceeding 20,000 by using commercially available membranes (S < 30). The isolated Li^+ product exhibits acceptable ion flow and current efficiency. Moreover, if novel membrane materials with definite selectivity

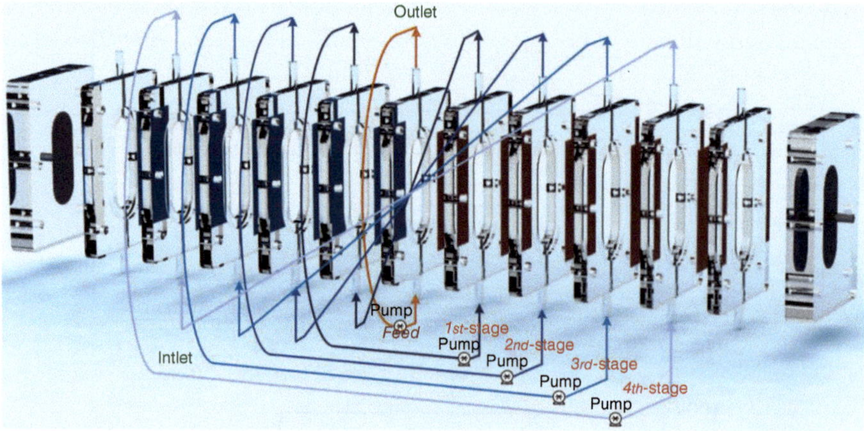

Figure 9.15 3D scheme of the ion-"distillation" membrane module and solution circulation manner in the system.

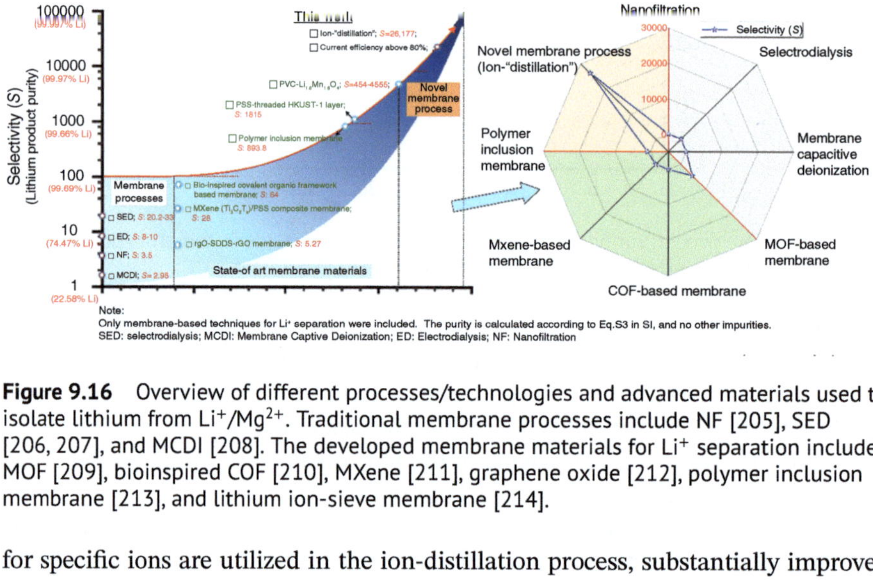

Figure 9.16 Overview of different processes/technologies and advanced materials used to isolate lithium from Li^+/Mg^{2+}. Traditional membrane processes include NF [205], SED [206, 207], and MCDI [208]. The developed membrane materials for Li^+ separation include MOF [209], bioinspired COF [210], MXene [211], graphene oxide [212], polymer inclusion membrane [213], and lithium ion-sieve membrane [214].

for specific ions are utilized in the ion-distillation process, substantially improved separation efficiency is envisaged. Therefore, the suggested approach offers an unprecedented avenue for the separation of diverse species, including isotopes (e.g. Li^6 and Li^7), chiral compounds (e.g. chiral amino acids), and metallurgical compounds (e.g. rare earth metals).

9.9 Fuel Cells

Fuel cells are electrochemical devices that can convert the chemical energy of fuel into electrical energy. Compared with conventional combustion-based engines, including electronics, housing power, power plants, and passenger vehicles in which fuels are burned and gases are expanded for operation, fuel cells can convert chemical energy directly into electrical energy that is independent of the Carnot cycle. Therefore, fuel cells present inherent advantages in energy efficiency with an electrical energy conversion efficiency as high as 60%, in comparison with 40% or less in the internal combustion engine. In addition, the almost zero noxious gas emissions, such as nitrogen dioxide and carbon dioxide, render fuel cells very promising power generation stations.

The reactions on fuel cells mainly include fuel oxidation and O_2 reduction that take place on the two electrodes and the transport of ions in the electrolyte solution. Figure 9.17 depicts the key components of a proton exchange membrane fuel cell stack, which consists of a membrane-electrode assembly (MEA), two bipolar plates, and two seals. The oxidative and reductive half-reactions are kept separated by one piece of proton exchange membrane (PEM). Specifically, hydrogen, a typical fuel, is fed into the anode, where it is oxidized, producing hydrogen ions and electrons. In contrast, oxygen, as a representative oxidizer, is subjected to the cathode, where the hydrogen ions from the anode accept electrons from the latter and react with the oxygen to produce water.

The difference among various energy levels at the electrodes is the voltage per unit cell. The amount of electric current available to the external circuit depends on the chemical activity and amount of the substances supplied as fuels. The current-producing process continues for as long as there is a supply of reactants because the electrodes and electrolytes of a fuel cell, unlike those in a regular battery, are designed to remain unchanged by chemical reaction. To be specified, the following electrode reactions happen.

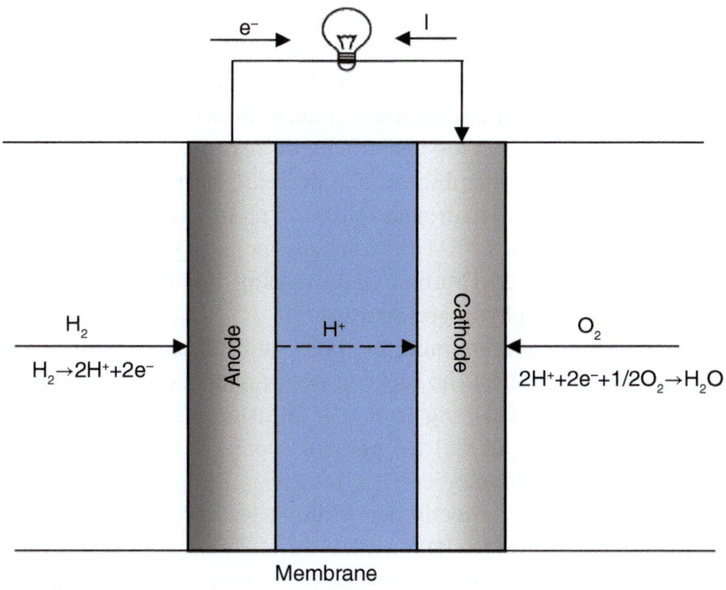

Figure 9.17 A schematic principle of a PEM fuel cell process.

On the anode side, reductive reactions occur,

$$H_2 \rightarrow 2H^+ + 2e^-$$

On the cathode side, oxidative reactions occur,

$$2H^+ + 2e^- + 1/2O_2 \rightarrow H_2O$$

According to the type of electrolyte, fuel cells can be mainly divided into proton exchange membrane fuel cells (PEMFCs), AEM fuel cells (AEMFCs), solid oxide fuel cells (SOFCs), and others. PEMs are commonly used in PEMFCs to transport protons and provide pathways for proton conduction between the anode and cathode, to prevent an electrical short circuit and to avoid the mixing of reactant gases. A desirable membrane for PEMFCs should present low resistance, high conductivity, and oxidation resistance [215, 216]. Unfortunately, the high price of the perfluorinated membrane, the high cost of the noble metal catalysts, the troublesome storage and transportation of hydrogen, and the increase in power density are the major reasons that limit the large-scale application of fuel cells [217]. To facilitate ion transport and reduce transport resistance, tremendous efforts have been dedicated to decreasing membrane thicknesses while incorporating reinforcements such as polytetrafluoroethylene (PTFE) to fabricate PEMs [218]. For instance, the thicknesses of the representative Nafion membranes decreased from the initial product Nafion™ 117 (177.8 μm) to Nafion 115 (127 μm), Nafion 1 135 (88.9 μm), Nafion 212 (50.8 μm), and Nafion 211 (25.4 μm). For another top-tier PEM manufacturer, GORE-SELECT® membranes have a much thinner thickness, with the lowest membrane thickness of approximately 8.5 μm for GORE-SELECT M765.08. The decrease in membrane thickness caused several drawbacks, such as poor membrane strength, accelerated crossover of hydrogen, and unsatisfactory membrane physical durability. Tang et al. [219] found that the fatigue strength or the safety limit of the cyclic stress for the Nafion NR111 membrane is ~1.5 MPa, which is only 1/10 of the tensile strength of the membrane. The cyclic stress and dimensional change induced by the swelling of the membranes are pronounced and are also the major reasons for the mechanical degradation and failure of the membrane. The decrease in membrane thickness could exacerbate the crossover of hydrogen across the membrane. For instance, Balci et al. [220] investigated various PEM thicknesses, including 25.4, 50.8, 127, and 183 μm, on hydrogen crossover and found that decreasing the PEM thickness in MEA resulted in more pronounced hydrogen crossover.

In addition to the ion exchange membrane, the MEA, and its assembly manner are the foremost parameters that affect the fuel cell performance. Generally, MEA assembly technologies are also developing from gas diffusion electrodes (GDEs) and catalyst coating membranes (CCMs) to gradient/ordered MEAs and direct membrane deposition (DMD) technologies. For instance, Klingele et al. [221] reported the fabrication of MEA by directly depositing a Nafion® dispersion as a membrane layer on anode and cathode GDEs with a thinner thickness of 8–25 μm and obtained a peak power density of 4.07 W/cm^2 and a current density of 5 A/cm^2 at 0.6 V using pure oxygen as the cathode fuel. The exciting fuel cell performance is mainly attributed to the relatively thinner membrane and a strongly decreased contact resistance of the membrane and the proton conducting phase of the catalyst layer. It is worth mentioning that high-temperature proton exchange membrane fuel cells (HT-PEMFCs) are one of the most emerging directions for energy conversion owing to their improved catalyst CO tolerance, high energy conversion rate, and simple heat and water management [222]. HT-PEMFCs are expected to be the next generation of PEMFCs that will play a large role in hydrogen-fueled automobile vehicles and combined heat and power systems.

AEMFCs using AEMs are regarded as promising candidates for PEMFCs. AEMFCs allow the use of low-cost membranes and nonprecious metal catalysts and rapid redox kinetics under basic circumstances [223, 224]. However, the energy efficiency and service life of AEMFCs are associated with the properties of AEMs [225]. Therefore, high thermal stability, chemical stability, and high ionic conductivity are the main requirements for the AEM for AEMFCs. To date, the commercially available membranes for AEMFCs include A201 (Tokuyama, Japan), FAA3 (Fumatech, Germany), AEMION™ (Ionomer, Canada), SUSTAINION® (Dioxide Materials, USA), Xergy® (AEM-Pension, USA), etc. However, these membranes still have shortcomings in long-term stability under basic circumstances and unsatisfactory ion conductivity due to the low mobility of OH^-. Increasing the conductivity and alkaline stability are the primary directions for the fabrication of AEMs for AEMFCs. Notably, the fabrication of novel AEMs with an aryl ether-free backbone and highly stable ionic functional groups, such as nitrogen heterocyclic cations, via the superacid polymerization method is a promising strategy to overcome the "trade-off" effect for the conductivity and selectivity of the membrane. For instance, Wang et al. [226] reported the fabrication of an AEM based on a rigid, hydrophobic, and ether-bond-free aryl backbone with highly stable piperidinium cations. The partial substitution of piperidone with the high-reactivity monomer 2,2,2-trifluoroacetophenone enables a high-molecular-weight polymer with tunable IEC and satisfactory alkaline stability due to the usage of the piperidinium cations as the ionic functional groups. The membrane showed a high peak power density of 920 mW/cm^2 and operated stably at a constant current density of 500 mA/cm^2 for 300 h at 95 °C. Recently, Chen et al. [227] prepared an AEM with a high molecular weight alkyl block type polyarylpiperidine copolymer by regulating the IEC and water vapor permeability of the polymer. The prepared membrane exhibited a high OH^- conductivity of greater than 150 mS/cm, low alkali resistance, high durability in 1 M NaOH at 80 °C for more than 1 500 h, and a record high power density of 2.58 W/cm^2. Liu et al. [228] developed an AEM with a new type of oriented ferrocenium. The ion transport channel arranged in the direction of the membrane's transmutation surface has greatly improved the power output in the fuel cell and alkali stability and redox stability of the AEM. The AEM exhibited excellent thermal stability and durability at 120 °C and 40% relative humidity, and there was no significant loss of hydroxide conductivity in the longest reported alkali stability test of 4 320 h thus far. The conductivity and stability of the membrane were improved, creating a continuous output of 737 mW/cm^2 at 80 °C and 40% relative humidity.

However, compared with PEMFCs, AEMFCs have some limitations. The membrane electrode assembly is the core of fuel cells. A major limitation of the commercialization of AEMFCs is the preparation of the membrane electrode assembly. The membrane electrode assembly is commonly prepared by the catalyst-coated membrane (CCM) method, but there are certain technical difficulties in the large-sized preparation. In addition, there are few kinds of commercially available alkaline membranes, and they are very costly. The performances of alkaline membranes in the market still cannot meet the requirements of AEMFCs.

Due to the advantages of high modularity, wide power range, and variation of properties among different types, fuel cells have been used in a multitude of applications, including transportation, stationary, portable, and emergency backup power applications. The global fuel cell market size was estimated at USD 6.3 billion in 2022 and is expected to expand at a compound annual growth rate of 19.9% from 2023 to 2030.

The transportation industry is one of the main powerhouses in the advocation of renewable and sustainable sources of energy. Fuel cells allow the transportation industry near-zero harmful emissions and avoid compromising the efficiency of the vehicle's propulsion system. The hydrogen fuel-cell vehicle (HFCV) has been commercialized for end customers since 2015 by three different car companies: the Honda Clarity Fuel Cell, the Hyundai Nexo SUV, and the Toyota Mirai. By the end of 2022, the Mirai sedans had sold more than 11 357 vehicles in the U.S. market. The latest Toyota Mirai has three hydrogen fuel tanks with a driving range of approximately 850 km. However, the challenge for a fuel cell vehicle is that there is insufficient space for the power system and fast response and start-up times are required for car propulsion. In addition, the high cost of fuel cell systems results in expensive FCEVs, challenging market growth. Investment risks such as operational and capital costs and the underutilization of various FCEV facilities are likely to hinder market growth. Especially with the rapid development of electric vehicles based on lithium-ion batteries, HFCVs seem less competitive than electric cars for light-duty fuel cell electric vehicles. However, fuel cell vehicles have superiority under the circumstances of intensive power output, such as heavy load trucks, rail transport, and aircraft. For instance, Vision launched the world's first hydrogen fuel cell-based heavy-duty class-8 truck. The 8.4 m long by 2.9 m high truck runs on a hybrid hydrogen fuel cell/Li-ion battery drive system capable of providing a peak power of approximately 400 kW and a top speed of approximately 105 km/h at a total vehicle weight of more than 36 000 kg. Similarly, Airbus joined the FCHEA and unveiled its ambition to develop the world's first zero-emission commercial aircraft by 2035.

Recently, Guiver et al. [229] provided a perspective on the development of fuel cell vehicles in comparison with battery electric vehicles. The cost of fuel cell vehicles is currently higher than that of battery electric vehicles for short-range vehicles (below 322 km). Notably, if fuel cell vehicles are manufactured with high production capacity, the cost of fuel cell vehicles could be comparable to or even lower than that of electric vehicles, especially for long-range vehicles with driving ranges higher than 483 km. The authors also conclude that fuel cell vehicles are more appropriate for heavy-duty and long-distance transportation and special vehicles such as forklifts, as shown in Figure 9.18. For the fuel cell system for transportation propulsion

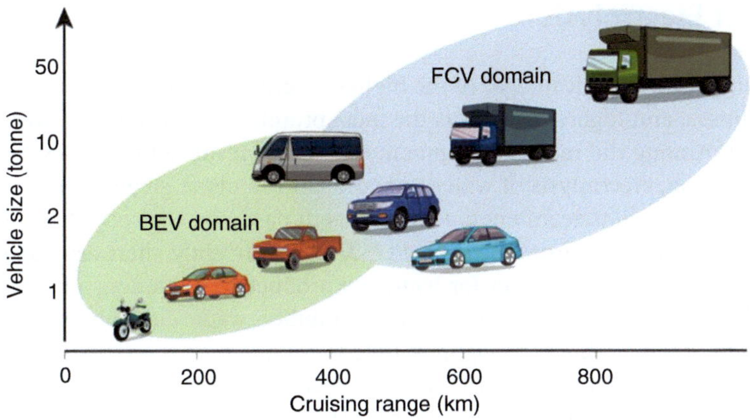

Figure 9.18 Expected application domains of battery electric vehicles and battery electric vehicles in future automotive transportation. Source: Jiao et al. [229]/Springer Nature.

applications, the power density is the key issue to be improved. The power density goals for next-generation proton exchange membrane fuel cells are to increase from the current 4 kW/l to the short-term target of 6 kW/l and the long-term target of 9 kW/l. For instance, Japan's New Energy and Industrial Technology Development Organization has set power density targets of 6 kW/l by 2030 and 9 kW/l by 2040.

Another important application field of fuel cells is stationary applications. Fuel cells play a particular role in the residential, commercial, and industrial stationary power generation sectors. Especially for a small, distributed power system, such as single or multiple home power generation, the fuel cell system combined with the energy from a heat cycle system is capable of providing reliable power for residential use. Using an inverter, the direct current from the fuel cell can be converted to an alternating current, which can be consumed in the household for lighting, heating, etc. Meanwhile, the heat discharged from the fuel cell stack can be used to warm the water for household use. The cogeneration of heat of power is more effective than the single production of electricity. According to the fuel cell technologies market report, the stationary fuel cell market accounts for approximately the majority of fuel cell shipments at the MW level. Notably, the dominant operation methods of fuel cells for stationary applications include the emergency backup power supply, remote-area power supply, and distributed power. Due to the unique advantages of high energy and power densities, high longevity, simple maintenance, large compactness, and easy operation, fuel cells are very promising alternatives for emergency backup power supply systems in telecommunications and hospitals, data centers, banks, and government agencies where the continuous supply of power is needed. The fuel cell backup power systems can be operated in a wide range of temperatures and humidity, such as operation in temperature ranging from −40 to 49 °C. Meanwhile, fuel cell systems are also believed to be a competitive power supply in grid-isolated locations, such as islands, deserts, forests, remote technical installations, holiday retreats, and remote research facilities. Similarly, fuel cells are feasible for shifting from large centralized power generation to decentralized distributed generation.

9.10 Water Electrolysis

Hydrogen energy, with the advantages of the highest energy mass density, environmental friendliness, and high efficiency, is the most promising alternative to fossil fuels [230, 231]. Among the rapid development of various advanced processes for hydrogen production, electrolysis of water to hydrogen is a clean green hydrogen technology for future mainstream energy development based on the specialties of eliminating carbon emissions from the source [232, 233]. Currently, there are three kinds of water electrolysis techniques for hydrogen production, including alkaline aqueous solution electrolysis, proton exchange membrane water electrolysis, and AEM water electrolysis, as shown in Figure 9.19 [234–236].

Alkaline aqueous solution electrolysis (alkaline electrolysis) is a mature technology for hydrogen production that uses aqueous KOH solution as the electrolyte and asbestos as the diaphragm in the electrolytic cell. This water electrolysis has been established for more than 100 years in the chemical industry for producing relatively pure hydrogen and chlorine. The first alkaline electrolyzers for hydrogen production were developed by Norsk Hydro (today NEL) in Norway, where inexpensive electricity from hydropower is available. Notably, one of the largest alkaline aqueous solution electrolysis systems has been installed in Assuan, Egypt, with a production capacity of 40 000 Nm^3/h and an electric input of 200 MW. This water electrolysis system demonstrated an energy consumption of 4.1 kWh/Nm^3 under the conditions of atmospheric pressure and 60–80 °C [237]. This technology has the advantages of low capital and operating costs because it allows the use of nonprecious metals such as Ni, Co, and Mn as catalysts [238]. However, the current efficiency of the technology is generally 60–80% because of the use of the asbestos membrane, which will cause the crossover of gases and even explode in the cell [239]. Lye loss, corrosion, high energy consumption, and other problems also exist in alkaline aqueous solution electrolytic cells. In addition, alkaline electrolysis has relatively low adaptability to renewable energy because it is difficult to instantly change the loading of alkaline electrolysis for hydrogen production and is incapable of rapidly adjusting the hydrogen production rate.

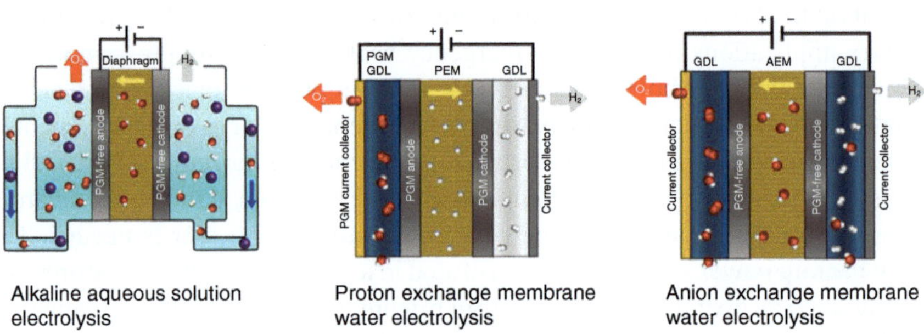

Figure 9.19 The main principles of alkaline aqueous solution electrolysis (AEL), proton exchange membrane water electrolysis (PEMWE), and anion exchange membrane water electrolysis (AEMWE) for hydrogen production. Source: Adapted from Refs. [234–236].

In contrast, proton exchange membrane water electrolysis (PEMWEO) and alkaline membrane water electrolysis (AEMWE) are easy to operate and capable of achieving a significant increase in hydrogen production efficiency at low temperatures [240, 241]. The primary functions of proton exchange membranes or alkaline membranes are to selectively transport H^+ or OH^- and to block the contact of the produced gas in the anode and cathode. The membrane is one of the key factors determining the efficiency of hydrogen production. Proton exchange membranes such as Nafion with good chemical stability, high proton conductivity, and excellent gas migration control are widely used to transport H^+. The PEMWE has high flexibility, is easy to operate, has high hydrogen production efficiency, has high current density, and matches well with photovoltaic power generation plants. The high efficiency is because protons can quickly respond to any change in the power input when they are transported through a solid polymer electrolyte membrane. In addition, a compact cell design and higher operational pressure could be achieved in a solid polymer electrolyte than in alkaline electrolysis. In recent decades, tremendous efforts have been dedicated to improving PEMWE performance. For example, Marshall et al. reported a current density of 1 A/cm^2, a very low cell voltage of 1.57 V with a high cell electrolysis efficiency of 94.4%. However, the use of Nafion and precious metals increases the cost of PEMs because it is necessary to use the precious metal Pt to reduce H^+ to generate H_2 in an acidic environment.

Recently, AEMWE has attracted intensive attention because hydrogen production in this electrolytic cell can avoid the use of noble metal catalysts. However, it is crucial to develop an alkaline membrane with low cost, high OH conductivity, and good stability in an alkaline environment [242, 243]. Furthermore, developing suitable alkaline ionomers as binders between anode and cathode catalysts in AEMWE and improving the electrolytic performance and durability of pure water AEMWE are challenges to be faced [244, 245]. The recent progress of AEMWE has made a significant contribution to the development of membrane materials. Generally, the commercial Sustainon® AEM in AEMWEs is able to obtain a current density of 1 A/cm^2 at 1.63 V in 1 M KOH at 60 °C [246]. To obtain a higher current density and longer stability in AEMWEs, many studies have been carried out to prepare novel membranes. Kim et al. [239] obtained an excellent current density of 2.7 A/ cm^2 at 1.8 V in pure water and a current density of 5.5 A/ cm^2 in 1 M KOH at 85 °C with high-performance AEMWEs based on alkylammonium polyphenylene AEMs and BTMA-polystyrene ionomers, but the cell durability was limited to below 170 h under 0.2 A/ cm^2. Soni et al. [247] developed a kind of AEM with a fully aromatic polymer consisting of only aromatic bonds such as benzene rings. This membrane showed high durability in alkaline hydrolysis with a high water electrolysis efficiency of 69–76% and a high current density of 1 A/cm^2. This AEM realized efficient hydrogen production in the case of a pure water supply. This was attributed to the high conductivity of OH^- ions, which were able to conduct at high speed even without using a KOH solution. It also remained stable in more than 100 h of tests at 80 °C, realizing low-cost, high-performance, and high-durability hydrogen production. Chen et al. [238] achieved a current density of 7.68 A/cm^2 at 2 V in 1 M KOH using poly(fluorenyl-*co*-aryl piperidinium) (PFAP)-based anhydrous cathode

AEMWEs, which exceed the state-of-the-art PEMWEs of 6 A/cm^2 current density at 2 V. The system demonstrated excellent high stability of more than 1000 h under a 0.5 A/cm^2 current density at 60 °C.

A membrane electrode assembly (MEA) for AEMWE was developed by a research team at the Korean Academy of Sciences (KIST). The MEA comprises an AEM, an electrode binder, and iron instead of titanium to replace the catalyst of the water electrolytic cell. It displayed high ionic conductivity and durability under alkaline conditions. The module was found to have 10 times longer life and 6 times better performance than the AEM, with a 20% performance advantage over the current use of PEMs and a 3000 times lower manufacturing cost than PEMWEs through testing. However, it had not been commercialized due to less than 100 h of continuous operation. Wan et al. [248] designed a novel three-dimensional (3D) ordered MEA based on a highly porous catalyst layer (CL) with unique vertical channels, an ultrathin membrane layer (ML), and 3D CL/ML interface structures nanoengineered for alkaline electrolyzers. The results showed that a high current density of 4.2 A/cm^2 could be achieved at 2.0 V in 1 M KOH. Additionally, the alkaline electrolyzer for pure water at 3.1 A/cm^2 was operated at 2.0 V, and it could be maintained for more than 600 h at a high current density of 1 A/cm^2. The energy conversion efficiency and durability were significantly improved.

9.11 Industrial Applications

Due to the inherent advantages of ion exchange membranes, substantial efforts have been dedicated to applying ion exchange membrane-based technologies such as DD, ED, SED, and BMED to industrial applications in pollution control, resource recovery, and clean production. In this section, we examine some typical applications of ion exchange membranes that have been developed in recent years.

9.11.1 Acid Recovery Using Diffusion Dialysis

The almost zero operating cost, easy operation, and long stability of DD technology have stimulated the large-scale application of this concentration-gradient driven membrane separation technology for acid recoveries, such as steel pickling, metallurgical acid stripping, aluminum foil, electroplating, and others. It is estimated that more than 100 000 m^2 of DD membranes are manufactured annually for acid recovery applications in China.

Copper hydrometallurgical processes are usually accomplished by the wet leaching of copper ores with sulfuric acid, as shown in Figure 9.20. Specifically, sulfuric acid was used to dissolve the copper ions, and the extractant was further introduced to isolate the copper in the organic phase. Subsequently, 170 g/l sulfuric acid was added as the stripping agent. This leaching and extraction copper extraction procedure discharged a large amount of lean copper solution with $H_2SO_4 > 190$ g/l, Fe of 6.3 g/l, and Cu of 32 g/l. The accumulation of iron will affect the following electrolysis performance on the cathode side; therefore, the system must be opened up for a

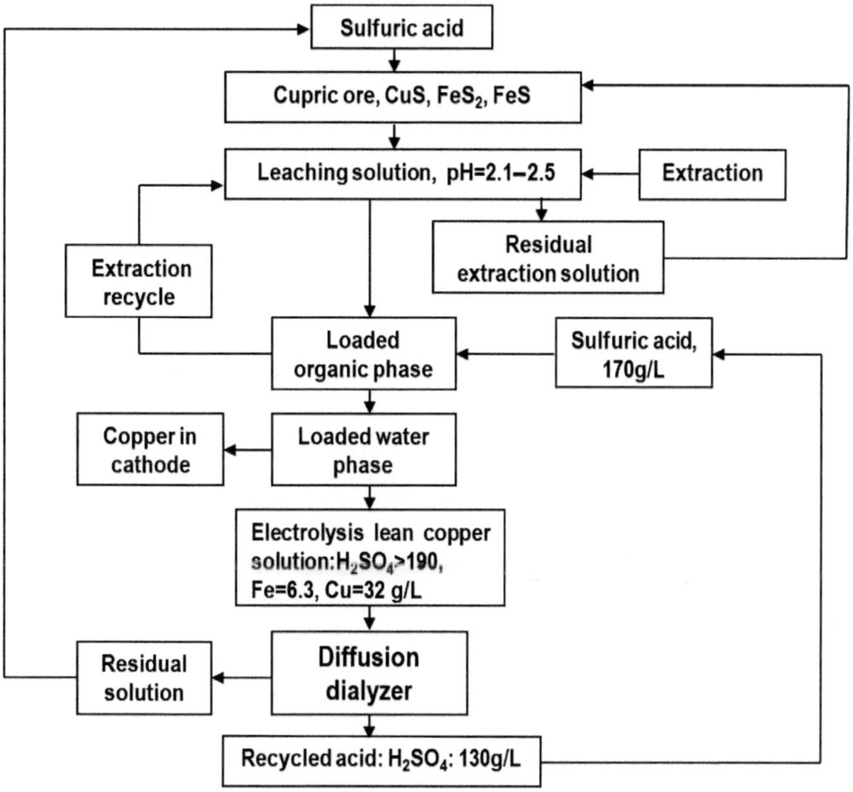

Figure 9.20 The main routes of diffusion dialysis for acid recovery from the electrolysis lean copper.

certain period of time. Generally, 1 ton of copper needs to discharge 3.5 tons of lean copper solution with a high content of sulfuric acid. Traditionally, the alkali neutralization method has been widely applied to tackle this waste acid. However, the neutralization method consumes a large amount of base and results in secondary pollution from the generation of solid waste. The DD method is a promising method for the recycling of acid in this field. Specifically, for one industrial DD stack with an effective membrane area of 0.8 m*1.6 m and a repeating unit of 400, the treatment capacity of the DD stack is approximately 6 tons per day. The recovery rate of this copper lean solution is higher than 80% with a metal ion rejection rate higher than 90%. Table 9.3 also tabulates the operating performance of DD for acid recovery from the copper lean solution. The DD system allowed the maximum recycling of approximately 200 tons of sulfuric acid with a recycled acid benefit of \$14 000 and a saved lime cost of \$44 000 a month (Figure 9.21).

Another major application field of the DD process is for steel pickling procedures. Steel pickling is an acid bath process that removes impurities, rust, and scale from the surface of steel. It is used before cold rolling or coating to make the steel easier to work with and paint. This acid pickling process generates a large amount of waste acid, which typically contains a variety of heavy metal ions with high

Table 9.3 The operating performance of diffusion dialysis for acid recovery from the copper lean solution.

Operation period	Recycling acid amount (tons)	Recycled acid benefits ($)	Saving lime amount (tons)	Saving lime cost ($)
Aug	134.22	8,418	148	17,698
Sep	226.57	14,209	249	44,084
Oct	198.80	12,468	219	26,213
Nov	204.59	12,831	225	26,977
Dec	85.86	5,385	94	11,321
Jan	170.76	10,709	188	22,516
Feb	110.19	6,910	121	14,529

Note: the prices of sulfuric acid and lime are $72.7/t and $62.9/t, respectively. During Aug, Dec, and Feb, diffusion dialysis was practically performed for approximately half a month.

Figure 9.21 Digit images of plate-and-frame diffusion dialysis for acid recovery from the electrolysis lean copper.

corrosivity and toxicity. At present, neutralization precipitation is widely adopted by small and medium-sized enterprises. In addition to the neutralization method, there are several feasible technologies that can realize the regeneration of useful resources from waste acid, such as decomposition, evaporation, crystallization, and membrane separation. Notably, the decomposition method is widely adopted in the industry for waste acid recovery. This waste acid recycling technique is mainly divided into three steps. In the first step, spent acid is cracked at a high temperature in a cracking furnace, $H_2SO_4 = SO_2 + 1/2O_2 + H_2O$. In the second step, SO_2 is converted into SO_3 by the catalyst in the converter, $SO_2 + 1/2\ O_2 = SO_3 + 99\ kJ/mol$. In the third step, sulfuric acid is regenerated through multistage absorption towers,

$SO_3 + H_2O = H_2SO_4 + 101$ kJ/mol. Generally, the decomposition method is feasible to produce a product acid of acceptable quality for use and sale. However, the cumbersome sequences, high capital cost, and inflexible integration capability have restricted the application of this method only in small- and medium-sized factories. DD is a perfect choice for this application with acid recovery in this field. According to the practical operating data from the factory, the recovery capacity of five DD stacks is approximately 25 tons/d with a total membrane area of 2 560 m^2. The acid concentrations in the diffusate and the dialysate chamber are approximately 1.8 mol/l and 0.16 mol/l, respectively, in comparison with 2.0 mol/l of acid in the feed. Meanwhile, the metal ion concentrations in the diffusate and the dialysate chamber are 4 and 38 g/l, respectively, in comparison with 55 g/l in the feed. The recovered HNO_3 and HF concentrations in the diffusate chamber are 8.5% and 1.2%, respectively. It is possible to annually recover 1 045 tons of HNO_3 and 175 tons of HF, thus saving approximately 550 tons of lime and 15 tons of coagulation agent.

9.11.2 Resource Recovery Using Electrodialysis

To date, the most dominant application field of the ion-exchange membrane separation process is the ED process. Generally, ED technology is applicable to wastewater treatment for realizing resource savings and protecting the environment. Notably, the current major applications of ED involve salt desalination and concentration as well as the separation of organic substances with inorganic salts. Compared with other pressure-driven desalination technologies, such as RO and nanofiltration, ion migration through the ion exchange membrane is not restricted by the osmotic pressure of the aqueous solution. As a result, the ED presents unique advantages for salt concentration in ZLD and WLC, both of which are highly recommended with stricter regulations for wastewater disposal. To achieve ZLD and WLC, there is usually a salt concentration process to minimize the volume of water for evaporation and crystallization, which are energy-intensive processes with high carbon emissions. Common salt concentration membrane technologies include various kinds of RO, such as DTRO, HPRO, FO, and ED. Generally, RO can concentrate the salt content to approximately 8–10 wt% NaCl because this process requires overcoming the osmotic pressure of the aqueous salt solution. FO is a relatively energy-efficient process, but the high energy consumption for the recovery of draw solutions has restricted the promotion of this technology to large-scale applications. In contrast, ED is a promising technology for the salt concentration to minimize the volume of wastewater for evaporation, thus reducing the required electricity in ZLD systems. There are a multitude of new applications of the ED technique for salt concentration in a variety of industrial fields, such as the coal chemical industry, electroplating, power generation industry, and paper and pulp.

A typical example of ED for the conversion of waste into resources is the implementation of ED for NaOH recovery from the wasted alkaline viscose fiber solution. Viscose fiber is a kind of regenerated cellulose film made from cellulose, such as cotton pulp and wood pulp, by special processes such as alkalization, sulfonation, and forming. Compared with traditional paper, regenerated cellulose film is flexible and transparent and presents several advantages, such as no static electricity, dustproof, and good kinking. The alkaline method is the predominant method for viscose fiber production. Specifically, the cellulose extracted from the wood pump reacts with 18% caustic soda to give alkali cellulose, which is further treated with carbon disulfide to split the cellulose molecules into smaller units. This preparation route generates a large amount of wasted alkaline viscose fiber wastewater, which normally contains 40 g/l NaOH and 50–150 g/l hemicellulose. Traditionally, this black viscose fiber liquor is subjected to the wastewater treatment system. Unfortunately, this wastewater is a great burden on conventional aerobic and anaerobic wastewater treatment systems due to the high alkaline content and antibacterial properties of hemicellulose. Subsequently, nanofiltration has been attempted to recover the alkali but has the shortcomings of high capital cost and poor stability of the membranes. The preparation of alkali-resistant nanofiltration is extremely challenging since most polymeric nanofiltration membranes are not tolerant under strong alkali circumstances, especially operation at high operating pressure. In addition, the nanofiltration technique does not have a very high alkali recovery rate (~75%) for this black viscose fiber liquor, and the fouling potential of the membrane is very high. Recently, ED has been successfully applied in this field, creating the possibility for waste minimization at maximum recovery of valuable components. A completely new valorization route for the wasted alkaline viscose fiber solution is developed, as depicted in Figure 9.22. Specifically, two ED systems are deployed; the first ED is for the recycling of NaOH from the black viscose fiber liquor, and the second ED is for the separation of xylose with the inorganic salt. Notably, the first ED can increase the NaOH recovery rate up to 95%, and the recycled NaOH is further treated by evaporation to obtain concentrated NaOH, which can be returned to the upstream cellulose extraction procedure. Meanwhile, the diluate stream after the first ED

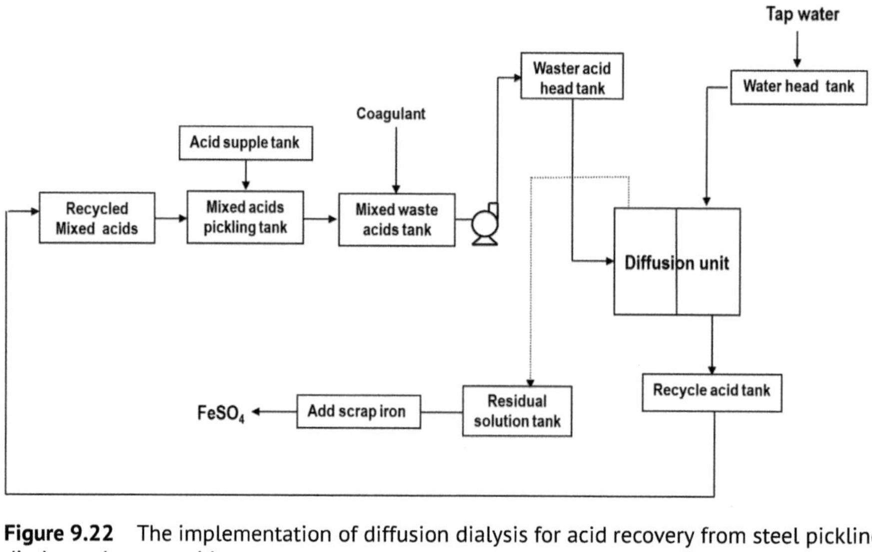

Figure 9.22 The implementation of diffusion dialysis for acid recovery from steel pickling discharged waste acid.

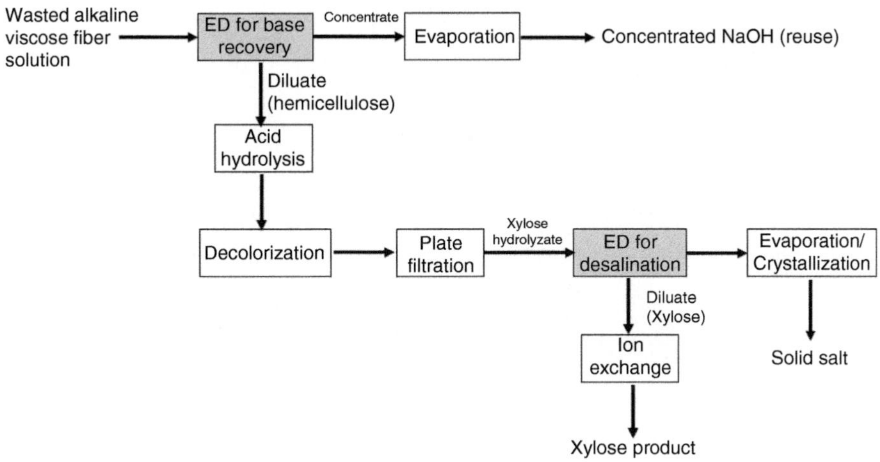

Figure 9.23 A schematic route for the recycling of xylose product from the wasted alkaline viscose film solution using electrodialysis.

mainly contains hemicellulose, which is further hydrolyzed with acid, decolorized, and filtered by plate filtration equipment to generate the xylose hydrolyzate. Then, the second ED is further introduced to isolate 10–20% xylose from 1.5–4.0% Na_2SO_4. The diluate stream after the ED treatment is further treated by an ion exchanger to obtain the xylose product, while the concentrate stream after the ED is further evaporated and crystallized to generate the solid salt. An industrial ED project was developed for the production of xylose from the wasted alkaline viscose fiber solution in 2019 at Tangshan, Hebei Province, China, as shown in Figure 9.23. The total area of the ion-exchange membrane was 13 800 m^2. It is possible to treat 21 000 tons of black viscose fiber liquor annually and produce approximately 18 000 m^2 tons of xylose. This is also the pioneering application of ED technology in the viscose fiber industry. Compared with conventional wastewater treatment methods, these kinds of ED technologies allow the conversion of waste into high value-added xylose, thus alleviating the burden on the environment. There are several major challenges to these novel applications. First, high alkalinity requires membranes to have high resistance to the base, and conventional functional groups such as quaternary ammonium are not appropriate. Second, the removal of a large amount of NaOH from the solution would decrease the solubility of hemicellulose, and it is particularly important to avoid the crystallization of organic dissolved matter during the ED process. Third, the temperature increased sharply during the ED operation, and it is imperative to cool the temperature below 35 °C for steady operation (Figure 9.24).

Wastewater mining is another major application field of ED technology. The use of ion exchange membranes can achieve the processing of mine and wastewater, allowing the simultaneous production of fresh water and solid salts, which are of commercial value. For instance, in the ceramic capacitor industry, it is necessary to provide alkaline conditions to facilitate the generation of TiO_2 precursors. The discharge of NH_4–N wastewater could cause eutrophication, produce offensive odors, and hinder the disinfection of the water supplier. Therefore, the treatment of wastewater with a

Figure 9.24 The digital images of the hydraulic pressure electrodialysis stacks for the production of xylose from the wasted alkaline viscose fiber solution.

high concentration of NH_4–N and a marginal amount of metal ions such as Ba^{2+} is a difficult problem. Traditionally, the air stripping method is the predominant method for high-concentration NH_4–N wastewater treatment. By mediating the solution pH ~10.5, the ammonium ions are converted into gaseous ammonia, and it is convenient to isolate it from the aqueous solution by air stripping. The air stripping method has high base consumption, undesirably increasing the running cost. The ED technique was performed for the recycling of NH4Cl salt from ceramic capacitor wastewater in Zibo, Shandong Province, China. A schematic route for the combination of nanofiltration, ED and RO for this wastewater treatment is depicted in Figure 9.25. Specifically, nanofiltration was used to remove the dissolved organic matter, and the permeate solution was fed into the ED unit for salt concentration. To increase the concentration of the salt, two-stage ED stacks were designed. The diluate steam from

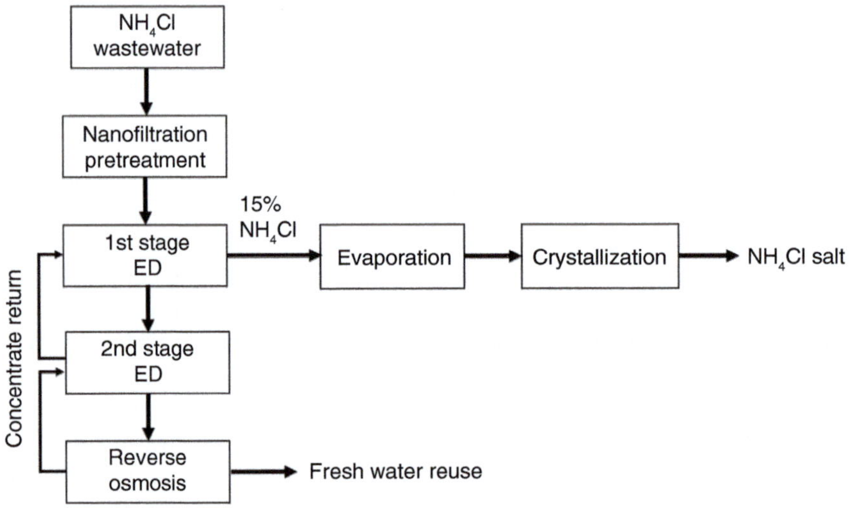

Figure 9.25 A schematic route for recycling NH_4Cl salt from ceramic capacitor wastewater.

Figure 9.26 Digital images of the hybrid membrane system including nanofiltration, electrodialysis, and reverse osmosis for the recycling of NH_4Cl salt from ceramic capacitor wastewater.

the 1st stage ED was fed to the 2nd ED stack, while the concentrate stream effluent from the 2nd stage was returned to the 1st stage. Meanwhile, the diluate stream from the 2nd stage ED was subjected to RO, while the retention stream of RO was returned to the 2nd stage ED. Therefore, the 1st-stage ED was always operated with a relatively high concentration of NH_4Cl, and the ED system was performed only for salt concentration and not for water production. The total area of the ion exchange membrane used for the ED project was approximately 17 280 m^2, with a total annual treatment capacity of approximately 450 000 tons of wastewater. Digital images of the hybrid membrane systems for NH_4Cl recovery from ceramic capacitor wastewater are provided in Figure 9.26. Notably, the salt recovery rate and the water reuse rate were higher than 98%, and it is possible to achieve complete wastewater mining.

The lead and zinc smelting acid production systems discharge a large amount of wasted acid. This wasted acid has a complex composition, such as 2.8% H_2SO_4, F^-, Cl^-, and metal ions. Traditionally, two major methods were performed for this wasted acid treatment. One is the neutralization method based on the use of chemical reagents, which leads to the appearance of secondary emissions. The other method is the evaporation method, which has high capital and operating costs. ED has advantages over traditional methods. A typical acid recovery route from the lead and zinc smelting acid production system is depicted in Figure 9.27. Specifically, the flue gas absorption liquid was first treated by sulfide precipitation to remove the metal ions. Subsequently, the liquid solution was further subjected to clarifier and fluoride resin to remove the suspended solid and fluoride ions. Thereafter, the waste acid is concentrated with ED to increase the H_2SO_4 content from the initial 2.8 to 10 wt%. The concentrated solution from the ED was further evaporated, and the HCl was stripped to obtain 70 wt% H_2SO_4. The diluate stream from the ED is further treated by RO to obtain fresh water. A practical ED plant was projected in

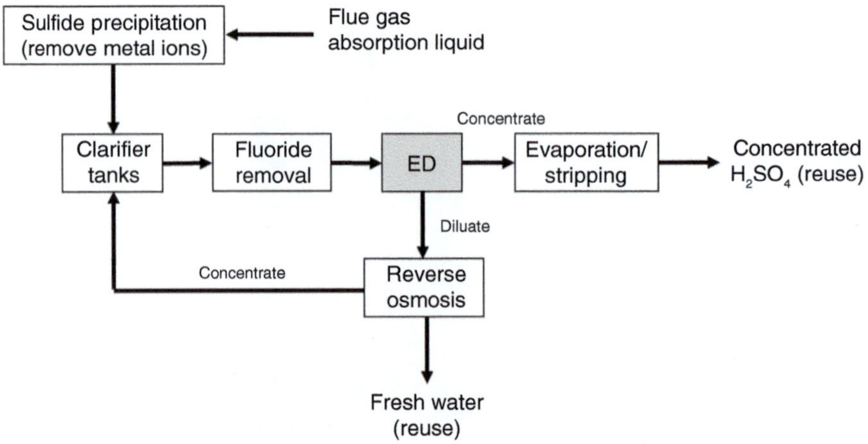

Figure 9.27 A schematic route for the recycling of H_2SO_4 using electrodialysis from lead and zinc smelting acid production systems.

Figure 9.28 Digital images of electrodialysis for the recycling of H_2SO_4 from lead and zinc smelting acid production systems.

Changsha, Hunan Province, China, in 2018, as shown in Figure 9.28. The total area of the ion exchange membrane was 8 160 m^2, which has a treatment capacity of 158 400 tons of wasted acid. In addition, the evaporation volume decreased from the initial 480 to 144 m^3/d while avoiding the generation of 13 300 tons/a of solid waste. Generally, this industrial ED system is capable of achieving the concurrent valorization of acid, water, and metals.

9.11.3 Clean Production Using Bipolar Membrane Electrodialysis

BMED overcomes the drawbacks of conventional separation and purification of organic acids techniques associated with low yields (due to product inhibition), low conversion efficiency, complex preparation routes, and high consumption of acid, base, and water. Therefore, there are tremendous industrial applications of this technology for cleaner organic production, lithium resource recovery, and waste salt valorization, especially with progress in bipolar membrane materials.

A dominant application of BMED is for the cleaner production of gluconic acid. Gluconic acid is a commonly used carboxylic acid that has numerous applications, such as food additives, calcium chelators, cement retarders, superplasticizers, and surface cleaners. Currently, microbial fermentation is the preponderant technique for the synthesis of gluconic acid due to the flexibility, robustness, and economy of automated fermentation platforms. During the fermentation process, saturated lime is usually supplied to the broth to maintain the pH near neutral and thus provide an optimal living environment for microbes and to precipitate the organic acid as Ca-salt for separation purposes. The subsequent acidification is an indispensable process because the precipitation process with lime makes the acid ions exist as their salts. Conventional acidification is employed by adding strong organic acids, such as H_2SO_4, but it is notorious for $CaSO_4$ sludge. This conventional process has several drawbacks. At first, it is not technologically advanced since it involves frequent filtration for separation and has a low productivity (low organic acid concentration, i.e. 0.1–2 mol/l) due to the product inhibitory effect. Furthermore, this process is not economically favorable. A substantial part of the cost is related to adding base and acid and disposal of the sludge. Finally, this process is not environmentally benign. The continuous input of external chemicals (the supplementation of acid and base in fermentation and acidification processes, respectively) and output of waste result in not only a waste of useful resources but also pollution of the environment. In comparison, BMED is a promising alternative to the Ca-salt precipitation route. In particular, protons dissociated by water splitting can be used to acidify the organic salt, abandoning the use of strong inorganic acid. Water-splitting hydroxyl ions can be used to adjust the broth pH, avoiding the use of lime. Therefore, BMED is of great significance for cleaner gluconic acid production with high industrial relevance. The pilot trial of the BMED technique for gluconic acid production was developed in Jiangxi Province, China. A schematic route of BMED technology for the cleaner production of gluconic acid is depicted in Figure 9.29. By taking advantage of the water dissociation in the bipolar membrane, the new preparation could abandon five procedures, including condensation, crystallization, drying, dissolving, and the majority of the ion exchanger. Compared with the conventional separation and purification route, the BMED method avoids the use of 90% of the ion exchanger capacity and saves 90% of the wastewater discharge. In 2016, a practical industrial BMED project was developed, with a total membrane area of 256 m^2 and a gluconic production capacity of 5 000 tons/a, as shown in Figure 9.30. The energy consumption for gluconic acid production ranged from 267 to 325 kWh/t, with a current efficiency in the range of 76–88%.

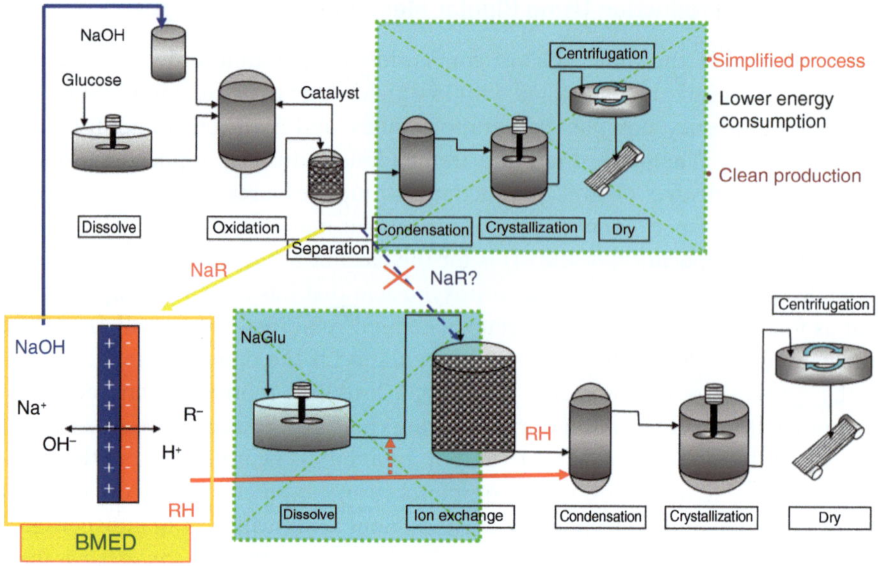

Figure 9.29 Bipolar membrane electrodialysis for the cleaner production of gluconic acid.

Figure 9.30 Digital images of bipolar membrane electrodialysis for cleaner production of gluconic acid.

References

1 Gueccia, R., Aguirre, A.R., Randazzo, S. et al. (2020). Diffusion Dialysis for separation of hydrochloric acid, Iron and zinc ions from highly concentrated pickling solutions. *Membranes (Basel)* 10 (6): 129.

2 Lin, X., Shamsaei, E., Kong, B. et al. (2016). Porous diffusion dialysis membranes for rapid acid recovery. *Journal of Membrane Science* 502: 76–83.

3 Luo, J., Wu, C., Wu, Y., and Xu, T. (2013). Diffusion dialysis of hydrochloric acid with their salts: effect of co-existence metal ions. *Separation and Purification Technology* 118: 716–722.

4 Zhang, X., Li, C., Wang, H., and Xu, T. (2011). Recovery of hydrochloric acid from simulated chemosynthesis aluminum foil wastewater by spiral wound diffusion dialysis (SWDD) membrane module. *Journal of Membrane Science* 384: 219–225.

5 Luo, J., Wu, C., Wu, Y., and Xu, T. (2011). Diffusion dialysis processes of inorganic acids and their salts: the permeability of different acidic anions. *Separation and Purification Technology* 78: 97–102.

6 Tongwen, X. and Weihua, Y. (2003). Industrial recovery of mixed acid (HF^+ HNO_3) from the titanium spent leaching solutions by diffusion dialysis with a new series of anion exchange membranes. *Journal of Membrane Science* 220: 89–95.

7 Oh, S.J., Moon, S.-H., and Davis, T. (2000). Effects of metal ions on diffusion dialysis of inorganic acids. *Journal of Membrane Science* 169: 95–105.

8 Yan, H., Xu, C., Wu, Y. et al. (2017). Integrating diffusion dialysis with membrane electrolysis for recovering sodium hydroxide from alkaline sodium metavanadate solution. *ACS Sustainable Chemistry & Engineering* 5: 5382–5393.

9 Stachera, D.M., Childs, R.F., Mika, A.M., and Dickson, J.M. (1998). Acid recovery using diffusion dialysis with poly(4-vinylpyridine)-filled microporous membranes. *Journal of Membrane Science* 148: 119–127.

10 Palaty, Z., Zakova, A., and Prchal, P. (2007). Continuous dialysis of carboxylic acids. Permeability of Neosepta-AME membrane. *Desalination* 216: 345–355.

11 Ran, J., Hu, M., Yu, D.B. et al. (2016). Graphene oxide embedded "three-phase" membrane to beat "trade-off" in acid recovery. *Journal of Membrane Science* 520: 630–638.

12 Xu, T.W. and Yang, W.H. (2001). Fundamental studies of a new series of anion exchange membranes: membrane preparation and characterization. *Journal of Membrane Science* 190: 159–166.

13 Yan, J.Y., Wang, H.Y., Fu, R. et al. (2022). Ion exchange membranes for acid recovery: diffusion dialysis (DD) or selective electrodialysis (SED)? *Desalination* 531.

14 Luo, J.Y., Wu, C.M., Xu, T.W., and Wu, Y.H. (2011). Diffusion dialysis-concept, principle and applications. *Journal of Membrane Science* 366: 1–16.

15 Wu, Y.H., Lin, H.Y., Zhang, G.C. et al. (2013). Non-charged PVA-SiO_2 hybrid membranes for potential application in diffusion dialysis. *Separation and Purification Technology* 118: 359–368.

16 Dong, Q.A., Liu, J.S., Song, L., and Shao, G.Q. (2011). Novel zwitterionic inorganic-organic hybrids: synthesis of hybrid adsorbents and their applications for Cu^{2+} removal. *Journal of Hazardous Materials* 186: 1335–1342.

17 Wu, Y.H., Wu, C.M., Li, Y. et al. (2010). PVA-silica anion-exchange hybrid membranes prepared through a copolymer crosslinking agent. *Journal of Membrane Science* 350: 322–332.

18 Hao, J.W., Wu, Y.H., Ran, J. et al. (2013). A simple and green preparation of PVA-based cation exchange hybrid membranes for alkali recovery. *Journal of Membrane Science* 433: 10–16.

19 Wu, C.M., Wu, Y.H., Xu, T.W., and Fu, Y.X. (2008). Novel anion-exchange organic-inorganic hybrid membranes prepared through sol-gel reaction and UV/thermal curing. *Journal of Applied Polymer Science* 107: 1865–1871.

20 Kim, D.H., Park, J.H., Seo, S.J. et al. (2013). Development of thin anion-exchange pore-filled membranes for high diffusion dialysis performance. *Journal of Membrane Science* 447: 80–86.

21 Emmanuel, K., Cheng, C.L., Erigene, B. et al. (2016). Imidazolium functionalized anion exchange membrane blended with PVA for acid recovery via diffusion dialysis process. *Journal of Membrane Science* 497: 209–215.

22 Vasil'eva, V., Goleva, E., Pismenskaya, N. et al. (2019). Effect of surface profiling of a cation-exchange membrane on the phenylalanine and NaCl separation performances in diffusion dialysis. *Separation and Purification Technology* 210: 48–59.

23 Sun, F.J., Wu, C.M., Wu, Y.H., and Xu, T.W. (2014). Porous BPPO-based membranes modified by multisilicon copolymer for application in diffusion dialysis. *Journal of Membrane Science* 450: 103–110.

24 You, X.Q., Chen, J.Q., Teng, L. et al. (2022). Porous anion exchange membranes fabricated by phase inversion and in-situ modification for acid recovery. *Journal of Polymer Science* 60: 3054–3063.

25 You, X.Q., Chen, J.Q., Pan, S. et al. (2022). Piperazine-functionalized porous anion exchange membranes for efficient acid recovery by diffusion dialysis. *Journal of Membrane Science* 654: 120560.

26 Lin, X.C., Kim, S.J., Zhu, D.M. et al. (2017). Preparation of porous diffusion dialysis membranes by functionalization of polysulfone for acid recovery. *Journal of Membrane Science* 524: 557–564.

27 Lin, X.C., Shamsaei, E., Kong, B. et al. (2016). Fabrication of asymmetrical diffusion dialysis membranes for rapid acid recovery with high purity (vol 3, pg 24000, 2015). *Journal of Materials Chemistry A* 4: 8478.

28 Ul Afsar, N., Erigene, B., Irfan, M. et al. (2018). High performance anion exchange membrane with proton transport pathways for diffusion dialysis. *Separation and Purification Technology* 193: 11–20.

29 Pan, J.F., He, Y.B., Wu, L. et al. (2015). Anion exchange membranes from hot-pressed electrospun QPPO-SiO_2 hybrid nanofibers for acid recovery. *Journal of Membrane Science* 480: 115–121.

30 Pan, J.F., Ge, L., Lin, X.C. et al. (2014). Cation exchange membranes from hot-pressed electrospun sulfonated poly(phenylene oxide) nanofibers for alkali recovery. *Journal of Membrane Science* 470: 479–485.

31 Miao, J.B., Li, X.Y., Yang, Z.J. et al. (2016). Hybrid membranes from sulphonated poly(2, 6-dimethyl-1, 4-phenylene oxide) and sulphonated nano silica for alkali recovery. *Journal of Membrane Science* 498: 201–207.

32 Ji, W.G., Ul Afsar, N., Wu, B. et al. (2019). In-situ crosslinked SPPO/PVA composite membranes for alkali recovery via diffusion dialysis. *Journal of Membrane Science* 590.

33 Lin, J.Y., Huang, J.M., Wang, J. et al. (2021). High-performance porous anion exchange membranes for efficient acid recovery from acidic wastewater by diffusion dialysis. *Journal of Membrane Science* 624: 119116.

34 Liu, R., Wu, L., Pan, J.F. et al. (2014). Diffusion dialysis membranes with semi-interpenetrating network for alkali recovery. *Journal of Membrane Science* 451: 18–23.

35 Hu, K. and Dickson, J.M. (2009). In vitro investigation of potential application of pH-sensitive poly(vinylidene fluoride)-poly(acrylic acid) pore-filled membranes for controlled drug release in ruminant animals. *Journal of Membrane Science* 337: 9–16.

36 Choudhury, A., Patnaik, P., Mondal, R. et al. (2023). GO-anchored imidazolium based cross-linked composite anion exchange membranes for the enhancement in acid recovery via diffusion dialysis. *Chemical Engineering Journal* 451: 138901.

37 Cheng, C.L., Li, P., He, Y.B. et al. (2019). Branched polyvinyl alcohol hybrid membrane for acid recovery via diffusion dialysis. *Chemical Engineering & Technology* 42: 1180–1187.

38 Wang, C., Wu, C.M., Wu, Y.H. et al. (2013). Polyelectrolyte complex/PVA membranes for diffusion dialysis. *Journal of Hazardous Materials* 261: 114–122.

39 Wang, C.W., Liang, Y.X., Miao, J.B. et al. (2019). Preparation and properties of polyvinyl alcohol (PVA)/mesoporous silica supported phosphotungstic acid (MS-HPW) hybrid membranes for alkali recovery. *Journal of Membrane Science* 592: 117388.

40 Mao, F.L., Zhang, G.C., Tong, J.J. et al. (2014). Anion exchange membranes used in diffusion dialysis for acid recovery from erosive and organic solutions. *Separation and Purification Technology* 122: 376–383.

41 Luo, J.Y., Wu, C.M., Wu, Y.H., and Xu, T.W. (2010). Diffusion dialysis of hydrochloride acid at different temperatures using PPO-SiO_2 hybrid anion exchange membranes. *Journal of Membrane Science* 347: 240–249.

42 Wu, Y.H., Luo, J.Y., Yao, L.L. et al. (2012). PVA/SiO_2 anion exchange hybrid membranes from multisilicon copolymers with two types of molecular weights. *Journal of Membrane Science* 399: 16–27.

43 Wu, Y.H., Hao, J.W., Wu, C.M. et al. (2012). Cation exchange PVA/SPPO/SiO_2 membranes with double organic phases for alkali recovery. *Journal of Membrane Science* 423: 383–391.

44 Liang, Y.X., Huang, X.N., Yao, L.Z. et al. (2021). Regulation of polyvinyl alcohol/sulfonated nano-TiO_2 hybrid membranes interface promotes diffusion dialysis. *Polymers* 13.

45 Irfan, M., Ul Afsar, N., Wang, Y.M., and Xu, T.W. (2018). Investigation of key process parameters in acid recovery for diffusion dialysis using novel (MDMH-QPPO) anion exchange membranes. *Journal of the Taiwan Institute of Chemical Engineers.* 93: 405–413.

46 Kim, D.H., Park, H.S., Seo, S.J. et al. (2014). Facile surface modification of anion-exchange membranes for improvement of diffusion dialysis performance. *Journal of Colloid and Interface Science* 416: 19–24.

47 Wu, Y.H., Luo, J.Y., Zhao, L.L. et al. (2013). QPPO/PVA anion exchange hybrid membranes from double crosslinking agents for acid recovery. *Journal of Membrane Science* 428: 95–103.

48 Pan, J., Zhu, L., Han, J.J., and Hickner, M.A. (2015). Mechanically tough and chemically stable anion exchange membranes from rigid-flexible semi-interpenetrating networks. *Chemistry of Materials* 27: 6689–6698.

49 Kim, K., Jung, B.K., Ko, T. et al. (2018). Comb-shaped polysulfones containing sulfonated polytriazole side chains for proton exchange membranes. *Journal of Membrane Science* 554: 232–243.

50 Li, C.R., Wang, G.H., Yu, D.B. et al. (2019). Cross-linked anion exchange membranes with hydrophobic side-chains for anion separation. *Journal of Membrane Science* 581: 150–157.

51 Chen, Y.B., Thorn, M., Christensen, S. et al. (2010). Enhancement of anhydrous proton transport by supramolecular nanochannels in comb polymers. *Nature Chemistry.* 2: 503–508.

52 Li, N., Wang, C., Lee, S.Y. et al. (2011). Enhancement of proton transport by nanochannels in comb-shaped copoly(arylene ether sulfone)s. *Angewandte Chemie-International Edition* 50: 9158–9161.

53 Fu, R., Wang, H.Y., Yan, J.Y. et al. (2022). Asymmetric bipolar membrane electrodialysis for acid and base production. *AIChE Journal* 69 (3): e17957.

54 Tongwen, X. and Weihua, Y. (2001). Sulfuric acid recovery from titanium white (pigment) waste liquor using diffusion dialysis with a new series of anion exchange membranes—static runs. *Journal of Membrane Science* 183: 193–200.

55 Xu, T., Liu, Z., Huang, C. et al. (2008). Preparation of a novel hollow-fiber anion-exchange membrane and its preliminary performance in diffusion dialysis. *Industrial and Engineering Chemistry Research* 47: 6204–6210.

56 Zhang, X., Li, C., Wang, X. et al. (2012). Recovery of hydrochloric acid from simulated chemosynthesis aluminum foils wastewater: an integration of diffusion dialysis and conventional electrodialysis. *Journal of Membrane Science* 409–410: 257–263.

57 Zhuang, J.-X., Chen, Q., Wang, S. et al. (2013). Zero discharge process for foil industry waste acid reclamation: coupling of diffusion dialysis and electrodialysis with bipolar membranes. *Journal of Membrane Science* 432: 90–96.

58 Hussain, A., Yan, H.Y., Ul Afsar, N. et al. (2022). Acid recovery from molybdenum metallurgical wastewater via selective electrodialysis and nanofiltration. *Separation and Purification Technology* 295: 121318.

59 Wang, H., Wu, C., Wu, Y. et al. (2011). Cation exchange hybrid membranes based on PVA for alkali recovery through diffusion dialysis. *Journal of Membrane Science* 376: 233–240.

60 Gu, J., Wu, C., Wu, Y. et al. (2012). PVA-based hybrid membranes from cation exchange multisilicon copolymer for alkali recovery. *Desalination* 304: 25–32.

61 Xiao, X., Wu, C., Cui, P. et al. (2011). Cation exchange hybrid membranes from SPPO and multi-alkoxy silicon copolymer: preparation, properties and diffusion dialysis performances for sodium hydroxide recovery. *Journal of Membrane Science* 379: 112–120.

62 Zhu, Q., Li, Y.Y., Qian, Q.L. et al. (2022). A sulfonated ultramicroporous membrane with selective ion transport enables osmotic energy extraction from multiform salt solutions with exceptional efficiency. *Energy & Environmental Science* 15: 4148–4156.

63 Yip, N.Y. and Elimelech, M. (2012). Thermodynamic and energy efficiency analysis of power generation from natural salinity gradients by pressure retarded osmosis. *Environmental Science & Technology* 46: 5230–5239.

64 Lee, K.P., Arnot, T.C., and Mattia, D. (2011). A review of reverse osmosis membrane materials for desalination—development to date and future potential. *Journal of Membrane Science* 370: 1–22.

65 Chou, S., Wang, R., Shi, L. et al. (2012). Thin-film composite hollow fiber membranes for pressure retarded osmosis (PRO) process with high power density. *Journal of Membrane Science* 389: 25–33.

66 Post, J.W., Hamelers, H.V., and Buisman, C.J. (2008). Energy recovery from controlled mixing salt and fresh water with a reverse electrodialysis system. *Environmental Science & Technology* 42: 5785–5790.

67 Veerman, J., Saakes, M., Metz, S., and Harmsen, G. (2011). Reverse electrodialysis: a validated process model for design and optimization. *Chemical Engineering Journal* 166: 256–268.

68 Tufa, R.A., Pawlowski, S., Veerman, J. et al. (2018). Progress and prospects in reverse electrodialysis for salinity gradient energy conversion and storage. *Applied Energy* 225: 290–331.

69 Liu, F., Schaetzle, O., Sales, B.B. et al. (2012). Effect of additional charging and current density on the performance of capacitive energy extraction based on Donnan potential. *Energy & Environmental Science* 5: 8642–8650.

70 Brogioli, D., Zhao, R., and Biesheuvel, P. (2011). A prototype cell for extracting energy from a water salinity difference by means of double layer expansion in nanoporous carbon electrodes. *Energy & Environmental Science* 4: 772–777.

71 Olsson, M., Wick, G.L., and Isaacs, J.D. (1979). Salinity gradient power: utilizing vapor pressure differences. *Science* 206: 452–454.

72 Jia, Z.J., Wang, B.G., Song, S.Q., and Fan, Y.S. (2014). Blue energy: current technologies for sustainable power generation from water salinity gradient. *Renewable & Sustainable Energy Reviews* 31: 91–100.

73 Pattle, R.E. (1954). Production of electric power by mixing fresh and salt water in the hydroelectric pile. *Nature* 174: 660.

74 Vermaas, D.A., Saakes, M., and Nijmeijer, K. (2011). Power generation using profiled membranes in reverse electrodialysis. *Journal of Membrane Science* 385: 234–242.

75 Hong, J.G. and Chen, Y.S. (2015). Evaluation of electrochemical properties and reverse electrodialysis performance for porous cation exchange membranes with sulfate-functionalized iron oxide. *Journal of Membrane Science* 473: 210–217.

76 Cho, D.H., Lee, K.H., Kim, Y.M. et al. (2017). Effect of cationic groups in poly(arylene ether sulfone) membranes on reverse electrodialysis performance. *Chemical Communications* 53: 2323–2326.

77 Lee, S.Y., Jeong, Y.J., Chae, S.R. et al. (2016). Porous carbon-coated graphite electrodes for energy production from salinity gradient using reverse electrodialysis. *Journal of Physics and Chemistry of Solids* 91: 34–40.

78 He, Z.L., Gao, X.L., Zhang, Y.S. et al. (2016). Revised spacer design to improve hydrodynamics and anti- fouling behavior in reverse electrodialysis processes. *Desalination and Water Treatment.* 57: 28176–28186.

79 Weinstein, J.N. and Leitz, F.B. (1976). Electric-power from difference in salinity - dialytic battery. *Science* 191: 557–559.

80 Brauns, E. (2009). Salinity gradient power by reverse electrodialysis: effect of model parameters on electrical power output. *Desalination* 237: 378–391.

81 Dlugolecki, P., Ogonowski, P., Metz, S.J. et al. (2010). On the resistances of membrane, diffusion boundary layer and double layer in ion exchange membrane transport. *Journal of Membrane Science* 349: 369–379.

82 Fontananova, E., Zhang, W.J., Nicotera, I. et al. (2014). Probing membrane and interface properties in concentrated electrolyte solutions. *Journal of Membrane Science* 459: 177–189.

83 Kim, H.K., Lee, M.S., Lee, S.Y. et al. (2015). High power density of reverse electrodialysis with pore-filling ion exchange membranes and a high-open-area spacer. *Journal of Materials Chemistry A.* 3: 16302–16306.

84 Bohinc, K., Kralj-Iglic, V., and Iglic, A. (2001). Thickness of electrical double layer. Effect of ion size. *Electrochimica Acta* 46: 3033–3040.

85 Tedesco, M., Cipollina, A., Tamburini, A. et al. (2015). A simulation tool for analysis and design of reverse electrodialysis using concentrated brines. *Chemical Engineering Research & Design* 93: 441–456.

86 Veerman, J., Saakes, M., Metz, S.J., and Harmsen, G.J. (2010). Reverse electrodialysis: evaluation of suitable electrode systems. *Journal of Applied Electrochemistry* 40: 1461–1474.

87 Scialdone, O., Albanese, A., D'Angelo, A. et al. (2013). Investigation of electrode material – redox couple systems for reverse electrodialysis processes. Part II: experiments in a stack with 10–50 cell pairs. *Journal of Electroanalytical Chemistry* 704: 1–9.

88 Guler, E., Zhang, Y.L., Saakes, M., and Nijmeijer, K. (2012). Tailor-made anion-exchange membranes for salinity gradient power generation using reverse Electrodialysis. *ChemSusChem* 5: 2262–2270.

89 Guler, E., Elizen, R., Vermaas, D.A. et al. (2013). Performance-determining membrane properties in reverse electrodialysis. *Journal of Membrane Science* 446: 266–276.

90 Zhang, Z., Sui, X., Li, P. et al. (2017). Ultrathin and ion-selective Janus membranes for high-performance osmotic energy conversion. *Journal of the American Chemical Society* 139: 8905–8914.

91 Zhu, X.B., Zhou, Y.H., Hao, J.R. et al. (2017). A charge-density-tunable three/two-dimensional polymer/graphene oxide heterogeneous nanoporous membrane for ion transport. *ACS Nano* 11: 10816–10824.

92 Pawlowski, S., Rijnaarts, T., Saakes, M. et al. (2017). Improved fluid mixing and power density in reverse electrodialysis stacks with chevron-profiled membranes. *Journal of Membrane Science* 531: 111–121.

93 Guler, E., Elizen, R., Saakes, M., and Nijmeijer, K. (2014). Micro-structured membranes for electricity generation by reverse electrodialysis. *Journal of Membrane Science* 458: 136–148.

94 Vermaas, D.A., Saakes, M., and Nijmeijer, K. (2011). Doubled power density from salinity gradients at reduced intermembrane distance. *Environmental Science & Technology* 45: 7089–7095.

95 Veerman, J., Saakes, M., Metz, S.J., and Harmsen, G.J. (2009). Reverse electrodialysis: performance of a stack with 50 cells on the mixing of sea and river water. *Journal of Membrane Science* 327: 136–144.

96 Dlugolecki, P., Dabrowska, J., Nijmeijer, K., and Wessling, M. (2010). Ion conductive spacers for increased power generation in reverse electrodialysis. *Journal of Membrane Science* 347: 101–107.

97 Zhu, X.P., He, W.H., and Logan, B.E. (2015). Reducing pumping energy by using different flow rates of high and low concentration solutions in reverse electrodialysis cells. *Journal of Membrane Science* 486: 215–221.

98 Gurreri, L., Tamburini, A., Cipollina, A. et al. (2016). Flow and mass transfer in spacer-filled channels for reverse electrodialysis: a CFD parametrical study. *Journal of Membrane Science* 497: 300–317.

99 Frilette, V.J. (1956). Preparation and characterization of bipolar ion-exchange membranes. *The Journal of Physical Chemistry* 60: 435–439.

100 Donnan, F.G. (1995). Theory of membrane equilibria and membrane-potentials in the presence of non-dialyzing electrolytes – a contribution to physical-chemical physiology (Reprinted from Zeitshrift Fur Elektrochemie Und Angewandte Physikalische Chemie, Vol 17, Pg 572, 1911). *Journal of Membrane Science* 100: 45–55.

101 Sarkar, S., Sengupta, A.K., and Prakash, P. (2010). The Donnan membrane principle: opportunities for sustainable engineered processes and materials. *Environmental Science & Technology* 44: 1161–1166.

102 Cwirko, E.H. and Carbonell, R.G. (1990). A theoretical-analysis of Donnan Dialysis across charged porous membranes. *Journal of Membrane Science* 48: 155–179.

103 Wisniewski, J. and Rozanska, A. (2006). Donnan dialysis with anion-exchange membranes as a pretreatment step before electrodialytic desalination. *Desalination* 191: 210–218.

104 Rozanska, A. and Wisniewski, J. (2009). Modification of brackish water composition by means of Donnan dialysis as pretreatment before desalination. *Desalination* 240: 326–332.

105 Ho, C.C., Jan, D.S., and Tsai, F.N. (1993). Membrane diffusion-controlled kinetics of ionic transport. *Journal of Membrane Science* 81: 287–294.

106 Beck, A. and Ernst, M. (2015). Kinetic modeling and selectivity of anion exchange in Donnan dialysis. *Journal of Membrane Science* 479: 132–140.

107 Boubakri, A., Helali, N., Tlili, M., and Ben Amor, M. (2014). Fluoride removal from diluted solutions by Donnan dialysis using full factorial design. *Korean Journal of Chemical Engineering* 31: 461–466.

108 White, N., Misovich, M., Yaroshchuk, A., and Bruening, M.L. (2015). Coating of Nafion membranes with polyelectrolyte multilayers to achieve high monovalent/divalent cation Electrodialysis Selectivities. *ACS Applied Materials & Interfaces* 7: 6620–6628.

109 Ahmad, M., Yaroshchuk, A., and Bruening, M.L. (2020). Moderate pH changes alter the fluxes, selectivities and limiting currents in ion transport through polyelectrolyte multilayers deposited on membranes. *Journal of Membrane Science* 616: 118570.

110 Tapiero, Y., Rivas, B.L., Sanchez, J., and Avila, E.G. (2018). Donnan dialysis assisted by interpenetrating polymer networks for chromium ion transport in aqueous media. *Journal of the Chilean Chemical Society* 63: 3902–3913.

111 Bryjak, M. and Duraj, I. (2013). Anion-exchange membranes for separation of borates by Donnan dialysis. *Desalination* 310: 39–42.

112 Pessoa-Lopes, M., Crespo, J.G., and Velizarov, S. (2016). Arsenate removal from sulphate-containing water streams by an ion-exchange membrane process. *Separation and Purification Technology* 166: 125–134.

113 Nasser, I.I. and Ahmed, C. (2015). Application of Donnan Dialysis for cyanide removal from aqueous solutions. *International Journal of Environmental Research* 9: 505–510.

114 Tor, A. (2007). Removal of fluoride from water using anion-exchange membrane under Donnan dialysis condition. *Journal of Hazardous Materials* 141: 814–818.

115 Zhao, B., Zhao, H.Z., and Ni, J.R. (2010). Modeling of the Donnan dialysis process for arsenate removal. *Chemical Engineering Journal* 160: 170–175.

116 Chen, C., Dong, T., Han, M.Y. et al. (2020). Ammonium recovery from wastewater by Donnan dialysis: a feasibility study. *Journal of Cleaner Production* 265: 121838.

117 Shashvatt, U., Amurrio, F., Portner, C., and Blaney, L. (2021). Phosphorus recovery by Donnan dialysis: membrane selectivity, diffusion coefficients, and speciation effects. *Chemical Engineering Journal* 419: 129626.

118 Sudoh, M., Kamei, H., and Nakamura, S. (1987). Donnan dialysis concentration of cupric ions. *Journal of Chemical Engineering of Japan* 20: 34–40.

119 Aliaskari, M. and Schafer, A.I. (2021). Nitrate, arsenic and fluoride removal by electrodialysis from brackish groundwater. *Water Research* 190: 116683.

120 Nezungai, C.D. and Majozi, T. (2016). Optimum synthesis of an electrodialysis framework with a background process—I: a novel electrodialysis model. *Chemical Engineering Science* 147: 180–188.

121 Tado, K., Sakai, F., Sano, Y., and Nakayama, A. (2016). An analysis on ion transport process in electrodialysis desalination. *Desalination* 378: 60–66.

122 McGovern, R.K., Zubair, S.M., and Lienhard, V.J.H. (2014). The benefits of hybridising electrodialysis with reverse osmosis. *Journal of Membrane Science* 469: 326–335.

123 Jiang, C.X., Wang, Y.M., Zhang, Z.H., and Xu, T.W. (2014). Electrodialysis of concentrated brine from RO plant to produce coarse salt and freshwater. *Journal of Membrane Science* 450: 323–330.

124 Mohan, S.V., Butti, S.K., Amulya, K. et al. (2016). Waste biorefinery: a new paradigm for a sustainable bioelectro economy. *Trends in Biotechnology* 34: 852–855.

125 McCarty, P.L., Bae, J., and Kim, J. (2011). Domestic wastewater treatment as a net energy producer-can this be achieved? *Environmental Science & Technology* 45: 7100–7106.

126 Li, W.W., Yu, H.Q., and Rittmann, B.E. (2015). Chemistry: reuse water pollutants. *Nature* 528: 29–31.

127 Lu, L., Guest, J.S., Peters, C.A. et al. (2018). Wastewater treatment for carbon capture and utilization. *Nature Sustainability* 1: 750–758.

128 Al-Karaghouli, A. and Kazmerski, L.L. (2013). Energy consumption and water production cost of conventional and renewable-energy-powered desalination processes. *Renewable and Sustainable Energy Reviews* 24: 343–356.

129 Zhou, Y., Yan, H., Wang, X. et al. (2018). Electrodialytic concentrating lithium salt from primary resource. *Desalination* 425: 30–36.

130 Yan, H.Y., Wang, Y.M., Wu, L. et al. (2019). Multistage-batch electrodialysis to concentrate high-salinity solutions: process optimisation, water transport, and energy consumption. *Journal of Membrane Science* 570: 245–257.

131 Jiang, C., Chen, B., Xu, Z. et al. (2022). Ion-"distillation" for isolating lithium from lake brine. *AICHE Journal* 68 (6): e17710.

132 Zhao, J., Liu, D., Zhu, Y. et al. (2022). A review of forest carbon cycle models on spatiotemporal scales. *Journal of Cleaner Production* 339: 130692.

133 Hao, Z., Barecka, M.H., and Lapkin, A.A. (2022). Accelerating net zero from the perspective of optimizing a carbon capture and utilization system. *Energy & Environmental Science* 15: 2139–2153.

134 Digdaya, I.A., Sullivan, I., Lin, M. et al. (2020). A direct coupled electrochemical system for capture and conversion of CO_2 from oceanwater. *Nature Communications* 11: 4412.

135 Willauer, H.D., DiMascio, F., Hardy, D.R., and Williams, F.W. (2014). Feasibility of CO_2 extraction from seawater and simultaneous hydrogen gas generation using a novel and robust electrolytic cation exchange module based on continuous electrodeionization technology. *Industrial & Engineering Chemistry Research* 53: 12192–12200.

136 Lin, J., Lin, F., Chen, X. et al. (2019). Sustainable management of textile wastewater: a hybrid tight ultrafiltration/bipolar-membrane electrodialysis process for resource recovery and zero liquid discharge. *Industrial & Engineering Chemistry Research* 58: 11003–11012.

137 Kumar, A., Phillips, K.R., Cai, J. et al. (2019). Integrated valorization of desalination brine through NaOH recovery: opportunities and challenges. *Angewandte Chemie (International Ed. in English)* 58: 6502–6511.

138 Tongwen, X. (2001). Development of bipolar membrane-based processes. *Desalination* 140: 247–258.

139 Strathmann, H., Krol, J.J., Rapp, H.-J., and Eigenberger, G. (1997). Limiting current density and water dissociation in bipolar membranes. *Journal of Membrane Science* 125: 123–142.

140 Shen, J., Yu, J., Liu, L. et al. (2014). Synthesis of quaternary ammonium hydroxide from its halide salt by bipolar membrane electrodialysis (BMED): effect of molecular structure of ammonium compounds on the process performance. *Journal of Chemical Technology & Biotechnology* 89: 841–850.

141 Shen, J., Hou, Z., and Gao, C. (2017). Using bipolar membrane electrodialysis to synthesize di-quaternary ammonium hydroxide and optimization design by response surface methodology. *Chinese Journal of Chemical Engineering* 25: 1176–1181.

142 Geng, T., Teng, X.-G., Li, Q.-X. et al. (2013). Synthesis and properties of long-chain quaternary ammonium hydroxides. *Chinese Chemical Letters* 24: 494–496.

143 Yao, L., Qiu, Y., Zhao, Y. et al. (2020). A continuous mode operation of bipolar membrane electrodialysis (BMED) for the production of high-pure choline hydroxide from choline chloride. *Separation and Purification Technology* 233: 116054.

144 Novalic, S., Kongbangkerd, T., and Kulbe, K.D. (1997). Separation of gluconate with conventional and bipolar electrodialysis. *Desalination* 114: 45–50.

145 Pinacci, P. and Radaelli, M. (2002). Recovery of citric acid from fermentation broths by electrodialysis with bipolar membranes. *Desalination* 148: 177–179.

146 Zhang, X., Li, C.R., Wang, Y.M. et al. (2011). Recovery of acetic acid from simulated acetaldehyde wastewaters: bipolar membrane electrodialysis processes and membrane selection. *Journal of Membrane Science* 379: 184–190.

147 Iyyappan, J., Baskar, G., Gnansounou, E. et al. (2019). Recent advances in microbial production of malic acid from renewable byproducts. *Reviews in Environmental Science and Bio/Technology* 18: 579–595.

148 Ma, H.Z., Yue, S.Y., Li, H.G. et al. (2019). Recovery of lactic acid and other organic acids from food waste ethanol fermentation stillage: feasibility and effects of substrates. *Separation and Purification Technology* 209: 223–228.

149 Wang, Y.M., Wang, X.L., Yan, H.Y. et al. (2020). Bipolar membrane electrodialysis for cleaner production ofN-methylatedglycine derivative amino acids. *AICHE Journal* 66.

150 Chen, T., Bi, J., Ji, Z. et al. (2022). Application of bipolar membrane electrodialysis for simultaneous recovery of high-value acid/alkali from saline wastewater: an in-depth review. *Water Research* 226: 119274.

151 Wang, H.Y., Yan, J.Y., Fu, R. et al. (2022). Bipolar membrane electrodialysis for cleaner production of gluconic acid: valorization of the regenerated base for the upstream enzyme catalysis. *Industrial and Engineering Chemistry Research* 61: 7634–7644.

152 Innocent, C., Kameche, M., Xu, F.N., and Pourcelly, G. (2003). Electrodialysis in water-ethanol solutions: application to the acidification of organic salts. *Desalination* 154: 9–15.

153 Noguchi, M., Nakamura, Y., Shoji, T. et al. (2018). Simultaneous removal and recovery of boron from waste water by multi-step bipolar membrane electrodialysis. *Journal of Water Process Engineering* 23: 299–305.

154 Tang, Y.P., Luo, L., Thong, Z., and Chung, T.S. (2017). Recent advances in membrane materials and technologies for boron removal. *Journal of Membrane Science* 541: 434–446.

155 Bunani, S., Arda, M., Kabay, N. et al. (2017). Effect of process conditions on recovery of lithium and boron from water using bipolar membrane electrodialysis (BMED). *Desalination* 416: 10–15.

156 Pourcelly, G., Tugas, I., and Gavach, C. (1994). Electrotransport of sulphuric acid in special anion exchange membranes for the recovery of acids. *Journal of Membrane Science* 97: 99–107.

157 Jury, W.A. and Henry, V.J. (2005). The role of science in solving the world's emergingwater problems. *PNAS* 102: 15715–15720.

158 Peñate, B. and García-Rodríguez, L. (2012). Current trends and future prospects in the design of seawater reverse osmosis desalination technology. *Desalination* 284: 1–8.

159 Voutchkov, N. (2011). Overview of seawater concentrate disposal alternatives. *Desalination* 273: 205–219.

160 Liang, J., Deng, A., Xie, R. et al. (2013). Impact of seawater reverse osmosis (SWRO) product remineralization on the corrosion rate of water distribution pipeline materials. *Desalination* 311: 54–61.

161 Eisaman, M.D., Alvarado, L., Larner, D. et al. (2011). CO_2 separation using bipolar membrane electrodialysis. *Energy & Environmental Science* 4: 1319–1328.

162 Jiang, C., Zhang, Y., Feng, H. et al. (2017). Simultaneous CO_2 capture and amino acid production using bipolar membrane electrodialysis (BMED). *Journal of Membrane Science* 542: 264–271.

163 Wang, Q., Jiang, C., Wang, Y. et al. (2016). Reclamation of aniline wastewater and CO_2 capture using bipolar membrane Electrodialysis. *ACS Sustainable Chemistry & Engineering* 4: 5743–5751.

164 Nagasawa, H., Yamasaki, A., Iizuka, A. et al. (2009). A new recovery process of carbon dioxide from alkaline carbonate solution via electrodialysis. *AICHE Journal* 55: 3286–3293.

165 Jiang, C., Li, S., Zhang, D. et al. (2019). Mathematical modelling and experimental investigation of CO_2 absorber recovery using an electro-acidification method. *Chemical Engineering Journal* 360: 654–664.

166 Eisaman, M.D., Alvarado, L., Larner, D. et al. (2011). CO_2 desorption using high-pressure bipolar membrane electrodialysis. *Energy & Environmental Science* 4.

167 van der Vegt, N.F.A., Wilhelm, F.G., Punt, I. et al. (2001). Optimisation strategies for the preparation of bipolar membranes with reduced salt ion leakage in acid-base electrodialysis. *Journal of Membrane Science* 182: 13–28.

168 Pourcelly, G. (2002). Electrodialysis with bipolar membranes: principles, optimization, and applications. *Russian Journal of Electrochemistry* 38: 919–926.

169 Wilhelm, F.G., Punt, I., van der Vegt, N.F.A. et al. (2002). Asymmetric bipolar membranes in acid-base electrodialysis. *Industrial & Engineering Chemistry Research* 41: 579–586.

170 Wen, W.-F., Wang, J., Zhong, C.-Y. et al. (2022). Direct production of lithium nitrate from the primary lithium salt by electrodialysis metathesis. *Journal of Membrane Science* 654: 120555.

171 Chen, Q.-B., Tian, Z., Zhao, J. et al. (2022). Near-zero liquid discharge and reclamation process based on electrodialysis metathesis for high-salinity wastewater with high scaling potential. *Desalination* 525: 115390.

172 Camacho, L.M., Fox, J.A., and Ajedegba, J.O. (2017). Optimization of electrodialysis metathesis (EDM) desalination using factorial design methodology. *Desalination* 403: 136–143.

173 Zhong, C.-Y., Lv, Y.-P., Wen, W.-F. et al. (2022). Sustainable production of lithium acetate by bipolar membrane electrodialysis metathesis. *ACS Sustainable Chemistry & Engineering* 10 (18): 6045–6056.

174 Wang, B., Qin, L., Mu, T. et al. (2017). Are ionic liquids chemically stable? *Chemical Reviews* 117: 7113–7131.

175 Zhang, T., Doert, T., Wang, H. et al. (2021). Inorganic synthesis based on reactions of ionic liquids and deep eutectic solvents. *Angewandte Chemie (International Ed. in English)* 60: 22148–22165.

176 Weaver, K.D., Vrikkis, R.M., Van Vorst, M.P. et al. (2012). Structure and function of proteins in hydrated choline dihydrogen phosphate ionic liquid. *Physical Chemistry Chemical Physics* 14: 790–801.

177 Tateishi-Karimata, H., Nakano, M., and Sugimoto, N. (2014). Comparable stability of Hoogsteen and Watson-crick base pairs in ionic liquid choline dihydrogen phosphate. *Scientific Reports* 4: 3593.

178 Rottiers, T., Van der Bruggen, B., and Pinoy, L. (2017). Synthesis and transport of impurities in electrodialysis metathesis: production of choline dihydrogen phosphate. *Journal of Membrane Science* 541: 550–557.

179 Rottiers, T., De la Marche, G., Van der Bruggen, B., and Pinoy, L. (2015). Co-ion fluxes of simple inorganic ions in electrodialysis metathesis and conventional electrodialysis. *Journal of Membrane Science* 492: 263–270.

180 Chen, Q.-B., Ren, H., Tian, Z. et al. (2019). Conversion and pre-concentration of SWRO reject brine into high solubility liquid salts (HSLS) by using electrodialysis metathesis. *Separation and Purification Technology* 213: 587–598.

181 Li, P.-F., Chen, Q.-B., Wang, J. et al. (2021). Converting softening nanofiltration brine into high-solubility liquid salts (HSLS) via electrodialysis metathesis: effect of membrane type. *Separation and Purification Technology* 267: 118619.

182 Sheveleva, O.G., Rupcheva, V.A., and Poilov, V.Z. (2016). Production and properties of double potassium-ammonium sulfate. *Russian Journal of Applied Chemistry.* 89: 29–33.

183 Aagli, A., Tamer, N., Atbir, A. et al. (2005). Conversion of phosphogypsum to potassium sulfate. *Journal of Thermal Analysis and Calorimetry* 82: 395–399.

184 Khlissa, F., M'nif, A., and Rokbani, R. (2004). Application of the conductimetry to the study of the transformation of KCl and Na_2SO_4 into K_2SO_4 between 5 and 30 °C. *Chemical Engineering and Processing: Process Intensification* 43: 929–934.

185 Jannet, D.B., M'Nif, A., and Rokbani, R. (2004). Natural brine valorisation: application of the system K^+, Mg^{2+}/cl^-, $SO_4^{2-}//H_2O$ at 25°C observations on kinetics formation of K_2SO_4 and $K_2SO_4.MgSO_4.6H_2O$. *Desalination* 167: 319–326.

186 Sharma, P.P., Gahlot, S., Rajput, A. et al. (2016). Efficient and cost effective way for the conversion of potassium nitrate from potassium chloride using electrodialysis. *ACS Sustainable Chemistry & Engineering* 4: 3220–3227.

187 Zhang, X., Han, X., Yan, X. et al. (2019). Continuous synthesis of high purity KNO_3 through electrodialysis metathesis. *Separation and Purification Technology* 222: 85–91.

188 Wang, X., Du, Y., Liu, J. et al. (2022). Modeling and simulation of continuous electrodialysis metathesis process for conversion of Na_2SO_4 to K_2SO_4. *Desalination* 528: 115605.

189 Zhang, X., Wang, X., Liu, X. et al. (2015). Conversion of potassium chloride into potassium sulfate by four-compartment electrodialysis: batch operation process. *Industrial & Engineering Chemistry Research* 54: 11937–11943.

190 Jaroszek, H. and Dydo, P. (2018). Potassium nitrate synthesis by electrodialysis-metathesis: the effect of membrane type. *Journal of Membrane Science* 549: 28–37.

191 Han, X., Yan, X., Wang, X. et al. (2018). Preparation of chloride-free potash fertilizers by electrodialysis metathesis. *Separation and Purification Technology* 191: 144–152.

192 Trivedi, J.S., Bhadja, V., Makwana, B.S. et al. (2016). Sustainable process for the preparation of potassium sulfate by electrodialysis and its concentration and purification by a nanofiltration process. *RSC Advances* 6: 71807–71817.

193 McCabe, W.L., Smith, J., and Harriott, P. (1995). *Unit Operations of Chemical Engineering*, 7e.

194 Strathmann, H. (1981). Membrane separation processes. *Journal of Membrane Science* 9: 121–189.

195 Barrer, R.M. (1977). Membrane separation. *Nature* 265: 566.

196 Koh, D.Y., McCool, B.A., Deckman, H.W., and Lively, R.P. (2016). Reverse osmosis molecular differentiation of organic liquids using carbon molecular sieve membranes. *Science* 353: 804–807.

197 Marchetti, P., Jimenez Solomon, M.F., Szekely, G., and Livingston, A.G. (2014). Molecular separation with organic solvent nanofiltration: a critical review. *Chemical Reviews* 114: 10735–10806.

198 Sholl, D.S. and Lively, R.P. (2016). Seven chemical separations to change the world. *Nature* 532: 435–437.

199 Volpin, F., Badeti, U., Wang, C. et al. (2020). Urine treatment on the international space station: current practice and novel approaches. *Membranes* 10: 327.

200 Lin, J.C.-T. and Livingston, A.G. (2007). Nanofiltration membrane cascade for continuous solvent exchange. *Chemical Engineering Science* 62: 2728–2736.

201 Pruitt, J.M., Carter, L., Bagdigian, R.M., and Kayatin, M.J. (2015). Upgrades to the ISS water recovery system. 45th International Conference on Environmental Systems

202 Carter, D.L., Tobias, B., and Orozco, N.Y.2013). Status of ISS water management and recovery. 43rd International Conference on Environmental Systems. p. 3509.

203 Xu, T. (2005). Ion exchange membranes: state of their development and perspective. *Journal of Membrane Science* 263: 1–29.

204 Zhang, Y., Paepen, S., Pinoy, L. et al. (2012). Selectrodialysis: fractionation of divalent ions from monovalent ions in a novel electrodialysis stack. *Separation and Purification Technology* 88: 191–201.

205 Li, X., Mo, Y., Qing, W. et al. (2019). Membrane-based technologies for lithium recovery from water lithium resources: a review. *Journal of Membrane Science* 591: 117317.

206 Nie, X.-Y., Sun, S.-Y., Sun, Z. et al. (2017). Ion-fractionation of lithium ions from magnesium ions by electrodialysis using monovalent selective ion-exchange membranes. *Desalination* 403: 128–135.

207 Liu, H., Jiang, Y., Ding, J. et al. (2018). Surface layer modification of AEMs by infiltration and photo-cross-linking to induce monovalent selectivity. *AIChE Journal* 64: 993–1000.

208 Shi, W., Liu, X., Ye, C. et al. (2019). Efficient lithium extraction by membrane capacitive deionization incorporated with monovalent selective cation exchange membrane. *Separation and Purification Technology* 210: 885–890.

209 Guo, Y., Ying, Y., Mao, Y. et al. (2016). Polystyrene sulfonate threaded through a metal–organic framework membrane for fast and selective lithium-ion separation. *Angewandte Chemie, International Edition* 55: 15120–15124.

210 Bing, S., Xian, W., Chen, S. et al. (2021). Bio-inspired construction of ion conductive pathway in covalent organic framework membranes for efficient lithium extraction. *Matter* 4: 2027–2038.

211 Wang, H., Lu, Z., Wu, Y. et al. (2021). A Lamellar MXene (Ti_3C_2Tx)/PSS composite membrane for fast and selective lithium-ion separation. *Angewandte Chemie International Edition* 133 (41): 22439–22443.

212 Zhao, Y., Shi, W., Van der Bruggen, B. et al. (2018). Graphene oxide: tunable nanoscale interlayer of graphene with symmetrical polyelectrolyte multilayer architecture for lithium extraction (Adv. Mater. Interfaces 6/2018). *Advanced Materials Interfaces* 5: 1870025.

213 Kazemzadeh, H., Karimi-Sabet, J., Towfighi Darian, J., and Adhami, A. (2020). Evaluation of polymer inclusion membrane efficiency in selective separation of lithium ion from aqueous solution. *Separation and Purification Technology* 251: 117298.

214 Zhu, G., Wang, P., Qi, P., and Gao, C. (2014). Adsorption and desorption properties of Li+ on PVC-$H_{1.6}Mn_{1.6}O_4$ lithium ion-sieve membrane. *Chemical Engineering Journal* 235: 340–348.

215 Zhang, B. and Edwards, B.J. (2015). Modelling proton conductivity in perfluorosulfonate acid membranes. *Journal of the Electrochemical Society* 162: F1088–F1095.

216 Kundu, P.P. and Dutta, K. (2016). Hydrogen fuel cells for portable applications. *Compendium of Hydrogen Energy* 111–131.

217 Xiao, F., Wang, Y.C., Wu, Z.P. et al. (2021). Recent advances in electrocatalysts for proton exchange membrane fuel cells and alkaline membrane fuel cells. *Advanced Materials* 33: e2006292.

218 Shi, S.W., Weber, A.Z., and Kusoglu, A. (2016). Structure/property relationship of Nafion XL composite membranes. *Journal of Membrane Science* 516: 123–134.

219 Tang, H.L., Shen, P.K., Jiang, S.P. et al. (2007). A degradation study of nation proton exchange membrane of PEM fuel cells. *Journal of Power Sources* 170: 85–92.

220 Beste Balci, K.N., Kii, T., Kageyama, M., and Balci, B. (2021). Hydrogen crossover through a Nafion membrane and a membrane electrode assembly of PEFC at high temperature. *ECS Meeting Abstracts* p. MA2021-02 1076.

221 Klingele, M., Breitwieser, M., Zengerle, R., and Thiele, S. (2015). Direct deposition of proton exchange membranes enabling high performance hydrogen fuel cells. *Journal of Materials Chemistry A* 3: 11239–11245.

222 Haider, R., Wen, Y.C., Ma, Z.F. et al. (2021). High temperature proton exchange membrane fuel cells: progress in advanced materials and key technologies. *Chemical Society Reviews* 50: 1138–1187.

223 Qian, Y., Huang, L., Pan, Y. et al. (2018). Dependency of migration and reduction of mixed $Cr_2O_7^{2-}$, Cu^{2+} and Cd^{2+} on electric field, ion exchange membrane and metal concentration in microbial fuel cells. *Separation and Purification Technology* 192: 78–87.

224 Mandal, M. (2020). Recent advancement on anion exchange membranes for fuel cell and water electrolysis. *ChemElectroChem* 8: 36–45.

225 Pan, Z.F., An, L., Zhao, T.S., and Tang, Z.K. (2018). Advances and challenges in alkaline anion exchange membrane fuel cells. *Progress in Energy and Combustion Science* 66: 141–175.

226 Wang, J.H., Zhao, Y., Setzler, B.P. et al. (2019). Poly(aryl piperidinium) membranes and ionomers for hydroxide exchange membrane fuel cells. *Nature Energy* 4: 392–398.

227 Chen, N., Hu, C., Wang, H.H. et al. (2021). Poly(alkyl-terphenyl piperidinium) ionomers and membranes with an outstanding alkaline-membrane fuel-cell performance of 2.58 W cm(−2). *Angewandte Chemie (International Ed. in English)* 60: 7710–7718.

228 Liu, X., Xie, N., Xue, J. et al. (2022). Magnetic-field-oriented mixed-valence-stabilized ferrocenium anion-exchange membranes for fuel cells. *Nature Energy* 7: 329–339.

229 Jiao, K., Xuan, J., Du, Q. et al. (2021). Designing the next generation of proton-exchange membrane fuel cells. *Nature* 595: 361–369.

230 Park, C.H., Lee, S.Y., Hwang, D.S. et al. (2016). Nanocrack-regulated self-humidifying membranes. *Nature* 532: 480–483.

231 Li, D., Motz, A.R., Bae, C. et al. (2021). Durability of anion exchange membrane water electrolyzers. *Energy & Environmental Science* 14: 3393–3419.

232 Buttler, A. and Spliethoff, H. (2018). Current status of water electrolysis for energy storage, grid balancing and sector coupling via power-to-gas and power-to-liquids: a review. *Renewable and Sustainable Energy Reviews* 82: 2440–2454.

233 Miller, H.A., Bouzek, K., Hnat, J. et al. (2020). Green hydrogen from anion exchange membrane water electrolysis: a review of recent developments in critical materials and operating conditions. *Sustainable Energy & Fuels* 4: 2114–2133.

234 Li, C. and Baek, J.-B. (2021). The promise of hydrogen production from alkaline anion exchange membrane electrolyzers. *Nano Energy* 87: 106162.

235 Cha, M.S., Park, J.E., Kim, S. et al. (2020). Poly(carbazole)-based anion-conducting materials with high performance and durability for energy conversion devices. *Energy & Environmental Science* 13: 3633–3645.

236 Li, K., Yu, S., Li, D. et al. (2021). Engineered thin diffusion layers for anion-exchange membrane electrolyzer cells with outstanding performance. *ACS Applied Materials & Interfaces* 13: 50957–50964.

237 Noris Gallandat, K.R. and Züttel, A. (2017). An analytical model for the electrolyser performance derived from materials parameters. *Journal of Power and Energy Engineering* 34–49.

238 Chen, N., Paek, S.Y., Lee, J.Y. et al. (2021). High-performance anion exchange membrane water electrolyzers with a current density of 7.68 A cm−2 and a durability of 1000 hours. *Energy & Environmental Science* 14: 6338–6348.

239 Li, D., Park, E.J., Zhu, W. et al. (2020). Highly quaternized polystyrene ionomers for high performance anion exchange membrane water electrolysers. *Nature Energy* 5: 378–385.

240 Wan, L., Xu, Z., and Wang, B. (2021). Green preparation of highly alkali-resistant PTFE composite membranes for advanced alkaline water electrolysis. *Chemical Engineering Journal* 426: 131340.

241 Saba, S.M., Müller, M., Robinius, M., and Stolten, D. (2018). The investment costs of electrolysis – a comparison of cost studies from the past 30 years. *International Journal of Hydrogen Energy* 43: 1209–1223.

242 Wan, L., Xu, Z., Wang, P. et al. (2022). Dual regulation both intrinsic activity and mass transport for self-supported electrodes using in anion exchange membrane water electrolysis. *Chemical Engineering Journal* 431: 133942.

243 Liu, L., Chu, X., Liao, J. et al. (2018). Tuning the properties of poly(2,6-dimethyl-1,4-phenylene oxide) anion exchange membranes and their performance in H_2/O_2 fuel cells. *Energy & Environmental Science* 11: 435–446.

244 Chu, X., Shi, Y., Liu, L. et al. (2019). Piperidinium-functionalized anion exchange membranes and their application in alkaline fuel cells and water electrolysis. *Journal of Materials Chemistry A* 7: 7717–7727.

245 Ahmed Mahmoud, A.M. and Miyatake, K. (2018). Optimization of the pendant chain length in partially fluorinated aromatic anion exchange membranes for alkaline fuel cells. *Journal of Materials Chemistry A* 6: 14400–14409.

246 Kaczur, J.J., Yang, H., Liu, Z. et al. (2018). Carbon dioxide and water electrolysis using new alkaline stable anion membranes. *Frontiers in Chemistry* 6: 263.

247 Soni, R., Miyanishi, S., Kuroki, H., and Yamaguchi, T. (2020). Pure water solid alkaline water electrolyzer using fully aromatic and high-molecular-weight poly(fluorene-alt-tetrafluorophenylene)-trimethyl ammonium anion exchange membranes and ionomers. *ACS Applied Energy Materials* 4: 1053–1058.

248 Wan, L., Xu, Z., Xu, Q. et al. (2022). Overall design of novel 3D-ordered MEA with drastically enhanced mass transport for alkaline electrolyzers. *Energy & Environmental Science*.

Index

Ion Exchange Membranes: Design, Preparation, and Applications,
First Edition. Tongwen Xu and Yaoming Wang.

f

g

h

i

l

m

n

p

t

u

w

z